Ayurveda in the New Millennium

Ayurveda in the New Millennium

Emerging Roles and Future Challenges

Edited by
D. Suresh Kumar

CRC Press
Taylor & Francis Group
Boca Raton London New York

CRC Press is an imprint of the
Taylor & Francis Group, an **informa** business

First edition published 2021
by CRC Press
6000 Broken Sound Parkway NW, Suite 300, Boca Raton, FL 33487-2742

and by CRC Press
2 Park Square, Milton Park, Abingdon, Oxon, OX14 4RN

Library of Congress Cataloging-in-Publication Data
Names: Suresh Kumar, D., 1949- editor.
Title: Ayurveda in the new millennium : emerging roles and future
challenges / edited by D. Suresh Kumar.
Description: 1. | Boca Raton, FL : CRC Press, 2021. | Includes
bibliographical references and index.
Identifiers: LCCN 2020032988 | ISBN 9780367279547 (hardback) | ISBN
9780429298936 (ebook)
Subjects: LCSH: Medicine, Ayurvedic. | Herbs--Therapeutic use.
Classification: LCC R605 .A8873 2021 | DDC 615.5/38--dc23
LC record available at https://lccn.loc.gov/2020032988

ISBN: 978-0-367-27954-7 (hbk)
ISBN: 978-0-429-29893-6 (ebk)

Typeset in Times
by Deanta Global Publishing Services, Chennai, India

आयुर्वेदो रक्षति रक्षितः

Āyurvēdō rakṣati rakṣitaḥ

He who protects Ayurveda shall, in turn, be protected by It.

Contents

Preface.....ix
About the Editor.....xiii
Contributors.....xv

Chapter 1 What We Learn from the History of Ayurveda.....1
N.K.M. Ikbal, D. Induchoodan and D. Suresh Kumar

Chapter 2 Manufacture of Ayurvedic Medicines – Regulatory Aspects.....21
V. Remya, Alex Thomas and D. Induchoodan

Chapter 3 Industrial Manufacture of Traditional Ayurvedic Medicines.....41
Nishanth Gopinath

Chapter 4 Quality Control of Ayurvedic Medicines.....71
V. Remya, Maggie Jo Alex and Alex Thomas

Chapter 5 Scientific Rationale for the Use of Single Herb Remedies in Ayurveda.....101
S. Ajayan, R. Ajith Kumar and Nirmal Narayanan

Chapter 6 Biological Effects of Ayurvedic Formulations.....135
G.R. Arun Raj, Kavya Mohan, R. Anjana, Prasanna N. Rao, U. Shailaja and Deepthi Viswaroopan

Chapter 7 Evidence Building in Ayurveda: Generating the New and Optimizing the Old Could Be Strategic.....161
Sanjeev Rastogi, Arindam Bhattacharya and Ram Harsh Singh

Chapter 8 Conservation – A Strategy to Overcome Shortages of Ayurveda Herbs.....175
S. Noorunnisa Begum and K. Ravikumar

Chapter 9 Lessons to Be Learnt from Ayurveda: Nutraceuticals and Cosmeceuticals from Ayurveda Herbs.....199
Prachi Garodia, Sosmitha Girisa, Varsha Rana, Ajaikumar B. Kunnumakkara and Bharat B. Aggarwal

Chapter 10 Ayurveda in the West.....223
Atreya Smith

Chapter 11 Ayurveda Renaissance – *Quo Vadis*?.....241
D. Suresh Kumar

Index.....273

Preface

Ayurveda or "the sacred knowledge of longevity" has been practiced in India and many Asian countries since time immemorial. Names of the celestial physicians, Aśvinikumāra (*Dasra* and *Nāsatya*) appear in the documents excavated from Boghaz Koyi in the Cappadocia region of present-day Turkey. Similarly, the discovery of the Ayurveda text *Nāvanītakam* from a ruined Buddhist monastery near Kuchar in Chinese Turkestan indicates the popularity of Ayurveda in that faraway land. In India, this medical system occupied a lofty position as a result of royal patronage.

Western medicine was introduced into India by the Portuguese in the 16th century. Alfonso de Albuquerque, the Portuguese commander and conqueror of Goa, founded the *Hospital Real* (Royal Hospital) in 1510, which was converted in 1842 into the School of Medicine and Surgery. Although it was the Portuguese who introduced Western medicine into India, it was the British who later established and consolidated both its practice and study in the subcontinent. The history of Western medicine in India is almost exclusively that of the development of medicine during British rule. The report of the committee (1943) headed by Sir Joseph William Bhore was put into practice in independent India. Consequently, Western medicine became the dominant medical system of the country.

Interest in Ayurveda started growing all over the world in the late 1970s. To encourage national and international efforts to develop and implement primary healthcare throughout the world, the World Health Organization convened the International Conference on Primary Health Care (6–12 September 1978) at Alma Ata, in the former Soviet republic of Kazakhstan. This conference adopted the famous Alma Ata Declaration, which called on member nations to formulate national policies, strategies and plans to launch and sustain primary healthcare. The member states were especially encouraged to mobilize their own national resources. The Western world was thus inspired to study in depth the various traditional medical systems of the world. Ayurveda was an important one among them, having a sound theoretical basis.

The enthusiasm generated by spiritual organizations, such as the Chinmaya Mission, International Sivananda Yoga Vedanta Center, International Society for Krishna Consciousness (I.S.K.C.O.N.), Isha Foundation, Maharishi Foundation, Osho Foundation and the Art of Living International Center, was also a reason for the newfound interest in Ayurveda. In Germany, Austria and Switzerland, Ayurveda is one of the fastest-growing complementary and alternative medical systems.

In spite of the growing interest in Ayurveda, a closer look at the system prevailing in India reveals that it is suffering from many shortcomings. The theory of Ayurveda is based on unique concepts, which are very different from those of Western medicine. This fundamental difference between the two systems is ignored in contemporary Ayurveda to such an extent that Ayurveda is taught and practiced along the lines and with the help of Western medicine. Many stumbling blocks have to be removed if Ayurveda is to cater to the needs of a wider audience. This book attempts to identify such areas. Much progress has been made since the Alma Ata Declaration in understanding the theoretical constructs of Ayurveda. This book reviews that progress as well.

Chapter 1, entitled *What We Learn from the History of Ayurveda*, traces a few of the missions undertaken by ancient personalities to rediscover Ayurveda, and presents for the first time some evidence that is very valuable from the perspective of the history of medicine. The chapter objectively unravels the genesis of Unani medicine, the popular medicine of the Muslim society of northern India. What Ayurveda and Unani learned from each other is also highlighted. The chapter concludes with a call to revive the culture of inquisitiveness and questioning for advancing the Ayurveda renaissance.

The various regulations concerning the commercial manufacture of ayurvedic medicines are generally known only to those related to the industry. Not much published literature is available on the regulatory aspects of the manufacture of ayurvedic medicines. Chapter 2 describes the various steps to be undertaken to obtain a manufacturing license or loan license, good manufacturing practices, an inspection of the facility, labeling of products, the shelf-life of ayurvedic medicines, punishment for violation of the Drugs and Cosmetics Act (1940) and regulatory problems that require solutions.

Chapter 3 deals with the industrial manufacture of traditional ayurvedic medicines. The various steps involved, from procurement and pre-processing to industrial production of important dosage forms like decoctions, powders, medicinal lipids, fermented liquids, pills and pastes, are described in this chapter.

Many quality control methods are employed at present for assuring the quality of ayurvedic medicines. These methods are applied for the detection of adulterants, authentication of herbs, evaluation of raw materials and quality control of finished formulations. Chapter 4 discusses quality control in general and some problems, such as heavy metals in ayurvedic pills, souring of fermented liquids and irrational substitution of ingredients in formulae.

Ayurveda employs many single herbs in the treatment of diseases. Chapter 5 presents scientific rationales for the use of some such single herb remedies for a variety of ailments like epilepsy, irregular fever, rheumatoid arthritis, urinary stones, jaundice, dysentery, heart disease, wounds, hemorrhoids and dysfunctional bleeding.

Ayurveda makes use of several classes of medicine in the treatment of diseases. Expressed juices, decoctions, spirituous liquors, medicinal lipids, electuaries, pastes, powders and pills are the important ayurvedic dosage forms. Instructions for preparing these dosage forms are available in *Śārṅgadhara Samhita*, composed in the 13th century A.D. and considered to be the authoritative text on ayurvedic pharmacy. Ayurvedic medicines bring about a wide range of effects in the body. Chapter 6 deals with the various pharmacological and biological effects of ayurvedic formulations.

Evidence building in Ayurveda is a long-sought need, requiring careful planning and efforts to bridge the ancient wisdom with emerging knowledge. The observational strength of Ayurveda needs to be utilized in generating trusted evidence and filling the gaps of knowledge. Chapter 7 offers some suggestions for the optimization of ancient wisdom and generation of knowledge using newer technology.

Medicinal plants are globally valuable sources of herbal products, and they are disappearing at high speed. Chapter 8 provides an overview of the flora of the country, detailed insight into Indian medicinal plants and different methods to conserve the important, threatened, traded medicinal plants. It also illustrates the sustainable use of medicinal plant resources to provide a reliable reference for their conservation and sustainable use. This chapter also provides examples and success stories of conservation of some of the important threatened medicinal plants with restricted distribution.

Almost all therapeutic agents used in orthodox medicine are chemical compounds that are synthesized, costing billions of dollars, and exhibiting minimum benefits. These drugs also exhibit numerous side-effects. In view of these side-effects of modern medicines, Ayurveda has the potential to treat most chronic diseases using herbs. Chapter 9 highlights the nutraceuticals and cosmeceuticals derived from ayurvedic herbs that are inexpensive and highly efficacious with minimal side-effects.

Ayurveda was first presented to the Western public as a form of "wellness" and not as a medical system. Westerners are mainly exposed to "Wellness Ayurveda" through the media and spa-type treatments. Confusion between wellness and traditional medicine is increased by a lack of clarity from the Indian regulatory bodies and an outdated 70-year-old medical syllabus that is continuously being thrust upon Western governments. Chapter 10 describes what ails Ayurveda in the West and how to bring order to the scene.

Though the Ayurveda renaissance has sparked an interest in understanding the theoretical constructs of Ayurveda, many stumbling blocks have to be removed if Ayurveda is to cater to the needs of a wider audience. This is the major theme of Chapter 11. The botanical identity of herbs, identification of formulae with fewer ingredients, the discovery of new dosage forms, the substitution of plant parts, pharmaceutics of ayurvedic medicines, objective ways of teaching Ayurveda and the diagnosis of diseases on the basis of principles of Ayurveda are the major topics discussed in this chapter.

Many persons helped in the writing of this book by offering valuable help. They include Antony Selvaraj, Bangalore; T.S. Salim, Cherai; Migy John, Koratty; Dr J. Radhakrishnan, Oachira; Muhammed Murshid, Pulamanthole, Rishikesh and Kottayam; and Prof. R. Ramakumaran Thampuan, Trippunithura. The photographs related to ayurvedic medicine manufacturing were kindly provided by Nagarjuna Herbal Concentrates Ltd, Thodupuzha (www.nagarjunaayurveda.com) and Rajah Healthy Acres, Koottanad, Kerala (www.ayurvedichospital.com). I am thankful to Dr D. Suresh Baburaj, Ooty, for permitting the reproduction of the photographs of medicinal plants from his personal collection. All the graphics in this book were prepared by Alex Thomas, CIPLA, Mumbai.

The quality of the chapters of this book was enhanced by their revision based on the valuable comments received from experts who read them critically. In this regard, I thank Prof. Kenneth G. Zysk, University of Copenhagen, Denmark; Prof. V.K. Joshi, Emeritus Professor, Banaras Hindu University, Varanasi; Dr A. Kumaran, National Institute for Interdisciplinary Science and Technology (N.I.I.S.T.), Trivandrum; Prof. Robert Verpoorte, Leiden University, Netherlands; Dr P. Jayamurthy, N.I.I.S.T., Trivandrum; Dr A.P. Raman Nambudiri, Divakara Concepts, Coimbatore; Dr Job Thomas, The Arya Vaidya Pharmacy (Coimbatore) Ltd, Trivandrum; Dr D. Suresh Baburaj, formerly of the Centre for Medicinal Plants Research in Homoeopathy (C.C.R.A.S.), Ooty; Prof. Wandee Gritsanapan, Phyto Product Research, Bangkok; Neelam Toprani, Padmashri Naturals Inc. Vancouver and Dr Gary C. Beauchamp, Monell Chemical Senses Center, Philadelphia, for reviewing, respectively, Chapters 1–11. I am thankful to Renu Upadhyay, Shikha Garg and Jyotsna Jangra of Taylor and Francis Books India Pvt Ltd, New Delhi, for answering many of my queries promptly and helping in the successful completion of this project.

As editor of this book I am indebted to all the contributors for providing manuscripts in time. The contributors and the editor have the best intentions and efforts in producing this book. We hope that, despite any shortcomings, it will be a useful guide to take Ayurveda to greater heights in the new millennium. This book will be of interest to exponents of Ayurveda and all branches of traditional and alternative medicine.

About the Editor

D. Suresh Kumar was born on 21 September 1949, in the southern Indian province of Kerala, where he received his early education. He obtained a B.Sc. degree in zoology from the University of Kerala (1969) and secured M.Sc. (1972) and Ph.D. degrees (1977) from Banaras Hindu University, Varanasi. His doctoral thesis was on the hormonal control of oxidative metabolism in reptiles. Thereafter, he spent two years as a postdoctoral fellow in the Department of Biological Sciences, University of Aston in Birmingham, England, investigating the pancreatic physiology of the rainbow trout. He returned to India in 1980 and joined the Department of Zoology, University of Calicut, Kerala, as pool officer in the scientist pool of the Council of Scientific and Industrial Research, New Delhi. During his stay there, a chance encounter with some religious persons introduced him to the study of Ayurveda, the traditional medical system of India. He undertook a survey of the state of Ayurveda in the province and published his findings in provincial and national weeklies. In 1986, he joined the International Institute of Ayurveda, Coimbatore, as a research officer in the Department of Physiology. From 1986 to 2003, he carried out research on various aspects of Ayurveda. In collaboration with Dr Y.S. Prabhakar, presently at C.D.R.I. Lucknow, he proposed the first mathematical model for the ayurvedic concept of *tridōṣa* in the disease state. He also offered a novel definition for the ayurvedic class of medicine *arka*, based on his study of the Sanskrit text *Arkaprakāśa*. In 2003, he joined Sami Labs Ltd, Bangalore, as a senior scientist in the R&D laboratory. He spent several years in the company working on various aspects of new product development. From 2012 to 2016, he worked as head of the R&D laboratory of the Ayurveda consortium, Confederation for Ayurveda Renaissance Keralam Ltd, Koratty, Kerala. In January 2016, he joined Cymbio Pharma Pvt Ltd, Bangalore as Head of New Product Development, and since May 2020 he has worked as an Ayurveda consultant. He is the author of *Herbal Bioactives and Food Fortification: Extraction and Formulation* published by CRC Press (2016). Email: dvenu21@yahoo.com.

Contributors

Bharat B. Aggarwal
Director
Inflammation Research Center
San Diego, California

S. Ajayan
Ashtangam Ayurveda Vidyapeedham and Chikitsalayam
Palghat, India

Maggie Jo Alex
Aruna College of Pharmacy
Tumkur, India

R. Anjana
K.M.C.T. Ayurveda Medical College
Manassery, India

S. Noorunnisa Begum
Centre for Conservation of Natural Resources
University of Trans-disciplinary Health Sciences and Technology
Bangalore, India

Arindam Bhattacharya
Ministry of External Affairs
Government of India
New Delhi, India

Prachi Garodia
Integrated Health and Research Center
Ashland, Oregon

Sosmitha Girisa
Cancer Biology Laboratory and DAICENTER
Indian Institute of Technology Guwahati
Guwahati, India

Nishanth Gopinath
Nagarjuna Herbal Concentrates Ltd
Thodupuzha, India

N.K.M. Ikbal
Vaidyaratnam Ayurveda Medical College and Hospital
Thrissur, India

D. Induchoodan
Rudraksha Ayurvedic Holistic Centre
Thrissur, India

R. Ajith Kumar
Vishnu Ayurveda College
Shoranur, India

Ajaikumar B. Kunnumakkara
Cancer Biology Laboratory and DAICENTER
Indian Institute of Technology Guwahati
Guwahati, India

Kavya Mohan
Rajeev Institute of Ayurvedic Medical Sciences and Research Centre
Hassan, India

Nirmal Narayanan
Vishnu Ayurveda College
Shoranur, India

G.R. Arun Raj
S.D.M. College of Ayurveda and Hospital
Hassan, India

Varsha Rana
Cancer Biology Laboratory and DAICENTER
Indian Institute of Technology Guwahati
Guwahati, India

Prasanna N. Rao
S.D.M College of Ayurveda and Hospital
Hassan, India

Sanjeev Rastogi
State Ayurvedic College and Hospital
Lucknow, India

K. Ravikumar
Centre for Conservation of Natural Resources
University of Trans-disciplinary Health Sciences and Technology
Bangalore, India

V. Remya
Vaidyaratnam Oushadhasala Pvt Ltd
Thrissur, India

U. Shailaja
S.D.M College of Ayurveda and Hospital
Hassan, India

Ram Harsh Singh
Faculty of Ayurveda
Institute of Medical Sciences
Banaras Hindu University
Varanasi, India

Atreya Smith
Montreux 1
Switzerland

Alex Thomas
Cipla Ltd, Vikhroli
Mumbai, India

Deepthi Viswaroopan
Parul Institute of Ayurveda
Vadodara, India

1 What We Learn from the History of Ayurveda

N.K.M. Ikbal, D. Induchoodan and D. Suresh Kumar

CONTENTS

1.1 Introduction 2
1.2 Learning from Animals 2
1.3 Evolution of Medical Systems 2
1.4 India's Gift to the World 3
  1.4.1 Dṛḍhabala's Mission 4
  1.4.2 The Legacy of Vāgbhaṭa 4
1.5 Ayurveda and Tamil Medicine 7
1.6 The Puzzling Dosage Form Arka 8
1.7 Interaction between Ayurveda and Islamic Medicine 9
  1.7.1 Origin of Greco-Arabic Medicine 9
  1.7.2 Indian Influence on Greco-Arabic Medicine 9
  1.7.3 Emergence of Ūnāni 10
1.8 The Influence of Ayurveda on Ūnāni 11
  1.8.1 Development of Khamīra 11
  1.8.2 Development of Mā'jun 11
  1.8.3 Development of Kuṣṭā 12
  1.8.4 Acceptance of New Plants into Formulary 12
1.9 How Ayurveda Benefited from Greco-Arabic Medicine 12
  1.9.1 Pulse Examination 12
  1.9.2 Description of New Disease Entities 12
    1.9.2.1 Snāyukarōga 12
    1.9.2.2 Munnātakhyārōga 12
    1.9.2.3 Vardhma 12
  1.9.3 Acceptance of New Plants into Ayurveda 12
  1.9.4 Adoption of New Dosage Forms 13
  1.9.5 Interest in Ūnāni 13
1.10 Some Curiosities in Ayurveda Theory 14
  1.10.1 Prabhāva (Special Potency) 14
  1.10.2 Gender Bias 14
  1.10.3 Futuristic Provisions in Ayurvedic Pathogenesis 15
    1.10.3.1 Omission of Tenets 15
    1.10.3.2 Addition of Explanation 16
    1.10.3.3 Introduction of New Concept 16
1.11 Conclusion 16
References 17

1.1 INTRODUCTION

Scientific evidence shows that disease is older than the human race. Studies of ancient fossils demonstrate that arthritis was widespread among a wide range of medium- and large-sized mammals like aardvarks, anteaters, bears and gazelles (Magner 2005). According to some scientists, many modern human infectious diseases arose during the Neolithic period or shortly afterwards due to close contact with domestic animals and their pathogens (Pearce-Duvet 2006; Wolfe et al. 2007; Trueba and Dunthorn 2012; Trueba 2014). Studies of the ancient remains of human bodies confirm that many diseases prevalent today affected ancient human populations as well. Ancient skeletons show signs of diseases like hydrocephalus, spina bifida, congenital clubfoot, Paget's disease, osteoporosis, rickets, osteomalacia, acromegaly, microencephaly and achondroplasia (Sigerist 1951; Steinbock 1976; Campillo 1983; Priorischi 1995a).

1.2 LEARNING FROM ANIMALS

Primitive man learned the rudiments of the healing arts from the animals around him. Nonhuman primates, and especially chimpanzees are considered to constitute "living relatives" of *Homo sapiens*. Their behavior can be interpreted to present an approximate view of the behavior of a common ancestor of man (Fábrega Jr 1997). Chimpanzees in the wild occasionally lick each other's wounds, and take care of each other's infirmities like their human counterparts. Chimpanzees have also been recorded squeezing pus from abscesses and removing foreign bodies from each other's eyes (Priorischi 1995b).

Animals are also known to eat medicinal plants to cure diseases or to prevent their appearance. For example, Phillips-Conroy (1986) suggested that the leaves and nutritious and tasty berries of *Balanites aegyptiaca* were eaten by the baboon, *Papio hamadryas*, as a prophylactic agent against schistosomiasis. This plant was observed to be a regular part of their diet along the Awash River in Ethiopia, where schistosomiasis was very much prevalent. Similarly, chimpanzees around the Gombe Stream and Mahale Mountains National Parks, Tanzania swallow *Aspilia mossambicensis* and *Aspilia pluriseta* slowly and without chewing (Wrangham and Nishida 1983; Huffman and Wrangham 1994). A report of a chemical analysis of the plant (Rodriguez et al. 1985) and another report on the plant *Lippia plicata* eaten in a similar manner at Mahale confirmed the hypothesis of Wrangham and Nishida (1983).

The chimpanzees of Mahale, Tanzania are also reported to chew the pith of ethnomedically important *Vernonia amygdalina*, suck out and swallow the astringent, bitter-tasting juice, spitting out the fibrous remnants, when they exhibit signs of lethargy, loss of appetite and irregularity of bodily excretions. The symptoms displayed by the sick chimpanzees are the same shown by people throughout tropical Africa (Huffman and Seifu 1989).

1.3 EVOLUTION OF MEDICAL SYSTEMS

The highly developed nervous system of man favored the recognition of changes in one's own body and perception of behavioral changes in other individuals of the same group (Fábrega 1997). This ability, coupled with the wealth of information gleaned from the behavior of wild animals and his own experiences, formed the basis of the medical system of primitive man. In their search for food, early human beings inevitably encountered toxic or therapeutic plants, depending on the amount ingested and the physique of the person. Herbal medicine can be said to have developed alongside adventures in determining the diet, and it can be inferred that herbal medicine is an offshoot of nutrition (Conway 2011). The primitive knowledge thus gained was expanded upon by learning and social transmission (Hart 2005, 2011). This gave rise to various forms of traditional healing, broadly called folk (traditional) medicine. Traditional medicine is the sum total of the knowledge, skills and practices based on the beliefs and experiences indigenous to various cultures used in the maintenance of health as well as in the prevention, diagnosis, improvement or treatment of physical and mental illness (Anonymous 2013). Folk medical practices do not have a theoretical foundation and are based on empirical knowledge.

Since the days of Descartes and the Renaissance, science has taken a clear path in its analytical evaluation of nature. This approach is rooted in the assumption that complex problems can be solved by dividing them into smaller, simpler, and thus more controllable, units. As the processes are "reduced" into more basic units, this approach is also called "reductionism". It has been the predominant paradigm of science over the past two centuries (Ahn et al. 2006). Using principles of pharmacognosy, solvent extraction of phytocompounds, pharmacology, pharmacokinetics and pharmacodynamics, Western medicine evaluates the efficacy of folk medical practices. Modern investigations have demonstrated the effectiveness of these forms of phytotherapy (Ramzan and Li 2015).

On the other hand, Ayurveda explains life processes using a doctrine based on the six schools of Hindu philosophy, namely, *nyāya*, *vaiśēṣika*, *sāmkhya*, *yōga*, *mīmāmsa* and *vēdānta* (Dasgupta 1997). Ayurveda states that the body is made up of *pañcabhūta* or the five primordial elements *pṛdhvi* ("earth"), *ap* ("water"), *tējas* ("fire"), *vāyu* ("air") and *ākāśa* ("sky"). The ability of the *pañcabhūta* to modulate life processes under the influence of a driving force (*ātma*) is denoted by the collective term *tridōṣa*, which consists of *vāta*, *pitta* and *kapha*. The body is said to be in a state of health when the *tridōṣa* exist in a steady state. Destabilization of the *tridōṣa* causes the appearance of diseases. Herbs and other medicinal substances are identified in Ayurveda on the basis of five characteristics – *rasa* (taste), *guṇa* (qualities), *vīrya* (potency), *vipāka* (post-digestive taste) and *prabhāva* (specific action). These are the ayurvedic counterparts of pharmacological characteristics (Murthy 2017a). Ayurvedic pharmacy makes use of many herbs used in Indian cuisine. Examples are asafetida (*Ferula asafoetida*), black pepper (*Piper nigrum*), coriander (*Coriandrum sativum*), cumin seeds (*Cuminum cyminum*), curry leaves (*Murraya koenigii*), bitter gourd (*Momordica charantia*), drumstick (*Moringa oleifera*), fenugreek (*Trigonella foenum-graecum*), ginger (*Zingiber officinale*), mustard (*Brassica juncea*), inflorescence and pseudostem of banana (*Musa paradisiaca*), onion (*Allium cepa*) and turmeric (*Curcuma longa*) (Kumar 2018). It is quite possible that, in the early stage of its development, Indian medical and culinary traditions worked hand-in-hand with each other (Conway 2011).

Thus, folk medicine, Western medicine and Ayurveda are three different approaches toward health and disease. For example, the tribal communities of India use the plant *Adhatoda vasica* in several ways to treat respiratory diseases. Flowers, root powder, decoctions of leaves or juice of leaves mixed with goat milk are administered for conditions like cough, cold, fever, bronchitis, malarial fever, asthma, tuberculosis and pneumonia (Arjariya and Chaurasia 2009; Desale et al. 2013). *Adhatoda vasica* is an important plant in Ayurveda, used in the treatment of coughs and other respiratory disorders. Formulations like *Vāśāriṣṭam* (Vaidyan and Pillai 1986), *Vāśā ghṛtam*, *Vāśā khaṇḍa kūśmāṇḍakam*, *Vāśā khaṇḍam* and *Khaṇḍakādyō lauham* are indicated in these conditions (Mishra 1983). Based on the information from Ayurveda, Amin and Mehta (1959) isolated for the first time vasicinone, a compound with bronchodilatory activity, in crystalline form from the leaves of *Adhatoda vasica*. One of the derivatives of vasicinone is bromhexine or bromhexine hydrochloride (N-cyclohexyl-N-methyl-(2-amino-3,5-dibromo-benzyl) amine hydrochloride). It was first introduced into the market in 1963 and has been used in the treatment of a variety of respiratory diseases (Zanasi et al. 2017). When used on the basis of different principles, the same medicinal plant is identified with different medical systems.

1.4 INDIA'S GIFT TO THE WORLD

Ayurveda is India's gift to the world. In the remote past when pen, pencil, paper, printing technology and data-processing systems were non-existent, ascetics conceived a comprehensive medical doctrine, put it to clinical testing and compiled voluminous works dealing with diagnosis and treatment of diseases. Though texts like *Khāranādi* and *Nimitantram* are mentioned in latter-day compendia, they are not available anymore. *Caraka Samhita*, *Suśruta Samhita*, *Aṣṭāṅgasamgraha* and *Aṣṭāṅgahṛdaya* are now the mainstay of Ayurveda.

As Ayurveda is "sacred knowledge of life" (Gode and Karve 1959) it was intended to be used wisely for the welfare of the world. Therefore, it was traditionally taught only to aspirants who had the required qualities. Great importance was attached to ethical values. This is reflected in the narration on the qualities of an aspiring student of Ayurveda. In the *Vimānasthāna* of *Caraka Samhita*, it is stated unambiguously that the aspirant should be of noble nature, humble, modest, good-natured, having no greed for wealth, compassionate to all creatures and endowed with good memory (Sharma 1981a). Because of the continuous training imparted by a sage master, a physician acquired spiritual wisdom and perfect knowledge of Ayurveda by the time he completed his studies. It is no wonder that such training equipped the physician to serve society effectively as well as Ayurveda. It also instilled in him the moral strength to undertake arduous missions. The lives of Dṛḍhabala and Vāgbhaṭa speak of this.

1.4.1 Dṛḍhabala's Mission

The oldest classical Ayurveda work in Sanskrit is the *Caraka Samhita*, believed to have been composed between 100 B.C. and 200 A.D. In olden days, textbooks were customarily memorized by students and scholars. It so happened that by A.D. 500 *Caraka Samhita* became a difficult-to-find text and the few manuscripts that survived were full of gaps. Reproduction of *Caraka Samhita* was attempted by Dṛḍhabala, a resident of *Pañcanada*, in present-day Punjab region (Maas 2010). He travelled extensively in India, met many of the scholars who had memorized the *Caraka Samhita* in its entirety, and transcribed the various versions. He compared the text stanza by stanza and edited the entire *Caraka Samhita*. The present-day *Caraka Samhita* contains 17 chapters of *Cikitsāsthāna*, the whole of *Kalpasthāna* and *Siddhisthāna* composed by Dṛḍhabala himself. Thus was born the present-day *Caraka Samhita*, which is a creative revision of its forerunner, the *Agnivēśa Tantra* (Valiathan 2006).

1.4.2 The Legacy of Vāgbhaṭa

The most perilous mission for the benefit of Ayurveda was the one undertaken by a young student of Ayurveda, who later became immortal under his pen name, Vāgbhaṭa (the one who uses words as his soldiers). Interestingly, this legend is narrated only in the Malayalam literary work *Aitihyamāla* (Garland of legends), compiled by Kottarathil Sankunni and first published in 1909 (Sankunny 1982).

With the conquest of Sind by Mohammed-bin-Qasim in 712 A.D. (Asif 2016), Muslim supremacy in that region increased. Slowly the medical profession also came under their domination. All the healers of the region were Muslim and most of the Sanskrit texts of Ayurveda had disappeared by that time. There were no physicians among Hindus of that region, as the Muslim *hakim*s (physicians) refused to accept Hindu pupils. Saddened by this state of affairs, members of the priestly class held a conclave by the Sindhu (Indus) river. They decided to depute an intelligent and valiant youth to recover the appropriated sacred knowledge of life. The young son of Simha Gupta was chosen for this mission. He was instructed to disguise himself as a Muslim youth, impress the best *hakim* and become his pupil. He was assured that a *yajña* (sacrificial rite) would begin that very day and would be concluded only when his mission came to an end.

Young Vāgbhaṭa could identify a learned *hakim*, who interviewed the young man to ascertain whether he had the required qualities. The *hakim* was impressed by the knowledge and mental faculties of the newcomer. He accepted him as a pupil. The young man came to his master every morning and after studying for the day, left in the evening, saying that he had a family to look after on the other side of the river. This went on for many months and the master was very happy with his pupil. Then one day the master expressed his willingness to teach him, if he could come back after having supper at home. Vāgbhaṭa, who was eager to complete his studies as soon as possible and escape without betraying his identity, happily accepted the offer.

The *hakim* lived in a palatial manor and the evening session was conducted in his chamber. The special training in the evening was a prerogative offered only to Vāgbhaṭa, who mastered most of the medical books in a short time. The *hakim* would explain late into the night the intricacies of Ayurveda and Vāgbhaṭa would listen to him very attentively. Then one night the master felt some uneasiness in his leg. He requested his pupil to sit on the bed and massage his leg for some time. Vāgbhaṭa quickly obliged the master, who soon fell asleep because of the relief from the massage. Then a train of thought flashed through the young man's mind. "I was born in a noble family and have studied the four *vēda* and six *śāstra*. It is really my misfortune to hold the feet of this barbarian", he bemoaned, unbecoming of a true Hindu, who should view his master (guru) as an equal to his father and mother. The grief caused some teardrops to fall on the *hakim*'s leg. The *hakim* woke up from sleep and saw a grief-stricken Vāgbhaṭa, with tears rolling down his cheeks. The old man immediately realized that he had been fooled by an imposter. With a revengeful mind he slowly rose from the bed, hoping to catch hold of a sword placed on the wall. At the same time the quick-witted Vāgbhaṭa realized that he would soon meet his end if he did not escape. Telling himself that no harm would fall on him if there were a power called Almighty and if the four *vēda* were true, he jumped through a window in the room. He landed on the ground with no serious injury and took to his heels, escaping from his master and the guards.

The priests were holding the *yajña*, when suddenly Vāgbhaṭa appeared in front of them. He narrated to them all that had happened, culminating in his fleeing. They listened to his words and replied, "You still doubt if there is an Almighty and whether the four *vēda* are true. We do not want to associate ourselves with such a person. From now onwards you are an outcast". Vāgbhaṭa accepted the punishment and left the group. However, he started documenting all that he had learnt from the master, so that his hard work would not go in vain. Moreover, it would have been nearly impossible to usurp the knowledge from the Muslims a second time.

The first treatise that Vāgbhaṭa recreated is known as *Aṣṭāṅgasamgraha*. But as it was composed in prose and poetry, he thought students might find it hard to memorize the text. Therefore, he composed in poetic style *Aṣṭāṅgahṛdaya*, which is a summary of all the earlier Sanskrit works and which is more abridged than *Aṣṭāṅgasamgraha*. After offering these two texts at the feet of his mentors, Vāgbhaṭa left Sind forever (Sankunny 1982).

It is believed that after travelling all over northern India, Vāgbhaṭa finally arrived in Kerala, which was part of the Tamil country at that time. He spent the remaining years in the new-found home, disseminating the knowledge enshrined in his treatise. The *Aṣṭavaidya* tradition of Kerala is said to have begun with him. Indu and Jajjaṭa, famous commentators of *Aṣṭāṅgahṛdaya* are said to have been his disciples (Variar 1985, 1987). The *Aṣṭavaidyas* of Kerala chant every day the following *dhyāna śḷōka* during their daily worship of Vāgbhaṭa:

Lamba śmaśru kalāpamambuda nibha chāyādyutīm vaidyakānām
Antēvāsina indu jajjaṭa mukhān addhyāpayantam sadā
Āgulphāmala kañcukāñcita tanum lakṣyōpavītōjvalam
Kaṇḍhasthāgaru sāramañcita dṛśam vandē gurum vāgbhaṭām!

(I pay obeisance to my guru Vāgbhaṭa, who is long-bearded, having bright eyes and the complexion of clouds, who teaches his physician-disciples Indu and Jajjaṭa, wearing a white robe that flows up to the feet, and wearing around his neck, a garland made of sweet-smelling *aguru* wood!) (Figure 1.1) (Rishikesh, Vayaskara Illam, personal communication).

Members of the Pulamanthole Mooss family in northern Kerala believe that the sepulcher that exists in their estate and which is still venerated, contains the entombed body of Vāgbhaṭa (Figure 1.2). The two stone lamps over the sepulcher are lit every evening and special devotional service is held on all full moon days (Anonymous 2019).

FIGURE 1.1 Vāgbhaṭa teaching Indu and Jajjaṭa – an artist's impression (Photo courtesy of J. Suryanarayanan, Pattampi).

FIGURE 1.2 The sepulcher that is believed to contain the mortal remains of Vāgbhaṭa (Photo courtesy of Muhammed Murshid, Pulamanthole).

Some linguistic evidence is also available to adduce the impact of the personage of Vāgbhaṭa on the south Indian medical tradition. *Vākaṭam* is a word commonly employed in Tamil medical literature. *Piḷḷaippiṇi Vākaṭam* (Book of diseases of children) and *Kuḷantai Vākaṭam* (Book of diseases of children) are two examples (see Kasinathan 1955). According to the Tamil-English medical dictionary of T.V. Sambasivam Pillai, the word *Vākaṭam* means "a book that describes perfectly the measures to cure diseases" (Pillai 1931a). Dr. Asko Parpola, the Finnish Tamil scholar, is of the opinion that the word *Vākaṭam* is derived from the Sanskrit word Vāgbhaṭa (Dr. Asko Parpola, personal communication). Physicians of the *Aṣṭavaidya* tradition of Kerala refer to Vāgbhaṭa as *Vāhaṭa* or *Vāhaṭan* (Mooss 1982).

1.5 AYURVEDA AND TAMIL MEDICINE

This brings us to a discussion on the interrelationship between Ayurveda and Tamil medicine, popularly called Siddha medicine, presently being practiced predominantly in Tamil Nadu and marginally in Kerala, Karnataka and Andhra Pradesh (Reddy 1973; Kumar 1995). The Siddha system is said to have an ancient origin. Akattiyar (*Agastya* in Sanskrit) is believed to have conceived this system incorporating many elements of *Śaiva tantra* into the medical practices of the time (Pillai 1931b). The various legends associated with Akattiyar have prompted many authors to speculate on the history of Tamil medicine. Gurusironmani (1972, 1983) is of the opinion that the Siddha system must have had its origin around 5000 B.C. On the other hand, the legend in the commentary of the ancient work *Akapporuḷ* emboldens some to fix the date of the commencement of the first Tamil Sangam (*Caṅkam*) period to 9000 B.C. (Kumar 1995). However, such statements are hardly substantiated with authentic evidence.

Based on a comparative study of references to the "three faults" discussed in the Tamil work *Tirumantiram*, some Pāli medical texts and the treatises of Caraka and Suśruta, Scharfe (1999) concluded that the theoretical foundation of the medicine of the Tamil *cittars* denoted by the terms *muttōcam* and *mukkuṟṟam*, is based on the later development of the doctrinal principles found in *Caraka Samhita* and *Suśruta Samhita*.

Tirukkuraḷ, a philosophical treatise usually assigned to the 3rd century A.D., but may be actually from the 6th century, states in *kuraḷ*, 941 that imbalance of "the three starting with wind" causes diseases. It is significant that the word *nūlōr* (the learned books) implies the existence of a well-developed medical system anterior to the *Tirukkuraḷ* (Venkatraman 1990a).

Though no ancient Tamil medical texts are available to us, certain Sangam (*Caṅkam*) works (around 400 A.D.) like *Tirikaṭukam*, *Cirupañcamūlam* and *Ēlāti* are named after reputed medicinal formulae of Ayurveda. *Cilappatikāram* (approximately 450 A.D.) mentions that *āyuḷvētar* (experts in Ayurveda) used to live in the city of Kāviripūmpaṭṭiṇam next to priests and astrologers. Emperors of the *Cōḷa* dynasty (850–1280 A.D.) established and financed hospitals and medical schools. A stone inscription unearthed from Tiruvāvaṭuturai and dated to 1121 A.D. states that the students of such medical schools were taught *Aṣṭāṅgahṛdaya* and *Caraka Samhita* (Venkatraman 1990a).

Venkatraman is of the opinion that alchemy and tantrism, which are hallmarks of the Tamil Siddha system, owe their origin to influences from northern India. Around 1200 A.D. the Buddhist monasteries of Udantapura and Vikramaśila (in present-day Bihar) were destroyed by invaders, and the monks fled southwards carrying the esoteric knowledge with them. Nathism or the cult of Nāthasiddhas, which was already in existence in the north, rose to prominence following the downfall of Buddhist institutions in the Tamil country. These Nāthasiddhas are held to be directly responsible for the development of the Tamil Siddha system. The Tamil Siddha *Koṅkaṇar* states in *Koṅkaṇar Nāṭi* that pulse diagnosis was a gift from the Nāthasiddhas (Venkatraman,1990a). Similarly, *Caṭṭaimuni* states in *Caṭṭaimuni Nikaṇṭu*-1200 that alchemy and "the secret of arsenics" were propounded by the Nāthasiddhas (Venkatraman 1990b). The present-day Siddha system is an amalgam of Ayurveda and tantrism of the Nathasiddhas.

TABLE 1.1
Some Evidence From *Vaittiyacintāmaṇi*-800 Suggesting Its Ayurvedic Origins

Reference	Statement
*Pāṭal**1	I narrate in this *Vaittiyacintāmaṇi* the principles of *āyuruvētam* as narrated by the god Tanvantiri.
Pāṭal 3	Teraiyar passed it on to me and I, Yūki, record it hereby for the benefit of mankind. After studying the northern language and after receiving the blessings of the elders of that region, I hereby recite that knowledge.
Pāṭal 392	The knowledge which came from the northern language has been narrated by me without concealing any point.
Pāṭal 422	Akattiyar passed on the knowledge to Pulattiyar. He passed on this great *āyuruvētam* to sage Teraiyiar, who passed on this knowledge to me. I hereby narrate the characteristics of *piramēkam***.
Pāṭal 450	I, Yūkimuni, delved deep into the mystical *āyuruvētam*, compiled the knowledge and recorded it clearly and faithfully as the words of Siva.
Pāṭal 669	After studying the works in the northern language, I composed these verses in Tamil for the benefit of the world.
Pāṭal 743	The knowledge of healing is the foremost among the six bodies of knowledge (*cāttiram*). It is the knowledge of *āyuruvētam*. It is enshrined in 700,000 texts. I, Yūkimuni, composed this work after meticulous study of texts of the northern language and after receiving the blessings of the master.

*Verse
***pramēha* in Sanskrit

Majeed and Kumar (2008) reported their study of *Vaittiyacintāmaṇi*-800, an exceptional Tamil medical work devoted to nosology (classification of diseases). This work is attributed to the Tamil sage Yūkimuni. The author states at several places in the text that he composed the work in the Tamil language after studying in detail all the "northern texts" of *āyuruvētam*. It is obvious that the words "northern language" (*vaṭamoḷi*) and "northern texts" (*vaṭamoḷiyin nūlkaḷ*) denote respectively the Sanskrit language and the Sanskrit texts. Other evidence available in *Vaittiyacintāmaṇi*-800 is presented in Table 1.1.

All authoritative Sanskrit texts of mainstream Ayurveda follow a relatively rigid pattern of classification of diseases. The *Suśruta Samhita*, *Caraka Samhita*, *Aṣṭāṅgasamgraha* and *Aṣṭāṅgahṛdaya* follow more or less the same nosology. This orthodox pattern is evident in most of the ayurvedic literature even up to present times. However, in *Vaittiyacintāmaṇi*- 800, one observes a heterodox approach to the classification of many diseases like *kācam*, *kāmālai*, *kirāṇi* and *makōdaram* (*kāsa*, *kāmila*, *grahaṇi* and *mahōdaram* in Sanskrit). This indicates that Tamil medicine was not constrained by the rigors of ayurvedic dogmatism. Socio-historical factors prevalent in older times might have contributed to this development (Majeed and Kumar 2008).

1.6 THE PUZZLING DOSAGE FORM ARKA

Ayurveda employs several classes of medicine in the treatment of diseases. According to the medieval text *Śārṅgadhara Samhita*, considered today to be the authoritative work on ayurvedic pharmacy, expressed juices (*svarasa*), decoctions (*kvātha*), spirituous liquors (*āsava*, *ariṣṭa*), oils (*taila*), clarified butter (*ghṛta*), electuaries (*lēhya*), pastes (*kalka*), powders (*cūrṇa*) and pills (*guḷika*) are the important classes of medicines (Murthy 2017b). The Sanskrit treatise *Arkaprakāśa*, attributed to King Rāvaṇa of *Rāmāyaṇa* fame, describes yet another variety called *arka* (Kumar 1992a).

The equipment recommended in the preparation of *arka* is very similar to an alembic-like still used in distillation. However, closer examination of the text reveals recipes of many *arka* involving blood, milk, honey, clarified butter and ingredients having little volatile components, like rock-salt, bamboo manna, lac, cuttlefish bone, sulfur, soot from chimneys of houses and so on. Interesting findings emerged from a comparison of the five single drug and 12 compound drug formulations described in *Arkaprakāśa* with similar recipes of *Cakradattam*, composed in the 11th century by Cakrapāṇidatta (Sharma 1994). Not even one of these 17 identical medicines described in *Cakradattam* involves distillation. The author of *Arkaprakāśa* cautions that one should never administer an *arka* that has developed a foul odor. Fumigation of such malodorous *arka* with sulfur is recommended to make them "powerful and noble like the sun". These and several other pieces of evidence suggest that an *arka* may be a pooled mixture of the hydrodistillate fraction representing the volatile compounds and the aqueous extract remaining in the distillation vessel. Comparison of the chemical and pharmacological profiles of *arka* and decoctions of single and compound formulations can be useful in ascertaining the correct identity of the dosage form *arka* (Kumar 1992a).

1.7 INTERACTION BETWEEN AYURVEDA AND ISLAMIC MEDICINE

1.7.1 Origin of Greco-Arabic Medicine

According to historians, Arabic medicine of the pre-Islamic period was of an empirical nature (Hamarneh 1990). After the capitulation of Alexandria in 642 A.D., Greek medicine was introduced to the Arab world and it started influencing the native medical practices. The works of Hippocrates and Galen were warmly received by Arab physicians. The erudite scholar Abu Yusuf Ya'qub bin Ishaq al-Kindi (died ca. 871 A.D.) composed more than 20 treatises on medicine. One of them was specifically on Hippocratic medicine. The writings of Abu Bakr Muhammed ibn Zakariyya al-Razi (850–925 A.D.) were also greatly influenced by Greek medical literature (El-Gammal 1993).

With the founding of the Abbasid caliphate in Baghdad, greater attention was paid to Hellenistic medicine. Al-Ma'mun, the most liberal among the Abbasids, welcomed intellectuals to his court. In 833 A.D. he founded the famous *Bayt-al-Hikma* (House of Wisdom) which had an important influence on the transmission of ancient learning to the Islamic world and to stimulate a burst of intellectual activity (Anawati 1970). In this prestigious institution scholars were engaged on a full-time basis to translate medical works into Arabic. By the 850s almost all of Galen's works had been rendered into Arabic (El-Gammal 1993). Greco-Arabic medicine was thus born out of a synthesis of Greek and Arab medicine.

1.7.2 Indian Influence on Greco-Arabic Medicine

A significant influx of Hindu thought into Arabia took place during the reign of the liberal caliph, Harun al-Rashid (786–809 A.D.). Under his patronage many Sanskrit texts were translated into Arabic. The first Indian to make his mark was Manka, appointed in the Royal Barmecides Hospital in Baghdad. Well-versed in Persian and Arabic, this saintly person translated many Sanskrit medical texts into Arabic. Ibn Dhan and Saleh-bin-Bhela were two other famous Indian physicians of Baghdad (Keswani 1974).

In *Uyūn-al-anbā fi tabāqat-al-atibbā*, the 12th book of history of physicians, Ibn Abi Usaybia (died 1270 A.D.) gives a full list of some Indian works studied by Arabs. The title of one of them is *Bdān* or *Ndān*. Usaybia mentions that characteristics of 404 diseases are described in this work, without indicating their treatment. This points toward the *Mādhavanidāna* (Mueller 1880). Madhava's treatise is also described in a similar manner by the Arab historian Ibn Wadih-al-Yaqubi (850 A.D.) (Meulenbeld 1974).

Greco-Arabic medicine owes much to pioneers like Ali ibn Sahl Rabban-al-Tabari who became secretary to Prince Mazyar ibn Qarin in the Persian province of Tabaristan. In 850 A.D. he completed the book *Kitāb Firdaus al-Hikma* (The Paradise of Wisdom). It contains a mixture of rational and magical observations of nature and concludes with a discussion of Ayurveda. Al-Tabari had depended upon Persian and Arabic translations of the treatises of Caraka, Suśruta, Vāgbhaṭa and Mādhavakara, as he mentions *Jrk*, *Susrud*, *Ashtanghrdy* and *Ndān* (El-Gammal 1993; Meulenbeld 1974). Roşu (1988) states that *Kitāb Firdaus al-Hikma* contains details of a *yantra* representing a magic square of the order three, originally found in *Vṛnda*'s *Siddhayōga* (900 A.D.)

1.7.3 Emergence of Ūnāni

Though the Muslim presence in India is said to have begun with the military expedition of Mahmud of Ghazna (1014 A.D.), evidence suggests that the interaction between Hindus and Muslims began much earlier. Sayyid Sulayman Nadvi remarks that during the caliphate of Umar (636 A.D.), the governor of Bahrain attacked Thana (Bombay) and later Bahruch and Daybu on the Gujarat coast (Ahmad 1988). Greco-Arabic medicine reached India with these visitors. It is said that the new system of medicine was not easily accepted on account of the temperament of the people and the relatively superior nature of Ayurveda. Therefore, a hybrid amalgam of Greco-Arabic medicine and Ayurveda was slowly developed. This new medical system later came to be known as *Ūnāni Tibb* or *Ūnāni* medicine. *Ūnāni* was enriched by the addition of many therapeutic measures of Ayurveda, especially the use of medicinal plants, many of which had not then reached the Arab world (Verma and Keswani 1974a, 1974b).

Under the patronage of Muslim rulers, scholars translated many Sanskrit texts into Arabic or composed *Ūnāni* treatises borrowing profusely from Ayurveda. Zia Muhammed Mubarak, a courtier of Muhammed Tughlaq (1525–1351 A.D.) composed *Majmā-e-Ziāyi* (Collections of Zia), which had a separate chapter on medicine as prescribed by Nāgarjunā and other sages of India (Verma and Keswani 1974b).

Firus Shah Tughlaq (1351–1388 A.D.) who succeeded Muhammed Tughlaq was himself an accomplished physician. He had a special interest in ophthalmology and is reputed to have designed an eye ointment which had the skin of black snake as an important ingredient (Verma and Keswani 1974b). This reminds us of a similar collyrium recommended by Vāgbhaṭa in the *Uttarasthāna* of *Aṣṭāṅgahṛdaya* (Upadhyaya 1975a).

Firus Shah's court physicians compiled a medical text called *Tibb-e-Firus Shāhi* (Medicine of Firus Shah), which reportedly describes the treatment of many diseases that were not mentioned in *Al-Qānūn* of Avicenna (Verma and Keswani 1974b). *Ūnāni* was greatly patronized by Sultan Mahmud Shah of Gujarat (1458–1511 A.D.), who ordered the founding of a special department for translating Arabic and Sanskrit medical works into Persian. Muhammed bin Ismail Asavale Asili translated Vāgbhaṭa's *Aṣṭāṅgahṛdaya*. It is known as *Tibb-e-Mahmūdi* (Medicine of Mahmud) or *Shifa-e-Mahmūdi* (Cure of Mahmud) (Verma and Keswani 1974b).

In 1512 A.D. Behwa bin Khawas Khan, an *amir* of Sikander Shah Lodhi (1489–1517 A.D.) completed the compilation of a medical text called *Madan-ul-Shifā Sikander Shāhi*. This voluminous treatise was based on authoritative Ayurveda texts and the first chapter, like the *Sūtrasthāṇa* of Sanskrit medical works, discusses the fundamental principles of treatment (Verma and Keswani 1974b).

The Deccan disintegrated after the decline of the Bahmani kingdom and five princely states came into existence. The Adil Shahi dynasty of Bijapur was established in 1489 A.D. by Yusuf Adil Shah. During the reign of Ibrahim Adil Shah II, his courtier Muhammed Qasim Hindu Shah, alias Firishta, composed the medical text *Dastūr-ul-Atibbā* or *Ikhtiyārāt-e-Qāsimi* (1590 A.D.). This work deals with Ayurveda. In the preamble to the book Firishta states that he embarked upon this project to introduce Ayurveda to his Muslim friends. He was apparently impressed by the well-founded theories of Ayurveda, the practice of which seemed strange at the outset (Verma and Keswani 1974b).

Babar (1526–1530 A.D.), the founder of the Moghul dynasty had many great physicians in his court. The most respected of them was Yusuf bin Muhammed bin Yusuf. He gleaned information on hygiene, general principles, diseases, diagnosis and therapeutics from Ayurveda and composed several books. He is credited with the production of a composite and integrated medical system by amalgamating Greco-Arabic and ayurvedic medical thought. The important texts composed by Muhammed bin Yusuf are *Jami-ul-Fawāid* (Collection of Benefits), *Fawāid-ul-Akhyār* (Benefits of the Best), *Qasidā fi Hifz-ul-Sihhat*, *Riyāz -ul-Adwiya* (Garden of Remedies), *Tibb-e-Yūsufi* (Medicine of Yusuf) and *Ilāj-ul-Amrāz* (Verma and Keswani 1974b).

Aurangazeb's reign (1658–1707 A.D.) created an atmosphere conducive to the popularization of *Ūnāni* (Rizvi 1975). A famous physician of his court, Muhammed Akbar Arzani, produced about eight Persian medical compilations. One of them, *Tibb-e-Hindi* (Medicine of the Hindus) deals with drugs of the ayurvedic formulary (Verma and Keswani 1974b). The Moghul period was marked by the translation of most of the medical texts written in Arabic into Persian, as Persian was the court language of the time. By the time Aurangazeb ascended the throne, all Arabic texts used in the *Ūnāni* system were available in Persian. In the 19th century many of these works were translated into Urdu, the popular language of the Muslims of northern India (Verma and Keswani 1974b).

1.8 THE INFLUENCE OF AYURVEDA ON ŪNĀNI

1.8.1 Development of Khamīra

Khamīra or medicated spirituous liquors were developed on the lines of *āsava* and *ariṣṭa* of Ayurveda. The Moghul nobility had an aversion to drinking bitter decoctions of herbs and the Persian physicians of the court circumvented this problem by developing *khamīra*s and making the medicines more palatable (Said 1978a). *Khamīra*s are usually prepared by making decoctions of herbs and reducing the volume by one tenth. To this are added citric acid (*sat limun*), sodium benzoate (*nitrun bunjavi*) and honey. Sometimes clarified butter (ghee) is also added "to effect lubrication and to destroy dryness". It may be remembered that Sanskrit medical texts advise the *āsava* and *ariṣṭa* to be prepared in earthen pots, the inside of which are smeared with ghee and sometimes scented with fragrant fumes. *Khamīra*s are usually named after the principal ingredient. For example, *Khamīra-e- Abrēṣam* has *abrēṣam mugharāz* (cocoons of *Bombyx mori*) as the major ingredient. It is said that the idea of fermenting decoctions and honey was first suggested by the medieval Turkish physician Najab-al-Din-Samarqandi (died 1222 A.D.) (Said 1978a). By virtue of its mode of preparation, a *khamīra* is preserved for a long time and its absorption into the body is also faster. The rationale behind the selection of ingredients is vindicated by the observation that *Khamīra-e-Abrēṣam* is a proven cardiotonic medicine (Siddiqui 1964; Nazmi et al. 2011).

1.8.2 Development of Mā'jun

Many electuaries or *mā'jun* were also developed by *Ūnāni* physicians. Examples are *Mā'jun Jograj, Gujul Mā'jun, Mā'jun-e-Hamāl Alāwi Khāni, Mā'jun Rāh al-Mumimin, Mā'jun Shir Dagārd Wāli* and *Mā'jun Kalkakanāj*. Hakim Azad Khan, who composed the text *Muhit-i-Āzam* is credited with the designing of many electuaries. Though the *Ūnāni* physicians had taken cues from Ayurveda, they ingeniously formulated many novel *mā'jun* which have few parallels in ayurvedic pharmacy. An example is *Mā'jun Murawwāh al-Arwāh*, which has more than hundred ingredients, including exotic ones like camel milk cheese (*mayāshuir A'rābi*), dried turtle eggs (*baizā sāṅg puṣṭ khuṣk kiyā huā*), mongoose flesh (*ibn irs*), sparrow brain (*maghz sar kuñjāṣk*) and so on. A cursory look at the list of ingredients reveals the acceptance into *Ūnāni* of drugs from several countries (Said 1978a).

1.8.3 Development of Kuṣṭā

A *kuṣṭā* is the *Ūnāni* equivalent of an ayurvedic *bhasma*, which is a calcined mineral or metal. The word *kuṣṭā* is derived from the Persian equivalent *kuṣṭān* which means "killed or conquered" (Aziz et al. 2002). The material to be calcined is ground in the juice of appropriate herbs, put in a pit of dried cow dung pats and set on fire. The recipes of many of the *kuṣṭās* were formulated in India (Said 1978a).

1.8.4 Acceptance of New Plants into Formulary

The *Ūnāni* formulary was enriched by the inclusion of many plants from Ayurveda. Ali (1990) has identified 210 such plants. In the majority of cases the *Ūnāni* names are Persianized Sanskrit words. Examples are *Bish* (*Aconitum ferox*, *Viṣa* in Sanskrit), *Wuz* (*Acorus calamus*, *Vaca* in Sanskrit), *Mothoo* (*Cyperus rotundus*, *Musta* in Sanskrit) and so on.

1.9 HOW AYURVEDA BENEFITED FROM GRECO-ARABIC MEDICINE

1.9.1 Pulse Examination

It is often said that the technique of pulse examination (*nāḍiparīkṣa*) is a later addition to Ayurveda, possibly from Greco-Arabic medicine (Majumdar 1971; Upadhyaya 1986). The cardinal evidence for such a line of argument is the observation that *Śārṅgadhara Samhita* is the first ayurvedic text to mention this topic (Murthy 2017c).

1.9.2 Description of New Disease Entities

1.9.2.1 Snāyukarōga

Snāyukarōga or dracunculiasis is described for the first time in Vṛnda's *Siddhayōga*. This disease was already recognized by Greco-Arabic physicians (Said 1978b; Meulenbeld 1981; Sharma 1981b; Litvinov 1991). Arab physicians repeatedly mentioned the view of the Greek author Soranus (2nd century A.D.) that the "little snakes" or *drakontia* found in the disease were nerves and not animals. Subsequently, this parasitic infestation was called *Al-irk-al-madani*. The ayurvedic physicians also gave a similar Sanskrit name, *Snāyukarōga*, to the disease, alluding to its similarity with sinews or tendons (Macdonell 1893). Trimalla (17th century A.D.) later classified it in his *Yōgataraṅgiṇi*, on the basis of the *tridōṣa* doctrine (Meulenbeld 1981).

1.9.2.2 Munnātakhyārōga

This curious entity is an affliction of penis mentioned for the first time in Śamkarā's *Vaidyavinōda Samhita* of 17th century. The name of the disease was adapted from Greco-Arabic medicine (Meulenbeld 1981).

1.9.2.3 Vardhma

The name of this disease is said to have a Greek origin and was first mentioned in Vṛnda's *Siddhayōga*. Sharma is of the opinion that it is *lymphogranuloma venereum* (Sharma 1981b).

1.9.3 Acceptance of New Plants into Ayurveda

As *Ūnāni* medicine became popular in the country, ayurvedic physicians had an opportunity to study the medicinal value of many drugs used by the *hakims*. Consequently, many of these were accepted into the ayurvedic system. Table 1.2 lists some such plants. The pellitory root *Anacyclus pyrethrum* known in Arabic as *aqarqarha* (Tauheed et al. 2017) was given many Sanskrit names like *Ākārakarabha*, *Akarkarha*, *Ākallaka*, *Akarkara* and so on. The Marathi, Gujarati, Telugu, Tamil,

TABLE 1.2
Some Drugs Which Ayurveda Borrowed From *Ūnāni*

Sl. No.	Latin Name	Sanskrit Name	Texts Which Mention The Drug
1	*Acacia arabica*	*Babbūla*	*Rājamārtāṇḍam*
2	Ambergris	*Agnijara*	*Dhanvantari Nighaṇṭu*
3	*Anacyclus pyrethrum*	*Ākārakarabha*	*Śārṅgadhara Samhita*
4	*Blepharis edulis*	*Kāmavṛddhi*	*Rājanighaṇṭu*
5	*Cassia angustifolia*	*Sanāya*	*Arkaprakāśa*
6	*Hyoscyamus niger*	*Pārasīkayavāni*	*Śārṅgadhara Samhita*
7	*Lawsonia inermis*	*Mehandi*	*Arkaprakāśa*
8	*Lepidium sativum*	*Candraśūra*	*Bhāvaprakāśa*
9	*Papaver somniferum*	*Ahiphēna*	*Gadanigraha*
10	*Pistacia lentiscus*	*Mastagi*	*Gadanigraha*
11	*Plantago ovata*	*Īśvarabōla*	*Siddhabhēṣajamaṇimāla*
12	*Quercus infectoria*	*Māyāphala*	*Rājamārtāṇḍam*
13	*Smilax china*	*Cōpacīni*	*Bhāvaprakāśa*

Kannada and Malayalam names of this plant are respectively *Akarkara*, *Akarkaro*, *Akkalkara*, *Akkirākāram*, *Akkalakari* and *Akkikaruka*, suggesting adaptations of the Arabic name. Similarly, *Isaphgol* (*Plantago ovata*) was given Sanskrit names like *Īśvarabōla and Īṣaḍgōḷa.* The use of opium also increased during medieval times (Kumar 1992b).

1.9.4 Adoption of New Dosage Forms

Several types of *Ūnāni* preparations like *guḷkhaṇḍ*, *malham* and *sarbat* were accepted by Ayurveda. Krishnarama Bhatta's *Siddhabhēṣajamaṇimāla* (1896 A.D.) testifies to this (Sarma 1912). *Guḷkhaṇḍ* is a sweet, jam-like preserve made of rose petals. It is prepared by mixing rose petals with sugar. The word *guḷkhaṇḍ* is derived from the Persian *gul*, meaning rose, and *khand*, which means sugar. *Malham* is an oil-based, thick preparation. It was first introduced into Ayurveda by the text *Yōgaratnākara*. The term *malham* later transformed into *malahara* (Sinduja et al. 2018).

Sarbat is a soft drink made of white sugar, candied sugar, honey and jaggery dissolved in water. However, in *Ūnāni*, *sarbat* means a concentrated drink made from decoction of herbs or fruit juices by adding sugar. The invention of *sarbat* is attributed to the ancient Greek philosopher and mathematician Pythagoras. *Ūnāni* physicians considered *sarbat* as a convenient dosage form to administer extracts of bitter herbs (Ahmad et al. 2016).

1.9.5 Interest in Ūnāni

Scholars of Ayurveda who realized the usefulness of *Ūnāni* medicine made attempts to introduce the system to ayurvedic physicians. The pioneer in this line was Mahadeva Deva, whose *Hikmatprakāśa* (1773 A.D.) described in Sanskrit the principles of *Ūnāni*, properties of drugs and many useful formulae. Mahadeva Deva later composed another text, *Hikmatpradīpa*, and both the works were later utilized by Mouktika in the composition of *Vaidyamuktāvali*. Some more texts were published in modern times. Notable among them are *Ūnāni Siddhayōgasamgraha*, *Ūnāni Dravyaguṇavijñān* and *Ūnāni Dravyaguṇādarś*, written in Hindi by Vaidyaraj Hakim Daljit Singh (Kumar 1992b).

The synergy between Ayurveda and Greco-Arabic medicine is a glaring example of cross-cultural interaction. Being experimentalists, the Greco-Arabic physicians found it fascinating to study the many facets of the unique medical system they came across in India. As a result of their efforts, they were able to accept several positive aspects of Ayurveda, without sacrificing the tenets of their own system. Similarly, ayurvedic physicians also had an opportunity to observe from close quarters the style of working of *Ūnāni hakims*. Many hospitals, like the one established at Etawah by Nawab Khair-Andesh Khan, where *hakims* and *vaids* worked side by side, helped the latter to assess objectively the usefulness of *Ūnāni* (Verma and Keswani 1974b). A large body of useful information was thus incorporated into ayurvedic practice.

1.10 SOME CURIOSITIES IN AYURVEDA THEORY

1.10.1 Prabhāva (Special Potency)

The term *prabhāva* means special potency. In ayurvedic pharmacology, a drug is composed of five basic elements (*pañcamahābhūta*). A drug is expected to act by virtue of its taste (*rasa*), property (*guṇa*), potency (*vīrya*), metabolic effect (*vipāka*), or by special potency called *prabhava*. Action (*karma*) of a drug in the human system depends on the above set of pharmacological attributes in varying proportions (Upadhyaya 1975b).

Taste is that which is specifically appreciated by the tongue. These are six in number. Property refers to general properties a drug can possess, one or many among the 20 properties attributed. Potency is a score of the ability of a drug for a specific action. There are two potencies: Hot (*uṣṇa*), which is catabolic, and cold (*śīta*), which is anabolic.

Metabolic effect is derived from the final action of a drug within the system. There are three effects identified in this category viz. sweet (*madhura*), sour (*amḷa*) and acrid (*kaṭu*). These are not the taste (*rasa*) mentioned earlier, but are termed by the "*rasa*" nomenclature, since they have actions similar to *rasa* (Upadhyaya 1975b).

Prabhāva on a different note, is truly unpredictable and cannot be numbered or counted. It is held to be a variant of potency (*vīrya*) that is unpredictable, unlike the other properties. When two drugs are of similar taste, property, potency and metabolic effect, if one of them exhibits some special action which the other drug does not have, it is due to special potency (*prabhāva*) of the former. For example, milk and ghee (clarified butter) have the same properties. Ghee is carminative (*dīpana*) in action, and milk is not. This is *prabhāva* or special potency of the ghee in a human system. The term *prabhāva* is a measure of relative action (Upadhyaya 1975b).

On a closer review, it can be commented that the concept and nomenclature of *prabhāva* can be seen as a case where ancient authors of Ayurveda could not explain the actions (*karma*) of many drugs based on the principles *of rasa-guṇa-vīrya-vipāka*. This situation might have necessitated introducing the term *prabhāva*, a dominant theory of uncertainty or unpredictability. To put it simply, the concept of *prabhāva* is a window for future generations of ayurvedists to think beyond, work more on and find out possible explanations of the actions of drugs with definiteness and precision. Ironically, even contemporary ayurvedists are also quite contented with the term *prabhāva*, not striving to advance further.

Added to the imbroglio, there is still another category of drugs in Ayurveda called "*vicitrapratyayārabdha dravya*" (Upadhyaya 1975b). Drugs, on a broad labeling, are tagged as standard or strange. The standard drug is the one which is predictable and acts according to its pharmacological categories of *rasa-guṇa-vīrya-vipāka-prabhāva*. Strange drugs do not obey the framework of the rules of pharmacology. In considering drugs acting on "strange" (unknown?) causes, the term *vicitra-pratyaya-ārabdha* is introduced. This is also a curious case of vagueness where windows are left open to look for plausible explanations for the actions of drugs used in Ayurveda for treating diseases.

1.10.2 Gender Bias

As the knowledge of longevity progressed gradually, Ayurveda was divided into eight branches, which are *kāyacikitsa* (internal medicine), *śalya* (surgery), *śālākya* (knowledge of diseases of supra-clavicular region), *kaumārabhṛtya* (pediatrics, obstetrics and gynecology), *agadatantra* (toxicology), *bhūtavidya* (knowledge of subtle beings or spirits), *rasāyana* (promotive therapy) and *vājīkaraṇa* (virilification) (Sharma 1981c). Among them *vājīkaraṇa* is solely intended for the treatment of men. *Aṣṭāṅgahṛdaya* advocates several therapeutic measures and many medicinal formulations for improving stamina, vigor and sexual drive of men. This therapy is believed to impart great physical strength and improve libido. Interestingly, no such measures or medicines intended for women are mentioned anywhere in Ayurveda, even though women are equally important in the furtherance of the human race. This indicates that all the classical texts of Ayurveda were composed in a male-dominant society.

The imprint of gender bias is evident in the theory and nosology of Ayurveda. The human body is said to be formed of seven *dhātu* (tissue elements) which are *rasa*, *rakta*, *māmsa*, *mēdas*, *asthi*, *majja* and *śukra*. It is implied that all of these seven *dhātu* exist in women as well. However, while discussing the "lowering" of *śukra dhātu*, *Aṣṭāṅgahṛdaya* mentions delayed ejaculation of seminal fluid, emission of blood-stained semen, pricking pain in scrota and burning sensation in phallus, as signs of this "lowered" state. Increased interest in women and *śukḷāśmari* (hardened semen?) are the only signs mentioned in the context of "high" state of *śukra dhātu* (Upadhyaya 1975c). These are discernible only in men. Signs related to women are totally lacking in these descriptions.

In classical Ayurveda texts, importance is always given to a son, rather than a daughter. For example, *Caraka Samhita* describes procedures for begetting a fair-complexioned son, with reddish brown eyes, or a sky-complexioned son with red eyes, or one with black, soft and long hair (Sharma 1981d). Many medicinal formulations are reputed to bless the user with a son, as in the case of *Mahāmāyūra ghṛta* (Sharma 1983).

Much evidence suggesting gender bias is available in the *Nidānasthāna* (section dealing with classification of diseases) of *Aṣṭāṅgahṛdaya*. Pricking pain in the phallus (*mēdrārti*, *mēhanayōstōdaḥ*), heaviness of the phallus (*mēdragauravam*) and edema in the phallus (*mēdraśōpham*) are the signs of various types of dysuria (*mūtrāghāta*). *Sīvaniruk* (pain in perineal area) and *muṣkaśvayathu* (edema in scrota) are also mentioned as the signs of some urinary diseases. These anomalies need to be corrected.

1.10.3 Futuristic Provisions in Ayurvedic Pathogenesis

The methodology of understanding the stages of pathogenesis in ayurvedic parlance is termed *rōganidāna*. The stages of evolution of diseases are termed *samprāpti*. There are general concepts (*sarvarōganidāna*) and specific details on individual diseases and their sub-types. The patterns of describing the evolution of various diseases followed by Caraka, Suśruta and Vāgbhaṭa in their respective treatises are similar in most instances. Among this celebrated triad of authors (*bṛhatrayi*), Vāgbhaṭa (who followed both Caraka and Suśruta) meticulously endorsed the basics of previous authors, omitted certain vital postulates and added explanations for hitherto unknown new groups into disease categories in the text *Aṣṭāṅgahṛdaya*.

1.10.3.1 Omission of Tenets

Rōganidāna, or etiopathogenesis in ayurvedic parlance, is depicted showcasing five attributes namely *nidāna* (etiology), *pūrvarūpa* (premonitory symptoms), *rūpa* (symptoms), *upaśaya* (trial and error) and *samprāpti* (pathogenesis). It is a curious case to see that Vāgbhaṭa did not follow the original idea of *ṣaṭkriyākāla* (six specific, yet distinct stages of development of a disease in terms of perturbed *dōṣa*) suggested by Suśruta. But a lesser attribute for select terms of vitiation and mitigation (*caya*, *prakōpa*, *praśama*) can be seen mentioned in terms of climatic impacts on *dōṣa* (Upadhyaya 1975d). The concept of the progressive stages of *sañcaya*, *prakōpa*, *prasara*, *sthānasamśraya*, *vyakti* and *bhēda* projected by Suśruta is not upheld by Vāgbhaṭa, though he endorses most of the anatomical and surgical tenets scattered through Suśruta's treatise.

1.10.3.2 Addition of Explanation

On a critical review of the chapters of *nidānasthāna* in *Aṣṭāṅgahṛdaya* by Vāgbhaṭa, it becomes imperative that the last chapter introduces a new disease category *vātaśōṇita*, placed after the highly structured chapter called *vātavyādhi*, describing diseases with dominance of *vātadōṣa* (Upadhyaya 1975e). The disease *vātaśōṇita* is mentioned in *Caraka Samhita* with the basic attributes. Despite the nomenclature *vātaśōṇita*, the prefix of *vāta* to *śōṇita* (blood) indicates the crucial role of *vāta*. Vāgbhaṭa did not include this disease entity in the previous chapter of *vātavyādhi* or diseases of *vāta*! The precedence of naming the disease as depicted in the second chapter *raktapitta* – a group of bleeding disorders – was intentional in recognizing the dominant role of *pitta* over *rakta* in that context. On the same scale, *vātaśōṇita* qualifies to be included in the *vātavyādhi* chapter. But Vāgbhaṭa opted to add a new chapter, introducing a new disease spectrum – *vātaśōṇita*. Looking at the nature of disease, it becomes obvious that the role of *śōṇita* (*rakta*) is primary and decisive over *vāta*. The previous generation of *vaidyā*s in Kerala preferred to mention the disease as *raktavāta*!

1.10.3.3 Introduction of New Concept

In the second segment of the chapter *vātaśōṇita nidānam*, Vāgbhaṭa established the group of diseases laced with a new concept – *āvaraṇa*, meaning encompassing or enveloping, originally postulated by Caraka (Upadhyaya 1975e). It argues that when a disease is manifested with overlapping symptoms applying the equation of *dōṣa* involvement, a new method can be adopted with the idea of considering one *dōṣa* to be covering or concealing the other. There can be a case of *vāta* vitiated but encircled by *kapha*, making the symptoms of *vāta* less impressive! This concept was later expanded, extrapolating *dōṣa* sub-types (15 in all) and also the *dhātu* variants into the picture. The idea of *āvaraṇa* was imagined through multiple propositions, and Vāgbhaṭa finally concluded the discussion by authorizing the physician to derive an appropriate equation of *dōṣa* involvement in each clinical situation considering its merits. Vāgbhaṭa concluded that enumerating *āvaraṇa* types can be endless, as each context can be specific in this regard to be concluded on a logical basis. It is quite clear that the new principle of *āvaraṇa* in the domain of pathogenesis according to Ayurveda is open-ended – neither precise nor finite.

On examination, it makes sense to understand that the discussion on many diseases (not described in ayurvedic classics) can be complex and ill-fitting when viewed in the framework of specific disease types with popular names as categorized in the *nidāna* chapters. Elaborating the concept of *āvaraṇa* was a window opened to allow thinking out of bounds. Vāgbhaṭa's stride was unprecedented – the first step of identifying the *vātaśōṇita* disease types as separate from *vātavyādhi*, further proceeding to explore the concept of *āvaraṇa* to explain diseases that cannot be included in the foregoing regular groups of diseases and to expand the scope of including new diseases that would emerge in future!

1.11 CONCLUSION

Ayurveda had always been dynamic. The knowledge of Ayurveda that originated from Brahma passed through Dakṣa Prajāpati, Indra, Bharadvāja or Atri and finally reached Ātrēya. Ātrēya discussed the topic of medicine with scholar-sages in different conclaves organized in various parts of ancient India and conceived the basic principles of the "sacred knowledge of life". His six illustrious disciples – Agnivēśa, Bhēla, Jatūkarṇa, Parāśara, Hārīta and Kṣārapāṇi composed medical treatises. However, *Agnivēśa Tantra*, penned by Agnivēśa, is the most notable and popular one, representing the Ātrēya school of medicine. Later it was refined and enlarged by Caraka when it began to be known as *Caraka Samhita*. *Caraka Samhita* was once again revised by Dṛḍhabala. The extant text of the *Caraka Samhita* is the *Agnivēśa Tantra* refined by Caraka and later revised by Dṛḍhabala (Sharma 1981c). *Caraka Samhita* and *Suśruta Samhita* facilitated the transition of Ayurveda from a mystical tradition to a clinically sound, evidence-based knowledge guided by etiology, detailed observations and therapeutic measures.

Vāgbhaṭa reclassified these *samhitā*s and presented a concise version with many additions from contemporary practice. His *Aṣṭāṅga Samgraha* and *Aṣṭāṅgahṛdaya* demonstrate the importance of revising the texts in line with changing contexts. Thus, an era of rediscovering Ayurveda can be said to have begun with Vāgbhaṭa. This was followed by important treatises like *Śārṅgadhara Samhita*, *Cikitsāsamgraha* by Vaṅgasēna and *Bhāvaprakāśa* of Bhāvamiśra. *Śārṅgdhara Samhita* of the 13th century documents the use of opium and highlights the importance of urine examination in diagnosis of diseases. Interestingly, the word *rōṭṭika*, forerunner of the name of the flat bread *rōṭṭi* popular in northern India was mentioned for the first time in *Bhāvaprakāśa*. Similarly, *phiraṅgarōga* (syphilis) and its remedy *Dvīpāntaravaca* or *Cōpacīni* (*Smilax china*) also appear for the first time in this work (1558–1559) (Keswani 1974).

An important aspect to be learnt from the study of the history of Ayurveda is its indefatigable spirit of curiosity, questioning and experimentation. The culture of inquisitiveness and questioning is very important for the progress of any medical system. And Ayurveda is no exception. However, unfortunately that culture has nearly disappeared from the contemporary Ayurveda world. Most of the ayurvedic physicians firmly believe in the oft-repeated lines from the last chapter of *Aṣṭāṅgahṛdaya*:

> *Idam āgama siddhatvāt pratyakṣa phaladarśanāt*
> *Mantravat samprayōktavyam na mīmāmsyam kathañcana*

(The teachings of *Aṣṭāṅgahṛdaya* born from the *vēda* bring forth discernible results. Therefore, this work can be used just like a *mantra*. One should not doubt that it will ever fail). They maintain that there is no need for any research and development in Ayurveda, as it is already perfect. Nevertheless, it is evident that many of the intricacies of Ayurveda can be unraveled only by using the tools of modern science. For example, the cardinal aspects of drug action like *rasa*, *guṇa*, *vīrya*, *vipāka* and *prabhāva* can be better understood with the help of molecular biology and pharmacology. Deeper understanding of these characteristics can facilitate the inclusion of newer herbs, fruits and vegetables into the ayurvedic formulary.

Aṣṭāṅgahṛdaya states that the six taste modalities can form 63 combinations and that they are of paramount importance in treatment of diseases (Upadhyaya 1975f). While elaborating on the *tridōṣa*, the text states that the *tridōṣa* also can exist in numerous combinations. However, for the sake of convenience, only 63 combinations need to be considered (Upadhyaya 1975d). The V.P.K. code proposed by Prabhakar and Kumar (1993) can be applied to the quantification of disease states, which are nothing but expressions of the *tridōṣa*. The cross-matching of 63 combinations of taste modalities with 63 combinations of *tridōṣa* states, when better understood, will revolutionize the diagnosis and treatment of diseases exclusively using the principles and parameters of Ayurveda. Research on these nuances will help rediscover Ayurveda.

REFERENCES

Ahmad, I., S. Shamsi, and R. Zaman. 2016. Sharbat: An important dosage form of unani system of medicine. *Medical Journal of Islamic World Academy of Sciences* 24: 83–88.

Ahmad, M. 1988. A critical study of *Arab-o-Hind ke T'aalluqat*. *Journal of Oriental Institute* 38: 343–54.

Ahn, A. C., M. Tewari, C. S. Poon, S. Russell, and R. S. Phillips. 2006. The limits of reductionism in medicine: Could systems biology offer an alternative? *PLoS Medicine* 3, no. 6: e208. doi:10.1371/journal.pmed.0030208.

Ali, M. 1990. Ayurvedic drugs in Unani materia medica. *Ancient Science of Life* 9, no. 4: 191–201.

Amin, A. H., and D. R. Mehta. 1959. A bronchodilator alkaloid (vasicinone) from *Adhatoda vasica* Nees. *Nature* 184, no. 17: 1317.

Anawati, G. 1970. Science. In *The Cambridge history of Islam, volume 2B: Islamic society and civilization*, ed. P. M. Holt, A. K. S. Lambton, and B. Lewis, 741–779. Cambridge: Cambridge University Press.

Anonymous. 2013. Introduction. In *WHO traditional medicine strategy 2014–2023*, 15–19. Geneva: WHO.

Anonymous. 2019. Vāgbhaṭa Samādhi. 2019. https://www.facebook.com/srdayurveda/posts/2303453639879509 (accessed July 21, 2019).

Arjariya, A., and K. Chaurasia. 2009. Some medicinal plants among the tribes of Chhatarpur District (MP) India. *Ecoprint: An International Journal of Ecology* 16: 43–50.

Asif, M. A. 2016. Frontier with the house of gold. In *A book of conquest: The Chachnama and Muslim origins in South Asia*, 23–46. Massachusetts: Harvard University Press.

Aziz, N., A. H. Gilani, and M. A. Rindh. 2002. Kushta(s): Unique herbo-mineral preparations used in South Asian traditional medicine. *Medical Hypotheses* 59, no. 4: 468–472.

Campillo, D. 1983. *La Enfermedad en la Prehistoria*, 15–23. Barcelona: Salvat (Barcelona).

Conway, P. 2011. Phytotherapy in context. In *The consultation in phytotherapy*, 1–38. London: Churchill Livingstone.

Dasgupta, S. 1997. General observations on the systems of Indian philosophy. In *A history of Indian philosophy*, Volume 1, 62–77. Delhi: Motilal Banarsidass.

Desale, M., P. Bhamare, P. Sawant, S. Patil, and S. Kamble. 2013. Medicinal plants used by the rural people of Taluka Purandhar, District Pune, Maharashtra. *Indian Journal of Traditional Knowledge* 12: 334–338.

El-Gammal, S. Y. 1993. The role of Hippocrates in the development and progress of medical sciences. *Bulletin of the Indian Institute of History of Medicine* 23: 125–136.

Fábrega, H. Jr. 1997. Origins of sickness and healing. In *Evolution of sickness and healing*, 29–74. Berkeley: University of California Press.

Gode, P. K., and C. G. Karve. 1959. *Prin. V.S. Apte's the practical Sanskrit-English dictionary*, Volume 3, 1496. Poona: Prasad Prakashan.

Gurusironmani, P. 1972. A short note on history of Siddha medicine. *Bulletin of the Institute of History of Medicine* 2: 78–80.

Gurusironmani, P. 1983. A chronological probe into Siddha system. *Bulletin of the Institute of History of Medicine* 13, no. 1–4: 34–37.

Hamarneh, S. K. 1990. Al-Zahrawi's Al-Tasrif, commemorating its millenary appearance. *Hamdard Medicus* 33, no. 2: 19–40.

Hart, B. L. 2005. The evolution of herbal medicine: Behavioural perspectives. *Animal Behaviour* 70: 973–989.

Hart, B. L. 2011. Behavioural defences in animals against pathogens and parasites: Parallels with the pillars of medicine in humans. *Philosophical Transactions of the Royal Society B* 366, no. 1583: 3406–3417.

Huffman, M. A., and M. Seifu. 1989. Observations on the illness and consumption of a possibly medicinal plant *Vernonia amygdalina* (Del.), by a wild chimpanzee in the Mahale Mountains National Park, Tanzania. *Primates* 30, no. 1: 51–63.

Huffman, M. A., and R. W. Wrangham. 1994. Diversity of medicinal plant use by chimpanzees in the wild. In *Chimpanzee cultures*, ed. P. G. Heltne, and L. A. Marquardt, 129–148. Cambridge: Harvard University Press.

Kasinathan, R. 1955. *Piḷḷaippiṇi Vākaṭam*, 1–96. Chennai: Government Oriental Manuscripts Library and Research Centre.

Keswani, N. H. 1974. Medical heritage of India. In *The science of medicine and physiological concepts in Ancient and Medieval India*, ed. N. H. Keswani, 1–49. New Delhi: All-India Institute of Medical Sciences.

Kumar, D. S. 1992a. On the identity of *arka*, an ayurvedic class of medicines. *Journal of the European Ayurvedic Society* 2: 54–59.

Kumar, D. S. 1992b. Glimpses of interaction between Ayurveda and Unani. *Aryavaidyan* 6: 109–116.

Kumar, D. S. 1995. Some preliminary observations on the chronology of Tamil medical texts. In *Glimpses of Indian ethnopharmacology*, ed. P. Pushpangadan, U. Nyman and V. George, 77–84. Trivandrum: Tropical Botanic Garden and Research Institute.

Kumar, D. S. 2018. Advances in food fortification with phytonutrients. In *Food biofortification technologies*, ed. Saeid, Agnieszka, 161–243. Boca Raton: CRC Press.

Litvinov, S. K. 1991. How the U.S.S.R. rid itself of Dracunculiasis. *World Health Forum* 12, no. 2: 217–219.

Maas, P. A. 2010. On what became of the *Carakasaṃhitā* after *Dṛḍhabala's* revision. *E Journal of Indian Medicine* 3: 1–22.

Macdonell, A. A. 1893. *A Sanskrit-English dictionary*, 367. London: Longmans, Green and Co.

Magner, L. N. 2005. Paleopathology and paleomedicine. In *A history of medicine*, 1–23. Boca Raton: Taylor & Francis.

Majeed, M., and D. S. Kumar. 2008. Classification of diseases in the Tamil medical work *Vaittiyacintāmaṇi* 800 of Yukimuni. I: Introduction. *Traditional South Asian Medicine (Halle-Wittenberg)* 8: 65–76.

Majumdar, R. C. 1971. Medicine. In *A concise history of science in India*, ed. D. M. Bose, S. N. Sen, and B. V. Subbarayappa, 213–273. New Delhi: Indian National Science Academy.

Meulenbeld, G. J. 1974. *The Mādhavanidāna and its chief commentary: Chapters 1–10*, 1–27. Leiden: E.J. Brill.

Meulenbeld, G. J. 1981. Developments in traditional Indian nosology: The emergence of new diseases in post-classical times. *Curare* 4: 211–216.

Mishra, B. 1983. *Raktapitta cikitsa* (Treatment of hemorrhages of obscure origin). In *Cakradatta*, 118–129. Varanasi: Chowkhamba Sanskrit Series Office.

Mooss, N. S. 1982. Identification and cultivation of medicinal plants mentioned in Ayurvedic classics. *Ancient Science of Life* 1, no. 4: 224–228.

Mueller, A. 1880. Arabische Quellen zur Geschichte der indischen Medizin. *Zeitscrift der Deutschen Morgenländischen Gesellschaft* 34: 465–556.

Murthy, K. R. S. 2017a. Prathama Khaṇḍa (First section), Chapter 2. In *Śārṅgadhara Samhita*, Volume 10–14. Varanasi: Chaukhamba Orientalia.

Murthy, K. R. S. 2017b. Madhyama Khaṇḍa (Middle section). In *Śārṅgadhara Samhita*, 49–184. Varanasi: Chaukhamba Orientalia.

Murthy, K. R. S. 2017c. *Prathama Khaṇḍa* (first section) Chapter 3. In *Śārṅgadhara Samhita*, 14–16. Varanasi: Chaukhamba Orientalia.

Nazmi, A. S., S. J. Ahmad, A. Rashikh, M. Akhtar, K. K. Pillai, and A. K. Najmi. 2011. Protective effects of Khamira Abresham Hakim Arshad Wala, a unani formulation against doxorubicin-induced cardiotoxicity and nephrotoxicity. *Toxicology Mechanisms and Methods* 21, no. 1: 41–47.

Pearce-Duvet, J. M. 2006. The origin of human pathogens: Evaluating the role of agriculture and domestic animals in the evolution of human disease. *Biological Reviews of the Cambridge Philosophical Society* 8, no. 3: 369–382.

Phillips-Conroy, J. E. 1986. Baboons, diet, and disease: Food selection and schistosomiasis. In *Current perspectives in primate social dynamics*, ed. D. M. Taub, and F. A. King, 287–304. New York: Van Nostrand Reinhold.

Pillai, T. V. S. 1931a. *Tamil-English dictionary of medicine, chemistry, botany and allied sciences*, Volume 5, 1036. Madras: The Research Institute of Siddhar's Science.

Pillai, T. V. S. 1931b. Introduction. In *Tamil-English dictionary of chemistry, botany and allied sciences*, Volume 1, 1–114. Madras: Institute of Siddhars' Sciences.

Prabhakar, Y. S., and D. S. Kumar. 1993. A model to quantify disease state based on the ayurvedic concept of *tridōṣa*. *Bulletin of the Indian Institute of History of Medicine* 23: 1–19.

Priorischi, P. 1995a. Primitive and naturalistic medicine. In *A History of medicine: Primitive and ancient medicine*, Volume 1, 9–73. Omaha: Horatio Press.

Priorischi, P. 1995b. Introduction. In *A history of medicine: Primitive and ancient medicine*, Volume 1, 1–7. Omaha: Horatio Press.

Ramzan, I., and G. Q. Li. 2015. Phytotherapies - Past, present, and future. In *Phytotherapies: Efficacy, safety, and regulation*, ed. I. Ramzan, 1–17. NJ: John Wiley & Sons, Inc.

Reddy, D. V. S. 1973. History of Siddha medicine: Need for further detailed studies. *Bulletin of the Institute of History of Medicine* 3: 182–185.

Rizvi, S. A. A. 1975. The Muslim ruling dynasties. In *Cultural history of India*, ed. A. L. Basham, 245–265. Oxford: Clarendon Press.

Rodriguez, E., M. Aregullin, T. Nishida et al. 1985. Thiarubine-A, a bioactive constituent of *Aspilia* (Asteraceae) consumed by wild chimpanzees. *Experientia* 41, no. 3: 419–420.

Roşu, A. 1988. *Mantra* and *yantra* in Indian medicine and alchemy. *Ancient Science of Life* 8, no. 1: 20–24.

Said, M. 1978a. Traditional medicine in the service of health. *Proceedings of the 26th international congress of history of medicine*, Plodiv, Bulgaria (August 20–25).

Said, M. 1978b. Eastern medicine in a changing world. *Proceedings of the international Finnish archipelago conference*, Turku, Finland (August 28–September 1), 1–14.

Sankunny, K. 1982. 25. Vāgbhaṭācāryar. In *Aitihyamāla*, Volumes 162–166. Kottayam: Kottarathil Sankunny Memorial Committee.

Sarma, N. 1912. *Hikmatprakāśa, 1–208*. Bombay: Khemraj Srikrishnadas.

Scharfe, H. 1999. The doctrine of the three humors in traditional Indian medicine and the alleged antiquity of Tamil Siddha medicine. *Journal of the American Oriental Society* 119, no. 4: 609–629.

Sharma, P. V. 1981a. *Rōgabhiṣagjitīya vimānam* (Specific features of therapeutics of diseases). In *Caraka Samhita*, Volume 1, 350–394. Varanasi: Chaukhamba Orientalia.

Sharma, P. V. 1981b. *Āyurvēd kā vaijñānik itihās*, 244–256. Varanasi: Chaukhamba Orientalia.

Sharma, P. V. 1981c. Introduction. In *Caraka Samhita, Volume 1*, v–xxxii. Varanasi: L Chaukhamba Orientalia.

Sharma, P. V. 1981d. Jātisūtrīyam Śārīram (Principles of procreation). In *Caraka Samhita*, Volume 1, 461–488. Varanasi: Chaukhamba Orientalia.

Sharma, P. V. 1983. *Tṛmarmīya cikitsitam* (Treatment of disorders of three vital organs). In *Caraka Samhita*, Volume 2, 419–455. Varanasi: Chaukhamba Orientalia.

Sharma, P. V. 1994. *Cakradatta*, 1–731. Varanasi: Chaukhamba Orienalia.

Siddiqui, H. H. 1964. Effect of *Khamira abresham arshadwala* on serum cholesterol levels in rabbits. *Planta Medica* 12, no. 4: 443–445.

Sigerist, H. E. 1951. *A history of medicine*, Volume 1, 45–48. New York: Oxford University Press.

Sinduja, V., R. Salve, and L. Priya. 2018. A brief review on *Shweta Malahara*. *World Journal of Pharmaceutical Research* 7: 1584–1588.

Steinbock, R. T. 1976. *Paleopathological diagnosis and interpretation*, 253–273. Springfield: Charles C. Thomas.

Tauheed, A., and Ali, A. Hamiduddin. 2017. Aqarqarha (*Anacyclus pyrethrum* DC.): A potent drug in *unani* medicine: A review on its historical and phyto-pharmacological perspective. *Journal of Pharmaceutical and Scientific Innovation* 6, no. 1: 22–28.

Trueba, G. 2014. The origin of human pathogens. In *Confronting emerging zoonoses*, ed. A. Yamada, L. H. Kahn, B. Kaplan, T. P. Monath, J. Woodall, and L. Conti, 3–11. Tokyo: Springer Japan.

Trueba, G., and M. Dunthorn. 2012. Many neglected tropical diseases may have originated in the paleolithic or before: New insights from genetics. *PLOS Neglected Tropical Diseases* 6, no. 3: e1393. doi:10.1371/journal.pntd.0001393.

Upadhyaya, S. D. 1986. *Nāḍivijñāna (Ancient Pulse Science)*, 15–46. New Delhi: Chaukhamba Sanskrit Pratishthan.

Upadhyaya, Y. 1975a. *Timirapratiṣēdham* (Treatment of cataract). In *Aṣṭāṅgahṛdaya*, 492–499. Varanasi: Chaukhamba Sanskrit Sansthan.

Upadhyaya, Y. 1975b. *Dravyādivijñānīyam* (Knowledge of matter). In *Aṣṭāṅgahṛdaya*, Volumes 79–82. Varanasi: Chowkhamba Sanskrit Sansthan.

Upadhyaya, Y. 1975c. *Dōṣadivijñānīyam* (Knowledge of *dōṣa*). In *Aṣṭāṅgahṛdaya*, 85–90. Varanasi: Chowkhamba Sanskrit Sansthan.

Upadhyaya, Y. 1975d. *Dōṣabhēdīyam* (Combinations of *tridōṣa*). In *Aṣṭāṅgahṛdaya*, 90 –96. Varanasi: Chowkhamba Sanskrit Sansthan.

Upadhyaya, Y. 1975e. *Vātaśōṇita nidānam* (Pathogenesis of *vātaśōṇita*). In *Aṣṭāṅgahṛdaya*, Volumes 280–284.Varanasi: Chowkhamba Sanskrit Sansthan.

Upadhyaya, Y. 1975f. *Rasabhēdīyam* (Combinations of taste). In *Aṣṭāṅgahṛdaya*, Volumes 82–85. Varanasi: Chowkhamba Sanskrit Sansthan.

Vaidyan, A. K. V., and A. S. G. Pillai. 1986. Ariṣṭayōgaṅgaḷ. In *Sahasrayōgam*, 255–271. Alleppey: Vidyarambham Publishers.

Valiathan, M. S. 2006. Caraka and his legacy. In *The legacy of Caraka*, i–xvi. Hyderabad: Orient Longman Pvt Ltd.

Variar, P. R. 1985. The Ayurvedic heritage of Kerala. *Ancient Science of Life* 5, no. 1: 54–64.

Variar, P. R. 1987. Ayurveda in Kerala. In *Encyclopaedia of Indian literature, volume 1*, Volumes 311–313, ed. A. Datta. New Delhi: Sahitya Akademi.

Venkatraman, R. 1990a. Features of the Tamil siddha cult. In *A history of the Tamil siddha cult*, 76–165. Madurai: Ennes Publications.

Venkatraman, R. 1990b. Origin of the Tamil siddha cult. In *A history of the Tamil siddha cult*, 23–41. Madurai: Ennes Publications.

Verma, R. L., and N. H. Keswani. 1974a. The physiological concepts of Ūnāni medicine. In *The science of medicine and physiological concepts in Ancient and Medieval India*, ed. N. H. Keswani, 145–163. New Delhi: All-India Institute of Medical Sciences.

Verma, R. L., and N. H. Keswani. 1974b. Ūnāni medicine in medieval India- Its teachers and texts. In *The science of medicine and physiological concepts in Ancient and Medieval India*, ed. N. H. Keswani, 127–142. New Delhi: All-India Institute of Medical Sciences.

Wolfe, N. D., C. P. Dunavan, and J. Diamond. 2007. Origins of major human infectious diseases. *Nature* 17, no. 7142: 279–283.

Wrangham, R. W., and T. Nishida. 1983. *Aspilia* spp. leaves: A puzzle in the feeding behavior of wild chimpanzees. *Primates* 24, no. 2: 276–282.

Zanasi, A., M. Mazzolini, and A. Kantar. 2017. A reappraisal of the mucoactive activity and clinical efficacy of Bromhexine. *Multidisciplinary Respiratory Medicine* 12: 7. doi:10.1186/s40248-017.

2 Manufacture of Ayurvedic Medicines – Regulatory Aspects

V. Remya, Alex Thomas and D. Induchoodan

CONTENTS

2.1 Emergence of Regulatory Activity 22
2.2 Early Days of the Ayurveda Industry 22
2.3 Drugs and Cosmetics Act, 1940 22
2.3.1 The Schedules 23
2.3.2 The Rules 23
2.4 Other Applicable Acts 24
2.4.1 The Pharmacy Act, 1948 24
2.4.2 The Drugs and Magic Remedies (Objectionable Advertisements) Act, 1954 24
2.4.3 The Narcotic Drugs and Psychotropic Substances Rules Act, 1985 24
2.4.4 The Poisons Act, 1919 24
2.4.5 The Drugs (Price Control) Order, 1995 24
2.4.6 The Trade Marks Act, 1999 24
2.4.7 The Biological Diversity Act, 2002 25
2.5 How to Begin Manufacture of Ayurvedic Medicines 25
2.5.1 Licensing 25
2.5.2 Loan License 26
2.5.3 Applying for a Manufacturing License 26
2.5.3.1 Conditions of License 26
2.5.3.2 Conditions of Loan License 27
2.5.4 Proof of Effectiveness 27
2.5.4.1 Classical Formulations 27
2.5.4.2 Proprietary Formulations 27
2.5.5 Records of Raw Material Used 28
2.5.6 G.M.P. Certification 28
2.5.6.1 Principles of the Guidelines 28
2.5.6.2 G.M.P. Guidelines 28
2.5.6.3 Inspectors for Conducting G.M.P. Inspection 29
2.5.6.4 Duties of the Inspectors 29
2.5.6.5 Procedure for Collection of Drug Sample by Inspector 30
2.5.6.6 Joint Inspection of Testing Laboratories 31
2.6 Punishment for Violation of the Act 31
2.7 Cancellation or Suspension of Licenses 32
2.8 Labeling of Products 32
2.9 Shelf Life of Ayurvedic Medicines 32
2.10 Good Clinical Practices 34

2.11 Regulatory Problems That Require Solution 35
2.11.1 Monitoring of A.Y.U.S.H. G.C.P. Guidelines 35
2.11.2 Revision of First Schedule 35
2.11.3 Regulation of Change in Dosage Form 36
2.11.4 Regulation of Complementary Products 36
2.11.5 Labeling of Products 36
2.11.6 Revision of Schedule E1 37
2.11.7 Effective Implementation of the Act 37
2.12 Conclusion 38
References 38

2.1 EMERGENCE OF REGULATORY ACTIVITY

Manufacture of medicinal products was not a regulated activity in the developed world in the early days. Then industrial-scale production and mass transportation were applied to medicine manufacturing, and customers could no longer be assured of the quality of products. Consequently, the adulteration of food and drugs became a problem. During the Mexican War (1846–1848) it was found that American troops were supplied with substandard imported medicinal products. This resulted in the passing of the first federal law dealing with medicinal products – the Drug Importation Act of 1848. The unhygienic conditions prevailing in the meat packing industry gave rise to public outcry about the safety of food, and also of the quality of medicinal substances. The death of a dozen children in 1902 from contaminated vaccines forced legislators to pass the Biologics Control Act of 1902, which called for the licensing of biological products and their production in licensed facilities. The Pure Food and Drug Act of 1906 prohibited the mislabeling and adulteration of medicinal products and introduced the U.S. Pharmacopeia and the National Formulary as official standards. This ushered in an era of regulated industrial manufacture of medicines (Dumitriu 1997).

2.2 EARLY DAYS OF THE AYURVEDA INDUSTRY

Up to the first half of the 19th century, a number of households produced and distributed ayurvedic medicines. The production and distribution were not based on principles of modern business management. The production of medicine was concentrated in and around the *vaidya*'s residence, and the service and production costs were not clearly distinguished. Ayurveda was active in most of the Indian villages, as modern medical systems had not made inroads into rural India. The demand for ayurvedic medicines shot up in the mid-19th century when *vaidyas* responded to the spread of epidemics like cholera and smallpox. While responding to these problems in the 1880s, bold attempts were made by some *vaidyas* to shift from household production to bulk production (Harilal 2009).

The first initiatives in large-scale medicinal production were seen in the late-19th century in Bengal. Vaidya Gangadhar Ray in Bengal was inspired by the increasing demand for ayurvedic medicines. He set up a large-scale manufacturing unit in 1884 called N.N. Sen and Company (Gupta 1976). Vaidyaratnam P.S. Variar started the industrial manufacture of ayurvedic medicines in Kerala, in southern India. These attempts later spread to other parts of the country. Manufacturing in Ayurveda passed from small-scale physician outlets to petty/cottage production, and later to the industrial scale, emerging as a competing alternative to the biopharmaceutical market (Harilal 2009).

2.3 DRUGS AND COSMETICS ACT, 1940

The Government of India enacted the Drugs and Cosmetics Act, 1940, to regulate the import, manufacture, distribution and sale of drugs and cosmetics (Anonymous 2016a). This Act was initially intended for chemical drugs. However, in 1969, a separate chapter relating to Ayurveda, Siddha and

Unani drugs was inserted by Act 13 of 1964. The laws are more or less similar to those intended for pharmaceuticals. This Act was again modified in 1983, 1987, 1994 and 2002. The Act ensures that the drugs and cosmetics sold in India are safe, are effective and conform to quality standards. It has four Schedules and five Rules related to the Ayurveda, Siddha and Unani systems. The Drugs and Cosmetics Rules, 1945, contains provisions for the classification of drugs under various Schedules and the guidelines for the storage, sale, display and prescription of each scheduled drug (Saroya 2016a).

2.3.1 The Schedules

The First Schedule came into effect on 1 February 1969. It lists the standard scriptures of Ayurveda to be followed for manufacturing Ayurveda, Siddha and Unani medicines. Fifty-seven books of Ayurveda (with additions in 1987, 1994, 2002) are listed in this Schedule (Table 2.1). The Second Schedule came into force on 15 September 1964. It set the standards to be complied with for manufacturing medicines. Schedule E (1) lists the poisonous substances under the Ayurvedic, Siddha and Unani Systems of Medicine differentiated into drugs of vegetable, animal and mineral origin. The fourth Schedule, Schedule T, deals with Good Manufacturing Practices (G.M.P.) for Ayurvedic, Siddha and Unani medicines. The government has made it mandatory under Schedule T of the Drugs and Cosmetics Act 1940, for all manufacturing units to adhere to G.M.P. (Anonymous 2016a).

2.3.2 The Rules

Part XVI deals with manufacturing of Ayurvedic, Siddha or Unani medicines. It details the procedures to acquire licenses, loans for establishing a manufacturing unit and also the identification of raw materials and their purity. Part XVIA is about the approval of institutions for carrying out tests on Ayurvedic, Siddha and Unani medicines and raw materials used in their manufacture. Part XVII defines labeling, packing and limits of alcohol in these products. Part XVIII lists government analysts and Inspectors for Ayurvedic, Siddha or Unani medicines. Part XIX sets standards for Ayurvedic, Siddha or Unani medicines (Anonymous 2016a).

TABLE 2.1
The First Schedule

- *Abhinava cintāmaṇi*
- *Arkaprakāśa*
- *Ārōgya Kalpadruma*
- *Ārya Bhiṣak*
- *Aṣṭāṅgahṛdaya*
- *Aṣṭāṅga Samgraha*
- *Āyurvēda Cintāmaṇi*
- *Āyurvēda Kalpadruma*
- *Āyurvēda Prakāśa*
- *Āyurvēda Ratnākar*
- *Āyurvēda Samgraha*
- *Ayurvēda Sāra Samgraha*
- The Ayurveda Formulary of India
- The Ayurvedic Pharmacopoeia of India
- *Baṅgasēna*
- *Basavarājīyam*
- *Bhaiṣajya Ratnāvali*
- *Bhārat Bhaiṣajya Ratnākara*
- *Bhāvaprakāśa*
- *Bhēlasamhita*
- *Bṛhat Nighaṇṭu Ratnākara*
- *Cakradatta*
- *Caraka Samhita*
- *Dravyaguṇa nighaṇṭu*
- *Gadanigraha*
- *Kāśyapasamhita*
- *Kūpipakva Rasāyana*
- *Nighaṇṭu Ratnākara*
- *Rasa Chandāmśu*
- *Rasamañjari*
- *Rasāmṛta*
- *Rasa Pradīpika*
- *Rasarāja Sundara*
- *Rasaratna Samuccaya*
- *Rasatantra Sāra Siddha Prayōga Samgraha* – Part I
- *Rasa Taraṅgiṇi*
- *Rasayōga Ratnākara*
- *Rasayōgasāgra*
- *Rasayōga Samgraha*
- *Rasēndra Sārasamgraha*
- *Sahasrayōga*
- *Śārṅgadhara Samhita*
- *Sarvarōga Cikitsā Ratnam*
- *Sarvayōga Cikitsāratnam*
- *Siddha Bhaiṣajya Maṇimāla*
- *Siddhayōga Samgraha*
- *Suśruta Samhita*
- *Vaidya Chintāmaṇi*
- *Vaidyaka Chikitsā Sāra*
- *Vaidyaka Śabda Sindhu*
- *Vaidya Jīvan Viśvanāthacikitsa*
- *Vṛndacikitsa*
- *Yōga Chintāmaṇi*
- *Yōga Ratnākara*
- *Yōgaratna samgraha*
- *Yōga Taraṅgiṇi*

2.4 OTHER APPLICABLE ACTS

In addition to the Drugs and Cosmetics Act, 1940, a few other important Acts are also applicable to the manufacture of Ayurvedic medicines in India.

2.4.1 The Pharmacy Act, 1948

Passed in 1948, the Pharmacy Act was amended in 1959 and in 1976. Under the provisions of this Act, the Government of India established the Pharmacy Council of India. Pharmacy courses conducted in the country are subject to the approval of the Council under each State Government. A State Pharmacy Council has to be established, and it has the responsibility to maintain a register of pharmacists of the concerned state. The Central Council makes regulations as to the standard of education required for qualification as a pharmacist and prescribes the nature and period of study to be undergone by the pharmacists (Anonymous 2019a).

2.4.2 The Drugs and Magic Remedies (Objectionable Advertisements) Act, 1954

This Act prohibits a person from advertising any drug which suggests that the drug can be used for the prevention of miscarriage or conception in women, the maintenance or improvement of the ability of a person to indulge in sexual activity, the correction of menstrual disorders and the diagnosis, treatment or prevention of venereal diseases and so on. It also prohibits giving false information or making claims regarding the true character of a drug (Anonymous 2019b).

2.4.3 The Narcotic Drugs and Psychotropic Substances Rules Act, 1985

This Act makes provisions for the control of operations related to narcotic drugs and psychotropic substances. Psychotropic substance means any substance, natural or synthetic or other material, included in the list of psychotropic substances under the Act. The Act provides for severe punishments to those who contravene its provisions (Anonymous 2019c).

2.4.4 The Poisons Act, 1919

The Poisons Act, 1919, restricts the importation, possession and sale of specified poisons mentioned under the Act. It provides for the collection of excise duty on medical and toilet preparations containing narcotics or narcotic drugs such as alcohol and opium. The list of ayurvedic preparations which can be consumed as alcoholic beverages is specified in the Act (Anonymous 2019d).

2.4.5 The Drugs (Price Control) Order, 1995

The Drugs (Price Control) Order, 1995, replaces the earlier Order of 1987. It is applicable more to synthetic medicines. The Government of India, under the Act, has the power to fix leader prices for the medicines. These leader prices shall operate as the ceiling sale prices for every manufacturer. The Government has the power to fix the manufacturer's price, retail price and trade commission of certain formulations. Those who violate the provisions of the order are prosecuted under the order and the provisions of the Essential Commodities Act, 1955 (Anonymous 2019e).

2.4.6 The Trade Marks Act, 1999

Under this Act, brands of any distinctive character that are capable of distinguishing some goods or services of a manufacturer from another shall not be registered. Similarly, a brand mark shall not be registered if it deceives or confuses the public, contains any matter that hurts the religious feelings of any section of citizens, contains obscene matter or if its use is prohibited under the Emblems and Names (Prevention of Improper Use) Act, 1950 (Anonymous 2019f).

2.4.7 The Biological Diversity Act, 2002

Foreign biotechnology companies trying to exploit India's traditional medical systems have to bear in mind the provisions of the Biological Diversity Act, 2002. This Act stipulates that a country which wants to use the biological resources of another country should obtain the prior consent of the latter. It prohibits the transfer of Indian genetic material to agencies outside the country without the approval of the Government of India. Biological resources include plants, animals, micro-organisms or parts thereof and their genetic material. However, the Act does not include human genetic material (Anonymous 2019g).

Under the Biological Diversity Act, a three-tier institutional mechanism operates, firstly at the apex level, then at the state levels and thirdly at local levels. The apex authority (the National Biodiversity Authority) should be contacted by all foreign nationals, organizations or associations seeking access to any of India's biological resources. Similarly, Indian nationals will have to inform the Authority before collecting India's biological resources. Further, Indian nationals should obtain the prior permission of the Authority to transfer the results related to research on any biological resource to foreigners. However, Indian *vaidyas* (ayurvedic physicians) and *hakims* (Unani physicians) will have free access to biological resources for use within the country. Whoever contravenes or abets the contravention of the provisions of the Act shall be punishable with imprisonment for a term which may extend to five years, or with a fine which may extend to one million rupees, and where the damage caused exceeds one million rupees, such fine may be commensurate with the damage caused, or with both (Anonymous 2019g).

The main objective of the Act is to protect Indian herbal wealth from over-exploitation by Western countries. Instead of closing the doors to foreign biotechnology companies, the Government of India has come up with a profit-sharing mechanism to ensure that India benefits from the commercialization of Indian herbal wealth (Anonymous 2019g).

2.5 HOW TO BEGIN MANUFACTURE OF AYURVEDIC MEDICINES

Obtaining a drug manufacturing license is mandatory in the Republic of India for manufacturing ayurvedic medicines for sale inside the country, as well as for export. The manufacturing license is renewable. Every state government appoints licensing authorities. The Director of the Department of Ayurveda or A.Y.U.S.H. of the state is the licensing authority. Applications are usually filed through a District Ayurvedic Officer or the Drugs cell of the A.Y.U.S.H. directorate. Some of the procedures may differ from state to state (Saroya 2016b), but the format is unchangeable.

The manufacture of ayurvedic medicines is governed by the rules of the Drugs and Cosmetics Act, 1940. All classical and proprietary formulations have to be cleared by the Authority before commencing commercial production. Rules 151–155 B of the Act relate to the manufacture and regulation of licensing of ayurvedic medicines. Rules 156–160 specify the conditions of licensing, and Rules 160 A–160 J are concerned with approval of testing laboratories and regulation of testing and quality control.

2.5.1 Licensing

If a company manufactures ayurvedic medicines on more than one set of premises, a separate application shall be made and a separate license shall be obtained in respect of each such set of premises. An application for the grant or renewal of a license to manufacture for sale any ayurvedic medicines shall be made in Form 24-D to the Licensing Authority along with a fee of 1000 rupees. Subject to the conditions of Rule 157 being fulfilled, a manufacturing license shall be issued by the Licensing

Authority in Form 25-D. The license shall be issued within a period of three months from the date of receipt of the application. An original license in Form 25-D or a renewed license in Form 26-D, unless suspended or cancelled earlier, shall be valid for a period of five years from the date of its issue or renewal (Anonymous 2014a).

2.5.2 Loan License

If a unit does not have its own manufacturing facilities, it can avail them from another unit which has those facilities. This arrangement is called loan licensing. An application for the grant or renewal of a loan license shall be made in Form 24-E to the Licensing Authority along with a fee of 600 rupees. A loan license shall be issued in Form 25-E. An original loan license in Form 25-E or a renewed loan license in Form 26-E, unless suspended or cancelled earlier, shall be valid for a period of five years from the date of its issue or renewal (Anonymous 2014a).

Before a manufacturing license or loan license is granted, the applicant shall comply with the following conditions:

i) The manufacture shall be carried out in such hygienic conditions as given in Schedule T of Rule 157 under the Drugs and Cosmetics Act, 1940
ii) The manufacture shall be carried out under the direct supervision of technical staff including at least one person who possesses (a) a degree or diploma in Ayurveda or Ayurvedic pharmacy conferred by a university or government or (b) a graduate in pharmacy, pharmaceutical chemistry, chemistry or botany of a university or government with experience of at least two years in the manufacture of ayurvedic drugs or (c) a *vaidya* registered in a State register of indigenous systems of medicine, having an experience of at least four years in the manufacture of the medicines

2.5.3 Applying for a Manufacturing License

The application is to be supported by blueprints of the manufacturing facility as per specifications laid down in Rule 157 pertaining to G.M.P.; a list of technical staff with qualifications; a list of formulations for which the license is sought, indicating the name and reference book in the case of classical medicines, and a complete details of ingredients, quantities and method of preparation in the case of proprietary medicines; a list of machinery and equipment, including those used in the testing laboratory and an application fee, which may vary from state to state. In addition to these the applicant has to submit several documents and affidavits related to ownership of the company and the premises in which the manufacturing facility is located (Saroya 2016b).

The manufacturing facility should have an office (100 sq. feet), a crude drug store (150 sq. feet), a finished medicine store (150 sq. feet), a laboratory (150 sq. feet), rejected medicine store (100 sq. feet), a *cūrṇa* room (200 sq. feet), a furnace (200 sq. feet), and a packaging area (Saroya 2016b).

2.5.3.1 Conditions of License

A license issued in Form 25-D shall be subject to the following further conditions also:

(i) The licensee shall maintain proper records of the details of manufacture and of the tests carried out by him regarding the raw materials and finished products
(ii) The licensee shall allow an Inspector appointed under the Act to enter any premises where the manufacture of ayurvedic medicine is carried out. He should permit the Inspector to inspect the premises, to take samples of the raw materials as well as finished products and to inspect the records maintained under the Rules
(iii) The licensee shall maintain an Inspection Book in Form-35 to enable an Inspector to record his impressions and the defects noticed (Anonymous 2014a)

2.5.3.2 Conditions of Loan License

A loan license in Form 25-E shall be subject to the following conditions as well:

(i) The license in Form 25-E shall be deemed to be cancelled or suspended, if the license owned by the licensee in Form 25-D whose manufacturing facilities have been availed of by the licensee is cancelled or suspended
(ii) The licensee shall comply with the provisions of the Act and of the Rules and with such further requirements if any, as may be specified in any Rules subsequently made under Chapter IV-A of the Act, provided that where such further requirements are specified in the Rules, these would come into force four months after publication in the Official Gazette
(iii) The licensee shall maintain proper records of the details of manufacture and of the tests carried out by him regarding the raw materials and finished products
(iv) The licensee shall allow an Inspector appointed under the Act to inspect all registers and records maintained under these Rules and shall furnish to the Inspector such information as he may require for the purpose of ascertaining whether the provisions of the Act and the Rules have been observed
(v) The licensee shall maintain an Inspection Book in Form-35 to enable an Inspector to record his impressions and the defects noticed (Anonymous 2014a)

2.5.4 Proof of Effectiveness

2.5.4.1 Classical Formulations

Traditional ayurvedic formulations with ingredients, dosage form and indications according to authoritative texts of First Schedule do not require any safety study. However, evidence of effectiveness is required from authoritative texts. Traditional ayurvedic formulation with a change in dosage form also does not require safety studies. Nevertheless, evidence of effectiveness is required from published literature. Traditional ayurvedic formulations for new indication also do not require a safety study. However, evidence of effectiveness needs to be provided from published literature and pilot studies (Anonymous 2014a).

2.5.4.2 Proprietary Formulations

Formulations based on R.&D. efforts and textual rationale fall into this category. The class also includes nutritive formulations, cosmetics and formulations based on aqueous, hydro-alcoholic and other extracts. Proprietary formulations with ingredients not from Schedule E (I) and based on textual rationale from authoritative texts do not require any safety study, but evidence of the effectiveness of ingredients needs to be supported by published literature and a pilot study. Proprietary formulations with any of the ingredients from Schedule E (I) and based on rationales from authoritative texts also do not require any safety study. However, effectiveness needs to be proved from published literature and a pilot study (Anonymous 2014a).

Extract-based formulations fall into five varieties. Only references from authoritative texts are required for formulations manufactured with aqueous extracts and based on authoritative texts. Formulations manufactured with aqueous extracts as per authoritative texts, but with new indications, do not require a safety study. However, proof of effectiveness is required. Evidence of the effectiveness of medicines manufactured with hydro-alcoholic extracts and based on textual information is required on a case-to-case basis. Safety study is required for formulations manufactured using specific hydro-alcoholic extracts and intended for a new indication. Proof of effectiveness is also required. The license requirements for medicines manufactured with various organic solvents are stringent. Data on acute toxicity, chronic toxicity, mutagenicity and teratogenicity are required in such cases. The effectiveness of these products needs to be proved with published literature and clinical studies (Anonymous 2014a).

2.5.5 Records of Raw Material Used

Each licensed manufacturer of ayurvedic medicines shall keep a record of raw materials used by the manufacturing unit in the preceding financial year in the proforma document given in Schedule TA. The information shall be submitted to the State Drug Licensing Authority and to the National Medicinal Plants Board by 30 June of the succeeding financial year (Anonymous 2014a).

2.5.6 G.M.P. Certification

To ensure the quality of products, the Government of India has made it mandatory for all Ayurveda medicine manufacturing units to adopt good manufacturing practices. G.M.P. Certification came into force as per the Gazette Notification No. GSR 561(E) dated 23 June 2000. The certificate will be issued to those manufacturers who comply with the requirements of G.M.P. as laid down in Schedule T of Rule 157 of the Drugs and Cosmetics Act, 1940. Those manufacturers who had registered their units prior to the date of the Gazette were asked to obtain the certificate within two years from that date. The certificate of Good Manufacturing Practices to manufacturers of ayurvedic medicines shall be issued for a period of five years to licensees who comply with the requirements of Good Manufacturing Practices (G.M.P.) for Ayurveda, Siddha and Unani drugs as laid down in Schedule T (Anonymous 2014a).

2.5.6.1 Principles of the Guidelines

The G.M.P. guidelines set by Government of India follow the following basic principles:

i) All manufacturing processes are clearly defined and regulated. All critical processes are validated to safeguard consistency and compliance with specifications
ii) Manufacturing processes are controlled, and any deviations from them are evaluated. Changes that have a bearing on the quality of the medicines are validated
iii) Instructions and standard operating procedures (S.O.P.s) are written in clear and unambiguous language
iv) Operators are trained to carry out procedures and document them properly
v) Records are prepared properly during manufacture and ensure that all the steps defined in the procedures and S.O.P.s are in fact taken and that the quality and quantity of the medicines are as expected. Deviations from them are investigated and documented
vi) Records of manufacture that trace the complete history of a batch are retained in a comprehensible and accessible form. A system is available for recalling any batch of medicines from sale
vii) Complaints about marketed medicines are examined, the causes of complaints are investigated and appropriate corrective measures are taken to prevent recurrence (Anonymous 2014b)

2.5.6.2 G.M.P. Guidelines

The main objectives behind insisting on the requirements of G.M.P. certification are to ensure the quality of raw materials used in production, to upgrade the standards and quality of production, to improve the acceptability of medicines manufactured and to make it compulsory to document all the different stages of the manufacturing process.

According to the guidelines set by A.Y.U.S.H., raw materials used in the manufacture of medicines are to be authentic, of good quality and devoid of contamination. The manufactured medicines should be of high quality, as evidenced by stringent quality control measures (Anonymous 2014b). However, under the Indian Medicine Central Council Act, 1970, registered *vaidyās* who prepare their own medicines to dispense to their patients and are not selling such medicines in the market are exempted from the purview of G.M.P. (Saroya 2016c).

According to G.M.P. guidelines the factory should have well-defined areas for receiving and storing raw material, manufacturing process areas, quality control section, finished goods store, office and storage areas for rejected raw materials and finished goods. The factory building should be designed in such a way to facilitate the production of medicines under hygienic conditions. It should be free of cobwebs and have provisions to prevent the entry of insects and rodents. It should be well-ventilated and have a proper drainage system, fire safety measures and space for drying raw materials and in-process medicines (Anonymous 2014a, 2014c).

The management should ensure that all workers employed in the factory are free from contagious diseases. The clothing of the workers should consist of clean uniforms suitable for the nature of the work and the climate. Provision should be made for separate lavatories for men and women, and such lavatories are to be located at places separated from the processing rooms. All workers are to be subjected to a medical examination at the time of employment, and thereafter once a year by a qualified physician, and the records maintained (Anonymous 2014a, 2014c).

For carrying out the manufacturing activity, depending on the size of the operation and the nature of product manufactured, suitable manual or semi-automatic machinery should be available at the premises. Proper S.O.P.s for cleaning, maintaining and operation of every machine should be available to the workers. Batch manufacturing records, distribution records and records of market complaints should be maintained. The manufacturing unit is also expected to keep books on regulatory affairs, such as *The Ayurvedic Pharmacopoeia of India (Part I & II)*, *The Ayurvedic Formulary of India (Part I & II)* and the *Drugs and Cosmetics Act, 1940* (Anonymous 2014a, 2014c).

Every licensee should provide a facility for quality control of the products in his own premises or through a Government-approved testing laboratory. The tests should be conducted according to *The Ayurvedic Pharmacopoeia of India*. The quality control section should test all raw materials, monitor in-process quality checks and assure the quality of medicines produced and released to the finished goods store. The G.M.P. guidelines give the specifications of the quality control section (Anonymous 2014a; 2014c).

2.5.6.3 Inspectors for Conducting G.M.P. Inspection

To ensure that all ayurvedic medicine manufacturing units in the country comply with the requirements of G.M.P., the Government of India appoints Inspectors, whose powers, duties, conditions, limitations or restrictions subject to which such powers and duties may be exercised, are prescribed in the Drugs and Cosmetics Act, 1940 and the Rules thereunder. The Drugs and Cosmetics Rules 49 and 67 stipulate that a person with any of the following qualifications can be appointed as an Inspector:

(i) A degree in Pharmacy/Pharmaceutical Sciences/Medicine with specialization in Clinical Pharmacology or Microbiology from a University who shall have undergone practical training in the manufacture of ayurvedic medicines
(ii) A degree in Ayurveda conferred by a University or State Government or a Statutory Faculty, Council or Board of Indian Systems of Medicine recognized by the Central Government or the State Government for this purpose
(iii) A diploma in Ayurveda granted by a State Government or an Institution recognized by the Central Government or a State Government for this purpose

However, the Rules make it clear that any person having a financial interest in the manufacture or sale of any medicine cannot be appointed as an Inspector, in spite of meeting the above requirements (Anonymous 2014b).

2.5.6.4 Duties of the Inspectors

The Inspectors may inspect any premises wherein any ayurvedic medicine is being manufactured or any premises wherein any ayurvedic medicine is being sold, stocked, exhibited, offered for sale or distributed. He can take samples of any ayurvedic medicine which is being manufactured or being sold or is stocked or exhibited or offered for sale, or is being distributed. He can also collect samples from any person who is in the course of conveying, delivering or preparing to deliver such ayurvedic medicines to a purchaser or a consignee (Anonymous 2014b).

At all reasonable times, with such assistance, if any, as he considers necessary, he can search any person, who he believes has secreted about his person any ayurvedic medicine in respect of which an offence under Chapter IV-A of the Drugs and Cosmetics Act, 1940, has been, or is being, committed. He can also enter and search any place in which he has reason to believe that an offence under Chapter IV-A of the Drugs and Cosmetics Act, 1940, has been, or is being committed. He can stop and search any vehicle, vessel or other conveyance which he has reason to believe is being used for carrying any ayurvedic medicine in respect of which an offence under Chapter IV-A of the Drugs and Cosmetic Act, 1940, has been, or is being, committed, and order in writing the person in possession of the ayurvedic medicines in respect of which the offence has been, or is being, committed, not to dispose of any stock of such ayurvedic medicines for a specified period not exceeding 20 days, or, unless the alleged offence is such that the defect may be removed by the possessor of the medicines, seize the stock of such medicines and any substance or article by means of which the offence has been or is being committed or which may be employed for the commission of such offence (Anonymous 2014b).

The Inspector can examine any record, register, document or any other material object found with any person, or in place, vehicle, vessel or other conveyance referred as above and seize the same, if he has reason to believe that it may furnish evidence of the commission of an offence punishable under the Act or the Rules made thereunder. For the stock of any ayurvedic medicines or any record, register, document or any other material object seized by him, the Inspector may issue a receipt on Form 16 of Drugs and Cosmetics Rules, 1945 (Anonymous 2014b).

He can require any person to produce any record, register or other document relating to the manufacture for sale, distribution, stocking, exhibition for sale, offer for sale or distribution of any ayurvedic medicines in respect of which he has reason to believe that an offence under Chapter IV-A of the Drugs and Cosmetics Act, 1940 has been, or is being, committed. The provisions of the Code of Criminal Procedure, 1973 shall apply to any search or seizure under Chapter IV-A of the Drugs and Cosmetics Act, 1940 as they apply to any search or seizure made under the authority of a warrant issued under section 94 of the said Code (Anonymous 2014b).

All records, registers or other documents seized or produced as above shall be returned to the person from whom they were seized within a period of 20 days of the date of such seizure or production, as the case may be, after copies thereof or extracts therefrom certified by that person, in such manner as may be prescribed, have been taken (Anonymous 2014b).

If any person intentionally obstructs an Inspector in the exercise of the powers conferred upon him under Chapter IV-A of Drugs and Cosmetics Act, 1940 or refuses to produce any record, register or other document when so required as mentioned above, he shall be punishable with imprisonment which may extend to three years, or with fine, or with both (Anonymous 2014b).

It shall be the duty of an Inspector to inspect not less than twice a year, all premises licensed for the manufacture of ayurvedic medicines within the area allotted to him and to satisfy himself that the conditions of the license and the provisions of the Act and the Rules made thereunder are being observed. He should send forthwith to the controlling authority after each inspection, a detailed report indicating whether or not the conditions of the license and the provisions of the Act and Rules are being observed. He should also collect samples of the medicines manufactured on the premises and send them for test or analysis in accordance with these Rules. He can also institute prosecutions in respect of violation of the Act and the Rules made thereunder (Anonymous 2014b).

2.5.6.5 Procedure for Collection of Drug Sample by Inspector

When an Inspector takes any sample of an ayurvedic medicine under Chapter IV-A of Drugs and Cosmetic Act, 1940, he shall tender its fair price and may require a written acknowledgment of the same. When the price tendered is refused, or when the Inspector seizes the stock of any ayurvedic medicine, he shall tender a receipt on Form 17A of Drugs and Cosmetics Rules, 1945 (Anonymous 2014b).

When an Inspector takes a sample of ayurvedic medicine for the purpose of test or analysis, he shall intimate such purpose in writing on Form 17 of Drugs and Cosmetics Rules, 1945 to the person from whom he takes it and, in the presence of such person, unless he willfully absents himself. The Inspector shall then divide the sample into three portions, seal them effectively, mark them suitably and permit such person to add his own seal and mark to all or any of the portions. When the ayurvedic medicine is packed in containers of small volume, instead of dividing a sample as aforesaid into three portions, the Inspector may collect three of the said containers after suitably marking and, if necessary, sealing them (Anonymous 2014b).

One portion or container he shall forthwith send to the Government Analyst for test or analysis. This sample shall be sent by registered post or by messenger, in a sealed package, enclosed with a memorandum on Form 18-A of Drugs and Cosmetics Rules, 1945. A copy of the memorandum and a specimen impression of the seal used to seal the package shall be sent by registered post or by messenger to the Government Analyst. On receipt of the package from the Inspector, the Government Analyst or an Officer authorized by him in writing on his behalf, shall open the package and shall also record the conditions of the seals on the package. He shall submit the second to the Court before which proceedings are instituted in respect of the ayurvedic medicine. He shall send the third portion to the person from whom he acquired the drug or cosmetic (Anonymous 2014b).

When an Inspector takes any action under clause (c) of section 22, he shall ascertain whether or not the ayurvedic medicine contravenes any of the provisions of section 33EEC. If it is ascertained that the medicine does not contravene such a provision, he shall forthwith revoke the order passed under the said clause and take such action as may be necessary for the return of the stock seized. When he seizes the stock of the ayurvedic medicine, he shall inform a Judicial Magistrate and take his orders as to the custody thereof. If the alleged contravention is such that the defect may be remedied by the possessor of the ayurvedic medicine, the Inspector shall, on being satisfied that the defect has been so remedied, forthwith revoke his order under the said clause. When an Inspector seizes any record, register, document or any other material object, he shall soon inform a Judicial Magistrate and take his orders as to the custody thereof (Anonymous 2014b).

2.5.6.6 Joint Inspection of Testing Laboratories

Before granting approval, the licensing authority shall cause the testing laboratory of the facility to be inspected jointly by the Inspectors appointed or designated by the Central Government and State Government for this purpose. They shall examine the premises and the equipment intended to be used for testing and verify the professional qualifications of the expert staff who may be employed by the laboratory. The laboratory shall allow the Inspectors to enter, with or without prior notice, the premises where testing is carried out and to inspect the premises, the equipment used and the testing procedures employed. The laboratory shall permit the Inspectors to inspect the registers and records maintained and shall furnish to them such information as is required for the purpose of ascertaining whether the provisions of the Act and Rules have been observed. On conclusion of the inspection the Inspectors shall forward to the licensing authority a detailed report of the results of the inspection (Anonymous 2014b).

2.6 PUNISHMENT FOR VIOLATION OF THE ACT

Chapter IVA of Drugs and Cosmetics Act, 1940, provides provisions for dealing with misbranded drugs, adulterated drugs, spurious drugs, regulation of manufacture for sale of ayurvedic drugs, inspection of manufacturing premises, analysis of products and so on. It also sets punishments for violations of the provision of the Act. Anyone who manufactures for sale or for distribution any ayurvedic medicine deemed to be adulterated or without a valid license shall be punishable with imprisonment for a term which may extend to one year and with a fine which shall not be less than 2000 rupees (Saroya 2016d).

Similarly, anyone who engages himself in the manufacture and sale of a spurious ayurvedic medicine shall be punishable with imprisonment for a term which shall not be less than one year, but which may extend to three years, and with a fine which shall not be less than 5000 rupees. Contravention of any other provisions of this Chapter shall be punishable with imprisonment for a term which may extend to three months and with a fine which shall not be less than 500 rupees. In extraordinary conditions the drug control department has the power to confiscate the property of the manufacturing facility (Saroya 2016d).

2.7 CANCELLATION OR SUSPENSION OF LICENSES

The Licensing Authority can cancel or suspend a manufacturing license, if the licensee has failed to comply with any of the conditions of the license or with any provisions of Drugs and Cosmetics Act, 1940, and the Rules made thereunder. The Licensing Authority may, after giving the licensee an opportunity to show cause why such an order should not be passed, by an order in writing stating the reasons, cancel a license issued under this Part or suspend it for such period as he thinks fit. A licensee whose license has been suspended or cancelled may appeal to the State Government within a period of three months from the date of receipt of the order. The State Government shall, after considering the appeal, take a decision on it (Anonymous 2014a).

2.8 LABELING OF PRODUCTS

Labeling and packing are mandatory for any ayurvedic medicine that is to be sold in the country. The label should provide clear information about different aspects of the product, like composition, dose, mode of administration and contra-indications. Improper labeling may lead to improper use. Rule 161 of Drugs and Cosmetics Act, 1940, provides guidelines for the labeling of ayurvedic products (Anonymous 2016a).

The label should bear the name of the drug as given in *The Ayurvedic Pharmacopoeia of India* or mentioned in the authoritative books included in the First Schedule. The words "Ayurvedic medicine" also need to be printed on it. The words "for external use only" are to be printed on the label if the medicine is for external application. If the medicine is for distribution to the medical profession as a free sample, the words "physician's sample, not to be sold" must be printed on the label. Every manufacturer is expected to declare on the label of the product a true list of all the ingredients used in its manufacture, quantity of each ingredient and a reference to the method of preparation, as detailed in the authoritative text specified in the First Schedule. If the list of ingredients is too large to be accommodated on the label, the same may be printed separately and enclosed with the packing. Reference to this is to be made on the label. The label should also carry the address of the manufacturer, manufacturing license number, quantity of medicine in each container, batch number, date of manufacture, date of expiry, mode of application and warning, if any (Anonymous 2016a).

Rule 161-A of Drugs and Cosmetics Act, 1940, gives guidelines for the labeling and packing of ayurvedic products for export. The labeling and packing should meet the requirements of the law of the country to which the product is exported. The other information to be included are: Name and address of manufacturer with manufacturing no., batch no. or lot no., date of manufacture, along with the date of expiry and the main ingredients with quantity, if required by the importing country (Anonymous 2016a).

2.9 SHELF LIFE OF AYURVEDIC MEDICINES

Till 2009, Rule No. 161 of the Drugs and Cosmetics Act, 1940, which deals with labeling, packing and limit of alcohol for ayurvedic medicines, had not made it obligatory to declare the shelf life of ayurvedic medicines on the label of the product. The rule insisted only on declaring the date of manufacture, for which purpose the date of completion of manufacture or the date of bottling or packing was considered sufficient. It was obligatory to declare on the label the other mandatory requirements like name of the product, license no., batch no., textual reference to the source of the formula and true list of ingredients (Gupta et al. 2017).

In 2009 the Department of A.Y.U.S.H., vide G.S.R. 764 (E) dated 15 October 2009, incorporated Rule no. 161-B referring to the shelf life of ayurvedic medicines into the Drugs and Cosmetics Act, 1940. The rule was further revised, vide G.S.R. No. 789 (E) dated 12 August 2016, by the Ministry of A.Y.U.S.H., amending Rule No. 161-B of the Drugs and Cosmetics Act, 1940, and Rules 1945, applicable to the shelf life of ayurvedic products (Gupta et al. 2017). The amended Rule made it mandatory to display conspicuously on the label of the container the date of expiry of the product, and after the said date of expiry, the medicine is not to be marketed, sold, or distributed. The following clauses were also included in the amended Rule:

a) Persons applying for license or renewal of license for manufacture of ayurvedic medicines shall submit to the State Licensing Authority scientific data pertaining to shelf life of the medicine based on real-time stability studies conducted in accordance with the guidelines prescribed in *The Ayurvedic Pharmacopoeia of India Part I, Volume-VIII*
b) Before granting license or renewal of license for ayurvedic medicine, the State Licensing Authority shall ensure the validity of the data submitted by the manufacturer in support of the claimed shelf-life of that medicine
c) At any time the State Licensing Authority may direct the manufacturer to provide samples of the medicine and any other related information and may share them with the Pharmacopeial Laboratory for Indian Medicine (P.L.I.M.), Ghaziabad for analysis or independent validation
d) When the manufacturer fails to comply with the direction of the State Licensing Authority under sub-rule (5), the license for the manufacturing of the medicine shall be suspended after giving a reasonable opportunity for being heard
e) Unless otherwise determined on the basis of scientific data, the shelf-life or date of expiry of an Ayurvedic medicine defined under clause (a) of section 3 of the Act shall be as given in Table 2.2

TABLE 2.2
Shelf Life Of Major Ayurvedic Dosage Forms*

Sl. No.	Dosage Form	Shelf Life
1	*Arka*	1 year
2	*Āsava, Ariṣṭa*	10 years
3	*Avalēha*	3 years
4	*Danta mañjan*	2 years
5	*Drāvaka, Lavaṇa, Kṣāra*	5 years
6	*Ghṛta*	2 years
7	*Guḷika*	3 years
8	*Kvātha cūrṇa, Cūrṇa*	2 years
9	*Lēpa*	3 years
10	*Taila*	3 years

*Anonymous 2016b.

2.10 GOOD CLINICAL PRACTICES

Considering the growing interest in Ayurveda, the Department of A.Y.U.S.H., Government of India, published *Good clinical practice guidelines for clinical trials in Ayurveda, Siddha and Unani medicine (GCP-ASU)*. The guidelines are intended to develop methodologies for evaluation of research and to provide appropriate methods to streamline the regulation and registration of products from Indian traditional medicine including Ayurveda. Department of A.Y.U.S.H. considers this document as a reference source for scientists of traditional medicine, registered medical practitioners, manufacturers and health authorities (Anonymous 2013).

However, these guidelines are drawn up on the basis of Western biomedical approaches, including randomized controlled trials (R.C.T.s), which are intended to provide necessary evidence of the efficacy of a traditional medicine like an ayurvedic product. They appear to be largely based on the standard G.C.P. guidelines proposed earlier by the Central Drugs Standard Control Organization (C.D.S.C.O.) (Patwardhan 2011). The C.D.S.C.O. is the National Regulatory Authority of India for Western medicine.

The new guidelines need to be revised in the backdrop of the debate on the ethics of R.C.T.s initiated by Miller and Joffe (2011) who suggest a solution to the acceptance of equipoise, which has assumed canonical status in research ethics and which holds that a patient may be enrolled ethically into a R.C.T. only when substantial uncertainty surrounds which of the trial treatments would most likely benefit him (Fries and Krishnan 2004). They suggest five conditions, which, when satisfied, show an R.C.T. should not be deemed necessary. They are listed below:

i) Compelling, usually mechanism-based, rationale favoring the efficacy of the new agent
ii) Evidence of large effect sizes on the basis of early clinical studies
iii) Well-understood, typically poor outcome with limited interpatient variability, given current therapy or supportive care
iv) Availability of one or more concurrent or historical control groups with characteristics similar to those of the patients to be enrolled in the proposed study
v) Use of a clinical or validated surrogate primary end point in the uncontrolled trial (Miller and Joffe 2011)

The criteria of Miller and Joffe (2011) can be extended to ayurvedic treatments. The first condition pertains to compelling, mechanism-based rationales favoring the efficacy of the new intervention. For correct diagnosis and treatment of diseases Ayurveda makes use of ten factors such as physiological constitution (*prakṛti*), tissue elements (*dhātu*), digestive efficiency (*agni*), temporal aspects of *tridōṣa* (*kālam*), age (*vaya*), physical strength (*balam*), place of residence (*dēśam*), homologation (*sātmya*), food (*āhāra*) and emotional status (*satvam*). These concepts provide well-positioned arguments in support of complex ayurvedic treatments. The second condition calls for evidence of large effect sizes on the basis of earlier clinical studies. Ayurveda has long-standing historical evidence of the efficacy of treatments judged in terms of its own basic principles. Application of Ayurveda to chronic disease satisfies the third criterion, as the effect size for cure using Western drug therapy is very low. Availability of concurrent or historical control groups with characteristics similar to those of the patients to be enrolled in the proposed study is the fourth criterion. Present populations receiving traditional treatments provide better clinical assessment than carefully selected members of some contemporary R.C.T. This satisfies the fourth condition. The theoretical basis of Ayurveda is sound and success of ayurvedic therapy can be assessed by its own parameters, thus justifying the fifth condition (Patwardhan 2011).

Significant success has been achieved in demonstrating the success of ayurvedic treatment of a chronic disease like rheumatoid arthritis (see Chapter 11). This shows that the rationale of ayurvedic therapy is acceptable to Western medical protocols. Therefore, regulators of the Ayurveda system in India should redesign the G.C.P. guidelines based on the principles of Ayurveda. Already there is an outcry against the pseudo-ethical principle of equipoise, with its inappropriate decision point, on which R.C.T.s are based. It is said to disregard patient autonomy, fails to protect patients on aggregate, ignores potential benefits to society and impedes medical progress (Fries and Krishnan 2004). The recent thinking is that trials of new treatments for life-threatening diseases that violate equipoise are both ethical and necessary for the development of evidence to support health policy decisions made on behalf of populations of patients (Miller and Joffe 2011).

2.11 REGULATORY PROBLEMS THAT REQUIRE SOLUTION

2.11.1 Monitoring of A.Y.U.S.H. G.C.P. Guidelines

Putting into practice the new A.Y.U.S.H. G.C.P. guidelines, in fact, stirs up a hornet's nest. For example, existing regulatory and licensing authority in Ayurveda is not competent to monitor the implementation of A.Y.U.S.H. G.C.P. guidelines because of a lack of infrastructure and human resources. Authorizing a Western medicine regulatory authority for this purpose is unscientific and unethical. A competent regulatory authority for quality assurance of ayurvedic product development needs to be formed with an advisory board comprising ayurvedic experts from various sectors like Ayurveda research and practice, drug manufacturing and marketing experts and members from administrative and academic sectors of government departments and ayurvedic institutions. This should be developed over a time period of 3–5 years.

Product development, according to A.Y.U.S.H. G.M.P. guidelines, calls for the investment of huge amounts of money and time which will not be affordable to most of the manufacturers of ayurvedic medicine. Lack of infrastructure, non-availability of clinical research experts in the Ayurveda sector and small market volume compared to Western medicine are deterrents.

Before formulating a better set of guidelines, the Department of A.Y.U.S.H. should conduct workshops in different parts of India involving experienced physicians from clinical practice, clinical research, Ayurvedic medicine manufacturing and medicine marketing. The new guidelines should be implemented step by step in a phased manner over a span of 8–10 years. The pertinent question here is: Why should we strive to formulate new products when there are numerous formulae in Ayurveda classical texts not requiring clinical trials? Classical texts describing such formulae should be included in the First Schedule.

2.11.2 Revision of First Schedule

An approved list of classical ayurvedic texts (First Schedule) does not cover important texts like *Vaidyamanōrama*, *Yōgāmṛtam*, *Cikitsāmañjari* and so on which have been widely used by Ayurveda exponents for many generations. As these texts are not included in the First Schedule, formulations from these texts are considered to be proprietary medicines according to A.Y.U.S.H. G.C.P. guidelines. A.Y.U.S.H. G.C.P. guidelines state that clinical trials are required for formulae outside the First Schedule, even if they are in traditional dosage forms like *kvātha*, *ariṣṭa*, *cūrṇa* or *avalēha*. The First Schedule should, therefore, be revised in consultation with physicians, manufacturers and academics before implementing new quality assurance rules for Ayurvedic proprietary medicines.

Interest in Ayurveda has been growing all over the world ever since the World Health Organization (WHO) convened the International Conference on Primary Health Care (6–12 September 1978) at Alma Ata, in the former Soviet Republic of Kazakhstan (Kumar 2016). This interest arises primarily from the belief that Ayurveda is a form of natural medicine utilizing herbs. Nevertheless, when one gets closer to Ayurveda, it becomes apparent that heavy metal-containing preparations are also freely employed. The First Schedule of Drugs and Cosmetics Act, 1940, lists 54 classical texts, out of which many deal with *rasaśāstra* or the art of treating diseases using mercurials. This is a self-contradiction. To erase the stigma attached to Ayurveda following the report of Saper et al. (2004), these *rasaśāstra* texts should be replaced with Sanskrit texts dealing with herbal Ayurveda. This calls for revision of the First Schedule.

2.11.3 Regulation of Change in Dosage Form

Ayurvedic practitioners traditionally used to prepare the same formula in different dosage forms for different conditions of the disease, even though the original formula might have been mentioned in only one dosage form in classical texts. An example is *Indukāntam kvātha*, the original formula described as a *ghṛta* in the text (Vaidyan and Pillai 1985). A.Y.U.S.H. guidelines limit the privilege of the physician or the manufacturer who manufactures the medicine for the physician by making clinical trials mandatory for any change in dosage form or dose. Ayurvedic medicines with the same formula, but in a different dosage form, should be considered under the classical medicine category only.

2.11.4 Regulation of Complementary Products

These days ayurvedic soaps are widely manufactured to protect the skin from dirt and germs. Chandrika, Medimix, Margo Neem Soap, Hamam Tulsi Soap and Himalaya Herbal Soap are some of them. They are considered proprietary ayurvedic products and their licensing, therefore, requires clinical trial reports, according to A.Y.U.S.H. G.C.P. guidelines. Even for the renewal of a license, phase 3 and 4 clinical trials are essential. In the larger interest of Ayurveda, such complementary products should be exempted from the ambit of A.Y.U.S.H. G.C.P. guidelines, and if necessary, only a pilot safety study should be insisted upon.

2.11.5 Labeling of Products

Bhalerao et al. (2010) carried out an interesting survey on the labeling information on 190 labels (101 classical and 89 proprietary formulations) of three ayurvedic pharmacies. The information given in them was checked against a set of quality criteria given in the Drugs and Cosmetics Act, 1940. The results were expressed as percentages. References to authoritative books as specified in the First Schedule of the Act were mentioned on only 90% of labels of the 101 classical formulations surveyed. 55% of labels of classical drugs and 88% of labels of proprietary drugs listed the ingredients. Only 20% of classical formulations' and 15% of proprietary formulations' labels mentioned warnings.

The authors observed that majority of the labels did not fulfill all the requirements that were mandatory as per the Act, and there was not even a single label which fulfilled all ten requirements specified by the Drugs and Cosmetics Act, 1940. The survey showed that nearly 45% of the classical formulations and 12% of proprietary formulations did not list the ingredients (Bhalerao et al. 2010). Similar findings were also reported by Mallick et al. (2012).

The Drugs and Cosmetics Act, 1940, states that mentioning the expiry date of a product is mandatory only in the case of ayurvedic medicines exported from the country (Anonymous 2002). It implies that it is not compulsory for the expiry date to be printed on containers of products marketed in India. This can be construed as permission to keep and use ayurvedic products for any number of years without a loss in quality. Nevertheless, *Śārṅgadhara Samhita* clearly specifies that all ayurvedic drugs should have an expiry date (Bhalerao et al. 2010). According to it every ayurvedic formulation has a specific *savīryatā avadhi*. It means that the medicinal constituents in the formulation are active only for a given period of time and that keeping those formulations beyond that period can render them ineffective. For example, a *cūrṇa* has a *savīryatā avadhi* of only four months, and if used beyond that period, the formulation may be ineffective (Bhalerao et al. 2010). Therefore, the printing of expiry date on labels should be made mandatory even for ayurvedic medicines sold inside the country.

The Drugs and Cosmetics Act, 1940, does not insist on processing details of *Bhasmas*. There are several methods for the preparation of one *bhasma* in ayurvedic texts (Anonymous 2003). Therefore, it is necessary to specify which method was used in the manufacture of the formulation in question. Considering the reports on the presence of heavy metals in *Bhasmas* (Saper et al. 2004, 2008; Kales and Saper 2009), it will be appropriate to also mention the heavy metals content on the product labels.

None of the labels specified potential interactions with food and synthetic medicine, as providing these details for ayurvedic drugs is not mandatory according to the Drugs and Cosmetics Act, 1940. It is widely believed that ayurvedic medicines are safe, as they are "natural". However, now it is known that many herbs have the ability to affect the CYP450 enzyme system. The most notable among them is St John's Wort (*Hypericum perforatum*), which reduces the plasma concentration of many drugs such as amitriptyline, digoxin, theophylline, non-steroidal anti-inflammatory drugs (N.S.A.I.D.s), and oral contraceptives (Izzo and Ernst 2001). St John's Wort also increases the activity of multidrug transporter P-glycoprotein efflux pumps in the duodenum. This mechanism has been implicated in its interactions with indinavir, digoxin, and cyclosporin. It also increases the metabolism of protease inhibitor indinavir, resulting in therapeutic failure. It has also been implicated in reducing the bioavailability of cyclosporine, an immunosuppressive agent given to transplant recipient patients (Mai et al. 2000; Milić et al. 2014; Bhadra et al. 2015).

Serious adverse effects have been encountered with warfarin due to herb-induced inhibition of its metabolizing enzymes. These effects include spontaneous post-operative bleeding, hematomas, hematemesis, melena, subarachnoid hemorrhage, subdural hematomas and thrombosis. The herbs that can cause such adverse reactions include *Allium sativum*, *Areca catechu*, *Boswellia serrata*, *Camellia sinensis* (tea), *Hibiscus sabdariffa* and *Zingiber officinale* (Milić et al. 2014; Bhadra et al. 2015; Izzo 2012; Izzo et al. 2016).

The common ayurvedic herb licorice (*Glycyrrhiza glabra*) contains glycyrrhizin and glycyrrhetinic acid. These phytocompounds are potent inhibitors of 11-β hydroxy steroid dehydrogenase, causing raised cortisol and increased mineralocorticoid activity, leading to hypertension and suppression of the renin-angiotensin aldosterone system. Licorice also interacts with some antihypertensives and antiarrhythmics (Serra et al. 2002). Therefore, manufacturers of ayurvedic medicines should also provide some information about the safety of the formulations (including dosage schedule and drug storage conditions) on the drug labels. The drug control department should enforce the laws pertaining to the marketing of ayurvedic medicines. These measures will prevent the irrational use of ayurvedic medicines.

2.11.6 Revision of Schedule E1

A list of potentially toxic ingredients including heavy metals has already been classified under Schedule E1 of the drugs and cosmetic rules as a special category (Anonymous 2016a). Nevertheless, it includes only 21 drugs from Ayurveda and has not been amended since 2010. In 2010 certain ingredients like red oxide of lead were also removed from the list, which previously contained 25 entries. The Drugs and Cosmetics Act, 1940, needs to be amended appropriately to include potentially toxic drugs in Schedule E1. This can be achieved with the help of an expert committee that can review the available recent literature and formulate the list of such ingredients to be appended to the list. Additionally, a separate document for the appropriate detoxification-cum-potentiation protocols for toxic drugs in recommended media is also required (Ilanchezhian et al. 2010). This is all the more important, as *śōdhana* procedures enhance the medicinal value of the herb in addition to removing toxic phytocompounds (Singh et al. 1985; Mitra et al. 2012; Acharya 2014).

2.11.7 Effective Implementation of the Act

Sahoo and Manchikanti (2013) carried out a questionnaire-based survey to assess the constraints faced by the Indian herbal medicine industry. Responses were collected from 150 companies through email, telephone and personal interviews. Insufficient regulatory guidelines, particularly guidelines for G.M.P.; non-implementation of good agricultural and collection practices; and weak implementation of the Drugs and Cosmetics Act, 1940 were mentioned as major impediments to the industry. The implementation of G.M.P. is decisive for the quality of herbal medicines. Most of the respondents suggested that apart from Schedule T, separate guidelines on quality control and quality assurance should be developed and that greater emphasis should be laid on documentation.

A major drawback of the Indian herbal industry is the implementation of the Drugs and Cosmetics Act, 1940, and its regulation. The survey revealed that only 107 of the 150 surveyed companies were G.M.P. compliant, even though G.M.P. compliance as per Schedule T of the Act has been mandatory since 2006. Further, survey responses revealed that the drug licensing authorities in various states of India interpret the Drugs and Cosmetics Act, 1940 differently. Consequently, the same medicine that is not permitted in one state is allowed to be manufactured in another state. The survey also identified non-uniformity in the drug registration timeline across states as a major problem. Development of unified protocols, defined timelines and specific guidelines defining the meetings with regulators may remove the anomalies related to state licensing authorities and establishing a unified system in the country (Sahoo and Manchikanti 2013).

2.12 CONCLUSION

In addition to the *bṛhatrayi* (greater triad) and *laghutrayi* (lesser triad), many Ayurveda texts have become popular over the centuries. *Vaidyamanōrama*, *Yōgāmṛtam* and *Cikitsāmañjari* are three such texts that are immensely popular in Kerala. Many unique and potent formulae unavailable in other texts are described in these works. Considering their inherent value, these and similar texts need to be included in First Schedule.

The fine prescribed for many violations of provisions of Drugs and Cosmetics Act, 1940, is very nominal. For example, if an Inspector commits offences like searching any place, vehicle or vessel without reasonable ground, or vexatiously and unnecessarily seizes any drug or cosmetic, or any substance or article, or any record, register, document or other material object, he shall be punishable with fine which may extend to 1000 rupees. Similarly, whoever manufactures for sale or for distribution, or sells, or stocks or exhibits or offers for sale any cosmetic other than a cosmetic referred to in clause (i) in contravention of any provision of Chapter 4 or any Rule made thereunder, shall be punishable with imprisonment for one year or with fine of 1000 rupees or with both. The penalty for non-disclosure of the name of the manufacturer is imprisonment for one year, or with a fine of 1000 rupees or with both (Anonymous 2016a). The levying of the fine is intended to act as a deterrent to violations. One thousand rupees is not a large sum in contemporary India. Therefore, the amounts of fine prescribed for many of the violations of the Act need to be revised for its implementation.

Insufficient regulatory guidelines and weak implementation of the Drugs and Cosmetics Act, 1940, are often mentioned as major impediments to Ayurveda industry. The existing regulatory and licensing authority in Ayurveda is not competent to monitor the implementation of A.Y.U.S.H. G.C.P. guidelines because of the lack of infrastructure and human resources. Entrusting this responsibility to the Western medicine regulatory authority is unscientific and unethical. A competent regulatory authority needs to be formed with an advisory board comprising ayurvedic experts, marketing experts, administrators and academics. According to industry sources, apart from Schedule T, separate guidelines on quality control and quality assurance should be developed and implemented effectively.

REFERENCES

Acharya, R. 2014. *Shodhana*: An Ayurvedic detoxification technique and its impact on certain medicinal plants. In *Conservation, cultivation and exploration of therapeutic potential of medicinal plants.* 1st edition, ed. M. Abhimanyukumar, M. Padhi, N. Srikanth, B. P. Dhar, and A. K. Mangal, 427–50. New Delhi: Central Council for Research in Ayurvedic Sciences.

Anonymous. 2002. Labelling, packing and limit of alcohol in ayurvedic (including Siddha) or Unani drugs, Part XVII: 161. In *Drugs and cosmetics act, 1940; together with drugs and cosmetic rules, 1945; drugs (prices control) order 1987*, 185–87. Lucknow: Eastern Book Company.
Anonymous. 2003. *The ayurvedic formulary of India. Part I, 2nd revised English edition*, 227–48. Delhi: The Controller of Publications.
Anonymous. 2013. *Good clinical practice guidelines for clinical trials in Ayurveda, Siddha and Unani medicine (GCP-ASU)*, 1–104. New Delhi: Department of AYUSH.
Anonymous. 2014a. Legal provisions for GMP certification. In *Guidelines for inspection of GMP compliance by Ayurveda, Siddha and Unani drug industry*, 11–55. New Delhi: Department of AYUSH.
Anonymous. 2014b. GMP: An overview. In *Guidelines for inspection of GMP compliance by Ayurveda, Siddha and Unani drug industry*, 1–10. New Delhi: Department of AYUSH.
Anonymous. 2014c. Stepwise points for conducting GMP inspection. In *Guidelines for inspection of GMP compliance by Ayurveda, Siddha and Unani drug industry*, 56–69. New Delhi: Department of AYUSH.
Anonymous. 2016a. The Drugs and Cosmetics Act, 1940 (23 of 1940) (As amended up to the 31st December 2016, Government of India, Ministry of Health and Family Welfare. http://www.cdsco.nic.in/writereaddata/2016Drugs%20and%20Cosmetics%20Act%201940%20&%20Rules%201945.pdf (accessed March 23, 2018).
Anonymous. 2016b. *The gazette of India, extraordinary Part-II, Section 3*. Sub-Section (i) No. 561. New Delhi.
Anonymous. 2019a. The Pharmacy Act, 1948. http://legislative.gov.in/sites/default/files/A1948-8.pdf (accessed July 24, 2019).
Anonymous. 2019b. The Drugs and Magic Remedies (Objectionable Advertisement) Act, 1954. https://indiacode.nic.in/bitstream/123456789/1412/1/A1954-21.pdf (accessed July 24, 2019).
Anonymous. 2019c. The Narcotic Drugs and Psychotropic Substances Rules Act, 1985. http://dor.gov.in/sites/default/files/Narcotic-Drugs-and-Psychotropic-Substances-Act-1985.pdf (accessed July 24, 2019).
Anonymous. 2019d. The Poisons Act, 1919. http://indiacode.nic.in/bitstream/123456789/5751/1/the_poisons_act%2C_1919.pdf (accessed July 24, 2019).
Anonymous. 2019e. The drugs (price control). Order 1995. http://fdaharyana.gov.in/actpdf/Drugs%20_Price%20Control_%20Order_1995_2.pdf (accessed July 24, 2019).
Anonymous. 2019f. The Trade Marks Act, 1999. http://www.ipindia.nic.in/writereaddata/Portal/IPOAct/1_43_1_trade-marks-act.pdf (accessed July 24, 2019).
Anonymous. 2019g. The Biological Diversity Act, 2002. http://nbaindia.org/uploaded/Biodiversityindia/Legal/31.%20Biological%20Diversity%20%20Act,%202002.pdf (accessed August 13, 2019).
Bhadra, R., K. Ravakhah, and R. K. Ghost. 2015. Herb-drug interaction: The importance of communicating with primary care physicians. *Australian Medical Journal* 8, no. 10: 315–9.
Bhalerao, S., R. Munshi, P. Tilve, and D. Kumbhar. 2010. A survey of the labeling information provided for ayurvedic drugs marketed in India. *International Journal of Ayurveda Research* 1, no. 4: 220–22.
Dumitriu, H. I., 1997. Historical overview. In *Good drug regulatory practices*, 3–21. Boca Raton: CRC Press.
Fries, J. F., and E. Krishnan. 2004. Equipoise, design bias, and randomized controlled trials: The elusive ethics of new drug development. *Arthritis Research and Therapy* 6, no. 3: R250–5.
Gupta, B. 1976. Indigenous medicine in nineteenth and twentieth century Bengal. In *Asian medical systems: A comparative study*, ed. C. Leslie, 368–77. CA: University of California Press.
Gupta, V., A. Jain, M. B. Shankar, and R. K. Sharma. 2017. Shelf life of ayurvedic dosage forms in regulatory perspectives. *International Journal of Advanced Ayurveda, Yoga, Unani, Siddha and Homeopathy* 6, no. 1: 360–9.
Harilal, M. S. 2009. "Commercialising traditional medicine": Ayurvedic manufacturing in Kerala. *Economic and Political Weekly* 44: 44–51.
Ilanchezhian, R., J. C. Roshy, and R. Acharya. 2010. Importance of media in *shodhana* (purification/processing) of some poisonous herbal drugs. *Ancient Science of Life* 30: 27–30.
Izzo, A. A. 2012. Interactions between herbs and conventional drugs: Overview of the clinical data. *Medical Principles and Practice* 21, no. 5: 404–28.
Izzo, A. A., and E. Ernst. 2001. Interactions between herbal medicines and prescribed drugs: A systematic review. *Drugs* 61, no. 15: 2163–75.
Izzo, A. A., S. Hoon-Kim, R. Radhakrishnan, and E. M. Williamson. 2016. A critical approach to evaluating clinical efficacy, adverse events and drug interactions of herbal remedies. *Phytotherapy Research: PTR* 30, no. 5: 691–700.
Kales, S. N., and R. B. Saper. 2009. Ayurvedic lead poisoning: An under-recognized, international problem. *Indian Journal of Medical Sciences* 63, no. 9: 379–81.

Kumar, D. S. 2016. Medical herbalism through the ages. In *Herbal bioactives and food fortification: Extraction and formulation*, 1–28. Boca Raton: CRC Press.

Mai, I., H. Krüger, K. Budde et al. 2000. Hazardous pharmacokinetic interaction of Saint John's wort (*Hypericum perforatum*) with the immunosuppressant cyclosporine. *International Journal of Clinical Pharmacology and Therapeutics* 38, no. 10: 500–2.

Mallick, A., A. Roy, A. Bose, and K. De. 2012. Assessment on the label of Ayurvedic classical and proprietary medicines in accordance with the Drugs and Cosmetics Act, 1940. *International Journal of Pharmaceutical Research and Bioscience* 1: 67–74.

Milić, N., N. Milosević, S. Golocorbin Kon, T. Bozić, L. Abenavoli, and F. Borrelli. 2014. Warfarin interactions with medicinal herbs. *Natural Product Communications* 9, no. 8: 1211–6.

Miller, G. G., and S. Joffe. 2011. Equipoise and the dilemma of randomized clinical trials. *The New England Journal of Medicine* 364, no. 5: 476–80.

Mitra, S., V. J. Shukla, and R. Acharya. 2012. Effect of purificatory measures through cow's urine and milk on strychnine and brucine content of Kupeelu (*Strychnos nuxvomica* Linn.) seeds. *The African Journal of Traditional, Complementary and Alternative Medicines* 9, no. 1: 105–11.

Patwardhan, B. 2011. Ayurveda GCP guidelines: Need for freedom from RCT ascendancy in favor of whole system approach. *Journal of Ayurveda and Integrative Medicine* 2, no. 1: 1–4.

Sahoo, N., and P Manchikanti. 2013. Herbal drug regulation and commercialization: An Indian industry perspective. *The Journal of Alternative and Complementary Medicine* 19, no. 12: 957–63.

Saper, R. B., S. N. Kales, J. Paquin et al. 2004. Heavy metal content of ayurvedic herbal medicine products. *Journal of American Medical Association* 292, no. 23: 2868–73.

Saper, R. B., R. S. Phillips, and A. Sehgal et al. 2008. Lead, mercury, and arsenic in US and Indian manufactured Ayurvedic medicines sold via the Internet. *Journal of the American Medical Association* 300, no. 8: 915–23.

Saroya, A. S. 2016a. Introduction to regulatory affairs. In *Regulatory and pharmacological basis of ayurvedic formulations*, 3–7. Boca Raton: CRC Press.

Saroya, A. S. 2016b. Ayurvedic drug manufacturing license. In *Regulatory and pharmacological basis of ayurvedic formulations*, 22–25. Boca Raton: CRC Press.

Saroya, A. S. 2016c. Good manufacturing practices for ASU medicines (Schedule-T). In *Regulatory and pharmacological basis of ayurvedic formulations*, 26–36. Boca Raton: CRC Press.

Saroya, A. S. 2016d. Drugs and Cosmetics Act and ayurvedic drugs. In *Regulatory and pharmacological basis of ayurvedic formulations*, 14–21. Boca Raton: CRC Press.

Serra, A., Uehlinger, D. E., Ferrari, P. et al. 2002. Glycyrrhetinic acid decreases plasma potassium concentrations in patients with anuria. *Journal of the American Society of Nephrology: J.A.S.N.* 13, no. 1: 191–6.

Singh, L. B., R. S. Singh, R. Bose, and S. P. Sen. 1985. Studies on the pharmacological action of aconite in the form used in Indian medicine. *Bulletin of Medico-Ethnobotanical Research* 6: 115–23.

Vaidyan, K. V. K., and A. S. G. Pillai, 1985. *Ghṛtayōgaṅgaḷ* (Formulae of clarified butters). In *Sahasrayōgam*, 339–96. Alleppey: Vidyarambham Publishers.

3 Industrial Manufacture of Traditional Ayurvedic Medicines

Nishanth Gopinath

CONTENTS

3.1 Introduction 42
3.2 Procurement of Herbal Raw Material.. 42
3.3 Storage of Raw Materials 42
3.4 Pre-Processing of Raw Materials 45
3.4.1 Cleaning of Herbs 45
3.4.2 Size Reduction of Herbs 47
3.5 Analysis of Raw Materials... 48
3.6 Sources of Medicinal Formulae 48
3.7 Standard Operating Procedures 48
3.8 Master Formula for Production 49
3.9 Manufacture of Various Dosage Forms... 50
3.9.1 Kvātha (Decoctions) 50
3.9.1.1 Textual Instructions 50
3.9.1.2 Industrial Manufacture.. 50
3.9.2 Cūrṇa (Medicinal Powders) 50
3.9.2.1 Textual Instructions 50
3.9.2.2 Industrial Manufacture.. 51
3.9.3 Ghṛta and Taila (Medicinal Lipids)... 51
3.9.3.1 Textual Instructions.. 51
3.9.3.2 Industrial Manufacture.. 52
3.9.4 Ariṣṭa and Āsava (Fermented Liquids) 56
3.9.4.1 Textual Instructions.. 56
3.9.4.2 Industrial Manufacture.. 56
3.9.5 Avalēha (Confections)... 57
3.9.5.1 Textual Instructions.. 57
3.9.5.2 Industrial Manufacture.. 59
3.9.6 Guḷika (Pills) 59
3.9.6.1 Textual Instructions.. 59
3.9.6.2 Industrial Manufacture 60
3.9.7 Lēpa (Paste) 61
3.10 Processing of Wastes 64
3.10.1 Vermicomposting 64
3.10.2 Briquetting of Solid Wastes.. 64
3.10.3 Stabilization of Black Cotton Soil 65
3.10.4 Recycling of Wastewater 66
3.11 Conclusion 66
References.. 67

3.1 INTRODUCTION

In the olden days, manufacture of ayurvedic medicines was carried out on a small scale, often by the physician himself, helped by his assistants. Herbs were mostly procured locally, with some exotic ones like guggul gum, *śilājit* and saffron coming from faraway places. Commercial production of ayurvedic medicines was first attempted by Vaidya Gangadhar Ray, a Kaviraja of Bengal. He was inspired by the increasing demand for ayurvedic medicines and set up a large-scale manufacturing unit in 1884. He called it N.N. Sen and Company (Gupta 1976). Vaidya Gangadhar Ray was followed by the physician- entrepreneur Dr S.K. Burman (1856–1907), who was affectionately called *Daktar Burman* by his clientele. In 1884, he set up the company Dabur (the name derived from the *Da* of *Daktar* and *Bur* of Burman) in a small house in Calcutta. After 133 years the company founded by him is today India's most trusted name and the world's largest manufacturer of ayurvedic medicines, with a portfolio of over 381 products (Anonymous 2018). By the turn of the 20th century, the demand for ayurvedic medicines had increased sufficiently to occupy a fair share in the country's medicine market (Kumar 2001).

Large-scale manufacture of traditional ayurvedic medicines was started in southern India by P.S. Variar (1869–1944), a visionary physician and philanthropist of the province of Kerala. The humble enterprise started by him in 1902 as a village clinic has now grown into a million-dollar business (Vellodi 1987). In recognition of the various services rendered by him to the development of Ayurveda, the colonial government of India conferred on him in 1933 the title of *Vaidyaratnam* (Gem of a physician). Arya Vaidya Sala was the first ayurvedic manufacturing facility in Kerala, processing ayurvedic drugs using modern methods. Moreover, P.S. Variar was the first person who introduced *kvātha* preserved in bottles without spoilage for any length of time. The success of Arya Vaidya Sala of Kottakkal inspired several others like Kesari Kuteeram Ayurveda Ouṣadhaśāla in Madras, Andhra Ayurveda Pharmacy Ltd, Sivagnana Siddha Vaidya Salai at Koilpatti and the Siddha Vaidya Agastya Āśramam in Tanjore (Kanagarathinam 2016). At present Kerala has the second largest number of ayurvedic manufacturing units (12% of total manufacturing units) next to Uttar Pradesh (Harilal 2009). Table 3.1 lists the important companies in India manufacturing traditional ayurvedic medicines.

3.2 PROCUREMENT OF HERBAL RAW MATERIAL

India's domestic herbal industry consists of 8610 licensed herbal units engaged in the manufacture of herbal formulations under different streams of Indian systems of medicine. They are grouped into large, medium, small and very small enterprises depending on their annual turnover. Analysis of their consumption data reveals that the industry consumes nearly 195,000 metric tons of crude drugs derived from 907 plant species. Of these, more than two-thirds are consumed by the large and medium units (Goraya et al. 2017b).

Most of these herbs are collected from forests, agricultural farms, fallow lands, sides of roads and railway tracks, banks of canals, ponds, lakes and waste lands (Goraya et al. 2017c). The trade of medicinal plants takes place in India in *mandi*s (large yards); the like of which in Neemuch, Unjha and Sojat are famous internationally for the trade of cultivated herbs like *aśvagandha* (*Withania somnifera*), isabgol (*Plumbago ovata*) and *mehendi* (*Lawsonia inermis*) (Figures 3.1–3.4). The one at Khari Baoli in Delhi is the largest spice market in Asia and the largest conventional crude drug *mandi* in the country (Goraya et al. 2017a). Medicinal herbs and other forest produce are offered for sale by some of the forest development corporations in the country. The major corporations are listed in Table 3.2.

3.3 STORAGE OF RAW MATERIALS

Herbs and other raw materials arriving from traders are stored in properly designated storage rooms under conditions that prevent contamination and deterioration (Figures 3.5 and 3.6). While dealing

TABLE 3.1
Leading Manufacturers of Traditional Ayurvedic Medicines

Sl. No.	Name of Company	Year of Establishment	URL
1	Dabur India Ltd	1884	www.dabur.com
2	Vasudeva Vilasam	1884	www.vasudeva.com
3	Arya Vaidya Sala	1902	www.aryavaidyasala.com
4	IMIS Pharmaceutical Pvt Ltd	1904	www.imispharma.com
5	The Sadvaidyasala Pvt Ltd	1913	www.sadvaidyasala.com
6	Shree Baidyanath Ayurved Bhawan Ltd	1918	www.baidyanath.com
7	S.N.A. Oushadhasala Pvt Ltd	1920	www.thaikatmooss.com
8	Sitaram Ayurveda Pharmacy Ltd	1921	www.sitaramayurveda.com
9	BIPHA Drug Laboratories Pvt Ltd	1929	www.bipha.com
10	Aryavaidya Nilayam	1930	www.avnarogya.in
11	Dhanwanthari Vaidyasala	1933	www.dhanwanthari.org
12	Ayurveda Rasashala	1935	www.ayurvedarasashala.com
13	Oushadhi	1939	www.oushadhi.org
14	S.D. Pharmacy	1939	www.sdpharmacy.com
15	Vaidyaratnam Oushdahasala Pvt Ltd	1941	www.vaidyaratnammooss.com
16	Arya Vaidya Pharmacy (Cbe) Ltd	1943	www.avpayurveda.com
17	IMPCOPS	1944	www.impcops.org
18	Kerala Ayurveda Ltd	1945	www.keralaayurveda.biz
19	Deseeya Ayurvedic Pharmacy	1946	www.deseeya.com
20	Santhigiri	1972	www.healthcare.santhigiriashram.org
21	Indian Pharmaceutical Company	1977	www.indianpharmaceutical.net
22	Alva Pharmacy	1981	www.alvasherbal.com
23	Everest Pharma	1981	www.everestayurveda.com
24	Dwaraka Ayurvedic Pharmaceuticals Ltd	1986	www.dwarakaayurveda.com
25	Kerala Sarvodaya Ayurveda Pharma & Research Center	1987	www.keralasarvodayaayurveda.com
26	Pankajakasthuri Herbals India (P) Ltd	1988	www.pankajakasthuri.in
27	Nagarjuna Herbal Concentrates Ltd	1989	www.nagarjunaayurveda.com
28	Patanjali Ayurved Ltd	2006	www.patanjaliayurved.org

with the storage of chamomile flowers, Boettcher and Guenther (2005) recommend certain storage conditions. Though described in the context of the storage of the German chamomile (*Matricaria chamomilla*), the desirable storage conditions and the characteristics of the storage room listed by Boettcher and Guenther (2005) are equally relevant to any other herb. Based on that report the following storage conditions can be adapted for ayurvedic herbs: moisture content of stored product: 8–10%; relative humidity of ambient air: <50%; room temperature: 25°C; lighting: largely in a darkened room; stock pests: free from living larvae and insects (Boettcher and Guenther 2005).

By storing dried herbs in these conditions, almost all quality deteriorations can be prevented or reduced. The entry of insects should be prevented by permanently covering all windows, walls, and ventilation openings with 1–2 mm mesh wide wire netting. All doors should be kept automatically closed. The continuous operation of electrical UV insect traps in the storage room is an effective measure. These will attract all insects flying in and kill them with the current. If pests are detected in the store, disinfestation must be carried out immediately. The stored products must be regularly inspected to determine their water content, external quality traits and infestation with stock pests. These checks, including the quality control analyses, must be repeated every 4 to 6 months. Stored herbs can be kept in this condition with minimal deterioration (Boettcher and Guenther 2005).

FIGURE 3.1 Mandi at Delhi – herbs in sacks. Photo courtesy: Janak Raj Rawal, Rawal Medherbs Consultants LLP, Delhi-110 052.

FIGURE 3.2 Mandi at Delhi – traders with samples of herbs. Photo courtesy: Janak Raj Rawal, Rawal Medherbs Consultants LLP, Delhi-110 052.

The characteristics of the storage room also influence successful storage. The store room should be cleaned and disinfected before products are taken into storage. It should be airy and cool, with a constant curve of temperature. The openings of buildings should be sealed with wire netting to prevent the entry of harmful insects, pests, birds and pets. Proper fire prevention measures should be taken. Toxic drugs are to be stored separately. To make the movement of stock easier, larger storage rooms should have mechanical stacking devices (Boettcher and Guenther 2005). Requirements of good manufacturing practices are generally implemented in the storage of herbs, raw materials and manufacture of the products (Anonymous 2014).

FIGURE 3.3 Mandi at Neemuch – heaps of herbs. Photo courtesy: Janak Raj Rawal, Rawal Medherbs Consultants LLP, Delhi-110 052.

FIGURE 3.4 Mandi at Neemuch – buyers inspecting herbs. Photo courtesy: Janak Raj Rawal, Rawal Medherbs Consultants LLP, Delhi-110 052.

3.4 PRE-PROCESSING OF RAW MATERIALS

3.4.1 Cleaning of Herbs

At the time of harvest or collection, most herbs are likely to contain contaminants. Ferrous and non-ferrous metals, filings, soil, stones, animal hair, bone and excreta are the common contaminants found in consignments of herbs. Therefore, it is necessary to perform operations of cleaning to ensure that herbs with uniformly high quality are made available for subsequent processing. It

TABLE 3.2
Major Sources of Forest Produce

Sl. No.	Organization	URL
1	Girijan Co-operative Corporation Ltd	www.apgirijan.com
2	Telangana Girijan Cooperative Corporation Ltd	www.twd.telangana.gov.in
3	M.P. State Minor Forest Produce Co-operative Federation Ltd	www.mfpfederation.org
4	Chhattisgarh State Minor Forest Produce (Trading & Development) Co-operative Federation Ltd	www.cgmfpfed.org
5	Gujarat State Forest Development Corporation Ltd	www.gsfdcltd.co.in
6	Andaman Plantations and Development Corporation Pvt Ltd	www.andamanplantations.com
7	Himachal Pradesh State Forest Development Corporation Ltd	www.hpforestco.in
8	Odisha Forest Development Corporation	www.odishafdc.com
9	West Bengal Forest Development Corporation Ltd	www.wbfdc.com

FIGURE 3.5 Sacks of raw materials stored in the herb store of an Ayurveda medicine manufacturing facility. Photo courtesy: Nagarjuna Herbal Concentrates Ltd, Kerala (www.nagarjunaayurveda.com).

is not possible to produce high-quality ayurvedic medicines from substandard raw materials, and, therefore, these cleaning procedures carried out in the beginning are a method of improving the quality of the raw material (Fellows 2000).

Wet cleaning is effective for removing soil from root crops. Dry cleaning procedures are used for raw materials that are smaller, like grains, seeds, nuts, rhizomes, fruits and their pericarps, which possess greater mechanical strength and have a lower moisture content. Dry procedures involve smaller and cheaper equipment than wet procedures. In addition to this, plant cleaning is easier. The major equipment used for dry cleaning is air classifiers, magnetic separators and separators based on screening. Classifiers use a moving stream of air to separate contaminants from foods by differences in their densities. They are used in the separation of heavy contaminants like stones and lighter contaminants like dust, leaves, stalks and husks. Metal fragments are common contaminants in consignments of herbs and pose a potential hazard in medicine manufacturing. Ferrous metals are removed by magnets or electromagnets (Fellows 2000).

Sieving or screening is used in the separation of pieces of herbs into two or more fractions on the basis of differences in size. Screens with fixed or variable apertures are used for size sorting. The screen may be stationary or rotating or vibrating. Fixed aperture screens are commonly used for sorting turmeric, nutmeg, cumin, black cumin and so on (Fellows 2000).

FIGURE 3.6 Sacks of herbs stored in the herb store of another manufacturing facility. Photo courtesy: Rajah Healthy Acres, Kerala (www.ayurvedichospital.com).

3.4.2 Size Reduction of Herbs

Size reduction or comminution is a process by which raw materials such as roots, leaves, bark, heartwood and seeds are reduced to powders that can pass through sieves of different mesh sizes. Pulverizing equipment is selected according to the material that is to be powdered. Chunks of heartwood are first minced before being fed into pulverizers (Figure 3.7). These mincing machines chop the wood lengthwise and crosswise, reducing it to small pieces, which are fed into hammer mills. Hammer-like solid metal blocks fixed to the shaft of the machine swing at high speed and powder the pieces of wood dropped into the chamber of the hammer mill. The number of hammers varies with different models. The particle size of the powder is dependent on the perforated screen that is fixed to the wall of the chamber (Kumar 2016).

Another equipment, called the knife mill, has sets of sharp knives fixed to a rotor. These knife mills are suitable for powdering leaves, roots, and barks. Herbs are also pulverized using a pin mill (teeth mill, impact mill). This equipment has a circular row of pins (teeth) fixed inside the chamber. Bits of herbs that are dropped into it are powdered through the beating, shearing, crushing and colliding that the material undergoes during the process (Kumar 2016).

Large-sized pin mills have a screw feeder with a variable speed drive, ensuring the uniform feeding of the herb into the mill. Other types of comminution equipment such as the shredding mill, ball mill, slow-speed attrition mill, micromill and air-swept mill are also used in the pulverization of herbs (Kumar 2016).

Different pulverization processes are applied to different herbs. Herbs with more stems and stalks need shredding or cutting mills, while hammer mills are suitable for hard and brittle materials like

FIGURE 3.7 Pulverizer for size reduction of herbs. Photo courtesy: Nagarjuna Herbal Concentrates Ltd, Kerala (www.nagarjunaayurveda.com).

resins, and leafy plants such as senna and *Adhatoda vasica*. Hard materials like cinnamon bark and dry ginger rhizomes can be pulverized in a two- or three-stage process combining cutting, shredding and pulverization in hammer or pin mills. Pulverization needs to be carried out carefully, so that the herb is not damaged due to the generation of heat (Anonymous 2001a; Kumar 2016).

3.5 ANALYSIS OF RAW MATERIALS

Herbs procured in bulk are subjected to macroscopic and microscopic evaluation for establishing their identity. Thereafter, physicochemical analyses like moisture content, foreign matter, ash value, TLC profile and so on are carried out. Other raw materials like clarified butter, sesame oil, coconut oil, castor oil, honey, jaggery and such others are also subjected to chemical tests. These tests are intended to ensure that all these raw materials conform to the specifications set in *The Ayurvedic Pharmacopoeia of India* (see Chapter 4 for details).

3.6 SOURCES OF MEDICINAL FORMULAE

Classical texts like *Caraka Samhita*, *Suśruta Samhita*, *Aṣṭāṅgasamgraha* and *Aṣṭāṅgahṛdaya* are veritable treasure houses of ayurvedic formulae. In addition to them manufacturers depend on latter-day compendia like *Śārṅgadhara Samhita* (Murthy 2017a), *Bhaiṣajya Ratnāvali* (Ayurvedacharya 1951) and *Sahasrayōga* (Vaidyan and Pillai 1985a), the latter being immensely popular in Kerala. To bring uniformity to industrial manufacture and improve the quality of medicines, the Government of India has published the *Ayurvedic Formulary of India* (Anonymous 2000, 2003a, 2011).

3.7 STANDARD OPERATING PROCEDURES

The Ayurveda drug manufacturing industry makes use of many standard operating procedures (S.O.P.s). An S.O.P. is a set of step-by-step instructions compiled by an organization to help workers carry out complex routine operations. S.O.P.s aim to achieve efficiency, quality output and uniformity of performance, while reducing miscommunication and failure to comply with industry

TABLE 3.3
Major SOPs Used in the Ayurveda Industry

Sl. No.	Routine Operation
1	Identification of suppliers of raw materials (vendor qualification)
2	Storage of raw materials
3	Cleaning of herbs
4	Drying of herbs
5	Comminution of herbs
6	Operation of drug boiler
7	Bottling of *kvātha*
8	Operation of evaporators used in the manufacture of medicated lipids
9	Cleaning of manufacturing equipment
10	Release of finished products
11	Waste management
12	Training of personnel
13	Audit by Q.A. department

regulations. The word *standard* implies that only one (standard) procedure is to be used for executing that function. An S.O.P. is always written with the user in mind. It follows the five Cs – clear, complete, concise, courteous and correct. SOPs detail the regularly recurring work processes that are to be conducted or followed within the organization. They document the way activities are to be performed to facilitate consistent conformance to technical and quality system requirements and to support data quality. They may describe, for example, fundamental and technical actions such as cleaning of herbs, comminution of herbs, cleaning validation of drug boiler and operation of *kvātha* bottling machine (Cook Jr. 1998). Some major S.O.P.s used in the Ayurveda industry are listed in Table 3.3.

3.8 MASTER FORMULA FOR PRODUCTION

The industrial production of ayurvedic medicines is carried out as a batch processing operation. A batch is a defined quantity of starting material processed in a single process or series of processes so that it could be expected to be homogeneous. The complete procedure for the manufacturing process is detailed in a master formula (M.F.), which describes the preparations to be made, the kind of equipment to be used, and the method to be followed (Chaloner-Larsson et al. 1997). The original approved M.F., with the signatures of the production and quality assurance (Q.A.) officials, is filed in a safe place. An approved copy of the M.F. is made available to the production department for every batch production run.

The entire process of the manufacture of the ayurvedic medicine of every batch is recorded in a batch manufacturing record (B.M.R.). A distinctive combination of numbers and/or letters is assigned to each batch and is used on labels and all documents pertaining to the batch. All aspects of the operation, such as the preparation of raw materials, the analysis of deionized water, sesame oil, coconut oil, honey, jaggery and so on, the cleaning of equipment, the analysis of raw materials and finished products and so on are carried out according to S.O.P.s, which are detailed written instructions that specify how a test or an administrative procedure is to be performed (Chaloner-Larsson et al. 1997). The B.M.R. contains the name of the product, dates and times of commencement and completion of production, name of the person responsible for each stage of production, initials of the operator of different steps of production and of the person who checked each of these operations, batch number, a record of in-process controls if any, and the initials of the person who carried them out, the yield of product obtained and detailed notes on special problems encountered, with authorization for any deviation from the M.F. (Anonymous 2008). On completion of the production process, the B.M.R. is returned to the Q.A., where it is stored in an appropriate place for a specific period of time according to A.Y.U.S.H. norms.

3.9 MANUFACTURE OF VARIOUS DOSAGE FORMS

3.9.1 Kvātha (Decoctions)

Kvātha is prepared by boiling the crude herbs in water for a stipulated period of time and reducing the volume to a specific quantity. Literally, the word *kvātha* means that which brings about normalcy to the body by maintaining the equilibrium of physiological factors and removing pathological processes. In addition to being an independent dosage form, *kvātha* are used in the preparation of other dosage forms like *ariṣṭa*, *guḷika*, *avalēha*, *rasakriya*, *ghṛta* and *taila*.

3.9.1.1 Textual Instructions

According to *Śārṅgadhara Samhita*, one *pala* (48 g) of coarsely powdered herbs is boiled in 16 parts of water in an earthen pot, which is heated over a mild fire till the volume is reduced to 1/8. The filtered fluid is called *śṛta*, *kvātha*, *kaṣāya* or *niryūha* (Murthy 2017a).

3.9.1.2 Industrial Manufacture

Kvātha prepared in earthen pots and having a shelf life of a maximum of two days have now become outdated. It has become now a standard procedure to manufacture *kvātha* industrially, concentrate the product further and extend the shelf life by adding approved preservatives. Instead of crushing the herbal raw materials manually, disintegrators are used. Though different extraction techniques like solvent extraction, counter-current extraction and supercritical fluid extraction methods are in vogue, mainly aqueous, alcoholic and hydroalcoholic extracts are used in ayurvedic medicine manufacturing (Anonymous 2001b).

The *kvātha* in question is manufactured according to its master formula. The master formula of *Amṛtōttaram kvātha* is given as an example (Table 3.4). Drug boilers are nowadays used for bulk manufacturing of *kvātha*. They include those in which the disintegrated materials are equally divided and first loaded into porous, stainless steel baskets which are immersed in water-filled boilers and boiling is carried out (Figures 3.8 and 3.9). First of all, the pre-processed raw materials are tested for their quality. They are weighed as per the stipulated quantity indicated in the formula and then loaded into the baskets or directly into the drug boilers. Deionized water added must always be above the level of the materials and heating is started.

Multiple extractions are carried out, and the filtrates are tested for the water-soluble extractive values, which are estimated as total soluble solids. The pooled dilute extract is further concentrated in machinery like forced circulation evaporators into which the *kvātha* is fed and passed across a heat source (Figure 3.10). The heat vaporizes the *kvātha* under vacuum. The vapor is removed from the solution, condensed and the *kvātha* concentrated. The desired concentration is measured in total soluble solids value based on which it is ensured that one dose of *kvātha* for a day is prepared out of 24 g of herbal raw material. Each dose of *Amṛtōttaram kvātha* is prepared out of *Zingiber officinale* (*Nāgara*) 4.00 g, *Terminalia chebula* (*Harītaki*) 8.00 g and *Tinospora cordifolia* (*Amṛta*) 2.00 g (Anonymous 2003b). A specific quantity of approved preservatives is added to the manufactured *kvātha*, and the product is bottled.

3.9.2 Cūrṇa (Medicinal Powders)

3.9.2.1 Textual Instructions

A fine powder of dry crude drugs sifted through cloth is called *cūrṇa*, *rajas* and *kṣōda* (Murthy 2017b).

TABLE 3.4
Some Elements of the Master Formula for the Production of *Amṛtōttaram Kvātha*

	Ingredients	
	Coarse powder of the following:	
	Zingiber officinale rhizome	35.200 kg
	Terminalia chebula pericarp	70.400 kg
	Tinospora cordifolia stem	140.800 kg
	Deionized water	3100.000 L
	Steps of the operation	
1.1	Check and certify that the drug boiler is clean as per S.O.P.	
1.2	Extraction I. Charge 246.400 kg of the herb powder into the drug boiler.	
1.3	Charge 1400 L of deionized water into the drug boiler. Ensure soaking.	
1.4	Heat to reflux and maintain for 3 h.	
1.5	Filter and collect the extract.	
1.6	Extraction II. Charge 1000 L of deionized water into the drug boiler and ensure soaking.	
1.7	Heat to reflux and maintain for 3 h.	
1.8	Filter and collect the extract.	
1.9	Extraction III. Charge 700 L of deionized water into the drug boiler and ensure soaking.	
1.10	Heat to reflux and maintain for 3 h.	
1.11	Filter and collect the extract.	
1.12	Combine the three filtrates and send for concentration.	
2 .0	Concentration in forced circulator evaporator.	
2.1	Check and certify that the equipment is clean as per S.O.P.	
2.2	Charge the combined filtrates and distil the water till the concentrate attains the T.S.S. value of 31.	
2.3	Filter and allow to cool.	
2.4	Add preservative and mix well.	
2.5	Bottle *Amṛtōttaram kvātha*.	
	Expected yield: 175.000 L.	

3.9.2.2 Industrial Manufacture

Crude drugs mentioned in the formula are cleaned and dried well. They are pulverized and sifted through a sieve. When the formula contains several herbs, each herb is powdered separately and sifted through a 120-mesh sieve. Each of these herb powders is weighed accurately and mixed together in a blender. Ingredients like rock-salt, sugar and camphor are powdered and blended with the other herb powders in the end. Asafetida and rock-salt may be roasted, powdered and mixed with the rest of the ingredients. Herbs like *Śatāvari* (*Asparagus racemosus*), *Guḍūci* (*Tinospora cordifolia*) and *Bhṛṅgarāj* (*Eclipta alba*) are recommended to be used when in a fresh condition. However, while manufacturing *cūrṇa*, these herbs are made into a paste, dried, powdered and added to the other ingredients. Thereafter, the finished product is packed in containers and labels affixed (Figure 3.11) (Anonymous 1978a). To prevent the *cūrṇa* from absorbing moisture, blending and packing operations are carried out in dehumidified atmosphere. The master formula of *Triphalā Cūrṇam* is given in Table 3.5 as an example (Anonymous 1978a).

3.9.3 Ghṛta and Taila (Medicinal Lipids)

3.9.3.1 Textual Instructions

Ghṛta and *taila* are prepared by boiling together one part of *kalka* (paste of herbs), four parts of clarified butter or oil and four parts of any decoction. The decoction is prepared by boiling one part of a coarse herb powder with four parts of water and reducing the volume by one-quarter. If the

FIGURE 3.8 A drug boiler used in the manufacture of *kvātha*. Photo courtesy: Nagarjuna Herbal Concentrates Ltd, Kerala (www.nagarjunaayurveda.com).

herbs are soft, four parts of water are taken. If the herbs are moderately hard, eight parts of water are taken, and, if very hard, 16 parts of water are taken. The volume is to be reduced by one-quarter (Murthy 2017e).

If the *ghṛta* or *taila* is prepared using only water, decoction, meat juice or fresh juice of herbs, their volumes should be respectively, four, six and eight times of that of the *kalka*. If *ghṛta* or *taila* is to be prepared with only water or decoction, the volume of fluid should be four times of the *kalka* or lipid material. If in the formulation, *kalka* consists of flowers, the flowers should be taken in one-eighth quantity of the lipid and water equivalent to four volumes of the lipid (Murthy 2017e).

The stage of filtering the product is determined on the basis of the texture of the sediment that forms when most of the water in the boiling mass is evaporated. When a significant volume of water is removed from the boiling mixture, it acquires a mud-like consistency. As evaporation progresses, sediments form. They are usually of three types – *mṛdu* (soft wax-like), *maddhyama* (hard wax) and *khara* (very hard). Heating beyond the *khara* stage will result in *dagdha pāka* or a charred stage (Figure 3.12). *Ghṛta* and *taila* filtered at this stage are unfit for use. *Ghṛta* or *taila* filtered at *mṛdu pāka* are suitable for nasal instillation (*nasya*) and *maddhyama pāka* is recommended for internal administration. *Khara pāka* is suitable only for formulations that are used in body massage (*abhyaṅga*). *Śārṅgadhara Samhita* states that *ghṛta* or *taila* should not be prepared in one day. They need to be prepared over several days, as a longer duration of preparation improves the quality of the medicine (Murthy 2017e).

3.9.3.2 Industrial Manufacture

Steam-jacketed stainless steel evaporators are employed for the large-scale manufacture of medicinal lipids (Figure 3.13). The lipids commonly used are sesame oil, coconut oil, castor oil and

FIGURE 3.9 Basket containing chopped and pulverized herbs. Photo courtesy: Nagarjuna Herbal Concentrates Ltd, Kerala (www.nagarjunaayurveda.com).

FIGURE 3.10 Forced circulator evaporator for concentrating *kvātha*. Photo courtesy: Nagarjuna Herbal Concentrates Ltd, Kerala (www.nagarjunaayurveda.com).

FIGURE 3.11 Packing of *cūrṇa*. Photo courtesy: Nagarjuna Herbal Concentrates Ltd, Kerala (www.nagarjunaayurveda.com).

TABLE 3.5
Some Elements of the Master Formula tor Production of *Triphalā Cūrṇam*

Ingredients

Fine powders (120 mesh sieve) of the following:

Emblica officinalis pericarp	166.666 kg
Terminalia belerica pericarp	166.666 kg
Terminalia chebula pericarp	166.666 kg

Check and confirm that moisture content is within the specified limits (N.M.T. 4%)

Steps of the operation

1.1	Check and certify that the blender is clean as per S.O.P.
1.2	Charge the herb powders into the blender.
1.3	Blend the powders for 45 minutes.
1.4	Transfer the blended powder into a clean stainless steel drum.
1.5	Check and certify that the packing machine is clean and in good condition as per S.O.P.
1.6	Pack the product in containers.
	Expected yield: 475 kg.

clarified butter. The lipid, fine paste of fresh or powdered herbs, required volumes of water, decoction, juice of herbs, buttermilk, milk or meat juice are poured into the evaporators according to the formula. The evaporator is heated over mild heat. Decoctions or buttermilk are evaporated in five days and juice of fresh herbs (*svarasa*) in three days. Milk and meat juices are evaporated in two days and one day, respectively. The boiling mass is stirred frequently with large ladles, to improve

FIGURE 3.12 Mud-like stage (a), soft wax stage (b), hard wax stage (c) and burnt sediment stage (d). Photo courtesy: D. Suresh Kumar.

FIGURE 3.13 Steam-jacketed evaporators employed in production of medicinal lipids. Note froth on surface. Photo courtesy: Nagarjuna Herbal Concentrates Ltd, Kerala (www.nagarjunaayurveda.com).

TABLE 3.6
Some Elements of ohe Master Formula for the Production of *Amṛtā Ghṛta*

	Ingredients	
	Clarified butter	50.00 L
	Decoction of *Tinospora cordifolia* derived from 100.00 kg stem	200.00 L
	Paste of *Zingiber officinale* dry rhizomes	8.33 kg
	Steps of the operation	
1.1	Check and certify that the steam-jacketed evaporator is clean as per S.O.P.	
1.2	Charge 50 L clarified butter and 200 L of decoction of *Tinospora cordifolia* stem into the evaporator.	
1.3	Suspend 8.33 kg of paste of *Zingiber officinale* dry rhizomes.	
1.4	Start heating the evaporator on medium heat.	
1.5	Stir the boiling mass frequently with ladles.	
1.6	Stop heating after 3 hours.	
1.7	Cover the evaporator.	
1.8	Repeat heating the evaporator next day for 3 hours. Stir frequently with ladles. Cover the evaporator.	
1.9	Repeat heating on the third day, with frequent stirring.	
1.10	Reduce the temperature when the sediment acquires mud-like consistency.	
1.11	Continue heating at lower temperature.	
1.12	Filter the *ghṛta* at *maddhyama pāka* into a clean and dry stainless steel drum Allow the product to cool.	
1.13	Check and certify that the bottling machine is in good condition as per S.O.P.	
1.14	Bottle the *ghṛta* in containers.	
	Expected yield: 45.00 kg.	

mixing of ingredients, encourage evaporation and to prevent sticking of solids to the bottom and sides of the evaporators (Anonymous 1978b). It is possible that during the long periods of continuous heating, cooling and stirring, many bioactive compounds in the herbs get incorporated into the lipid medium (see Chapter 11).

The *ghṛta* or *taila* is filtered at the appropriate stage. Certain ingredients like camphor, rock-salt and saffron are powdered and put in the vessel into which the medicinal lipid is filtered. This process is called *pātrapāka* (Anonymous 1978b). The finished product is packed in appropriate containers. The master formula of *Amṛtā ghṛta* (Anonymous 1978b) is provided in Table 3.6 to serve as an example.

3.9.4 Ariṣṭa and Āsava (Fermented Liquids)

3.9.4.1 Textual Instructions

Ariṣṭa and *āsava* are liquids prepared by keeping drugs in water for a long time, allowing it to ferment. *Ariṣṭa* are prepared by fermenting decoctions of herbs, whereas *āsava* are prepared in cold water and fresh juices of herbs. If not specified, the quantity of water to be used in the preparation of *ariṣṭa* is one *drōṇa* (12.288 kg). One *tōla* (4.800 kg) of jaggery is used for sweetening the medium. 2.400 kg honey and 480 g of herb powder are also to be added (Murthy 2017f).

3.9.4.2 Industrial Manufacture

The process of preparing *ariṣṭa* and *āsava* through fermentation is called *Sandhāna kalpana*, in which the mixture of powdered drugs, decoction, jaggery and honey is kept undisturbed for a specific period of time, usually 28 to 30 days. Fermentation generates alcohol, which facilitates the extraction of the active principles in powdered drugs. Moreover, the resultant product has good taste and flavor.

FIGURE 3.14 Modern brewing tanks in fermentation room. Photo courtesy: Nagarjuna Herbal Concentrates Ltd, Kerala (www.nagarjunaayurveda.com).

In earlier times when the process of preparing *ariṣṭa* and *āsava* was carried out in earthen pots, it was necessary to use a new pot each time a product was prepared. The vessel would be properly cleaned, and water was first boiled in the vessel. Absolute cleanliness was maintained, and the pot was fumigated using herbs like *Pippalī cūrṇa* (*Piper longum*) and the inside of the pot was then smeared with clarified butter before starting the fermentation process. When bulk manufacturing of ayurvedic products started to flourish, it became difficult to manage with earthen pots. New brewing vessels were introduced into the process. At present, brewing tanks with a storage capacity of even 10,000 liters are being used in the industrial manufacture of *ariṣṭa* and *āsava* (Figure 3.14). The smearing of clarified butter and fumigation with *Pippalī cūrṇa* are replaced with modern fumigation techniques.

In the traditional manufacture of *ariṣṭa*, the crude drugs mentioned in the formula are coarsely powdered and then made into *kvātha*, which contains the water-soluble bioactive compounds. The *kvātha* is then filtered and poured into the fermentation tank. According to the formula, either sugar or jaggery or a mix of these is dissolved in water, boiled, filtered and added to the *kvātha*. If honey is used, it is added as such to the liquid without heating. (In the case of *āsava*, fresh juices of herbs or fruits are used instead of *kvātha*.) Another group of herbs, collectively called *Prakṣēpa dravya*, is made into a fine powder and mixed with the contents of the brewing tank. After this, *Dhātakī puṣpa* (flowers of *Woodfordia fruticosa*), if included in the formula, is properly cleaned and added. This initiates the fermentation process. The brewing vessel is closed tightly so as to avoid any contact with external air and left undisturbed. The bioactive ingredients from the herbs get extracted into the medium during the fermentation process. The progress of fermentation is monitored by measuring at regular intervals the alcohol content, pH, specific gravity, reducing sugar and non-reducing sugar. After the specified period of fermentation, the vessel is opened, and the contents examined to ascertain whether the process of fermentation has been completed. The fermented liquid is filtered through a muslin cloth or clarified using a centrifuge. The filtered *ariṣṭa* or *āsava* is thereafter bottled (Anonymous 1978c) (Figure 3.15). The master formula of *Bhṛṅgarājāsavam* is presented to serve as an example (Table 3.7).

3.9.5 Avalēha (Confections)

3.9.5.1 Textual Instructions

Avalēha or *lēha* is a semi-solid dosage form prepared by evaporating a *kvātha* and adding jaggery, sugar or candied sugar. It is also processed sometimes with specific juices of herbs. Another name

FIGURE 3.15 Bottling of *ariṣṭa* and *āsava*. Photo courtesy: Nagarjuna Herbal Concentrates Ltd, Kerala (www.nagarjunaayurveda.com).

TABLE 3.7
Some Elements of the Master Formula for Production of *Bhṛṅgarājāsavam*

	Ingredients	
	Jaggery	33.33 kg
	Eclipta alba juice derived from 127.5 kg fresh herb	85.00 L
	Coarse powder of *Terminalia chebula* (pericarp)	2.67 kg
	Fine powder of the following herbs:	
	Piper longum (fruits)	0.056 kg
	Myristica fragrans (seeds)	0.056 kg
	Syzygium aromaticum (flower buds)	0.056 kg
	Cinnamomum zeylanicum (bark)	0.056 kg
	Elettaria cardamomum (seeds)	0.056 kg
	Cinnamomum tamala (leaves)	0.056 kg
	Mesua ferrea (flowers)	0.056 kg
	Steps of the operation	
1.1	Break the jaggery into pieces and dissolve in the juice of *Eclipta alba*.	
1.2	Filter the juice through 80 mesh sieve and discard the solids.	
1.3	Check and certify that the brewing tank is clean and in good condition as per S.O.P.	
1.4	Transfer the juice into a brewing tank.	
1.5	Add coarse powder of *Terminalia chebula* (pericarp) and stir well.	
1.6	Put the lid of the brewing tank and leave it in the brewing room.	
2.1	After 15 days open the brewing tank. Put the powders of the seven herbs into the brewed liquid.	
2.2	Stir well, close the brewing tank and leave for five more days.	
3.1	Check and certify that the Nutsche filter and vacuum pump are in good condition as per SOP.	
3.2	After 5 days, filter the brewed *Bhṛṅgarājāsavam* through Nutsche filter packed with hyflo, using a vacuum pump.	
3.3	Check and certify that the bottling machine is clean and in good condition as per S.O.P.	
3.4	Bottle the *Bhṛṅgarājāsavam*.	
	Expected yield: 100 L.	

of *avalēha* is *rasakriya*. *Avalēha* is usually administered along with milk, sugarcane juice, decoctions of grains, *pañcamūla* group of herbs or *vāsā* (*Adhatoda vasica*). When powder of herbs is used in the preparation of an *avalēha*, a four-fold quantity of sugar should be added. If jaggery is used, it should be double the quantity of powder. If any liquid is to be added, its quantity should be four times that of the powder. A properly prepared *avalēha* should fall like a rope if taken out of the

FIGURE 3.16 *Avalēha* being cooked in large steam-jacketed evaporators. Photo courtesy: Rajah Healthy Acres, Kerala (www.ayurvedichospital.com).

vessel and poured from a ladle. It will also sink in water. If rolled between fingers, it should carry the imprint of the fingers and assume various shapes. Properly prepared *avalēha* possesses good color, aroma and taste (Murthy 2017d).

3.9.5.2 Industrial Manufacture

Avalēha has ingredients like *kvatha* or other fluids, jaggery, sugar or candied sugar, powders or paste of herbs and clarified butter, oil or honey. First of all, jaggery, sugar or candied sugar is dissolved in the fluid and filtered to remove insoluble particles. This sweet liquid is poured into large steam-jacketed evaporators, like the ones used in the manufacture of medicinal lipids (Figure 3.16). The evaporator is then heated over moderate heat and stirred continuously with ladles. Heating is stopped when the cooked mass acquires the consistency of soft wax. A fine powder of herbs is slowly sprinkled over the mass, by sifting through a sieve. The mass is stirred vigorously and continuously so that the product turns soft and homogeneous. While the mass is still hot, clarified butter or oil, if mentioned in the formula, is added and mixed by stirring. Honey, if part of the formula, is added when the cooked mass is cool. After thorough mixing, the *avalēha* is bottled (Figure 3.17). The master formula of *Aśvagandhādi avalēha* serves as an example (Table 3.8). The *avalēha* should not be hard or appear like a thick fluid. It should be possible to roll the *avalēha* into small and soft balls. Growth of fungus or fermentation are signs of improper preparation of the *avalēha* (Anonymous 1978d).

3.9.6 Guḷika (Pills)

3.9.6.1 Textual Instructions

Guṭika, *vaṭi*, *vaṭika*, *varti*, *mōdaka*, *piṇḍi* and *guḍa* are the Sanskrit synonyms of this dosage form. It is prepared by two methods. Firstly, the powder of herbs can be cooked with jaggery, sugar or

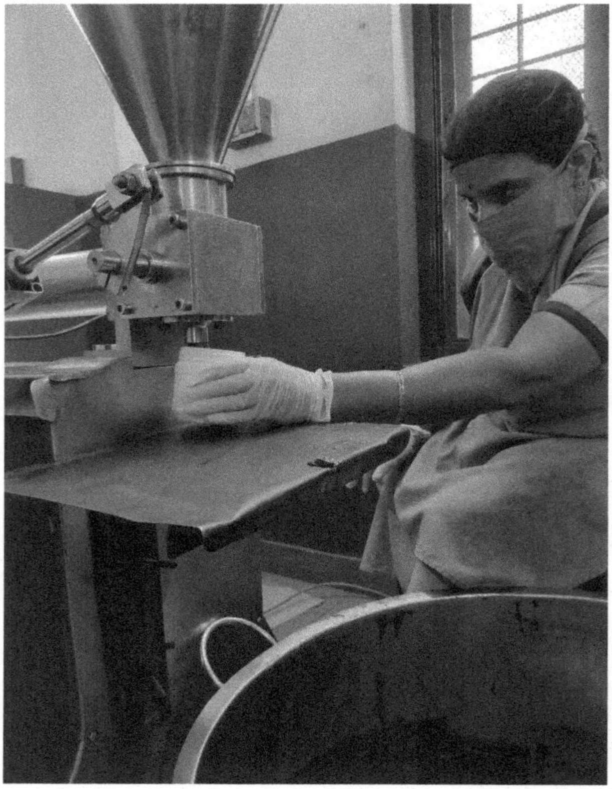

FIGURE 3.17 Bottling of *avalēha*. Photo courtesy: Rajah Healthy Acres, Kerala (www.ayurvedichospital.com).

guggul gum and then rolled into pills. Alternatively, the powder can be ground with any liquid. If sugar is to be added, its quantity should be four times that of the powder of herbs. The mixture is to be ground with twice the volume of water or other fluids like decoctions or juices of herbs, lemon juice, juice of pomegranate flowers or fruits, latex of *Jatropha curcas*, cow milk, breast milk, cow urine, goat urine, castor oil, honey, clarified butter or sesame oil. The grinding is carried out continuously for long periods lasting even 7 to 18 days. Finally, they are rolled into pills, dried under shade and stored (Vaidyan and Pillai 1985b; Murthy 2017c).

3.9.6.2 Industrial Manufacture

The herbs are cleaned and pulverized separately to fine powder that passes through an 80 mesh sieve. Weighed quantities of the herbs are then blended and placed on a granite grinding stone fitted with motorized pestles. The powder is ground to a soft paste by adding the specified liquids (Figure 3.18). If more than one liquid is to be used, they are added in succession. When the groundmass acquires the desired consistency, fragrant drugs like musk (*kastūri*) and camphor (*karpūra*), if mentioned in the formula, are added and ground again. Grinding is stopped when the mass does not stick to fingers when rolled. Pills are then rolled using a pill-rolling machine (Figure 3.19).

With the modernization of the manufacture of ayurvedic medicines, ayurvedic pills are also produced as tablets. In such case, the groundmass is dried in a tray dryer and pulverized. The resultant powder is put into a mass mixer. Required quantities of excipients are added to the powder, which is then converted into granules. The granulated material is dried in a fluid bed dryer. Tablets are

TABLE 3.8
Some Elements of the Master Formula for the Production of *Aśvagandhādi Avalēha*

	Ingredients	
	Fine powders (80 mesh sieve) of the following:	
	Withania somnifera (root)	28.32 kg
	Hemidesmus indicus (root)	28.32 kg
	Cuminum cyminum (fruit)	28.32 kg
	Smilax china (root tuber)	28.32 kg
	Vitis vinifera (dry fruit)	28.32 kg
	Elettaria cardamomum (seed)	3.54 kg
	Jaggery	200.00 kg
	Clarified butter	33.33 kg
	Deionized water	66.67 L
	Honey	66.67 kg
	Steps of the operation	
1.1	Break jaggery into pieces and put into a clean stainless steel drum.	
1.2	Add 66.667 L of deionized water, stir and dissolve it completely.	
1.3	Filter the fluid through 80 mesh sieve and remove the solids.	
1.4	Check and certify that the steam-jacketed evaporator is clean as per S.O.P.	
1.5	Pour the jaggery-water into the evaporator and start heating at low heat.	
1.6	When the jaggery-water becomes slightly viscous sprinkle, the fine powders of *Withania somnifera*, *Hemidesmus indicus*, *Cuminum cyminum*, *Smilax china*, *Vitis vinifera* and mix well by stirring with ladles.	
1.7	Continue stirring and heating till the *avalēha* becomes more viscous.	
1.8	Add fine powder of *Elettaria cardamomum* and mix well by stirring with ladles.	
1.9	Add clarified butter to the mass and mix.	
1.10	Confirm that the *avalēha* acquires the desired consistency.	
1.11	Allow the *avalēha* to cool.	
1.12	After the *avalēha* is cooled, add honey and mix well.	
1.13	Bottle the *avalēha*.	
	Expected yield: 490 kg	

later punched from the granulated material. The tablets are then packed in airtight containers. The master formula of *Bilvādi Gutika* serves as an example (Anonymous 1978e) (Table 3.9).

3.9.7 Lēpa (Paste)

Lēpa is a paste of soft consistency, intended for external application. The crude drugs are powdered finely, mixed with specific fluid and made into a paste for applying on the body. Water, cow urine, sesame oil, castor oil, coconut oil, honey, butter, sour buttermilk and clarified butter are the adjuvants generally used (Anonymous 1978f).

A common *lēpa* like *Kuṅkumādi lēpa* is manufactured in a simple way. First of all, *Kuṅkumādi taila* is prepared according to the formula in authentic text and using the equipment described earlier (see Section 3.9.3 on medicinal lipids). Thereafter, weighed quantities of *Kuṅkumādi taila* and beeswax (80:20) are placed in a clean and dry stainless steel vessel. The vessel is warmed at very low heat and the oil is stirred frequently. When the wax is molten completely, the product is allowed to cool to room temperature. Thereafter, it is filled in collapsible tubes (Figure 3.20) (Anonymous 1978f).

FIGURE 3.18 Grinding of *Bilvādi Gutika*. Photo courtesy: Rajah Healthy Acres, Kerala, (www.ayurvedichospital.com).

FIGURE 3.19 Pills rolling machine. Photo courtesy: Nagarjuna Herbal Concentrates Ltd, Kerala (www.nagarjunaayurveda.com).

TABLE 3.9
Some Elements of the Master Formula for the Production of *Bilvādi Gutika*

	Ingredients	
	Fine powders (80 mesh sieve) of the following:	
	Aegle marmelos (root)	1.155 kg
	Ocimum sanctum (flower)	1.155 kg
	Pongamia pinnata (seed)	1.155 kg
	Valeriana wallichi (root)	1.155 kg
	Cedrus deodara (heartwood)	1.155 kg
	Terminalia chebula (pericarp)	1.155 kg
	Terminalia bellirica (pericarp)	1.155 kg
	Emblica officinalis (pericarp)	1.155 kg
	Zingiber officinale (rhizome)	1.155 kg
	Piper nigrum (fruit)	1.155 kg
	Piper longum (fruit)	1.155 kg
	Curcuma longa (rhizome)	1.155 kg
	Berberis aristata (stem)	1.155 kg
	Goat urine 15.00 L	
	Steps of the operation	
1.1	Check and certify that the grinder is clean as per S.O.P.	
1.2	Load the herb powders into the grinder.	
1.3	Load the fresh goat urine into the grinder.	
1.4	Grind the powder till the urine completely gets absorbed into the paste, and the paste become non-sticky.	
1.5	Ensure that the paste achieved the required consistency.	
1.6	Check and certify that the pills rolling machine is clean as per S.O.P.	
1.7	Roll the paste into pills.	
1.8	Bottle the *gulika* in airtight containers.	
	Expected yield: 14.25 kg	

FIGURE 3.20 Packing of *Kuṅkumādi Lēpa*. Photo courtesy: Nagarjuna Herbal Concentrates Ltd, Kerala (www.nagarjunaayurveda.com).

3.10 PROCESSING OF WASTES

The Ayurveda industry leaves behind considerable quantities of plant residues. For example, a medium-sized company throws out about one ton of ayurvedic wastes daily in the process of manufacturing different formulations. After the preparation of medicines, the residual material is discarded as waste. These wastes are often dumped into heaps in open areas, resulting in water and soil pollution through their decay and release of leachate containing toxic substances, including metals. Sometimes these wastes are set on fire also. The burning of waste in open spaces also causes air pollution (Ndegwa and Thompson 2000). Therefore, there is a need for safer and eco-friendly technologies to manage these wastes. Several such technologies have been identified, and they can be employed for the effective management of wastes from the Ayurveda industry.

3.10.1 Vermicomposting

Scientific studies have established the utility of using earthworms as a treatment technique for numerous waste streams (Hand et al. 1988; Harris et al. 1990; Logsdon 1994). The action of earthworms in this process is physical, mechanical and biochemical. The physical and mechanical processes include substrate aeration and mixing, as well as actual grinding. The biochemical action is effected by microbial decomposition of the substrate in the intestines of the earthworms. Vermicomposting, thus, saves on all these unit operations. Hand et al. (1988) define vermicomposting as a low-cost technology for the processing or treatment of organic wastes.

Differing from traditional microbial waste treatment, vermicomposting results in the bioconversion of the waste stream into two useful products viz., the earthworm biomass and the vermicompost. The earthworm biomass can be processed into proteins (earthworm meal) or high-grade horticultural compost. Vermicompost is also considered an excellent product as it is homogenous, has desirable aesthetics, very low levels of contaminants and tends to hold more nutrients over a longer period without adversely affecting the environment (Ndegwa and Thompson 2000). A detailed account of vermicomposting technology is available (Nancarrow and Taylor 1998; Rajkhowa et al. 2005).

Pant and Yami (2008) carried out a laboratory experiment for the proper management of the solid wastes of Kathmandu valley in Nepal, generated by the Ayurveda industry, sugar mills (bagasse), wood mills, kitchens and vegetable and fruit markets. The experiment dealt with the decomposition of solid wastes through the action of the red earthworm (*Eisenia foetida*). The vermicomposting of the mixtures was carried out for 12 weeks. The vermicomposting of boiled, non-woody, Ayurveda industry waste was completed in ten weeks. The earthworms grown on Ayurveda industry wastes were found to be healthier than those found in other solid wastes. The study also showed that vermicompost obtained from Ayurveda industry wastes was found to be rich in N, P, K and organic matter. Vermicomposting seems to be an efficient method for the management of solid wastes emanating from the Ayurveda industry.

3.10.2 Briquetting of Solid Wastes

The ayurvedic medicine manufacturing industry gives out many waste products like herbal residues and solid wastes. From an eco-protection point of view it is better to convert these wastes into an ecofriendly and useful product like briquettes. Briquettes are biomass cylinders (50–80 mm diameter and 150 mm length) compressed at a high temperature, with a moisture content ranging between 10% and 20%. Other shapes like rectangular or prismatic are also frequent. In some cases, they have holes to improve their combustion. Briquettes may be composed of crushed and densified wood or composed of crushed, dried and molded charcoal under high pressure (Solano et al. 2016a).

The densification process consists of compaction or reduction of the raw material volume and sealing to ensure that the product remains in a stable, compacted state. Current regulations like ISO

17225 allow the use of specific additives to enhance and maintain compaction of briquettes. These additives contain starches like rice flour, cassava flour, mashed sweet potato, or molasses and gum arabic to give greater consistency to the resulting product (Solano et al. 2016b).

Briquettes are manufactured by compressing and compacting the raw material, Natural compaction of the raw material is achieved by compressing at high pressure, which raises the temperature leading to a Bakelized surface, which gives a glossy appearance and consistency to the briquettes. The typical briquette production process involves milling, drying and pressing or briquetting (Solano et al. 2016c). Machinery with differing specifications and capacities are employed in the manufacture of briquettes (Figures 3.21 and 3.22). Briquetting enables the valorization of waste material generated by the Ayurveda industry.

3.10.3 Stabilization of Black Cotton Soil

Black cotton soil is a black colored soil that covers about 30% of land area in India. It is suitable for growing cotton. The high content of montmorillonite bestows a high degree of expansiveness to black cotton soil. Due to its characteristics of high plasticity, excessive swelling, shrinkage and low strength when wet, this type of soil is regarded as unsuitable for construction (Oza and Gundaliya 2013). Many attempts have been made to stabilize this weak soil using industrial wastes, as black cotton soil is abundantly available. Fly ash, coconut coir fiber, crushed glass, H.D.P.E. waste fibers,

FIGURE 3.21 Briquette press used in compaction of powdered raw material. Figure of Jumbo Brq 9075 briquette press. Reproduced with kind permission of Jay Khodiyar Machine Tools, Rajkot, India (www.jaykhodiyar.com).

FIGURE 3.22 Briquettes made from feedstocks such as corn stover, switchgrass and prairie cord grass. Reproduced with permission from Karunanithy et al. (2012).

stone dust, lime and cement kiln waste have been used for this purpose (Oza and Gundaliya 2013; Varsha et al. 2017).

Varsha et al. (2017) carried out an interesting study on the suitability of waste material from the Ayurveda industry for improving the engineering properties of black cotton soil. They added 5%, 10% and 15% of sediment left from manufacture of *Kṣīrabala taila* to black cotton soil and studied the engineering properties of the modified soil, like unconfined compression, standard proctor compaction, the Atterberg limit and so on. The addition of the waste in varying percentages resulted in the decrease of maximum dry density and lowering of degree of expansiveness from 73.33% to 36.36%, thereby increasing the strength of the soil. The liquid limit of the soil also increased from 64.75% to 82.57%. The results of this study showed that black cotton soil can be stabilized using the inexpensive waste from Ayurveda industry.

3.10.4 Recycling of Wastewater

A large quantity of water is used in various operations in ayurvedic medicine manufacturing companies. Such units generate huge volumes of biodegradable wastewater during the processing of raw materials and the production of medicines. This wastewater is moderately rich in chemical oxygen demand (C.O.D.) and biochemical oxygen demand (B.O.D.) concentrations. It can be discharged only after proper treatment. Condenser waste from evaporators, chemical waste, spent liquors from fermentation operations, sewage, laboratory and floor washing waste contribute to the organic and inorganic matter in ayurvedic wastewater (Vanerkar et al. 2015). The impact of the wastewater on the environment needs to be reduced considerably by adopting efficient and eco-friendly water recycling methods.

Vermifiltration or lumbrifiltration is one such method, first advocated by the late Professor Jose Toha, at the University of Chile in 1992 (Li et al. 2008). Vermifiltration is a low-cost sustainable technology over conventional systems with immense potential for decentralization in rural areas (Taylor et al. 2003; Sinha et al. 2008). It was initially used to process organically polluted water using earthworms (Li et al. 2008). Introduction of earthworms into the system was an innovation to the conventional biofilter of wastewater treatment, and it created a new method of biological reaction through extending food chains, converting energy and transferring mass from the biofilm to the earthworms. Vermifiltration is found to be generally good for the treatment of swine wastewater (Li et al. 2008), municipal wastewater (Godefroid and Yang 2005), domestic wastewater (Taylor et al. 2003; Sinha et al. 2008) and wastewater from dairy industry (Telang and Patel 2013).

Das et al. (2015) developed a treatment method for ayurvedic liquid effluents by integrating microbial pre-treatment and vermifiltration. The ayurvedic effluent was at first pre-treated with a microbial consortium and later fed to a vermifiltration unit. Organic wastes and solids were ingested and absorbed by the earthworm's body wall and degraded. The process removed B.O.D., C.O.D., total dissolved solids (T.D.S.) and the total suspended solids (T.S.S.) from wastewater. There was no sludge formation during the process, and the resultant water was odor-free. The vermifiltered water exhibited a significant reduction in C.O.D. by 98.03%, B.O.D. by 98.43%, T.S.S. by 95.8%, T.D.S. by 78.66% and oil and grease by 92.58%. The vermifiltered water was clean and clean enough to be reused for irrigation. Vermifiltration is an eco-friendly method that can be applied to the ayurvedic medicine manufacturing industry.

3.11 CONCLUSION

Ayurvedic medicine manufacturing companies are mostly family-owned businesses. The origin of most of these companies can be traced back to a *vaidya* who used to prepare some medicines for dispensing to his patients. With the gradual acceptance of the medicines, these small units grew into fully-fledged companies. Many of them are now being run by third-generation owner–managers. The ownership pattern has helped the transfer of knowledge from one generation to the other, thereby enriching the knowledge base of these companies.

The production facilities of the ayurvedic manufacturers differ widely. They range from manual to semi-automated to fully manual modes of production. There are ultramodern factories where production is fully automated, and no hands touch the products until the machine-packed medicines are collected by personnel. Other manufacturers depend highly upon manual labor. In their workshop-like establishments herbs are ground by hand or with the help of simple machinery. In more traditional settings, vessels which can hold a few kilograms and wooden barrels having a capacity of 50 or 100 liters are used for making *kvātha* and *ariṣṭa*. Here workers stir liquids in bronze cauldrons placed over gas stoves and perform traditional boiling and cooking techniques. Large traditional manufacturing units employ workmen who skillfully process raw materials and half-finished preparations in what looks like a giant kitchen. This way of manufacturing is not limited to small or very small manufacturers. Even large companies manufacture some of their products this way.

The use of modern manufacturing equipment and laboratory tests for monitoring production processes has raised questions about the authenticity of mass-produced ayurvedic medicines. It is argued that modern production technologies affect the quality and efficacy of ayurvedic medicines. Medicines produced on a large scale are said to be of a lesser quality because production processes are speeded up and machines lack the skill of artisans. Manual, small-scale production is time-tested and in line with the rules for the production instructed in Sanskrit books of Ayurveda. Traditional techniques are elaborate and time consuming. Large manufacturers see these traditional methods, demands for materials and processing times as very unpractical and too time-consuming for their modern businesses (Narayana et al. 1997).

The star products of many large manufacturers are made in fully automated plants. Modern factories harbor fully automatic, computer-controlled production processes. Employees comply with contemporary hygienic standards, as is evident in their hair caps and white overalls. Production areas are sealed off to prevent contamination of raw materials and fungal and bacterial infections. These products are tested in laboratories, and the final products are stored in cool rooms. No hands touch the products as they move through various phases of manufacture. Process monitoring techniques adapted from modern food technology are applied. The scale of operations is very large. Apart from the scale of operations, the time taken for manufacturing also differs from those sticking to tradition. While an artisan manufacturer takes about a month to make a 100 liters of *Daśamūlāriṣṭa*, a large company needs only one day to produce a batch of 15,000 liters of the same product. To our knowledge no systematic evaluation has been made to determine if, and to what extent, modern production and monitoring techniques influence the efficacy of ayurvedic medicines. Without such a study, controversies about the appropriateness of modern technology will never end.

Most of the companies manufacture traditional products using their own master formulae and variations in processes, which are treated as jealously guarded secrets. For example, many herbs of the *daśamūla* group are now getting rare. To circumvent this problem, manufacturers resort to the substitution of herbs. This reduces the efficacy of products. Radha et al. (2014) procured seven brands of *Mahārāsnādi Kvātha* available in the Kerala market and compared several of their characteristics. The color of these seven brands, their pH, total dissolved solids, sodium benzoate content, H.P.T.L.C. profiles, microbial load and content of compound classes (total tannins, alkaloids, phenols and sterols) showed wide variation. The same is the case with almost all ayurvedic medicines available at present. This calls for standardization of manufacturing protocols. The Acts related to the manufacture of ayurvedic medicines, especially the Drugs and Cosmetics Act, 1940 need to be amended so as to incorporate provisions for correcting this problem.

REFERENCES

Anonymous. 1978a. *Cūrṇa*. In *The ayurvedic formulary of India, part I first edition*, 83–95. New Delhi: Ministry of Health and Family Planning.

Anonymous. 1978b. *Ghṛta*. In *Ayurvedic formulary of India, part-1 first edition*, 61–82. New Delhi: Ministry of Health and Family Planning.

Anonymous. 1978c. *Āsava* and *Ariṣṭa*. In *The ayurvedic formulary of India, part-I, first edition*, 1–17. New Delhi: Ministry of Health and Family Planning.

Anonymous. 1978d. *Avalēha* and *Pāka*. In *The ayurvedic formulary of India, part I, first edition*, 23–39. New Delhi: Ministry of Health and Family Planning.

Anonymous. 1978e. *Vaṭi* and *Guṭika*. In *The ayurvedic formulary of India, part I, first edition*, 139–54. New Delhi: Ministry of Health and Family Planning.

Anonymous. 1978f. Lepa. In *The ayurvedic formulary of India, part I, first edition*, 133–8. New Delhi: Ministry of Health and Family Planning.

Anonymous. 2000. *The ayurvedic formulary of India part II*, 1–425. New Delhi: Ministry of AYUSH.

Anonymous. 2001a. Processing of medicinal plants-Grinding, sieving and comminution. In *A guide to the European market for medicinal plants and extracts*, 60–1. London: Commonwealth Secretariat.

Anonymous. 2001b. Special introduction (extracts). In *Ayurvedic pharmacopoeia of India, part-1 volume 8*, XXIX–XIII. Delhi: The Controller of Publications.

Anonymous. 2003a. *The ayurvedic formulary of India part I (2nd revised edition)*, 1–488. New Delhi: Ministry of AYUSH.

Anonymous. 2003b. Kvātha cūrṇa. In *Ayurvedic formulary of India, part-1 second edition*, 51–62. Delhi: The Controller of Publications.

Anonymous. 2008. *Guidelines on good manufacturing practice for traditional medicines and health supplements*, 1st edition. Kuala Lumpur: National Pharmaceutical Control Bureau of Malaysia.

Anonymous. 2011. *The ayurvedic formulary of India part III*, 1–706. New Delhi: Ministry of AYUSH.

Anonymous. 2014. *Guidelines for inspection of GMP compliance by Ayurveda, Siddha and Unani Drug Industry*, 1–70. New Delhi: Department of AYUSH.

Anonymous. 2018. http://www.dabur.com/in/en-us/about/aboutus/history (accessed March 21, 2018).

Ayurvedacharya, R. S. 1951. *Bhaiṣajya Ratnāvali*. Varanasi: Chowkhamba Sanskrit Pusthakalaya.

Boettcher, H., and I. Guenther. 2005. Storage of the dry drug. In *Chamomile: Industrial profiles*, ed. R. Franke, and H. Schilcher, 211–20. Boca Raton: CRC Press.

Chaloner-Larsson, G., R. Anderson, and A. Egan. 1997. *A WHO guide to Good Manufacturing Practice (GMP) requirements. Part 1: Standard operating procedures and master formulae*, 1–111. Geneva: WHO.

Cook, Jr., J. L. 1998. Introduction. In *Standard operating procedures and guidelines*, 1–9. NJ: PennWell Publishing Company.

Das, D. C., M. Joseph, and D. Varghese. 2015. Integrated microbial-vermifiltration technique for ayurvedic industrial effluents. *International Journal of Engineering Research and General Science* 3: 338–46.

Fellows, P. 2000. Raw material preparation. In *Food processing technology: Principles and practice*, 83–97. Boca Raton: CRC Press.

Godefroid, B., and J Yang. 2005. Synchronous municipal sewerage-sludge stabilization. *Journal of Environmental Sciences* 17: 59–61.

Goraya, G. S., D. K. Ved, and V. Jishtu. 2017a. Supply of herbal raw drugs from wild collections. In *Medicinal plants in India: An assessment of their demand and supply*, ed. G. S. Goraya, and D. K. Ved, 83–105. New Delhi: N.M.P.B. - Ministry of AYUSH.

Goraya, G. S., D. K. Ved, K. Ravikumar, and R. S. Rawat. 2017b. Domestic trade of herbal raw drugs. In *Medicinal plants in India: An assessment of their demand and supply*, ed. G. S. Goraya, and D. K. Ved, 145–79. New Delhi: N.M.P.B. - Ministry of AYUSH.

Goraya, G. S., D. K. Ved, R. S. Rawat, and S. Gautam. 2017c. Consumption by domestic herbal industry. In *Medicinal plants in India: An assessment of their demand and supply*, ed. G. S. Goraya, and D. K. Ved, 19–37. New Delhi: N.M.P.B. - Ministry of AYUSH.

Gupta, B. 1976. Indigenous medicine in nineteenth and twentieth century Bengal. In *Asian medical systems: A comparative study*, ed. Charles Leslie, 368–78. CA: University of California Press.

Hand, P., W. A. Hayes, J. C. Frankland, and J. E. Satchell. 1988. The vermicomposting of cow slurry. *Pedobiologia* 31: 199–209.

Harilal, M. S. 2009. Commercialising traditional medicine: Ayurvedic manufacturing in Kerala. *Economic and Political Weekly* 44: 44–51.

Harris, G. D., W. L. Platt, and B. C. Price. 1990. Vermicomposting in a rural community. *BioCycle* 90: 48–51.

Kanagarathinam, D. V. 2016. *Physicians, print production and medicine in colonial South India (1867 -1933)*, 1–315. Doctor of Philosophy thesis submitted to Pondicherry University.

Karunanithy, C., Y. Wang, K. Muthukumarappan, and S. Pugalendhi. 2012. Physiochemical characterization of briquettes made from different feedstocks. *Biotechnology Research International*, 2012. doi:10.1155/2012/165202.

Kumar, A. 2001. The Indian drug industry under the Raj, 1860-1920. In *Health, medicine and empire: Perspectives on colonial India*, ed. B. Pati, and M. Harrison, 356–85. New Delhi: Orient Longman.

Kumar, D. S. 2016. Extraction of the bioactives. In *Herbal bioactives and food fortification: Extraction and formulation*, 63–127. Boca Raton: CRC Press.

Li, Y. S., P. Robin, D. Cluzeau et al. 2008. Vermifiltration as a stage in reuse of swine wastewater: Monitoring methodology on an experimental farm. *Ecological Engineering* 32, no. 4: 301–9.

Logsdon, G. 1994. Worldwide progress in vermicomposting. *BioCycle* 35: 63–5.

Murthy, K. R. S. 2017a. *Maddhyama Khaṇḍa* (Middle Section), Chapter 2. In *Śārṅgadhara Samhita*, 56–77. Varanasi: Chaukhamba Orientalia.

Murthy, K. R. S. 2017b. *Maddhyama Khaṇḍa* (Middle section). Chapter 6. In *Śārṅgadhara Samhita*, 84–101. Varanasi: Chaukhamba Orientalia.

Murthy, K. R. S. 2017c. *Maddhyama Khaṇḍa* (Middle section), Chapter 7. In *Śārṅgadhara Samhita*, 101–10. Varanasi: Chaukhamba Orientalia.

Murthy, K. R. S. 2017d. *Maddhyama Khaṇḍa* (Middle section). Chapter 8. In *Śārṅgadhara Samhita*, 111–5. Varanasi: Chaukhamba Orientalia.

Murthy, K. R. S. 2017e. *Maddhyama Khaṇḍa* (Middle Section). Chapter 9. In *Śārṅgadhara Samhita*, 115–36. Varanasi: Chaukhamba Orientalia.

Murthy, K. R. S. 2017f. *Maddhyama Khaṇḍa* (Middle Section). Chapter 10. In *Śārṅgadhara Samhita*, 136–45. Varanasi: Chaukhamba Orientalia.

Nancarrow, L., and J. H. Taylor. 1998. *The worm book*, 1–150. Berkeley: The Speed Press.

Narayana, D. B. A., B. Brindavan, and C. K. Katiyar. 1997. Herbal remedies: Through GMP/HACCP techniques. *The Eastern Pharmacist* 40: 21–28.

Ndegwa, P. M., and S. A. Thompson. 2000. Effects of C-to-N ratio on vermicomposting of biosolids. *Bioresource Technology* 75 no. 1: 7–12.

Oza, J. B., and P. J. Gundaliya. 2013. Study of black cotton soil characteristics with cement waste dust and lime. *Procedia Engineering* 51: 110–18.

Pant, S. R., and K. D. Yami. 2008. Selective utilization of organic solid wastes by earthworm (*Eisenia foetida*). *Nepal Journal of Science and Technology* 9: 99–104.

Radha, A., J. Sebastian, M. Prabhakaran, and D. S. Kumar. 2014. Observations on the quality of commercially manufactured ayurvedic decoction. *Mahārāsnādi Kvātha. Hygeia: Journal for Drugs and Medicines* 6: 74–80.

Rajkhowa, D. J., A. K. Gogoi, and N. T. Yaduraju. 2005. *Weed utilization for vermicomposting - success story*, 1–16. Jabalpur: National Research Centre for Weed Science.

Sinha, R. K., G. Bharambe, and U. Chaudhari. 2008. Sewage treatment by vermifiltration with synchronous treatment of sludge by earthworm: A low-cost sustainable technology over conventional systems with potential for decentralization. *Environmentalist* 28 no. 4: 409–20.

Solano, D., P. Vinyes, and P. Arranz. 2016a. Introduction to the energy use of wood and agricultural wastes. In *The biomass briquetting process - a guideline report*, 1–4. New York: UNDP-CEDRO.

Solano, D., P. Vinyes, and P. Arranz. 2016b. Local and international context for briquette use. In *The biomass briquetting process -a guideline report*, 5–8. New York: UNDP-CEDRO.

Solano, D., P. Vinyes, and P. Arranz. 2016c. Technical description. In *The biomass briquetting process -a guideline report*, 9–20. New York: UNDP-CEDRO.

Taylor, M., W. P. Clarke, and P. F. Greenfield. 2003. The treatment of domestic wastewater using small-scale vermicompost filter beds. *Ecological Engineering* 21 no. 2–3: 197–203.

Telang, S., and H. Patel. 2013. Vermi-biofiltration – A low cost treatment for dairy wastewater. *International Journal of Scientific and Engineering and Research* 4: 595–9.

Vaidyan, K. V. K., and A. S. G. Pillai. 1985a. *Sahasrayōgam, Vidyarambham*, 1–584. Alleppey: Vidyarambham Publishers.

Vaidyan, K. V. K., and A. S. G. Pillai. 1985b. *Guḷikā yōgaṅgaḷ* (Formulae of pills). In *Sahasrayogam*, 133–66. Alleppey: Vidyarambham Publishers.

Vanerkar, A. P., A. B. Fulke, S. K. Lokhande, M. D. Giripunje, and S. Satyanarayan. 2015. Recycling and treatment of herbal pharmaceutical wastewater using *Scenedesmus quadricuada*. *Current Science* 108: 979–83.

Varsha, R. A., M. Anvar, G. Aswini, K. U. Athulya, R. K. Ratheesh, and G. Unni. 2017. Stabilization of black cotton soil using ayurvedic industrial waste. *International Research Journal of Engineering and Technology* 4: 1770–3.

Vellodi, M. K. 1987. Vaidyaratnam P.S. Variar. *Aryavaidyan* 1: 8–13.

4 Quality Control of Ayurvedic Medicines

V. Remya, Maggie Jo Alex and Alex Thomas

CONTENTS

4.1 Introduction 72
4.2 Quality Control Measures in Ancient Times 72
4.3 Present State of Quality Control of Ayurvedic Medicines 73
4.4 Ayurvedic Pharmacopoeia of India 73
4.5 Factors That Influence the Quality of Ayurvedic Medicines 73
4.5.1 Good Agricultural and Collection Practices 74
4.5.2 Contamination 74
4.5.2.1 Contamination with Heavy Metals 74
4.5.2.2 Contamination with Micro-Organisms 75
4.5.2.3 Contamination with Pesticides 75
4.5.3 Adulteration and Substitution 76
4.5.4 Content of Bioactives 76
4.5.5 Stability 78
4.5.6 Processing Methods 79
4.6 Quality Control Methods 80
4.6.1 Macroscopic Evaluation 80
4.6.2 Microscopic Evaluation 81
4.6.3 Physicochemical Analysis 82
4.6.4 Fingerprinting 82
4.6.4.1 T.L.C. 83
4.6.4.2 H.P.L.C. 83
4.6.5 Application of Fingerprint Data 83
4.6.6 Similarity Evaluation of Fingerprints 84
4.6.7 Principal-Component Analysis (P.C.A.) 84
4.6.8 Application of Fingerprinting to Quality Control 84
4.6.8.1 Detection of Adulterants 84
4.6.8.2 Authentication of Herbs 84
4.6.9 Quality Control of Finished Formulations 85
4.7 Some Problems Related to Quality Control 85
4.7.1 Heavy Metals in Ayurvedic Pills 85
4.7.2 Souring of Ariṣṭa and Āsava 89
4.7.2.1 Quality of Water 89
4.7.2.2 Quality of Herbs 89
4.7.2.3 Use of Appropriate Brewing Vessel 89
4.7.2.4 Hygienic Design of Brewing Area 90
4.7.2.5 Disinfection 90
4.7.2.6 Use of Yeast 90
4.7.2.7 Temperature 90
4.7.2.8 Improvised Fermentation 90
4.7.3 Irrational Substitution 93
4.8 Conclusion 93
References 94

4.1 INTRODUCTION

When the expression "quality" is used in the context of a medicinal product, people usually think in terms of an excellent product that fulfills or exceeds their expectations. The quality of a product is dependent on several factors, like the raw materials used in its production, equipment used, the expertise of the personnel in charge of the operations, manufacturing process, analytical methods, packing and hygiene. Quality control is the application of techniques and measures to achieve and sustain at a constant rate the quality of a product or service. Quality improvement is the use of tools to continually make the product or service unbeatable (Besterfield 2013).

The quality of ayurvedic preparations is achieved through the development of procedures and practices designed to ensure conformity to specifications. Disciplines like botany, chemistry, pharmacology and pharmacognosy are vital to the implementation of these activities (Malone 1983; Dentali 2009). Factors like identification and collection of good constituent herbs, aspects of contamination, adulteration and substitution of herbs, the chemical profile of the herbs, manufacturing process, robust analytical methods and stability define the quality of these medicines (Yau et al. 2015).

4.2 QUALITY CONTROL MEASURES IN ANCIENT TIMES

Ayurveda was conceived and developed in a very remote era, when science, as we know it today, had no influence on the society. However, in their own modest way, the ancient masters made sure that the quality of the system was maintained. As the inclusion of the right herb is essential for the therapeutic quality of medicinal formulae, Ayurveda texts give advice on how to identify medicinal plants. For example, Caraka says that goatherds, shepherds, cowherds and other forest-dwellers know medicinal plants by their name and form (Sharma 1981). Suśruta also gives similar advice. He instructs that medicinal plants should be recognized and identified with the help of cowherds, hermits, huntsmen and others who trek the forest as well as with the help of those who collect and eat edible roots and fruits of the forests (Murthy 2017a). Therefore, a student or practitioner of Ayurveda in those days used to learn from these sources.

The physicians of ancient India were aware of the factors that can influence the quality of medicinal products. Instructions are available in Ayurveda texts regarding the procurement of herbs. For example, *Śārṅgadhara Samhita*, considered to be the best book on ayurvedic pharmacy, gives valuable advice on this subject. With purity of mind the physician himself should collect the herbs, and while doing so, he should avoid herbs growing on anthills, dirty places, marshy land, burial grounds and salty soil. He should discard herbs infested with worms or affected by fire and snow. The root bark is to be collected in the case of large roots, while the entire root should be collected in the case of smaller ones. While the outer bark of large trees like *Nyagrodha* (*Ficus benghalensis*) is collected, the heartwood of smaller trees like *Bījaka* (*Pterocarpus marsupium*) is preferred (Murthy 2017b).

Instructions are available on the choice of medicinal ingredients. As far as possible, only fresh herbs should be used in the preparation of medicines. However, exceptions to this rule are *Viḍaṅga* (*Embelia ribes*), *Kṛṣṇa* (*Piper longum*), *Guḍa* (jaggery), *Dhānya* (*Coriandrum sativum*), *Ājya* (clarified butter) and *Mākṣika* (honey). Their medicinal qualities improve with age (Murthy 2017b). Ingredients like *Guḍūci* (*Tinospora cordifolia*), *Kuṭaja* (*Holarrhene antidysenterica*), *Vāśa* (*Adhatoda vasica*), *Kūśmāṇḍa* (*Benincasa cerifera*), *Śatāvari* (*Asparagus racemosus*), *Aśvagandha* (*Withania somnifera*), *Sahacara* (*Barleria cristata*), *Śatapuṣpa* (*Anethus graveeolens*) and *Prasāriṇi* (*Uraria logopoiedes*) are always to be used in fresh condition (Murthy 2017b).

If *Candana* is included in a recipe of *cūrṇa*, *taila*, *ghṛta* or *āsava*, only *Śvēta candana* (*Santalum album*) should be used. However, in the preparation of *kvātha* or *lēpa* (pastes), *Rakta candana* (*Pterocarpus santalinus*) is traditionally used. Interestingly, *Śārṅgadhara Samhita* sheds light on the shelf-life of major dosage forms. *Cūrṇa* stays potent for two months, *guḷika* and *avalēha* for one year and *ghṛta* or *taila* for four months. *Āsava* is said to become more potent as they age (Murthy

2017b). The exact procedures of preparing various dosage forms are not explained in the classical Ayurveda texts, as they are passed down from one generation to another via the master–pupil tradition. However, *Śārṅgadhara Samhita* provides clear instructions to prepare *kvātha*, *avalēha*, *ghṛta*, *taila*, *guḷika* and *ariṣṭa- āsava* (Murthy 2017c).

4.3 PRESENT STATE OF QUALITY CONTROL OF AYURVEDIC MEDICINES

Stringent quality control (QC) measures are essential for exporting ayurvedic medicines to Europe and North America. Therefore, the Department of A.Y.U.S.H. (Government of India) assigned Pharmacopeial Laboratory for Indian Medicine, Ghaziabad, to develop protocols for testing ayurvedic products. The laboratory prepared a draft document, circulated it among various experts of the field and, after the consideration of their comments, published it as a volume entitled *Protocol for Testing Ayurvedic, Siddha & Unani Medicines* (Lohar 2008). This work provides analytical specifications of various ayurvedic dosage forms like *kvātha* (decoctions), *cūrṇa* (powders), *taila* (medicated oils), *ghṛta* (medicated clarified butter) and *lēhya* (electuaries). Additionally, methods for testing the parameters are also included.

4.4 AYURVEDIC PHARMACOPOEIA OF INDIA

The Ayurvedic Pharmacopoeia of India is a unique compendium of standards defining the quality, purity and strength of selected medicinal preparations that are manufactured, distributed and sold by licensed manufacturers in India. It was developed in two parts. Part I consists of monographs on medicinal substances of natural origin, and Part II includes selected compound formulations sourced from books listed in the First Schedule of the Drugs and Cosmetics Act, 1940. The first part of *The Ayurvedic Formulary of India* was published in 1978, and, thereafter, *The Ayurvedic Pharmacopoeia of India Part I, Volume I* was published in 1989 (Joshi et al. 2017). So far nine volumes of *The Ayurvedic Pharmacopoeia of India Part I* (Anonymous 2001a, 2001b, 2001c, 2004, 2006a, 2008a, 2000b, 2011a, 2016a) and four volumes of *The Ayurvedic Pharmacopoeia of India Part II* (Anonymous 2007, 2009a, 2010, 2017a) have been published. Three volumes of *The Ayurvedic Formulary of India* are also available (Anonymous 2000, 2003, 2011b). *The Ayurvedic Pharmacopoeia of India Part I* provides for each herb the following pharmacopeial information: Synonyms, names of the herb in regional languages, description, information on powder microscopy, identification by T.L.C., quantitative parameters, limits of contaminants, assay by H.P.L.C., storage conditions, labeling requirements and reference standards; and method of preparation, identification by T.L.C., and assay by H.P.L.C. of hydro-alcoholic and water extracts.

The Ayurvedic Pharmacopoeia of India Part II gives the following pharmacopeial information on medicinal formulations: Definition, composition and method of preparation of the formulation, description, identification by T.L.C., physicochemical parameters, the assay of the major bioactive compound, limits of contaminants, storage conditions, therapeutic uses and dose.

The Ayurvedic Formulary of India provides definitions, textual sources, methods of preparation, general precautions, characterization and preservation of various ayurvedic dosage forms like *āsava*, *ariṣṭa*, *arka*, *avalēha*, *pāka*, *kvātha cūrṇa*, *guggulu*, *taila*, *ghṛta*, *cūrṇa* and so on. Detailed formulae of these medicines are also given. This voluminous work provides lists of single drugs of animal, mineral and herbal origin, with detoxification-cum-potentiation protocols of the toxic ingredients. *The Ayurvedic Pharmacopoeia of India* and *The Ayurvedic Formulary of India* are useful to the Ayurveda industry of India for controlling the quality of crude drugs and finished products.

4.5 FACTORS THAT INFLUENCE THE QUALITY OF AYURVEDIC MEDICINES

These A.Y.U.S.H. specifications are valuable to assess the quality of the final products. However, considering the inherent complexity of naturally growing medicinal herbs, quality assurance of ayurvedic medicines calls for testing of herbal raw materials, their storage and validation of the manufacturing processes similar to those adopted for European herbal medicines (EMA 2011). Characteristics that define the quality of the herbal substance/preparations and herbal medicinal products should be identified. There are several critical aspects that need to be monitored.

4.5.1 Good Agricultural and Collection Practices

The Government of India published *Guidelines on Good Field Collection Practices for Indian Medicinal Plants* in 2009 to address issues associated with harvesting and post-harvest management of medicinal plants collected in the wild. The objective of these guidelines among several others is to ensure consistency in the quality of crude drugs having a bearing on the safety and efficacy of herbal formulations, minimize the direct and indirect negative impact on the environment due to wanton collection from the wild and ensure equitable returns to the herb collectors and other stakeholders (Anonymous 2009b). This document provides guidelines on compliance with regulatory requirements, harvesting of medicinal plant produce, post-harvest management, packing and storage of the produce and guidelines for the collection and post-harvest management of various categories of medicinal plant parts like roots, stem bark, wood, leaves, fruits, gums and so on. Adherence to these guidelines ensures sustainable harvesting of herbs used in Ayurveda.

Similarly, the Government of India has published *Good Agricultural Practices for Medicinal Plants* (Anonymous 2009c). Adapted from the *WHO Guidelines on Good Agricultural and Collection Practices* (G.A.C.P.), this document is intended to suit the policy framework on environment and health in India. It describes soil, climatic conditions, seeds, propagation material, precautions and crop management. Good agricultural practices can alleviate the problem of herb misidentification and ensure that the quality of the starting herbal ingredients is reproducible through good practices for cultivation, harvesting and post-harvesting processes. Contamination and product safety can also be controlled in this way.

4.5.2 Contamination

Several environmental, agricultural, urban and industrial factors, coupled with inadequate harvesting, storage and processing techniques, cause contamination in crude drugs, leading to a compromise in quality and efficacy of the products (Kneifel et al. 2002; Street et al. 2008). Many ayurvedic herbs are collected from wastelands and agricultural fields. Therefore, they may be contaminated with chemical pollutants, pesticide residues, bacteria and fungi. Contaminating micro-organisms and their toxins may lead to diseases, making the herbs hazardous for consumption (Govender et al. 2006; Van Vuuren et al. 2014). Major contributing factors to the growth of microbes are elevated temperatures and moisture (Hell and Mutegi 2011).

4.5.2.1 Contamination with Heavy Metals

Accumulation of heavy metals in medicinal herbs has been reported. Such contamination occurs when herbs grow in heavily polluted soil. Some plants accumulate heavy metals like cadmium, from regular soil (Ernst and Coon 2001; Blagojević et al. 2009; Gasser et al. 2009; Wenzig and Bauer 2009; Kabelitz 1998; Chizzola et al. 2003). However, ayurvedic herbs are relatively free from heavy metal contamination. Rao et al. (2011) examined the heavy metal content of *Emblica officinalis* fruits, *Terminalia chebula* fruits, *Terminalia belerica* fruits and *Withania somnifera* roots. Arsenic, cadmium, lead and mercury were found below detection limits in all the samples.

Sebastian et al. (2015) reported an extensive investigation of the heavy metal content of ayurvedic herbs. They procured samples of 34 ayurvedic herbs from various suppliers in the country. These herbs were collected from four agro-climatic zones. The heavy metal content of the dried and powdered herbs was analyzed using an Agilent 7700X inductively coupled mass spectrometer (ICP-MS) (Agilent Technologies, U.S.A.). Heavy metals were not detected in many of the samples, and the others contained the contaminants well below the limits set by the Government of India (Lohar 2008).

4.5.2.2 Contamination with Micro-Organisms

Harvested ayurvedic crude herbs may inadvertently contain micro-organisms. Herbs tend to have higher microbial contamination than chemically defined active substances, and the microbial population present may differ qualitatively and quantitatively. Therefore, attention should be paid to the microbiological quality of ayurvedic medicines. Some micro-organisms can alter the physicochemical characteristics of the product, affecting the product's quality. Constituents of the crude herb may be metabolized by the micro-organisms, leading to undesirable chemical changes. Micro-organisms may also cause changes in appearance, smell or taste and pH of the medicine. If the pH changes significantly in an ayurvedic *kvātha* containing a chemically ionizable preservative the efficacy of which is pH-dependent, like benzoic acid and sorbic acid, then the efficacy of the preservative may be diminished (Anonymous 2015).

During growth on herbal substrates, some molds produce mycotoxins. Some of these substances like aflatoxins and ochratoxin A can be secondary metabolites, while others like fumonisins may be hydrophilic. Mycotoxins can be formed during cultivation or wild growth of the plant or during the storage of the harvested crude herb. Some ayurvedic herbs like licorice root may be contaminated by ochratoxin A. This toxin is produced by *Aspergillus ochraceus*, *Penicillium verrucosum* and several other species of *Aspergillus* and *Penicillium*. Ochratoxin A is nephrotoxic and carcinogenic (Anonymous 2015).

Microbial contamination originates during the harvesting of the plant material or by the handling of the plant material or during human intervention, equipment, buildings, air ventilation systems and contamination during transportation. Minimizing contamination with micro-organisms and microbial toxins should be ensured ideally by monitoring and limiting these steps rather than by the use of decontamination methods (Anonymous 2015).

4.5.2.3 Contamination with Pesticides

During the different stages of collection or cultivation, storage, transport, distribution and processing of the herbs, pesticides and often fumigating agents are used against pests and unwanted deterioration. Thus, contaminants arising from pesticides or illegal use of unapproved pesticides during cultivation can also pose a threat to health (Yau et al. 2015). In a study that examined 36 crude Chinese herbs procured from large wholesale herb markets in South Korea, many were found to be contaminated with eight pesticides like benzene hexachloride, procymidone and endosulfan (Oh 2007). In another study, harvested wild plants were found to contain significantly higher levels of pesticides than cultivated samples, suggesting that local sources of industrial or agricultural pollution could be important factors contributing to contamination of herbs (Harris et al. 2011).

Rao et al. (2011) investigated the content of pesticide residues in fruits of *Emblica officinalis*, *Terminalia chebula* and *Terminalia belerica* and roots of *Withania somnifera*. Pesticide residues were found below detection limits in all the samples. However, somewhat contradictory results were reported by Rai et al. (2008). They estimated the levels of organochlorine pesticide residues like isoforms of H.C.H., D.D.T., D.D.D., D.D.E. and α-endosulfan, in constituent plants of the famous ayurvedic formula *Daśamūla* (*Aegle marmelos*, *Desmodium gangeticum*, *Gmelina arborea*, *Oroxylum indicum*, *Premna latifolia*, *Solanum indicum*, *Solanum surratense*, *Stereospermum chelonoides*, *Tribulus terrestris* and *Uraria picta*). The organochlorine pesticide residues viz., different metabolites of D.D.T., D.D.E., isomers of H.C.H. and α-endosulfan were checked in a total of 40 samples of single crude drugs. Although α-H.C.H. and γ-H.C.H. were present in almost all the samples, other pesticides were not detected in them. D.D.T. and D.D.E. were found only in two samples. The authors concluded that the estimation of pesticides is essential for crude drugs used in the preparation of ayurvedic medicines. As the study also reports high residual buildup of α-H.C.H. and γ-H.C.H., the authors suggest the judicious use of these pesticides and the choice of pesticides for use in agriculture (Rai et al. 2008). Considering the need to maintain the high quality of ayurvedic medicines, the Government of India has set limits for these contaminants (Table 4.1).

4.5.3 Adulteration and Substitution

With the depletion of medicinal plants in their natural habitat, many herbs used in Ayurveda are becoming rare. Plants like *Coscinium fenestratum*, *Gymnema sylvestre* and many others that were endemic to Indian forests have become endangered or threatened. This has given rise to widespread adulteration. Sometimes herbs are substituted with different parts of the same species. In the majority of cases, stems are used instead of roots. Thus, the roots of *Sida* species are substituted with their stems. The same is the case with *Rubia cordifolia* (Menon 2003). Adulterants of some threatened medicinal plants of Kerala are listed in Table 4.2.

4.5.4 Content of Bioactives

Ayurveda medicines are generally prepared using several crude drugs in combination. The celebrated Ayurveda tonic *Cyavanaprāśa* contains 42 herbal ingredients (Anonymous 1978a). It is well known that the chemical constituents in these herbs vary due to genetic factors, chemical races, hybridization, ecological factors (light, temperature, altitude, latitude, water, soil chemistry), ontogenic stage of harvest, cultivation, post-harvest treatment and storage (Kumar 2016).

Herbs contain numerous phytocompounds. A common herb like turmeric contains around 200 distinct phytocompounds (Balaji and Chempakam 2010). Even though there are major compounds among them, the minor compounds are also essential to bring forth the biological activities attributed to the herb. While reporting the antibacterial activity of *Daucus crinitus*, Bendiabdellah et al. (2013) observed that there is some evidence that minor components in essential oils have a critical

TABLE 4.1
Limits of Contaminants for Ayurveda, Siddha and Unani Products*

Parameter	Specification
Bacterial load	
Total bacterial count	1×10^5 CFU/gm
Yeast and mold	1×10^3 CFU/gm
Escherichia coli	Should be absent
Salmonella sp.	Should be absent
Pseudomonas aeruginosa	Should be absent
Staphylococcus aureus	Should be absent
Pesticide residue	
Organochloro group	Should be less than 1 ppm
Heavy metals	
Lead	10 ppm
Mercury	1 ppm
Arsenic	3 ppm
Cadmium	0.3 ppm
Aflatoxin	B1 – 0.5 ppm
	G1 – 0.5 ppm
	B2 – 0.1 ppm
	G2 – 0.1 ppm

*Adapted from Lohar 2008.

TABLE 4.2
Adulterants of Some Threatened Medicinal Plants of Kerala*

Sl. No.	Threatened Species	Part Used	Adulterant	Substituted Part
1	*Aegle marmelos*	Root	*Atlantia bengalensis*	Root
2	*Canarium strictum*	Gum	*Vateria indica*	Gum
3	*Hemidesmus indicus*	Root	*Decalepis hamiltoni*	Root
4	*Oroxylum malabaricum*	Root	*Ailanthus exelsa*	Root
5	*Gloriosa superba*	Rhizome	*Costus speciosus*	Rhizome
6	*Aquilaria malaccanses*	Heartwood	*Dysoxylum malabaricum*	Heartwood
7	*Piper nigrum*	Fruits	*Lantana camara*	Seeds
8	*Saraca asoka*	Bark	*Polyalthia longifolia*	Bark
9	*Pseudarthria viscida*	Root	*Solanum melongena*	Root
10	*Tribulus terrestris*	Fruit	*Pedalium murex*	Fruit

*Menon 2003.

part to play in antibacterial activity, possibly by producing a synergistic effect between other components, as in the case of sage (Marino et al. 2001), some species of *Thymus* (Lattaoui and Tantaoui-Elaraki 1994) and oregano (Paster et al. 1995).

Synergy is often stated to occur when an extract of a plant gives a greater or safer response than an equivalent dose of the major compound considered to be the "active" one. The improved antispastic effect shown by cannabis extract compared with an equivalent dose of tetrahydrocannabinol is an example of such an action (Baker et al. 2000; Williamson 2001). It may be possible that other undiscovered active compounds may be present. Such compounds present may be active against a range of targets, all of them contributing to the observed effect. This type of actions is called polyvalence, rather than synergy (Houghton 2009).

Synergism is strictly concerned with only one pharmacological function and not a range of activities resulting in a gross effect. Williamson (2001) discussed the role played by polyvalence in the scientific evidence for the claims often made for traditional herbal medicines, noting that the overall effect is greater, and sometimes different than might be predicted from the activities of the individual components (Houghton 2009).

A disease, in most instances, is not due to a single factor. Therefore, a single compound alone cannot successfully treat it, even if the disease and its causative factor are known. This awareness is now encouraging Western medicine to make changes in the approach toward diseases (Walker 2007). As a result, Western medicine now employs a cocktail of drugs against diseases like H.I.V. and cancer (Houghton 2009).

Chemical transformations can take place during the process of preparation of ayurvedic medicines. This possibility is strengthened by evidence from Kampo medicine. Saikosaponins a and d, which are constituents of the *Bupleurum* root, are converted into saikosaponin b during the preparation of the Kampo decoction (Arichi et al. 1979; Yamaji et al. 1984). Similarly, berberine-like alkaloids present in *Coptis* rhizome or bark of *Phellodendrum* and glycyrrhizin present in licorice root cross-react and the new product precipitates (Tomimori and Yoshimoto 1980; Noguchi et al. 1978). A similar precipitation reaction occurs between alkaloids and tannins (Nakajima et al. 1994).

These evidences support the ayurvedic contention that the whole herb and not any single isolated compound from it can cause a therapeutic effect, without provoking newer disorders. Therefore, the estimation of a single compound in any herb cannot be an indicator of its therapeutic activity. Thin-layer chromatography (T.L.C.) can be a useful and inexpensive technique to assess the quality of crude herbs used in the manufacture of ayurvedic medicines.

4.5.5 Stability

Questions on the quality, safety and efficacy (Q.S.E.) of ayurvedic medicines have been raised following their industrial manufacture. The Government of India has also recognized the medicinal use of ayurvedic products which conform to the three attributes of Q.S.E. However, assuring the Q.S.E. of an ayurvedic medicinal product at release does not ensure consistent therapeutic effects and or safety of the product during its shelf-life, as the therapeutic attributes of the product may be altered due to exposure to different environmental conditions during their storage. Therefore, systematic stability studies are to be carried out on all types of ayurvedic products to confirm their Q.S.E. during storage (Bansal et al. 2018).

Stability testing of ayurvedic products is more difficult and challenging when compared to that of a synthetic medicinal product. The active pharmaceutical ingredient of a synthetic product is well-defined qualitatively and quantitatively. Its content is directly related to the therapeutic effectiveness. Therefore, the actives in the formulation function as direct markers for stability testing of the product. Contrary to this, an ayurvedic product is a complex heterogeneous mixture of compounds like alkaloids, terpenoids, organic acids, saponins and so on. All these compounds are also prone to degradation under the influence of various environmental factors like temperature, light, air, moisture and pH. Therefore, it is possible that the contents of these phytocompounds change during its shelf-life, altering the Q.S.E. of the product (Bansal et al. 2018).

Several studies have been reported on the stability of herbal ingredients. For example, andrographolide present in the extract of *Andrographis paniculata* whole herb disintegrates on air oxidation (Garg et al. 2016). Similarly, lipid-soluble fractions like ar-turmerone, turmerone and curlone of dried rhizomes *Curcuma longa* are affected by light (Jain et al. 2007). Such evidence suggests that herbal materials can also undergo degradation. Moreover, the therapeutic actions of an herbal product are believed to be a function of the combined actions of chemically diverse bioactive compounds. Therefore, any change in the content of specific marker(s) during stability testing of an herbal product may not translate to a similar change in its therapeutic efficacy. Thus, monitoring of specific marker(s) during stability testing of herbal products may not form a sound basis for establishing their shelf-life (Bansal et al. 2018).

Stability studies of herbal products are often weakened by factors like the quality of the crude drug material used in the preparation of the product, use of different names of herbs, drying techniques, extraction methodologies, storage conditions, contamination of crude drugs, adulteration, substitution and improper agricultural practices (Bansal et al. 2018).

The selection of markers for assessment of the shelf-life of herbal products is the most challenging task during stability testing. A survey of available stability reports on herbal products reveals that different investigators have used different markers which are likely to disintegrate at higher temperatures for stability testing of herbal products (Livesey et al. 1999; Bernatoniene et al. 2011; Jiang et al. 2013). These reports suggest that there is great variability in susceptibility of different classes of markers to chemical change during shelf-life. Therefore, the phytocompound that is to be used as a marker in the stability study should be chosen judiciously, so as to obtain a reliable estimate of the stability of the product.

The United States' F.D.A. has included biological assay as one of the quality control parameters (Anonymous 2016b). In view of this regulatory recommendation, it is necessary to evaluate the shelf-life of an herbal product vis-à-vis chemical stability and biological activity. But so far only a few studies have been conducted considering both marker compound and biological activity (Srivastava and Gupta 2009; Akowuah and Zhari 2010; Patil et al. 2010; Srivastava et al. 2010). Therefore, shelf-life studies of ayurvedic medicinal products should include the evaluation of physical and chemical stabilities, as well as the intended biological activity using suitable *in vitro* and *in vivo* methods (Bansal et al. 2018).

Considering the importance of stability testing in ensuring the quality of ayurvedic medicines, the Central Council for Research in Ayurvedic Sciences, Ministry of A.Y.U.S.H., Government of India has published *General Guidelines for Drug Development of Ayurvedic Formulations* (Anonymous 2018a). It discusses, among other subjects, stability testing and shelf-life determination of ayurvedic medicines. It provides clear guidelines on the selection of batches, container and closure systems, specifications, testing frequency, storage conditions and evaluation.

Knowing the biologically active constituent in an ayurvedic herb may not be sufficient to predict its efficacy, as ayurvedic formulations are composed of several herbs. In such a situation it is not possible to attribute a particular biological activity to a set of active markers. The W.H.O.'s supplementary guidelines on good manufacturing practices for the manufacture of herbal medicines state that it is often not feasible to determine the stability of each active ingredient. Additionally, as the herbal material in its entirety is regarded as the active ingredient, a mere determination of the stability of the constituents with known therapeutic activity will not usually be sufficient (Anonymous 2006b). Chromatographic fingerprints allow the detection of changes occurring during the storage of a complex mixture of biologically active substances like an ayurvedic medicine. By comparison of appropriate chromatographic fingerprints, the content of bioactives in an ayurvedic medicine can be shown to be within the defined limits (Anonymous 2006b).

4.5.6 Processing Methods

Manufacture of ayurvedic medicines involves various processing methods like washing, drying, pulverizing, boiling in water, boiling together oils and plant juices or decoctions, fermentation and grinding. These processes can affect the phytochemical content of the herb(s). Zainol et al. (2009) studied the effect of different drying methods on the degradation of flavonoids in *Centella asiatica*. Leaves, roots and petioles of the herb were dried using hot air oven, vacuum oven and freeze-dryer. The study revealed the presence of a high concentration of flavonoids like naringin, rutin, quercetin, catechin, luteolin, kaempferol and apigenin in the plant parts of *C. asiatica*. Drying with a hot air oven resulted in the highest degradation of total flavonoids, followed by the vacuum oven and freeze-drying.

The medicinal quality of leaves of *Azadirachta indica* dried under shade, oven-dried at 45°C and at 70°C varied in final moisture content, color, crispness and in their phenolic content. Grinding was directly proportional to the crispness of the dried leaves. Finer particles were obtained from crisper leaves. The phenolics content was higher in powder obtained from shade-dried leaves compared to the oven-dried leaves at 45°C or at 70°C (Sejali and Anuar 2011).

Ayurvedic herbs are pulverized for preparing decoctions (*kvātha*) and medicinal powders (*cūrṇa*). Different comminution processes are adopted for different herbs. Herbs with more stems and stalks need shredding or cutting mills, whereas hammer mills are suitable for hard and brittle materials such as resins. Leafy plants and rhizomes can be pulverized in a two- or three-stage process involving cutting, shredding and pulverization in hammer or pin mills. Pulverization should be carried out with care so that the herbal material is not damaged due to the generation of heat (Anonymous 2001d).

Medicated oils (*taila*) and clarified butters (*ghṛta*) are prepared by boiling oil or clarified butter with plant juices or decoctions. A small quantity of paste of herbs is also added. The mixture is boiled over a slow fire, with frequent stirring. As the water content of the mixture reduces slowly, the oil-insoluble portion of the herbal material starts settling down. It first resembles mud, then turns into a mass resembling soft or hard wax (see Chapter 3). The product is filtered at this stage. Different oils are filtered at different stages and at the hard wax stage, the medicated lipid has a temperature of 105°C. The stage of filtration and the intensity of heating the mixture can affect the content of bioactives in the finished product.

Preparation of *ariṣṭa and āsava* involves fermentation. The purity of raw materials and water, hygienic design of equipment used for production, physical separation of high care areas where critical operations are carried out and in which barriers are raised to prevent the entry of micro-organisms from raw materials, people, air or utensils and effective cleaning and disinfection of equipment and facilities are required to ensure product quality (Baxter and Hughes 2001).

Cooking in fat (frying) and grinding on granite stone are two processes usually employed in the production of *avalēha* (electuaries) and pills (*guḷika*), respectively. For example, production of an *avalēha* like *Cyavanaprāśa* involves cooking of a paste of steamed gooseberry pericarps in clarified butter and sesame oil (Anonymous 1978a). Grinding is invariably adopted in the production of ayurvedic pills (vide Section 4.7.1). The various processing methods are critical to the quality of ayurvedic medicines. Therefore, in-process control tests are to be conducted routinely to monitor and assure consistency in quality of ayurvedic products.

4.6 QUALITY CONTROL METHODS

4.6.1 Macroscopic Evaluation

The first step in ensuring the quality of ayurvedic medicine is ascertaining the identity of the starting herbal raw materials. This is achieved by studying the macromorphology and micromorphology of the herbs. Shape, size, color, texture, surface characteristics, fracture characteristics, appearance of the cut surface, odor and taste of the sample of the herb are studied. Herbarium specimens and descriptions in pharmacopeial monographs are compared with the material (Anonymous 2008a; Carranza-Rojas et al. 2017) (Figure 4.1).

The incoming raw materials should be segregated into one of a group, like powders, woods, barks, leaves, flowers, seeds, fruits, whole plants, rhizomes or roots, and unorganized drugs such as beeswax, guggul gum, olibanum, vegetable oils, agar and opium, which have a generally uniform

FIGURE 4.1 Herbarium sheet image of *Datura stramonium* L. Reproduced with permission from Carranza-Rojas et al. 2017.

structure but are not composed of cells. By studying the pharmacognostical characteristics of the samples, a clear distinction can be made between the powders of minerals, starches and herbs (D'Amelio 1999).

4.6.2 Microscopic Evaluation

Microscopy techniques are invariably used to study the finer aspects of the herbs. Characteristics such as crystals, trichomes, palisade ratios, vessels, vein-islet number, starch granules, fibers and sclerenchyma are relied upon for effective identification of the herb (Figure 4.2). Chemical solutions like chloral hydrate solution, potassium hydroxide solution, ether–alcohol mixture, and a solution of chlorinated soda are used for dissolving the chemical inclusions in cells and isolating the tissue elements so that the botanical identity of the herbal material can be confirmed (D'Amelio 1999). Microscopic analysis is also valuable as an initial tool to detect adulterants and contaminants in crude herbal drugs.

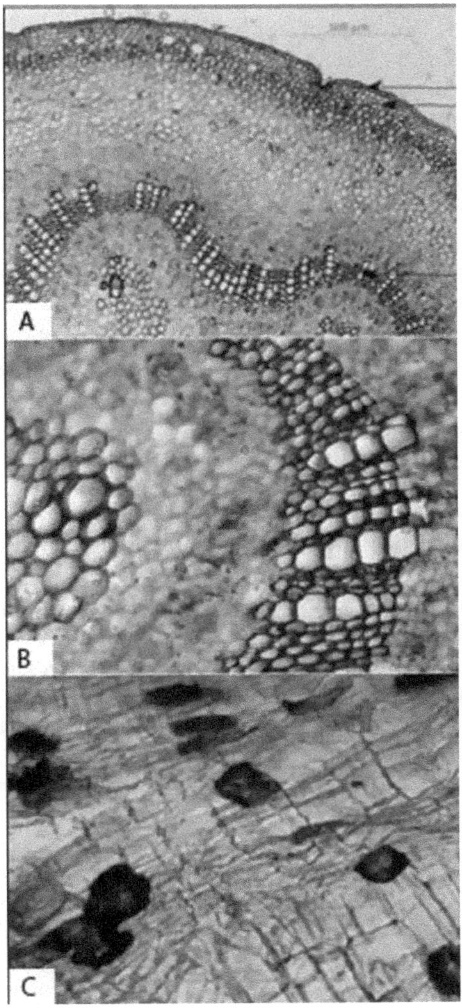

FIGURE 4.2 (a) Transverse section (T.S.) of *Alstonia scholaris* L. R. Br. stem showing wavy vascular bundle at ×10. (b) T.S. of *A. scholaris* stem showing central vascular bundle at ×40. (c) T.S. through the bark of *A. scholaris* showing cellular contents (arrows). Reproduced with permission from Ghosh et al. 2016.

4.6.3 Physicochemical Analysis

Several analytical tests are usually carried out to assess the quality of crude herbs and to detect adulterants and contaminants. They include moisture content, foreign matter, total ash, acid-insoluble ash, alcohol-soluble extractive and water-soluble extractive and T.L.C. profile (Anonymous 2008a). Physicochemical tests are also routinely carried out to evaluate the quality of finished ayurvedic medicines. The tests conducted on various traditional ayurvedic dosage forms are presented in Table 4.3. Descriptions of these test methods are available in Lohar (2008).

4.6.4 Fingerprinting

Chromatographic techniques are the most versatile tools for the analysis of herbs. They can be utilized for identification and authentication, as well as for the determination of various adulterants and for standardization of the product. Unlike macroscopic, microscopic and many molecular biological methods, they are not restricted to the crude herb. They can also be applied to finished ayurvedic preparations (Wenzig and Bauer 2009).

Chromatographic methods can be used for quality control of ayurvedic medicines, based on marker compounds and chromatographic fingerprint analysis. A marker compound can be any characteristic compound of a plant, whereas biomarkers are bioactive plant constituents, representing a plant's pharmacological activity (Techen et al. 2004). Generally, the quality of herbal medicines is judged on the basis of the analysis of one or more bioactive or characteristic compounds (Bauer 1998). This approach is widely used in the herbal medicine industry and also by many pharmacopoeias (Wenzig and Bauer 2009). To make chromatographic fingerprint analysis more meaningful, the suggestion has been made to match the fingerprint profile of an herbal preparation to the results of biological assays or clinical studies (Yuan and Lin 2000). The same approach can be adapted for ayurvedic medicines as well.

A chromatographic fingerprint of herbal medicine is a pattern representing the chemical characteristics of its ingredients. This pattern is obtained by analyzing chromatographically the pharmacologically active or characteristic chemical constituents in the preparation (Xie 2001; Ong 2002). The chromatographic profile should represent the herbal medicine being studied and should facilitate

TABLE 4.3
Tests Conducted on Major Traditional Ayurvedic Dosage Forms*

Decoctions (*kvātha cūrṇa*) and powders (*cūrṇa*)
Loss on drying at 105°C, total ash, acid-insoluble ash, water-soluble extractive, alcohol-soluble extractive, particle size (40–60 mesh for *kvātha cūrṇa*, 80–100 mesh for *cūrṇa*), identification using T.L.C./H.P.T.L.C. – with marker, tests for heavy metals, microbial contamination, tests for pesticide residues, tests for aflatoxins
Medicated oils and clarified butter (*taila, ghṛta*)
Weight/ml (in case of *taila*), the refractive index at 25°C, viscosity, iodine value, saponification value, acid value, peroxide value, identification using T.L.C./H.P.T.L.C. – with marker, tests for heavy metals, microbial contamination, tests for pesticide residues, tests for aflatoxins
Electuaries (*avalēha*)
Loss on drying at 105°C, total ash, acid-insoluble ash, total solids, fat content, reducing sugar, total sugar, identification using T.L.C./H.P.T.L.C. – with marker, tests for heavy metals, microbial contamination, tests for pesticide residues, tests for aflatoxins
Fermented liquids (*āsava, ariṣṭa*)
pH, specific gravity at 25°C, total solids, alcohol content, test for methanol, reducing sugar, non-reducing sugar, total acidity, identification using T.L.C./H.P.T.L.C. – with marker, tests for heavy metals, microbial contamination, tests for pesticide residues

*Lohar 2008.

the identification of similarity or differences between several such profiles. Thus, fingerprints can successfully reveal both "sameness" and "difference" between various samples. Herbal medicines can be accurately authenticated and identified even if the number and/or concentration of chemically characteristic constituents are not very similar in different samples of the medicine. Therefore, fingerprints can be handy tools to evaluate the quality of herbal medicines globally, considering the totality of the known and unknown constituents occurring in them (Montoro et al. 2012). T.L.C., high-performance thin-layer chromatography (H.P.T.L.C.), gas chromatography (G.C.) and high-performance liquid chromatography (H.P.L.C.) methods are commonly used for the development of chromatographic fingerprints of herbal medicines (Fan et al. 2006).

4.6.4.1 T.L.C.

Simplicity, versatility, high speed, specificity, sensitivity and simple sample preparation make T.L.C. an inexpensive and sensitive tool for quality control of herbal medicines. Coupled with image analysis and digital technologies developed in recent years, the evaluation of similarity between different samples is possible. T.L.C. is widely used for initial screening and semi-quantitative evaluation, very often combined with other chromatographic methods (Wenzig and Bauer 2009).

H.P.T.L.C. is an advanced version of T.L.C. The modern H.P.T.L.C. technique, combined with an automated sample application and densitometric scanning, is sensitive, completely reliable and suitable for qualitative and quantitative analysis. H.P.T.L.C. is a valuable tool for reliable identification, as it can generate chromatographic fingerprints that can be visualized and stored as electronic images. The advantages of H.P.T.L.C. include high sample throughput and low cost per analysis. Multiple samples and standards can be separated simultaneously and sample preparation is easier, as the stationary phase is disposable. In H.P.T.L.C. all steps of the T.L.C. process are computer-controlled (Srivastava 2011; Wagner et al. 2011).

4.6.4.2 H.P.L.C.

H.P.L.C. is widely applied for the analysis of herbal medicines because of its high separation capacity. It can be utilized to analyze almost all constituents of herbal products, provided an optimized procedure involving mobile and stationary phase is developed (Li et al. 2005). H.P.L.C. can be coupled to various detection techniques like ultraviolet diode array detection (D.A.D.), mass spectrometry (M.S.) and nuclear magnetic resonance (N.M.R.). These hyphenated techniques provide information on the structure of the compounds present in a chromatogram (Wenzig and Bauer 2009).

4.6.5 Application of Fingerprint Data

The establishment of a characteristic fingerprint chromatogram for the herbal medicine in question is a critical factor in the application of fingerprint data for judging its quality (Wong et al. 2004). Several factors like extraction method, detection instruments and operating conditions influence the generation of a good chromatographic fingerprint with phytoequivalence qualities. Meaningful fingerprints are obtained when these factors are optimized. Only then can the information provided by a chromatographic fingerprint be efficiently evaluated. Chemometric methods are employed to generate and evaluate chromatographic fingerprints (Liang et al. 2004).

The information content of a chromatographic fingerprint can be calculated by means of various approaches. Generally signal intensity, retention time, peak area and or peak height of each independent peak without overlapping are taken into consideration. This necessitates the identification of each single peak and the estimation of the noise and error level of a fingerprint. Calculation of information content becomes very complex when peaks overlap (Gong et al. 2003; Liang et al. 2004).

Shifts in chromatographic retention time interfere with fingerprint analysis. They are caused by successive degradation of the stationary phase, minor changes in the composition of the mobile phase, detector and other instrumental shifts, column overloading or interactions between analytes. To avoid erroneous results, these shifts need to be corrected before the evaluation of similarities and differences between chromatograms (Li et al. 2004a; Liang et al. 2004; Wenzig and Bauer 2009). Peak synchronization is achieved by several methods. A useful method is the addition of an internal standard (Liang et al. 2004). Retention time can be corrected mathematically using local least square analysis or spectral correlative chromatography (Li et al. 2004b; 2004c).

4.6.6 Similarity Evaluation of Fingerprints

A simple method for the evaluation of similarity is the calculation of correlation coefficient or congruence coefficient. This method has been successfully applied to the quality control of Chinese medicines. Yang et al. (2005) used an evaluation software for evaluating the similarity of H.P.L.C.-D.A.D. generated fingerprints of 56 *Hypericum japonicum* samples from six Chinese provinces. They used the correlation coefficient for similarity calculation, with a reference fingerprint representing the median of all chromatograms. The chromatogram with the highest correlation coefficient was selected as an authentic reference fingerprint (Wenzig and Bauer 2009).

4.6.7 Principal-Component Analysis (P.C.A.)

Principal-component analysis (P.C.A.) is also commonly used for comparison of chromatographic fingerprints and detection of variations. In P.C.A. the original multivariate data are projected on a set of orthogonal axes known as principal components, defining a sub-space of the original multivariate data space that maximally describes the variation contained within that multivariate data. The axes are arranged in descending order of the amount of variation in the original data. Each principal component has an associated loading and scores vector (Johnson et al. 2003). By extraction of useful information from object data, P.C.A. is able to construct a theoretical model that has a limited validity of principal components. When a fingerprint shows unexpected properties deviating from those of good major fingerprints or features matching with the theoretical major good fingerprint model, it is excluded from the model and considered to be different (Li et al. 2004b; Gemperline 2006).

4.6.8 Application of Fingerprinting to Quality Control

4.6.8.1 Detection of Adulterants

The fingerprinting technique is being applied to quality control of ayurvedic herbs. The bark of *Aśōka* (*Saraca asoca*) is the major ingredient of ayurvedic medicines indicated in dysfunctional bleeding. However, during a survey of major Indian crude drug markets, it was observed that almost all the samples of *Aśōka* were mixtures of *Saraca asoca*, *Saraca declinata* and *Polyalthia longifolia* (Khatoon et al. 2009). This practice might have originated as *Saraca asoca* is now considered to be an endangered species. Khatoon et al. (2009) collected stem bark of *Saraca asoca*, *Saraca declinata* and *Polyalthia longifolia* and subjected them to H.P.T.L.C. analysis. Using β-sitosterol and stigmasterol as marker compounds, they demonstrated that all three species could be differentiated easily.

4.6.8.2 Authentication of Herbs

Irshad et al. (2016) studied the T.L.C. profiles of commercial samples of *Śāṅkhupuṣpi* for which four different plants like *Convolvulus pluricaulis*, *Clitoria ternatea*, *Evolvulus alsinoides* and *Tephrosia purpurea* are used in different parts of the country. The authors procured commercial samples of *Śāṅkhupuṣpi* from eight herb markets of India viz., Lucknow, Delhi, Varanasi, Hisar, Jalandhar, Dehradun, Mumbai and Jaipur. The samples were subjected to T.L.C. fingerprinting. Caffeic acid, ferulic acid, β-sitosterol and lupeol were used as markers. The results provided interesting information. The Delhi, Jaipur and Hisar market samples comprised only *Convolvulus pluricaulis*. Nevertheless, samples procured from Lucknow and Varanasi were mixtures of *Convolvulus pluricaulis* and *Evolvulus alsinoides*. Sample from Jalandhar seemed to be a mixture of *Convolvulus pluricaulis* and *Tephrosia purpurea*. The Dehradun sample resembled *Evolvulus alsinoides* and *Tephrosia purpurea*. The Mumbai sample was entirely different from the other seven samples. This study shows that T.L.C. fingerprinting can be successfully applied to the quality control of ayurvedic herbs.

4.6.9 Quality Control of Finished Formulations

Terminalia arjuna is recommended in Ayurveda for the treatment of heart disease (Kumar and Prabhakar 1987). Recent studies report the efficacy of *Terminalia arjuna* in treating cardiac conditions such as anginal pain, palpitation, hypertension and ischemic heart disease (Dwivedi and Agarwal 1994; Bharani et al. 1995; Dwivedi and Jauhari 1997). On account of the growing interest in this herb and to satisfy the need for a quality control tool, Chitlange et al. (2009) compared the H.P.L.C. chromatograms of authentic *Arjuna cūrṇa* and three marketed formulations. The analysis of the retention time values and U.V. data revealed eight characteristic peaks in the chromatograms, which unambiguously confirmed the presence of the authentic crude herb used in the product (Figure 4.3). The fingerprints of six common peaks observed in chromatograms of sapogenins in the standardized formulation and marketed formulations could serve as a quality control parameter for the *Arjuna cūrṇa* formulation (Figure 4.4).

Pathyāṣaḍaṅgam kvātha is a classical ayurvedic medicine used in the treatment of cluster headache, migraine, upper respiratory diseases, earache and night blindness (Murthy 2017d). As there is a lack of information on the quality control of this medicine, Abraham et al. (2018) attempted H.P.T.L.C. fingerprinting of this medicine. Three batches of the *kvātha* were prepared according to standard procedures. H.P.T.L.C. analysis revealed that a mobile system consisting of toluene: ethyl acetate: formic acid (2.5: 2.0: 0.5) was suitable for the characterization of the *kvātha*. The analysis also generated an H.P.T.L.C. fingerprint with similarity in number, R_f, intensity and color of bands, indicating that the bioactives present in all the three batches were similar, thus helping to establish batch-to-batch consistency of the formulation (Figures 4.5 and 4.6). H.P.L.C. analysis of methanol extracts of three batches of *Pathyāṣaḍaṅgam kvātha* and andrographis *kvātha* was carried out along with andrographolide standard. H.P.L.C. analysis furnished a fingerprint with similar characteristics and showed that andrographolide is a suitable marker for standardization of the *kvātha*.

Sulaiman and Balachandran (2015) carried out the chemical profiling of *Amṛtōttaram kvātha* using liquid chromatography coupled with electrospray ionization mass spectrometry. *Amṛtōttaram kvātha* is an important ayurvedic formulation prepared using *Tinospora cordifolia* (stem), *Terminalia chebula* (pericarp) and *Zingiber officinale* (dry rhizome). It is the remedy of choice for fever. Lyophilized *Amṛtōttaram kvātha* was dissolved in deionized water and subjected to H.P.L.C. profiling using a H.P.L.C. system equipped with a photodiode array detector. The photodiode array chromatogram showed 11 major peaks. The unique H.P.L.C. fingerprint developed can be used as an analytical tool to ensure the quality of *Amṛtōttaram kvātha*.

4.7 SOME PROBLEMS RELATED TO QUALITY CONTROL

4.7.1 Heavy Metals in Ayurvedic Pills

Guḷika (pills) are an important dosage form used in Ayurveda. Nearly 30 *guḷika* are used in contemporary ayurvedic practice (Anonymous 2017b). Ayurvedic pills are prepared using a unique technology. Powders of herbs are mixed with various fluids like water, decoctions or juices of herbs, lemon juice, juice of pomegranate flowers or fruits, the latex of *Jatropha curcas*, cow milk, breast

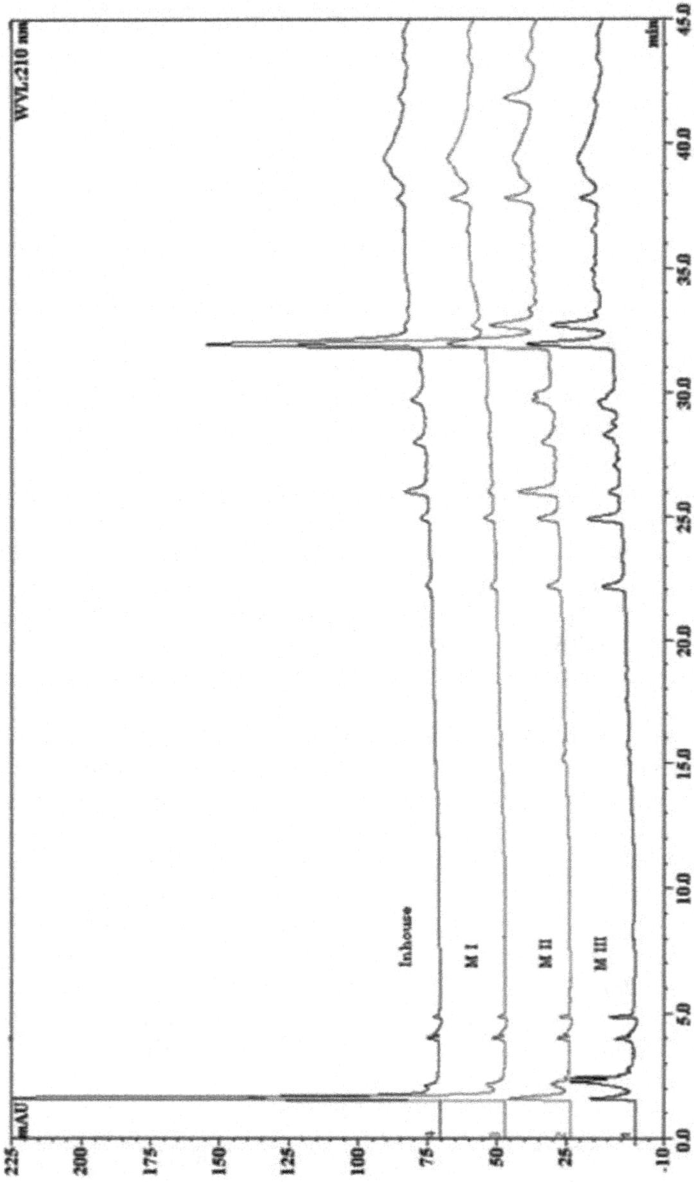

FIGURE 4.3 Overlay of chromatograms of in-house formulation and marketed formulations of *Arjuna cūrṇa*. Reproduced with permission from Chitlange et al. 2009.

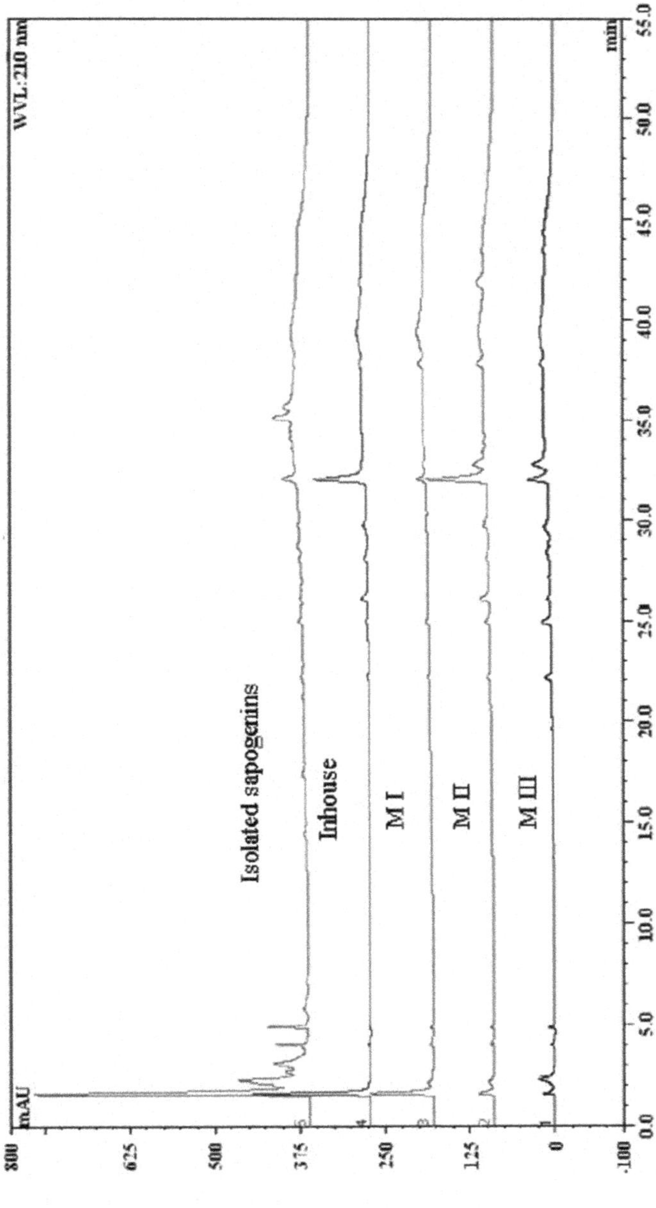

FIGURE 4.4 Overlay of chromatograms of isolated sapogenins, standardized in-house formulation and marketed formulations. Reproduced with permission from Chitlange et al. 2009.

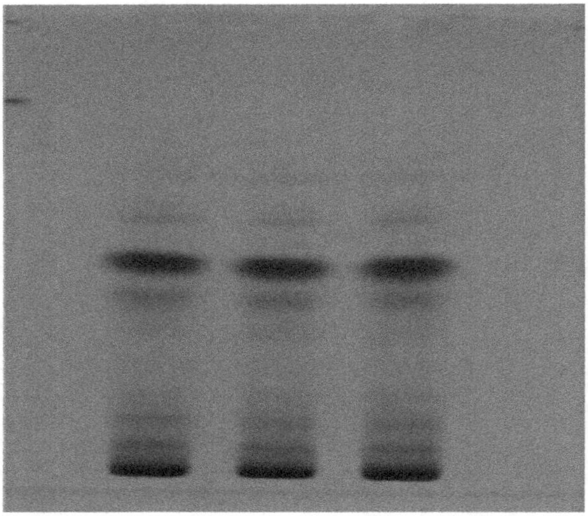

FIGURE 4.5 H.P.T.L.C. plate of three batches of *Pathyāṣaḍaṅgam kvātha* under U.V. 254 nm. Reproduced with permission from Abraham et al. 2018.

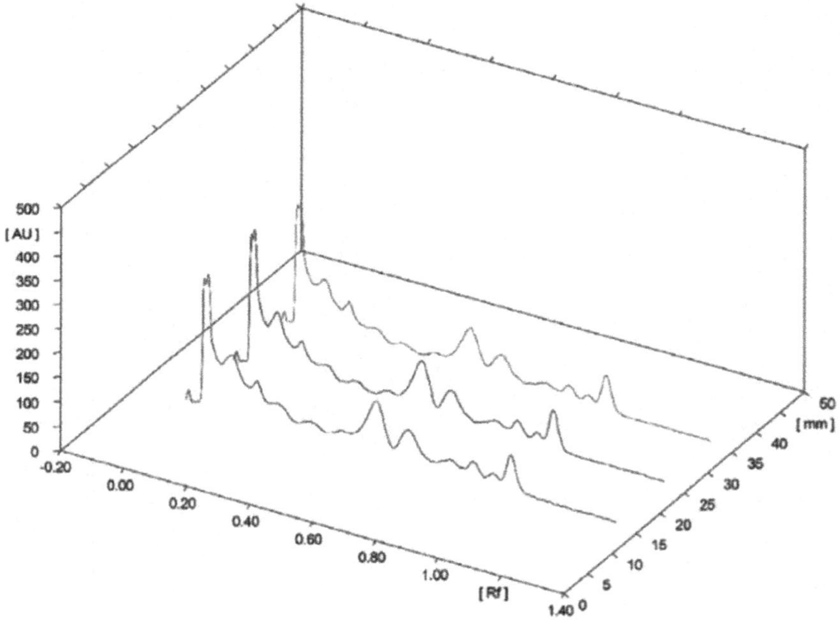

FIGURE 4.6 H.P.T.L.C. plate of three batches of *Pathyāṣaḍaṅgam kvātha* under 366 nm. Reproduced with permission from Abraham et al. 2018.

milk, cow urine, goat urine, castor oil, honey, clarified butter or sesame oil. The mass is ground on granite stone slabs using granite pestles. The grinding is carried out continuously for long periods, sometimes lasting even seven to 18 days (Vaidyan and Pillai 1985). Thereafter, the groundmass is rolled into pills of various sizes. Chemical transformations are bound to occur under such conditions, and these merit serious study.

Sebastian et al. (2015) carried out an extensive study of the heavy metal content of traditional ayurvedic medicines. The content of lead, arsenic, cadmium and mercury in 126 ayurvedic medicines manufactured by 32 companies was analyzed using inductively coupled plasma mass spectrometry (I.C.P.-M.S.). All the samples of *kvātha*, *cūrṇa*, *taila*, *ghṛta*, *lēhyam*, *āsava*, *ariṣṭa*, *drāvakam* and *lēpa* analyzed in the study conformed to the limits for heavy metals set by the Government of India. However, unlike the results reported by Sebastian et al. (2015), many traditional ayurvedic pills were reported to contain high levels of heavy metals (Thomas and Kumar 2018).

A probable source of heavy metal contamination in ayurvedic pills may be the traditional method of preparing them. Granite is basically an igneous rock and contains mercury. Its mercury content is nearly twice that of crustal material (McElroy 2010). Granite also contains, in addition to other heavy metals, 0.2–13.8 ppm of arsenic, 0.003–0.18 ppm of cadmium and 6–30 ppm of lead (Cannon et al. 1978; Nagajyoti et al. 2010). It needs to be ascertained whether such a grinding method influences the heavy metal content of ayurvedic pills.

4.7.2 Souring of Ariṣṭa and Āsava

Ayurvedic medicine manufacturers often complain that *ariṣṭa* and *āsava* turn sour. Uncontrolled fermentation often results in acidity and off-flavors in the product. As the manufacture of *ariṣṭa* and *āsava* is very similar to that of making wine and other fermented beverages, the Ayurveda industry can learn from brewery technology. Breweries adopt various strategies to control the factors that cause spoilage.

4.7.2.1 Quality of Water

Water is the principal ingredient used in breweries. Water contributing directly as an ingredient of brewed products is called brewing water. Brewing water is treated or adjusted to achieve the correct composition relevant to the product being brewed. There are many contaminating micro-organisms present in water. Therefore, usually a few key indicators of the presence of both harmful and harmless species are selected for analysis. Routinely, water is examined for *Escherichia coli* as an indicator of fecal contamination. If *E. coli* is present, it can be assumed that there is a possibility of other pathogenic organisms being present. Only water conforming to microbiological and chemical specification is to be used in brewing (Taylor 2006).

4.7.2.2 Quality of Herbs

Vermouth is an aromatized and fortified wine popular in Russia, Poland and other European countries. It is basically white wine fortified and infused with a proprietary recipe of different herbs, barks, seeds and fruit peels. Vermouth is very similar to ayurvedic *āsava*. Great care is taken in handling the herbs and spices, as their quality is affected by the harvesting and storage conditions. It is reported that the longer the dried products are stored before use, the poorer will be their flavor. Therefore, the manufacturers of vermouth ensure that the dried herbs and spices are used as fresh as possible (Panesar et al. 2009). A similar approach can be adopted in the manufacture of ayurvedic *ariṣṭa* and *āsava* as well. The quality of these products can be improved by using good quality herbs and jaggery devoid of microbial and chemical contamination.

4.7.2.3 Use of Appropriate Brewing Vessel

Micro-organisms attach to surfaces and develop biofilms (Donlan 2002). Biofilms resist sanitization during the washing process and allow bacteria to spread across the product (Srey et al. 2013). This is one of the major causes of spoilage of *ariṣṭa* and *āsava*. The threat of biofilms is apparent when wooden vats are used in the fermentation of *ariṣṭa* and *āsava*. It is difficult to disinfect wooden vats. Stainless steel tanks, on the other hand, are more hygienic and easier to clean. Moreover, biofilm production on stainless steel is less intense (Storgårds 2000). Wooden vats require seasoning, and even after seasoning there are chances of evaporation and diffusion of the fermenting fluid through the minute pores of wooden vats, affecting the total yield of the product. Moreover, there is the possibility of the entry of free oxygen into the fermenting medium through the minute pores of wooden vats, as a result of which alcohol may be converted into acetic acid by oxidation. Such *ariṣṭa* and *āsava* become too sour in taste and become unfit for therapeutic purposes (Hiremath and Joshi 1991). Therefore, stainless steel brewing tanks need to be used instead of wooden vats.

4.7.2.4 Hygienic Design of Brewing Area

High care areas where critical operations are carried out are to be separated physically. Barriers need to be raised to prevent the entry of micro-organisms from raw materials, people, air or utensils (Walter 2006). Contamination during the brewing process is one of the most prevalent problems, and, therefore, the building should be planned and constructed meticulously.

4.7.2.5 Disinfection

Fermentation of *ariṣṭa* and *āsava* is a very sensitive process. Therefore, the fermentation room, brewing tanks and bottling equipment should be disinfected periodically. The aim of disinfection is to reduce the surface population of viable micro-organisms after cleaning and to prevent microbial growth on surfaces during the idle time (Storgãrds. 2000).

The brewing tanks, walls and floors should be cleaned thoroughly and disinfected before the start of operations. The common disinfectants used in breweries are hydrogen peroxide-peracetic acid, acid iodophores, quaternary ammonium compounds and halogenated carbonic acid (Storgãrds 2000). Manzano et al. (2011) provided an example of a cleaning and disinfection protocol. Fermentation vessels were rinsed with water at 75°C and cleaned with a 6.25% basic detergent solution at 85°C for 20 minutes. After cleaning, the tank was rinsed with cold water, cleaned with a 6.25% acid detergent solution in cold water for 20 minutes and then rinsed with cold water. Microbial contamination of *ariṣṭa* and *āsava* can be considerably reduced by this approach.

4.7.2.6 Use of Yeast

Ayurvedic *ariṣṭa* and *āsava* are brewed traditionally using flowers of *Woodfordia fruticosa* as a source of inoculum. Kroes et al. (1993) investigated the significance of *W. fruticosa* flowers in the preparation of *Nimbāriṣṭa*. They noted that the flowers *per se* are not the source of alcohol-producing micro-organisms. But the flowers show considerable invertase activity, and this explains the alcohol content of this *ariṣṭa*. However, the use of *W. fruticosa* flowers as a source of inoculum gives varying results. A study of the alcohol content of more than 500 commercial samples of ayurvedic *ariṣṭa* and *āsava* showed that their alcohol content varied from 2% to 12.88% (Figure 4.7) (Thomas et al. 2016). According to *The Ayurvedic Pharmacopoeia of India*, *āsava* and *ariṣṭa* should contain not less than 5% and not more than 10% alcohol (Anonymous 2008d). To maintain the required amount of alcohol, it is better to adopt fermentation technology employed in the making of wine, including the use of yeast to cause fermentation.

4.7.2.7 Temperature

Temperature has a general accelerating effect on fermentation. At higher temperatures (35°C) the carbon dioxide dissolved in the fermentation medium gets converted to compounds like fusel oils (isoamyl alcohol and phenyl ethanol) and higher alcohols like 1-propanol, 1-butanol, 2-butanol, isobutanol, isoamyl alcohol and 1-hexanol which cause liver disease (Lachenmeier et al. 2008). The maximum specific growth rate of yeast is known to be at 25°C (Torija et al. 2003; Şener et al. 2007). Therefore, it is desirable to keep fermentation temperatures of *ariṣṭa* and *āsava* around 25°C.

4.7.2.8 Improvised Fermentation

For maintaining the high quality of the *ariṣṭa* or *āsava*, all the ingredients used in their production should be subjected to physicochemical and microbiological analysis. Only ingredients that conform to the specification should be accepted for production. Deionized water alone is to be used in production. Instead of wooden vats, it is better to use stainless steel fermentation tanks.

2.00	2.50	2.92					
3.00	3.05	3.23					
4.02	4.42	4.82					
5.00	5.20	5.36	5.60	5.88			
5.00	5.20	5.36	5.60	5.88			
6.00	6.10	6.20	6.22	6.25	6.40	6.58	6.96
7.20	7.24	7.25	7.48	7.60	7.62	7.76	
8.00	8.14	8.18	8.40	8.84	8.99		
9.00	9.10	9.12	9.32	9.40	9.50	9.80	
10.00	10.24	10.40	10.80				
11.08	11.60	11.68	11.88				
12.00	12.42	12.80	12.88				

FIGURE 4.7 Percentage of alcohol in commercial samples of *āsava* and *ariṣṭa*. Values in shaded areas conform to the pharmacopoeia range of alcohol. Reproduced from Thomas et al. 2016.

The fermentation tank should be housed in an aseptic, closed room maintained at a temperature of 25–30°C and relative humidity of 40–45% (Thomas et al. 2016).

The decoction of herbs or juices of herbs or fruits is to be prepared as instructed in the Sanskrit source of the formula. The specific gravity of the decoction or juice should be determined. Jaggery or honey is to be added to the decoction or juice. The specific gravity of water is 1, and when herbs are boiled in water, the resultant *kvātha* will have a slightly higher specific gravity. For example, *Vara Asanādi kvātha* has a specific gravity of 1.0185 (Ramachandra et al. 2012). When sugar is added to this *kvātha*, the specific gravity of the *kvātha* will increase by factor Y. Then specific gravity of the *kvātha* will be X + Y, where X is the specific gravity of *kvātha*. The required range of specific gravity, in respect to water, to yield 5–11% alcohol is 1 + 0.0388 (Ya) to 1 + 0.0799 (Yb). So increase in specific gravity is needed to get the desired percentage of alcohol = (X + Ya) to (X + Yb). An increase in 0.1° Brix increases specific gravity by 0.0004 and alcohol yield by 0.1% (Thomas et al. 2016).

Degrees Brix is the content of sugar in an aqueous solution of sugar. 1 g of sucrose dissolved in 100 g of water yields 1 Brix. Brix value is determined using a hydrometer or refractometer (Ribéreau-Gayon et al. 2006). Following the Brix determination, the acidity of the fermentation medium is to be checked and adjusted to the desired value, as described above. As fermentation is to begin, a baseline analysis for reducing sugar, titratable acidity, pH, acetic acid and volatile acidity is to be conducted. This will help to track the progress of fermentation (Thomas et al. 2016).

Dried activated yeast (D.A.Y.) is used to cause fermentation of the medium. D.A.Y. is added to warm water (35–40°C), suspended well and added to the fermentation medium. A dose of 10–20 g of D.A.Y./100 liters is generally recommended (Ribéreau-Gayon et al. 2006). The fermentation tank is to be left undisturbed for a specified period of time (Thomas et al. 2016).

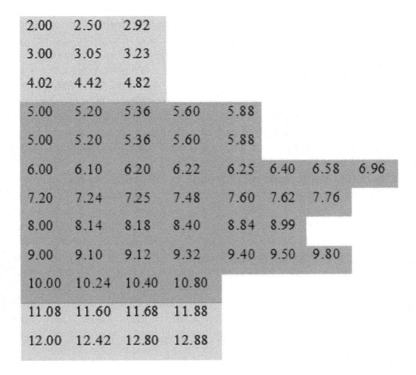

Several methods are used to monitor fermentation. Important among them is the counting of yeasts. After diluting the fermentation medium, the total number of yeast cells is counted under the microscope using a Malassez cell. The total cells enumerated in this way include both "dead" yeast and "live" yeast. The counting of "viable cells" is a better way to differentiate between the "dead" and "live" cells. When the diluted fermentation medium is placed in a solid nutritive medium, the viable yeast cells form a microscopic cluster. The number of viable yeast cells is enumerated by counting the colonies formed on this medium after nearly four days (Ribéreau-Gayon et al. 2006).

The rate of fermentation can also be tracked by estimating the amount of sugar consumed or the alcohol generated. As a relationship exists between the amount of alcohol generated and the initial concentration of sugar in the fermentation medium, the mass per unit volume (density) can give directly an approximate potential alcohol. The density and potential alcohol are usually marked on the stem of the hydrometer (Ribéreau-Gayon et al. 2006).

The monitoring of tank temperature daily during fermentation is indispensable to keeping a tab on fermentation. The tank temperatures are different in different areas of the tank. The temperature is highest in the cap and lowest at the bottom of the tank. The temperature is usually taken after a pumping-over, which agitates the fermenting medium and homogenizes the tank temperature. Average tank temperature can be measured in this way (Ribéreau-Gayon et al. 2006).

Cessation of fermentation is confirmed by the exact analysis of residual reducing sugar. Volatile acidity, acetic acid, titratable acidity and pH are measured. In addition to these tests, the *ariṣṭa* or *āsava* may be tested for the following parameters, like specific gravity at 20°C, total solids, alcohol content, methanol content, non-reducing sugar, T.L.C./H.P.T.L.C. identification, heavy metals, microbial contamination, specific pathogens and pesticide residues (Lohar 2008). A flowchart of the proposed protocol for the brewing of *āsava* and *ariṣṭa* according to principles of modern brewery technology is presented in Figure 4.8.

By using yeast, the alcohol content of these products can be maintained within the legal limits. Alcohol acts as a preservative and extends the shelf-life of these *ariṣṭa* and *āsava*. Yeast-aided fermentation also extracts a wide range of compounds from the herbs, as the menstruum undergoes a gradient of rising alcohol levels (Hiremath and Joshi 1991).

A gas chromatographic study of 12 brands of *Aśvagandhāriṣṭam* sold in Sri Lanka revealed that in addition to ethanol, the products contained methanol, n-propanol, iso butanol, amyl alcohol and

Chemical and Microbiological Analysis of Ingredients
(Ash value, extractive value, content of actives, microbial contamination, heavy metals, aflatoxins and pesticide residues)
↓
Preparation of Decocition/Extraction of Juice
(Specific gravity)
↓
Addition of Jaggery or Honey
(Degrees Brix)
↓
Baseline Analysis
(Reducing sugar, titratable acidity, pH, acetic acid and volatile acidity)
↓
Initiation of Fermentation
(Inoculation of Yeast)
↓
Tracking of Fermentation
(Counting yeasts, measurement of sugar and alcohol and monitoring temperature)
↓
Confirmation of Cessation of Temperature
(Residual reducing sugar)
↓
Analysis of Finished Product
(Titeratable acidity, pH, acetic acid, volatile acidity, alcohol, methanol, specific gravity, total solids, non-reducing sugar, TLC/HPLC, heavy metals, microbial contamination, and pesticide residues)
↓
Bottling of Finishing Product

FIGURE 4.8 Proposed protocol for brewing of *āsava* and *ariṣṭa* according to the principles of modern brewery technology. Reproduced from Thomas et al. 2016.

isopropyl alcohol (Weerasooriya and Liyanage 2006). The manufacturing technology of *ariṣṭa* and *āsava*, therefore, needs to be evaluated.

4.7.3 Irrational Substitution

Some instances of intentional addition of heavy metals to herbal formulae are also now known. An example is the addition of lead sulfide to *Puṣyānuga Cūrṇam*, indicated in dysfunctional uterine bleeding, piles, diarrhea and bloody stools. According to the formula described in *Caraka Samhita*, equal quantities of 25 herbs, including *Arjuna* (*Terminalia arjuna*) are to be powdered, mixed with honey and consumed (Sharma 1994a). The same formula is described in *Cakradattam* also (Sharma 1994b). In agreement with classical texts, the formula given in *The Ayurvedic Formulary of India* also recommends *Arjuna* (Anonymous 1978b). However, due to unknown reasons, several manufacturers in Kerala continue to substitute the bark of *Arjuna* with the similar-sounding mineral *añjana* (lead sulfide) (Anonymous 2018b). A recent analysis of ten market samples of *Puṣyānuga Cūrṇam* revealed that six of them contained 4% of lead sulfide (Dr N.P. Damodaran, International Institute of Ayurveda, Coimbatore, personal communication). In light of the toxicity of lead (Pearce 2007; Flora et al. 2012), such irrational substitutions should be prohibited.

4.8 CONCLUSION

In ancient times, physicians having comprehensive knowledge of *bhaiṣajya kalpana* (ayurvedic pharmacy) used to prepare the medicines themselves to treat their patients. Therefore, there was no difficulty in obtaining authentic medicines with desired therapeutic properties. However, in modern times, industrialization has compelled physicians to depend on commercially manufactured ayurvedic medicines. Thus, now there is the need for standardization of these medicinal preparations. The quality of ayurvedic medicines can be improved by utilizing herbs collected at appropriate seasons and stored under specified conditions. Certain toxic herbs like *Plumbago zeylanica* (root) and *Semecarpus anacardium* (seed) need to be subjected to detoxification-cum-potentiation processes (*śōdhana*), so that the appearance of undesirable side-effects can be prevented. Good manufacturing practices and standard processes are to be adopted in the manufacture of medicines, a responsibility that should rest with qualified and experienced personnel. The manufactured products should be packed and stored under specific conditions. The quality of crude drugs and finished products needs to be established using physicochemical tests and fingerprinting techniques.

An ayurvedic medicine is essentially a mixture of chemical compounds, and its quality can be assessed initially by the T.L.C. technique, using purified bioactive compounds as reference standards. Hiremath et al. (1993) demonstrated the suitability of applying this technique to the standardization of ayurvedic medicines. They standardized *Kuṭajāriṣṭa* using T.L.C., with conessine as a reference standard. T.L.C. fingerprinting of an ayurvedic product can be followed by H.P.L.C. fingerprinting. There is growing interest in the application of fingerprinting technology in assessing the quality of ayurvedic medicines. Garg et al. (2013) reviewed the published literature on H.P.T.L.C. and H.P.L.C. fingerprinting in Ayurveda. Most of the Ayurveda medicine manufacturing companies in India are small- to medium-sized business, and these two analytical technologies are quite suitable for their modest resources.

The safety of medicinal products is a major concern in modern healthcare. While it is popularly believed that ayurvedic medicines are harmless, the reality may be different. Most of the ayurvedic physicians believe that ayurvedic medicines can never evoke undesirable side-effects. It is stated in *Caraka Samhita* that even a strong poison can become an excellent medicine if administered well. However, even the most useful medicine can work like a poison, if handled carelessly (Thatte and Bhalerao 2008). The appearance of undesirable side-effects of ayurvedic medicines can be prevented by strengthening quality control in the ayurvedic medicine manufacturing sector and preventing the adulteration of such products. Administration of these medicines, according to the principles of Ayurveda, is also equally important. Quality control and good diagnostics should, therefore, go hand-in-hand.

REFERENCES

Abraham, A., S. Samuel, and L. Mathew. 2018. Phytochemical analysis of *Pathyashadangam kwath* and its standardization by HPLC and HPTLC. *Journal of Ayurveda and Integrative Medicine*. doi:10.1016/j.jaim.2017.10.011.

Akowuah, G. A., and I. Zhari. 2010. Effect of extraction temperature on stability of major polyphenols and antioxidant activity of *Orthosiphon stamineus* leaf. *Journal of Herbs, Spices and Medicinal Plants* 16, no. 3–4: 160–66.

Anonymous. 1978a. *Avalēha* and *Pāka*. In *The ayurvedic formulary of India, part 1, first edition*, 23–39. New Delhi: Ministry of Health and Family Planning.

Anonymous. 1978b. Cūrṇa. In *The ayurvedic formulary of India, part 1, first edition*, 83–95. New Delhi: Ministry of Health and Family Planning.

Anonymous. 2000. *The ayurvedic formulary of India part II*, 1–425. New Delhi: Ministry of AYUSH.

Anonymous. 2001a. *The ayurvedic pharmacopoeia of India part I, volume I*, 1–220. New Delhi Ministry of AYUSH.

Anonymous. 2001b. *The ayurvedic pharmacopoeia of India part I, volume II*, 1–249. New Delhi: Ministry of AYUSH.

Anonymous. 2001c. *The ayurvedic pharmacopoeia of India part I, volume III*, 1–312. New Delhi: Ministry of AYUSH.

Anonymous. 2001d. Processing of medicinal plants- Grinding, sieving and comminution. In *A guide to the European market for medicinal plants and extracts, Volumes 60–61*. London: Commonwealth Secretariat.

Anonymous. 2003. *The ayurvedic formulary of India part I (2nd revised edition)*, 1–488. New Delhi: Ministry of AYUSH.

Anonymous. 2004. *The ayurvedic pharmacopoeia of India part I, volume IV*, 1–215. New Delhi: Ministry of AYUSH.

Anonymous. 2006a. *The ayurvedic pharmacopoeia of India part I, volume V*, 1–290. New Delhi: Ministry of AYUSH.

Anonymous. 2006b. *Supplementary guidelines on good manufacturing practices for the manufacture of herbal medicines. World Health Organization technical report series, no. 937.* 85–106. Geneva: World Health Organization.

Anonymous. 2007. *The ayurvedic pharmacopoeia of India part II, volume I (Formulations)*, 1–280. New Delhi: Ministry of AYUSH.

Anonymous. 2008a. *The ayurvedic pharmacopoeia of India part I, volume VI*, 1–492. New Delhi: Ministry of AYUSH.

Anonymous. 2008b. *The ayurvedic pharmacopoeia of India part I, volume VII*, 1–157. New Delhi: Ministry of AYUSH.

Anonymous. 2008c. *Āsava* and *Ariṣṭa*. In *The ayurvedic pharmacopoeia of India Part II (Formulations)*, 1st edition, Volume II, 1–93. New Delhi: Ministry of AYUSH.

Anonymous. 2009a. *The ayurvedic pharmacopoeia of India part II, Volume II (Formulation)*, 1–322. New Delhi: Ministry of AYUSH.

Anonymous. 2009b. *Guidelines on good field collection practices for Indian medicinal plants*, 1–27. New Delhi: N.M.P.B.- Department of AYUSH.

Anonymous. 2009c. *Good agricultural practices for medicinal plants*, 1–20. New Delhi: N.M.P.B.- Department of AYUSH.

Anonymous. 2010. *The ayurvedic pharmacopoeia of India part II, volume III (Formulations)*, 1–291. New Delhi: Ministry of AYUSH.

Anonymous. 2011a. *The ayurvedic pharmacopoeia of India part I, volume VIII*, 1–250. New Delhi: Ministry of AYUSH.

Anonymous. 2011b. *The ayurvedic formulary of India part III*, 1–706. New Delhi: Ministry of AYUSH.

Anonymous. 2015. *Reflection paper on microbiological aspects of herbal medicinal products and traditional herbal medicinal products*, 1–22. London: European Medicines Agency.

Anonymous. 2016a. *The ayurvedic pharmacopoeia of India part I, Volume IX*, 1–146. New Delhi: Ministry of AYUSH.

Anonymous. 2016b. *Botanical drug development guidance for industry*, 1–30. MD: U.S. Department of Health and Human Services Food and Drug Administration Center for Drug Evaluation and Research (CDER).

Anonymous. 2017a. *The ayurvedic pharmacopoeia of India Part II, Vol. IV* (Formulations), 1–189. New Delhi: Ministry of AYUSH.

Anonymous. 2017b. Power of nature transcended to products. http://vaidyaratnammooss.com/pages.php?menu_id=2&submenu_id=34 (accessed November 9, 2017).

Anonymous. 2018a. *General guidelines for drug development of ayurvedic formulations*, 1–100. New Delhi: Ministry of AYUSH.

Anonymous. 2018b. http://www.aryavaidyasala.com/products.php?cate=4&cate_product=pushyanuga (accessed May 28, 2018).

Arichi, S., T. Tani, and M. Kubo. 1979. Studies on *Bupleuri radix* and Saikosaponin. (3) quantitative analysis of Saikosaponin of commercial *Bupleuri radix*. *Medical Journal of Kinki University* 4: 235–41.

Baker, D., G. Pryce, J. L. Croxford et al. 2000. Cannabinoids control spasticity and tremor in an animal model of multiple sclerosis. *Nature* 404, no. 6773: 84–7.

Balaji, S., and B. Chempakam. 2010. Toxicity prediction of compounds from turmeric (*Curcuma longa* L). *Food and Chemical Toxicology* 48, no. 10: 2951–9.

Bansal, G., J. Kaur, N. Suthar, S. Kaur, and R. S. Negi. 2018. Stability testing issues and test parameters for herbal medicinal products. In *Methods for stability testing of pharmaceuticals*, ed. S. Bajaj, and S. Singh, 307–33. New York: Humana Press.

Bauer, R. 1998. Quality criteria and standardization of phytopharmaceuticals: Can acceptable drug standards be achieved? *Drug Information Journal* 32, no. 1: 101–10.

Baxter, E. D., and P. S. Hughes. 2001. Assuring the safety of beer. In *Beer: Quality, safety and nutritional aspects*, 120–35. Cambridge: The Royal Society of Chemistry.

Bendiabdellah, A., M. E. A. Dib, N. Meliani et al. 2013. Antibacterial activity of *Daucus crinitus* essential oils along the vegetative life of the plant. *Journal of Chemistry* 2013. doi:10.1155/2013/149502.

Bernatoniene, J., R. Masteikova, J. Davalgiene et al. 2011. Topical application of *Calendula officinalis* (L.): Formulation and evaluation of hydrophilic cream with antioxidant activity. *Journal of Medicinal Plant Research* 5: 868–77.

Besterfield, D. H. 2013. Introduction to quality improvement. In *Quality improvement*, 9th edition, Volumes 1–4. NJ: Pearson Education, Inc.

Bharani, A., A. Ganguly, and K. Bhargava. 1995. Salutary effect of *Terminalia arjuna* in patients with severe refractory heart failure. *International Journal of Cardiology* 49, no. 3: 191–9.

Blagojević, N., B. Damjanović-Vratnica, V. Vukašinović-Pešić, and D. Đurović. 2009. Heavy metals content in leaves and extracts of wild-Growing *Salvia officinalis* from Montenegro. *Polish Journal of Environmental Studies* 18: 167–73.

Cannon, H. L., G. G. Connally, J. B. Epstein, J. G. Parker, I. Thornton, and G. Wixson. 1978. Rocks: Geological sources of most trace elements. *Geochemistry and Environment* 3: 17–31.

Carranza-Rojas, J., H. Goeau, P. Bonnet, E. Mata-Montero, and A. Joly. 2017. Going deeper in the automated identification of herbarium specimens. *BMC Evolutionary Biology* 17, no. 1: 181. doi:10.1186/s12862-017-1014-z.

Chitlange, S. S., P. S. Kulkarni, D. Patil, B. Patwardhan, and R. K. Nanda. 2009. High-performance liquid chromatographic fingerprint for quality control of *Terminalia arjuna* containing ayurvedic *churna* formulation. *Journal of AOAC International* 92, no. 4: 1016–20.

Chizzola, R., H. Michitsch, and C. Franz. 2003. Monitoring of metallic micronutrients and heavy metals in herbs, spices and medicinal plants from Austria. *European Food Research and Technology* 216, no. 5: 407–11.

D'Amelio, Sr., F. S. 1999. *Botanicals – a phytocosmetic desk reference*, 13–38. Boca Raton: CRC Press.

Dentali, S. 2009. Botanical preparations: Achieving quality products. In *Botanical medicine: From bench to bedside*, ed. R. Cooper, and F. Kronenberg, 3–12. New York: Mary Ann Liebert, Inc.

Donlan, R. M. 2002. Biofilms: Microbial life on surfaces. *Emerging Infectious Diseases* 8, no. 9: 881–90.

Dwivedi, S., and M. P. Agarwal. 1994. Antianginal and cardioprotective effects of *Terminalia arjuna*, an indigenous drug, in coronary artery disease. *Journal of Association of Physicians of India* 42, no. 4: 287–9.

Dwivedi, S., and R. Jauhari. 1997. Beneficial effects of *Terminalia arjuna* in coronary artery disease. *Indian Heart Journal* 49, no. 5: 507–10.

EMA. 2011. *Guideline on specifications: Test procedures and acceptance criteria for herbal substances, herbal preparations and herbal medicinal products/traditional herbal medicinal products*, 1–25. London: European Medicines Agency.

Ernst, E., and J. T. Coon. 2001. Heavy metals in traditional Chinese medicines: A systematic review. *Clinical Pharmacology and Therapeutics* 70, no. 6: 497–504.

Fan, X.-H., Y.-Y. Cheng, Z.-L. Ye, R. Lin, and Z. Qian. 2006. Multiple chromatographic fingerprinting and its application to the quality control of herbal medicines. *Analytica Chimica Acta* 555, no. 2: 217–24.

Flora, G., D. Gupta, and A. Tiwari. 2012. Toxicity of lead: A review with recent updates. *Interdisciplinary Toxicology* 5, no. 2: 47–58.

Garg, C., P. Sharma, S. Satija, and M. Garg. 2016. Stability indicating studies of *Andrographis paniculata* extract by validated HPTLC protocol. *Journal of Pharmacognosy and Phytochemistry* 5: 337–44.

Garg, S., A. Mishra, and R. Gupta. 2013. Fingerprint profile of selected ayurvedic churnas / preparations: An Overview. *Alternative and Integrative Medicine* 2: 125 doi:10.4172/2327- 5162.1000125.

Gasser, U., B. Klier, A. V. Kühn, and B. Steinhoff. 2009. Current findings on the heavy metal content in herbal drugs. *Pharmeuropa Scientific Notes* 1, no. 1: 37–50.

Gemperline, P. J. 2006. Principal component analysis. In *Practical guide to chemometrics*, 68–104. Boca Raton: CRC Press.

Ghosh, N., M. A. Veloz, N. Sherali et al. 2016. Microscopic characterization of some medicinal plants and elemental analysis of *Triphala* (three fruits) with anticarcinogenic properties. *Journal of Medicinal Herbs and Ethnomedicine* 2: 11–18.

Gong, F., Y.-Z. Liang, P.-S. Xie, and F.-T. Chau. 2003. Information theory applied to chromatographic fingerprint of herbal medicine for quality control. *Journal of Chromatography A* 1002, no. 1–2: 25–40.

Govender, S., D. Plessis-Stoman, T. G. Downing, and M. Van de Venter. 2006. Traditional herbal medicines: Microbial contamination, consumer safety and the need for standards. *South African Journal of Science* 102: 253–5.

Harris, E. S., S. Cao, and B. A. Littlefield et al. 2011. Heavy metal and pesticide content in commonly prescribed individual raw Chinese Herbal Medicines. *Science of the Total Environment* 409, no. 20: 4297–305.

Hell, K., and C. Mutegi. 2011. Aflatoxin control and prevention strategies in key crops of Sub-Saharan Africa. *African Journal of Microbiology Research* 5: 459–66.

Hiremath, S. G., and D. Joshi. 1991. Role of different containers and methods on alcoholic preparations with reference to *Kutajarista*. *Ancient Science of Life* 10, no. 4: 256–63.

Hiremath, S. G., D. Joshi, and A. B. Ray. 1993. TLC as a tool for standardisation of ayurvedic formulations with special reference to *Kutajarishta*. *Ancient Science of Life* 14, no. 3–4: 358–62.

Houghton, P. J. 2009. Synergy and polyvalence: Paradigms to explain the activity of herbal products. In *Evaluation of herbal medicinal products: Perspectives on quality, safety and efficacy*, ed. P. K. Mukherjee, and J. Houghton, 85–94. London: Pharmaceutical Press.

Irshad, S., P. K. Misra, A. K. S. Rawat, and S. Khatoon. 2016. Authentication of commercial samples of *Shankhupushpi* through physico-phytochemical analysis and TLC fingerprinting. *Indian Journal of Traditional Knowledge* 15: 646–53.

Jain, V., V. Prasad, R. Pal, and S. Singh. 2007. Standardization and stability studies of neuroprotective lipid soluble fraction obtained from *Curcuma longa*. *Journal of Pharmaceutical and Biomedical Analysis* 44, no. 5: 1079–86.

Jiang, B., F. Kronenberg, M. J. Balick, and E. J. Kennelly. 2013. Stability of black cohosh triterpene glycosides and polyphenols: Potential clinical relevance. *Phytomedicine* 20, no. 6: 564–9.

Johnson, K. J., B. W. Wright, K. H. Jarman, and R. E. Synovec. 2003. High speed peak matching algorithm for retention time alignment of gas chromatographic data for chemometric analysis. *Journal of Chromatography. Part A* 996, no. 1–2: 141–55.

Joshi, V. K., A. Joshi, and K. S. Dhiman. 2017. The Ayurvedic Pharmacopoeia of India, development and perspectives. *Journal of Ethnopharmacology* 197: 32–38.

Kabelitz, L. 1998. Heavy metals in herbal drugs. *European Journal of Herbal Medicine (Phytotherapy)* 4: 1–9.

Khatoon, S., S. Singh, S. Kumar, N. Srivastava, A. Rathi, and S. Mehrotra. 2009. Authentication and quality evaluation of an important Ayurvedic drug – Ashoka bark. *Journal of Scientific and Industrial Research* 68: 393–400.

Kneifel, W., E. Czech, and B. Kopp. 2002. Microbial contamination of medicinal plants – A review. *Planta Medica* 68, no. 1: 5–15.

Kroes, B. H., A. J. J. van den Berg, A. M. Abeysekera, K. T. D. de Silva, and R. P. Labadie. 1993. Fermentation in traditional medicine: The impact of *Woodfordia fruticosa* flowers on the immunomodulatory activity, and the alcohol and sugar contents of *Nimba arishta*. *Journal of Ethnopharmacology* 40, no. 2: 117–25.

Kumar, D. S. 2016. Bioactives from plants. In *Herbal bioactives and food fortification: Extraction and formulation*, 29–62. Boca Raton: CRC Press.

Kumar, D. S., and Y. S. Prabhakar. 1987. On the ethnomedical significance of the Arjun tree, *Terminalia arjuna* (Roxb.) Wight & Arnot. *Journal of Ethnopharmacology* 20, no. 2: 173–90.

Lachenmeier, D. W., S. Haupt, and K. Schulz. 2008. Defining maximum levels of higher alcohols in alcoholic beverages and surrogate alcohol products. *Regulatory Toxicology and Pharmacology* 50, no. 3: 313–21.

Lattaoui, N., and A. Tantaoui-Elaraki. 1994. Individual and combined antibacterial activity of the main components of three thyme essential oils. *Rivista Italiana* EPPOS 13:13–19.

Li, B.-Y., Y. Hu, and Y.-Z. Liang. 2004a. Spectral correlative chromatography and its application to analysis of chromatographic fingerprints of herbal medicines. *Journal of Separation Science* 27, no. 7–8: 581–8.

Li, B.-Y., Y. Hu, and Y.-Z. Liang. 2004b. Quality evaluation of fingerprints of herbal medicine with chromatographic data. *Analytica Chimica Acta* 514, no. 1: 69–77.

Li, B.-Y., Y Hu, and Y.-Z. Liang. 2004c. Spectral correlative chromatography and its application to analysis of chromatographic fingerprints of herbal medicines. *Journal of Separation Science* 27, no. 7–8: 581–8.

Li, W., Z. Chen, Y. Liao, and H. Liu. 2005. Separation methods for toxic components in traditional Chinese medicines. *Analytical Sciences* 21, no. 9: 1019–29.

Liang, Y.-Z., P. Xie, and K. Chan. 2004. Quality control of herbal medicines. *Journal of Chromatography. Part B* 812, no. 1–2: 53–70.

Livesey, J., D. V. C. Awang, J. T. Arnason, W. Letchamo, M. Barrett, and G. Pennyroyal. 1999. Effect of temperature on stability of marker constituents in *Echinacea purpurea* root formulations. *Phytomedicine* 6, no. 5: 347–9.

Lohar, D. R. 2008. *Protocol for testing Ayurvedic, Siddha & Unani medicines*, 1–200. Ghaziabad: Department of AYUSH.

Malone, M. H. 1983. The pharmacological evaluation of natural products: General and specific approaches to screening ethnopharmaceuticals. *Journal of Ethnopharmacology* 8, no. 2: 127–47.

Manzano, M., L. Iacumin, M. Vendrames, F. Cecchini, G. Comi, and S. Buiatti. 2011. Craft beer microflora identification before and after a cleaning process. *Journal of the Institute of Brewing* 117, no. 3: 343–51.

Marino, M., C. Bersani, and G. Comi. 2001. Impedance measurements to study the antimicrobial activity of essential oils from Lamiaceae and Compositae. *International Journal of Food Microbiology* 67, no. 3: 187–95.

McElroy, M. B. 2010. Coal: Origin, history, and problems. In *Energy perspectives, problems, and prospects*, 105–21. New York: Oxford University Press.

Menon, P. 2003. Adulteration and substitution of crude drugs. In *Conservation & consumption: A study on the crude drug trade in threatened medicinal plants in Thiruvananthapuram District, Kerala*, 37–42. Trivandrum: Centre for Development Studies.

Montoro, P., S. Piacente, and C. Pizza. 2012. Quality issues of current herbal medicines. In *Herbal medicines: Development and validation of plant-derived medicines for human health*, ed. G. Bagetta, M. Cosentino, M. T. Corasaniti, and S. Sakurada, 413–38. Boca Raton: CRC Press.

Murthy, K. R. S. 2017a. *Dravyasamgrahaṇīya Addhyāya* (Collection of ingredients). In *Suśruta Samhita*, 265–76. Varanasi: Chaukhambha Orientalia.

Murthy, K. R. S. 2017b. *Prathama Khaṇḍa* (Chapter 1). In *Śārṅgadhara Samhita*, 3–9. Varanasi: Chaukhamba Orientalia.

Murthy, K. R. S. 2017c. *Maddhyama Khaṇḍa* (Middle Section). In *Śārṅgadhara Samhita*, 51–184. Varanasi: Chaukhamba Orientalia.

Murthy, K. R. S. 2017d. Maddhyama Khaṇḍa (Chapter Two). In *Śārṅgadhara Samhita*, 56–77. Varanasi: Chaukhamba Orientalia.

Nagajyoti, P. C., K. D. Lee, and T. V. M. Sreekanth. 2010. Heavy metals, occurrence and toxicity for plants: A review. *Environmental Chemistry Letters* 8, no. 3: 199–216.

Nakajima, K., Y. Takeuchi, H. Taguchi, K. Hayashi, M. Okada, and M. Maruno. 1994. Physicochemical studies on decoctions of Kampo prescriptions. I. Transfer of crude drug components into the decoctions. *Chemical and Pharmaceutical Bulletin* 42, no. 10: 1977–83.

Noguchi, M., M. Kubo, T. Hayashi, and M. Ono. 1978. Studies on the pharmaceutical quality evaluation of the crude drug preparations used in oriental medicine kampoo part 1 precipitation reaction of the components of coptis rhizome and these of Glycyrrhiza root or rheum rhizome in decoction solution. *Shoyakugaku Zaashi* 32: 104.

Oh, C. H. 2007. Multi residual pesticide monitoring in commercial herbal crude drug materials in South Korea. *Bulletin of Environmental Contamination and Toxicology* 78, no. 5: 314–18.

Ong, E. S. 2002. Chemical assay of glycyrrhizin in medicinal plants by pressurized liquid extraction (PLE) with capillary zone electrophoresis (CZE). *Journal of Separation Science* 25, no. 13: 825–31.

Panesar, P. S., N. Kumar, S. S. Marwaha, and V. K. Joshi. 2009. Vermouth production technology – An overview. *Natural Product Radiance* 8: 334–44.

Paster, N., M. Menasherov, U. Ravid, and B. Juven. 1995. Antifungal activity of oregano and thyme essential oils applied as fumigants against fungi attacking stored grain. *Journal of Food Protection* 58, no. 1: 81–5.

Patil, D., M. Gautam, U. Jadhav et al. 2010. Physicochemical stability and biological activity of *Withania somnifera* extract under real-time and accelerated storage conditions. *Planta Medica* 76, no. 5: 481–8.

Pearce, J. M. S. 2007. Burton's line in lead poisoning. *European Neurology* 57, no. 2: 118–9.

Rai, V., P. Kakkar, J. Singh, C. Misra, S. Kumar, and S. Mehrotra. 2008. Toxic metals and organochlorine pesticides residue in single herbal drugs used in important ayurvedic formulation – "Dashmoola." toxic metals and organochlorine pesticides residue in single herbal drugs used in important ayurvedic formulation – "Dashmoola." *Environmental Monitoring and Assessment* 143, no. 1–3: 273–7.

Ramachandra, A. P., S. M. Prasad, S. M. S. Samarakoon, H. M. Chandola, C. R. Harisha, and V. J. Shukla. 2012. Pharmacognostical and phytochemical evaluation of *Vara Asanadi Kwatha. AYU* 33, no. 1: 130–5.

Rao, M. M., and A. K. Meena. 2011. Detection of toxic heavy metals and pesticide residue in herbal plants which are commonly used in the herbal formulations. *Environmental Monitoring and Assessment* 181, no. 1–4: 267–71.

Ribéreau-Gayon, P., D. Dubourdieu, B. Doneche, and A. Lonvaud. 2006. Conditions of yeast development. In *Handbook of enology, volume 1: The microbiology of wine and vinifications*, 2nd edition, 79–113. West Sussex: John Wiley & Sons Ltd.

Sebastian, J., A. Thomas, and D. S. Kumar. 2015. Heavy metals in ayurvedic herbs and traditional ayurvedic formulations- A study. *Aryavaidyan* 28: 101–11.

Sejali, S. N. F., and M. S. Anuar. 2011. Effect of drying methods on phenolic contents of neem (*Azadirachta indica*) leaf powder. *Journal of Herbs, Spices and Medicinal Plants* 17, no. 2: 119–31.

Şener, A., A. Canbaş, and M. Ü. Ünal. 2007. The effect of fermentation temperature on the growth kinetics of wine yeast species. *Turkish Journal of Agriculture and Forestry* 31: 349–54.

Sharma, P. V. 1981. Chapter 1, *Sūtrasthāna-Indriyasthāna*. In *Caraka Samhita*, Volume 1, 1–13. Varanasi: Chaukhamba Orientalia.

Sharma, P. V. 1994a. Chapter 30 *Cikitsāsthānam*. In *Caraka Samhita*, Volume 4, 219–25. Varanasi: Chaukhamba Orientalia.

Sharma, P. V. 1994b. *Asṛgdara Cikitsa* (Chapter on treatment of dysfunctional bleeding). In *Cakradatta*, Volumes 524–527. Varanasi: Chaukhamba Orientalia.

Srey, S., I. K. Jahid, and S. D. Ha. 2013. Biofilm formation in food industries: A food safety concern. *Food Control* 31, no. 2: 572–85.

Srivastava, J. K., and S. Gupta. 2009. Extraction, characterization, stability and biological activity of flavonoids isolated from chamomile flowers. *Molecular and Cellular Pharmacology* 1, no. 3: 138–47.

Srivastava, M. M. 2011. An overview of HPTLC: A modern analytical technique with excellent potential for automation, optimization, hyphenation and multidimensional applications. In *High-performance thin-layer chromatography (HPTLC)*, ed. M. M. Srivastava, 3–24. Heidelberg: Springer-Verlag.

Srivastava, P., H. N. Raut, H. M. Puntambekar, A. S. Upadhye, and A. C. Desai. 2010. Stability studies of crude plant material of *Bacopa monnieri* and its effect on free radical scavenging activity. *Journal of Phytological Research* 2: 103–9.

Storgårds, E. 2000. *Process hygiene control in beer production and dispensing*, 1–106. Vuorimiehentie, Finland: VTT.

Street, R. A., W. A. Stirk, and J. Van Staden. 2008. South African traditional medicinal plant trade – challenges in regulating quality, safety and efficacy. *Journal of Ethnopharmacology* 119, no. 3: 705–10.

Sulaiman, C. T., and I. Balachandran. 2015. Chemical profiling of an Indian herbal formula using liquid chromatography coupled with electro spray ionization mass spectrometry. *Spectroscopy Letters* 48, no. 3: 222–6.

Taylor, D. G. 2006. Water. In *Handbook of brewing*, ed. F. G. Priest, and G. G. Stewart, 91–137. Boca Raton: CRC Press.

Techen, N., S. L. Crockett, I. A. Khan, and B. E. Scheffler. 2004. Authentication of medicinal plants using molecular biology techniques to compliment conventional methods. *Current Medicinal Chemistry* 11, no. 11: 1391–401.

Thatte, U., and S. Bhalerao. 2008. Pharmacovigilance of ayurvedic medicines in India. *Indian Journal of Pharmacology* 40, no. Suppl 1: S10–12.

Thomas, A., and D. S. Kumar. 2018. Traditional ayurvedic tablets may contain heavy metals. *Hygeia: Journal of Drugs and Medicines* 9, no. 2. doi:10.15254/H.J.D.Med.9.2018.170.

Thomas, A., A. Radha, and D. S. Kumar. 2016. Some thoughts on improving the manufacturing process of Ayurvedic *ariṣṭa* and *āsava*. *Hygeia* 6: 11–18.

Tomimori, T., and M. Yoshimoto. 1980. Quantitative variation of glycyrrhizin in the decoction of Glycyrrhizae Radix mixed with other crude drugs. *Shoyakugaku Zaashi* 34: 138.

Torija, M. J., N. Rozès, M. Poblet, J. M. Guillamón, and A. Mas. 2003. Effects of fermentation temperature on the strain population of *Saccharomyces cerevisiae*. *International Journal of Food Microbiology* 80, no. 1: 47–53.

Vaidyan, K. V. K., and A. S. G. Pillai. 1985. *Guḷikāyōgaṅgaḷ* (Formulae of pills). In *Sahasrayōgam*, 133–66. Alleppey: Vidyarambham Publishers.

Van Vuuren, S., V. L. Williams, A. Sooka, A. Burger, and L. Van der Haar. 2014. Microbial contamination of traditional medicinal plants sold at the Faraday muthi market, Johannesburg, South Africa. *South African Journal of Botany* 94: 95–100.

Wagner, H., R. Bauer, D. Melchart, P. G. Xiao, and A. Staudinger. 2011. *Chromatographic fingerprint analysis of herbal medicines: Thin-layer and high performance liquid chromatography of Chinese drugs*, Volume 1, 1–1024. Wien: Springer-Verlag.

Walker, M. 2007. Drug evaluation in the 21st century. *Pharmaceutical Journal* 279, no. Supp.: B8.

Walter, V. E. 2006. Sanitation and pest control. In *Handbook of brewing*, ed. F. G. Priest, and G. G. Stewart, 629–53. Boca Raton: CRC Press.

Weerasooriya, W. M. B., and J. A. Liyanage. 2006. Identification of alcohols in selected arishta and asava used in ayurvedic medicine. In *Proceedings of the 62nd annual sessions of Sri Lanka association for the advancement of science*, Colombo, 11.

Wenzig, E. M., and R. Bauer. 2009. Quality control of Chinese herbal drugs. In *Evaluation of herbal medicinal products: Perspectives on quality, safety and efficacy*, ed. P. K. Mukherjee, and P. J. Houghton, 393–425. London: Pharmaceutical Press.

Williamson, E. M. 2001. Synergy and other interactions in phytomedicines. *Phytomedicine* 8, no. 5: 401–9.

Wong, S.-K., S.-K. Tsui, S.-Y. Kwan et al. 2004. Establishment of characteristic fingerprint chromatogram for the identification of Chinese herbal medicines. *Journal of Food and Drug Analysis* 12: 110–14.

Xie, P. S. 2001. A feasible strategy for applying chromatography fingerprint to assess quality of Chinese herbal medicine. *Traditional Chinese Drug Research and Clinical Pharmacology* 12: 141–69.

Yamaji, A., Y. Mareda, M. Oishi et al. 1984. Determination of saikosaponins in Chinese medicinal extracts containing Bupleuri radix. *Yakugaku Zaashi* 104, no. 7: 812–15.

Yang, L.-W., D.-H. Wu, W. Tang et al. 2005. Fingerprint quality control of Tianjihuang by high-performance liquid chromatography-photodiode array detection. *Journal of Chromatography. Part A* 1070, no. 1–2: 35–42.

Yau, W. P., C. H. Goh, and H. L. Koh. 2015. Quality control and quality assurance of phytomedicines: Key considerations, methods, and analytical challenges. In *Phytotherapies: Efficacy, safety, and regulation*, ed. I. Ramzan, 18–47. NJ: John Wiley & Sons.

Yuan, R., and Y. Lin. 2000. Traditional Chinese medicine: An approach to scientific proof and clinical validation. *Pharmacology and Therapeutics* 86, no. 2: 191–8.

Zainol, M. K. M., A. Abdul-Hamid, A. F. Bakar, and S. P. Dek. 2009. Effect of different drying methods on the degradation of selected flavonoids in *Centella asiatica*. *International Food Research Journal* 16: 531–7.

5 Scientific Rationale for the Use of Single Herb Remedies in Ayurveda

S. Ajayan, R. Ajith Kumar and Nirmal Narayanan

CONTENTS

5.1 Introduction 102
5.2 Evidence in Favor of Medicinal Uses 102
5.2.1 *Acorus calamus* L. 102
5.2.1.1 Sedative Action 102
5.2.1.2 CNS Depressant Action 102
5.2.1.3 Anti-Convulsant Action 104
5.2.2 *Allium sativum* L. 104
5.2.3 *Crataeva magna* (Lour.) DC 105
5.2.4 *Curcuma longa* L. 106
5.2.5 *Cyperus rotundus* L. 106
5.2.6 *Desmodium gangeticum* (L.) DC 107
5.2.7 *Evolvulus alsinoides* (L.) L. 108
5.2.8 *Glycyrrhiza glabra* L. 108
5.2.9 *Hygrophila auriculata* (K. Schum.) Heine 110
5.2.10 *Ipomoea mauritiana* Jacq 111
5.2.11 *Moringa oleifera* Lam. 111
5.2.12 *Piper longum* L. 112
5.2.12.1 Effect on Digestive Efficiency 112
5.2.12.2 Antibacterial Activity 113
5.2.12.3 Anti-Inflammatory Activity 113
5.2.12.4 Anti-Cancer Effect 113
5.2.13 *Plumbago zeylanica* L. 114
5.2.14 *Saraca asoca* (Roxb.) de Wilde 115
5.2.15 *Strychnos potatorum* L. f. 117
5.2.16 *Tabernaemontana divaricata* (L.) R.Br. ex Roem. & Schultes 118
5.2.17 *Terminalia arjuna* (Roxb. ex DC) Wight & Arn. 119
5.2.18 *Terminalia belerica* (Gaertn.) Roxb. 121
5.2.19 *Tinospora cordifolia* (Willd.) Miers ex Hook. f. & Thoms 122
5.2.19.1 In Ethnomedicine 123
5.2.19.2 Clinical Studies 123
5.2.19.3 Effects on Immunity 123
5.2.19.4 Antibacterial Activity 124
5.2.19.5 Anti-Inflammatory Activity 124
5.2.20 *Vitex negundo* L. 124
5.3 Role of Co-Administered Adjuvants 126
5.4 Conclusion 126
References 126

5.1 INTRODUCTION

Ayurveda employs single herbs and combinations of herbs in the treatment of diseases. Single herb remedies are usually administered as powders, pastes, decoctions, medicated clarified butter or incorporated into food matrices. Small quantities of adjuvants like honey, clarified butter, sesame oil or jaggery are administered along with the primary drug, most probably to enhance absorption and therapeutic effect. Ayurveda generally uses combinations of herbs to balance the *tridōṣa*-modulating properties of herbs and to minimize side-effects. Single herb remedies are unbalanced and therefore, there can be adverse effects. For example, *Aṣṭāṅgahṛdaya* recommends consumption of powder of dry ginger (*Zingiber officinale*) in warm water to cure *grahaṇi* (diarrhea) (Upadhyaya 1975a). But on account of its constipating property, called *grāhi* in Ayurveda (Upadhyaya 1975b), it is to be used with caution. Because of the inherent danger, knowledge of many single herb remedies was treated in the olden days as a jealously guarded secret. Single herb remedies are considered to be faster in action. Table 5.1 lists some such single herb remedies. These formulae have been collected from *Aṣṭāṅgahṛdaya* (Upadhyaya 1975c), *Cakradatta* (Mishra 1983), *Śārṅgadhara Samhita* (Murthy 2017) and *Vaidyamanōrama* (Mooss 1978, 1979).

5.2 EVIDENCE IN FAVOR OF MEDICINAL USES

5.2.1 *Acorus calamus* L.

Acorus calamus is a semi-aquatic rhizomatous perennial herb (Figure 5.1). It improves mental faculties and digestion. *Vaca*, *Ugragandha*, *Śataparvikā* and *Kṣudrapatrī* are some of its Sanskrit names (Warrier et al. 2007a). No clinical trials are reported for the anti-epileptic property of *Acorus calamus*. However, several studies lend support to the use of this plant in the treatment of epilepsy.

5.2.1.1 Sedative Action

The volatile oil isolated from *A. calamus* potentiated the sedative activity of pentobarbitone in mice. The active principle responsible for this activity was inferred to be present in the hydrocarbon fraction of the oil or in an oxygenated component in the oil (Dandiya et al. 1959a). The volatile fractions prolonged sleeping time in mice with pentobarbital, hexobarbital and ethanol. The sedation-potentiating activity was maximally present in the volatile fraction of the petroleum ether extract (Dandiya and Cullumbine 1959). Priming of mice with lysergic acid diethylamide partially prevented the hypnosis-potentiating action of the volatile oil (Dandiya et al. 1959b). Further studies on the potentiation of barbiturate-induced hypnosis in mice by the volatile oil showed that the potentiating action was antagonized by lysergic acid diethylamide as well as dibenzyline hydrochloride. These results suggest that the hypnosis-potentiating action might be mediated through serotonin and catecholamines (Malhotra et al. 1962). Further studies with the essential oil demonstrated tranquilizing action in rats, mice, cats and dogs. The oil inhibited monoamine oxidase at a higher dose (Dhalla and Bhattacharya 1968). *A. calamus* essential oil antagonized amphetamine-induced agitational symptoms and also inhibited the conditioned avoidance response in rats (Bhattacharya 1968). β-asarone produces sedative and hypothermic effects in rats. In one study, the alcohol extract of the rhizome showed sedative and analgesic properties. The essential oil, crude alcohol extract and aqueous extracts exhibited a depressant action in dogs (Bose et al. 1960).

5.2.1.2 CNS Depressant Action

Hazra and Guha (2003) studied the effect of ethanol extract of *A. calamus* on spontaneous electrical activity and monoamine levels of the rat brain. In treated animals, electrogram recording revealed an increase in α activity together with an increase in the norepinephrine level in the cerebral cortex. However, there was a decrease in their content in the midbrain and cerebellum. The serotonin level was increased in the cerebral cortex but decreased in the midbrain. In a similar way, the dopamine

TABLE 5.1
Some Single Herb Remedies Recommended in Ayurveda

Herb	Mode of Administration and Indication	Reference
Acorus calamus L.	Administration of powder of rhizome mixed with honey and followed by drinking of milk cures even chronic *apasmāra* (epilepsy)	Mishra (1983)179* Upadhyaya (1975c) 478
Allium sativum L.	Paste of cloves mixed with sesame oil cures *viṣamajvara* (irregular fever)	Mishra (1983) 29 Upadhyaya (1975c) 297
Hygrophila auriculata (K. Schum.) Heine	Regular consumption of root decoction and cooked leaves cures *vātaśōṇita* (rheumatoid arthritis), "just as compassion pacifies anger"	Upadhyaya (1975c) 424
Crataeva magna (Lour.) DC	Decoction of bark administered with jaggery cures *aśmari* (urinary stone) and *vastiśūla* (pricking pain in bladder) Paste of bark suspended in bark decoction cures *aśmari*	Mishra (1983) 285 Mishra (1983) 285
Curcuma longa L	*Ghṛta* (medicated clarified butter) prepared with decoction and paste of rhizomes cures *kāmala* (jaundice)	Mooss (1978) 68
Cyperus rotundus L.	Twenty tubers crushed and boiled in milk cure *āmātisāra* (dysentery with pain)	Mishra (1983) 51
Desmodium gangeticum (L.) DC	Root decoction prepared in milk cures *vāta* seated in heart (heart disease)	Upadhyaya (1975c) 417
Evolvulus alsinoides L.	*Kalka* (paste) of roots mixed with sugar, honey and clarified butter relieves within seven days *pariṇāmaśūla* (colic due to peptic ulcer)	Murthy (2017) 82
Glycyrrhiza glabra L.	Topical application of root powder mixed with clarified butter heals *sadyōvraṇa* (fresh wounds)	Mishra (1983) 352
Ipomoea digitata L.	Powder of tubers soaked in decoction of tubers, dried and administered with clarified butter and honey increases sexual vigor	Mishra (1983) 572
Moringa oleifera Lam.	Oral administration of juice of root bark with honey cures *antarvidradhi* (internal abscess)	Mishra (1983) 343
Piper longum L.	*Ghṛta* prepared with the fruit cures *pḷīha* (splenomegaly), *yakṛt* (hepatomegaly) and *agnimāndya* (lowering of abdominal fire)	Mishra (1983) 314
Plumbago zeylanica L.	Detoxified roots are to be ground to a paste and smeared inside a fresh earthen pot. Consumption of curd prepared in that pot cures *arśas* (hemorrhoids)	Mishra (1983)75-76 Mooss (1979) 148
Saraca asoca (Roxb.) de Wilde	Administration of bark decoction prepared in milk cures *asṛgdara* (dysfunctional bleeding)	Mishra (1983) 503
Strychnos potatorum L.f.	15 g of seeds ground into paste, suspended in buttermilk and administered with honey cures advanced *pramēha* (urinary diseases of polyuric nature)	Mooss (1978) 14
Tabernaemontana divaricata (L.) R.Br. ex Roem. & Schultes	All the 20 *pramēha* will be cured by eating muffin-like cakes prepared with dough made of ground leaves, sesame oil and rice	Mooss (1978) 17
Terminalia arjuna (Roxb. ex DC) Wight & Arn.	Administration of bark powder with clarified butter, milk or jaggery cures *hṛdrōga* (heart disease), *jīrṇajvara* (chronic fever) and *raktapitta* (hemorrhages of obscure origin)	Mishra (1983) 272
Terminalia belerica (Gaertn.) Roxb.	Powder of pericarp mixed with honey cures *hikka* (hiccup) and *śvāsa* (respiratory distress)	Mishra (1983) 151
Tinospora cordifolia (Willd.) Miers ex Hook. f. & Thoms.	Consumption of *ghṛta* prepared with decoction and paste of stem cures *vātaśōṇita* and *kuṣḍha* (leprosy and other skin diseases)	Mishra (1983) 218 Murthy (2017) 120
Vitex negundo L.	*Ghṛta* prepared with leaf juice cures *kāsa* (cough)	Mooss (1979) 74

*Number of the page on which the formula is described.

FIGURE 5.1 *Acorus calamus* (a), plant with spadix bearing flowers (b), dried rhizomes (c). Reproduced with permission from Dr D. Suresh Baburaj, Ooty.

level was increased in the caudate nucleus and midbrain but decreased in the cerebellum. Therefore, *A. calamus* seems to produce its depressive action by changing electrical activity and by differentially altering brain monoamine levels in different regions of the brain (Hazra and Guha 2003).

5.2.1.3 Anti-Convulsant Action

Acorus oil was given to adult albino mice one hour prior to the induction of convulsions. It successfully prevented seizures in maximal electroshock seizures test (Khare and Sharma 1982). α-asarone showed a tendency to offer protection against metrazol convulsions and modified electroshocks (Sharma et al. 1961). In a study using electroconvulsions, α-asarone increased the percentage mortality of animals treated with chlorpromazine, but not of those treated with reserpine (Dandiya and Sharma 1962; Dandiya and Menon 1963). The aqueous and alcohol extracts were found to reduce the severity of maximum electric shock-induced seizure in rats. Further, the extracts significantly increased the pentylenetetrazole-induced seizure latency (Manis et al. 1991). The essential oil showed a protective effect against electroshock seizures in rats (Madan et al. 1960). Anti-convulsant activity of *A. calamus* has been reported by Hazra et al. (2007) and Chandrashekar et al. (2013).

Bhat et al. (2012) studied the effect of detoxification of rhizomes of *A. calamus* using induced seizures model. Comparative anticonvulsant activity of crude and processed rhizomes was screened against a maximal electroshock (M.E.S.) seizure model to assess the effect of a classical purificatory procedure on the pharmacological action of the herb. Phenytoin was used as standard anti-epileptic drug for comparison. Pretreatment with both crude and processed samples exhibited significant anti-convulsant activity by decreasing the duration of the tonic extensor phase. Processed rhizomes significantly decreased the duration of convulsion and stupor phases of M.E.S.-induced seizures. The results obtained from this study clearly confirms the anti-convulsant activity of crude rhizomes of *A. calamus*. Subjecting the crude rhizomes to the detoxification procedure did not alter the efficacy of the rhizomes. Contrarily, the process enhanced the activity profile of the rhizomes.

5.2.2 *Allium sativum* L.

References to garlic in the Bible and the Koran reflect its importance to ancient civilizations as a spice and a healing herb. The name *Allium* is derived from the Greek root meaning "to avoid" because of its offensive smell (Block 2010). It is an important *rasāyana* herb in Ayurveda (Warrier et al. 2007a).

Organosulfur compounds, sulfides and disulfides are present in human bodies and the environment. The three sulfur-containing amino acids – cysteine, cystine and methionine – are important constituents of proteins, peptides, enzymes, vitamins and hormones. They can interact with or neutralize reactive oxygen species (R.O.S.) and play a vital role in signaling and sensing (Jones 2010). Thus they help to maintain healthy bodies (Goncharov et al. 2015, 2016).

A. sativum is a well-known representative of the *Allium* genus (family *Amaryllidaceae*) that contain *S*-alk (en)yl-l-cysteine sulfoxides (Munday 2012). Garlic has proven effect against an array of Gram-positive, Gram-negative and acid-fast bacteria, including *Salmonella*, *Escherichia coli*, *Pseudomonas*, *Proteus*, *Staphylococcus aureus*, *Klebsiella*, *Micrococcus*, *Bacillus subtilis*, *Clostridium*, *Mycobacterium* and *Helicobacter* (Bayan et al. 2014). Garlic extract showed activity against viral infections like influenza A and B, cytomegalovirus, rhinovirus, H.I.V., herpes simplex virus 1, herpes simplex virus 2, viral pneumonia and rotavirus.

Weber et al. (1992) made an attempt to identify the compounds responsible for the antiviral activity of *A. sativum*. Using direct pre-infection incubation assays, they determined the *in vitro* virucidal effects of fresh garlic extract, its polar fraction and the garlic associated compounds diallyl thiosulfinate (allicin), allyl methyl thiosulfinate, methyl allyl thiosulfinate, ajoene, alliin, deoxyalliin, diallyl disulfide and diallyl trisulfide. The virucidal activity was determined against selected viruses, including herpes simplex virus type 1, herpes simplex virus type 2, parainfluenza virus type 3, vaccinia virus, vesicular stomatitis virus and human rhinovirus type 2. Fresh garlic extract, in which thiosulfinates appeared to be the active components, was virucidal to all the tested viruses.

Influenza is an infectious disease caused by an influenza virus. The common symptoms include: high fever, runny nose, sore throat, muscle pain, headache, cough and tiredness. The influenza virus is constantly evolving and new antigenic variants give rise to epidemics and pandemics. The influenza virus is unique among respiratory tract viruses because of its considerable antigenic variations. These mutations make it extremely difficult to develop effective vaccines and drugs against the virus (Moghadami 2017).

Mehrbod et al. (2009) evaluated the antiviral activity of garlic extract against influenza virus in cell culture. Madin-Darby Canine Kidney cells were treated with effective minimal cytotoxic concentration of garlic extract and 100 $T.C.I.D._{50}$ (50% Tissue Culture Infectious Dose) of the virus during infection at different time periods. The viral titers were determined by hemagglutination and $T.C.I.D._{50}$ assays. The antiviral effect of the extract was studied at 1, 8 and 24 hours after treatment of the culture. R.N.A. extraction, reverse transcription-polymerase chain reaction and free band densitometry were performed to measure the amount of the viral genome synthesized at different times after treatment. It was observed that garlic extract with a good selectivity index has an inhibitory effect on the virus penetration and proliferation in cell culture. The experimental methods used for evaluating the antiviral activity of garlic extract demonstrated that this herb could be used as a potent antiviral agent. These observations are further strengthened by the report of Chavan et al. (2016) that *A. sativum* extracts can inhibit influenza A (H1N1) pdm09 virus by inhibiting viral nucleoprotein synthesis and polymerase activity.

5.2.3 *Crataeva magna* (Lour.) DC

Crateva magna (Varuna tree) is a small tree often found along streams. It bears red small tomato-sized berries. In Sanskrit language the word *Varuṇa* means "that which is earnestly desired by people". *Sētuvṛkṣa* (tree that fetters diseases) and *Mārutāpahā* (that which cures diseases arising from the destabilization of *vātā*) are the other Sanskrit names of the plant. The dried bark is used in Ayurveda for treating urinary stones and tumors (Prabhakar and Kumar 1990; Warrier et al. 2007b).

Deshpande et al. (1982) carried out a number of studies to evaluate the role of the stem-bark in the management of urinary calculi, prostatic hypertrophy, neurogenic bladder and chronic urinary tract infections. Cystometric studies were carried out on patients with prostatic hypertrophy and atonic bladder by means of a simple water cystometer. Decoction of the stem-bark (50 ml) twice a day was tested on 30 cases of prostatic hypertrophy with hypotonic bladder. The treatment provided marked relief of symptoms like incontinence, pain and retention of urine. The cystometric study showed increased bladder tone leading to an increase in the expulsive force of urination. The decoction improved the bladder tone in several cases of hypotonia and atonia which persist after prostatectomy. The drug also provided beneficial effects to neurogenic bladder. These effects were observable after three months of therapy (Deshpande et al. 1982).

After administration of the drug for a month it was observed that the excretion of urinary calcium was reduced to a great extent, while the excretion of sodium and magnesium increased significantly. On plotting the electrolyte values on a triangular graph, a shifting of values towards the non-lithogenic zone was observed. The drug is thus found to alter the relative proportions of calcium, magnesium and sodium which participate in calculus formation (Deshpande et al 1982).

Administration of 50 ml of bark-decoction twice a day to 46 patients with urinary stone caused a significant anti-urolithiatic action. Over a period of between 1 and 47 weeks, 28 patients were able to pass the stone, while 18 experienced symptomatic relief. The spontaneous passing of stones following *C. magna* therapy may be facilitated by the tonic contractile action of the drug on smooth muscle (Deshpande et al. 1982).

The lithotriptic action of the stem-bark decoction was re-examined by Singh et al. (1991), who administered the potion to patients suffering from calcium oxalate stones. After treatment for 12 weeks, a significant reduction in pain and dysuria was noted. There was also some reduction in the size of the stones (Singh et al. 1991).

5.2.4 *Curcuma longa* L.

C. longa or turmeric is a flowering plant belonging to the ginger family, Zingiberaceae. It is an important plant in Ayurveda. The Ayurveda lexicon *Abhidānamañjari* calls turmeric *Gauri* (of white complexion), *Haridra* (the yellow one), *Rajani* (night), *Pīta* (yellow), *Piṇḍa* (lump), *Kāñcani* (gold), *Strīvallabha* (liked by ladies) and *Varṇavati* (improver of complexion) (Warrier et al. 2007b).

The protective and curative effects of *C. longa* powder on CCl_4-induced hepatotoxicity in rats were investigated by Beji et al. (2019). Turmeric was administered before or after treatment with CCl_4. Results showed that the activities of aspartate aminotransaminase, alanine aminotransaminase and the levels of bilirubin and serum lipids were increased after CCl_4 treatment. Total protein, albumin levels and antioxidant enzyme activities were decreased. Turmeric administration, before or after CCl_4 treatment, significantly decreased the activities of marker enzymes and lipid levels in blood. Concomitantly, total protein and albumin contents were restored to nearly normal levels after turmeric administration. These effects were accompanied with an increase of antioxidant enzymes activities.

The hepatoprotective property of *C. longa* was further studied by Karamalakova et al. (2019) in a bleomycin-induced chronic hepatotoxicity model. Hepatic toxicity was induced by intraperitoneal injection of mice once daily with bleomycin for four weeks. *C. longa* extract was administered once a day for four weeks. The extract significantly protected the liver by decreasing plasma bilirubin, gamma glutamyl transpeptidase and lipid peroxidation levels. Bleomycin administration produced oxidative stress, leading to dysfunction in the antioxidant system and significant increase in R.O.S. production. *C. longa* extract provided significant hepatic protection by improving superoxide dismutase, catalase and malondialdehyde levels, and decreasing R.O.S., reducing membrane lipid peroxidation. Therefore, *C. longa* extract is a good therapeutic agent to treat chronic hepatotoxicity.

5.2.5 *Cyperus rotundus* L.

C. rotundus is a perennial plant, often growing in wastelands and in crop fields. The root system of a young *Cyperus* plant initially forms white, fleshy rhizomes in chains. Some rhizomes grow upward in the soil and form a bulb-like structure from which new shoots and roots grow. Some of these rhizomes grow horizontally or downward and form dark reddish-brown tubers or chains of tubers. In Ayurveda it is valued in the treatment of fevers and digestive disorders (Warrier et al. 2007b).

Tubers of *C. rotundus* contain an essential oil. Forty-three components were identified in the essential oil of tubers collected from South Africa. α-cyperone (11.0%), myrtenol (7.9%), caryophyllene oxide (5.4%) and β-pinene (5.3%) were major compounds in the oil (Lawal and Oyedeji 2009). *C. rotundus* essential oil was found to be significantly active against Gram-positive microorganisms (*Staphylococcus aureus* and *Streptococcus species*), moderately active against *Sarcina lutea*, *Bacillus subtilis* and the acid-fast *Mycobacterium phlei* (El-Gohary 2004). At 100% concentration the oil *C. rotundus* showed good activity against *Escherichia coli*, *Staphylococcus aureus*, *Bacillus subtilis* and *Pseudomonas aeruginosa* and less activity against *Micrococcus luteus* and *Klebsiella sp.* At low concentration the oil was also effective against *S. aureus* (Bisht et al. 2011).

The methanol extract of tubers of *C. rotundus* (250 and 500 mg/kg body weight) administered orally produced significant anti-diarrheal activity in castor oil-induced diarrhea in mice. When tested at 250 mg/kg, the petroleum ether fraction and residual methanol fraction were found to retain the activity, the latter being more active, compared to the control (Uddin et al. 2006).

Daswani et al. (2011) studied the anti-diarrheal activity of the decoction of *C. rotundus* tubers using representative assays of diarrheal pathogenesis. Antibacterial, antigiardial and antirotaviral activities were also studied. Effects on adherence of enteropathogenic *Escherichia coli* and invasion of enteroinvasive *E. coli* and *Shigella flexneri* to HEp-2 cells were evaluated to measure the effect on colonization. Effect on enterotoxins such as enterotoxigenic *E. coli*, heat-labile toxin, heat-stable toxin and cholera toxin was also assessed. The decoction showed antigiardial activity, reduced bacterial adherence to and invasion of HEp-2 cells and affected production of cholera toxin and action of heat-labile toxin. The decoction of *C. rotundus* seems to exert the anti-diarrheal action by mechanisms other than direct killing of the pathogen.

5.2.6 *Desmodium gangeticum* (L.) DC

Known as *Śāliparṇi*, *Sthirā*, *Tṛparṇi* and *Vidārigandhā*, *Desmodium gangeticum* is common on lower hills and plains throughout India (Figure 5.2) (Warrier et al. 2007d).

Kurian et al. (2005) tested the aqueous extract of *D. gangeticum* in isoproterenol-induced myocardial infarcted rats for hypocholesterolemic and antioxidant effects. After inducing myocardial infarction by isoproterenol, the aqueous extract of *D. gangeticum* root was orally administered daily for a period of 30 days. The activities of creatinine phosphokinase, lactate dehydrogenase, alkaline phosphatase and serum glutamate oxaloacetate transaminase increased in myocardial tissue, hepatic tissue and serum after induction of myocardial infarction. Pretreatment of the infarcted rats with the plant extract prevented the increase of these enzymes. The hypocholesterolemic effect of *D. gangeticum* was assessed by the concentration of total cholesterol, low density lipoprotein cholesterol, high density lipoprotein cholesterol and through the activities of the enzymes HMG CoA reductase and lecithin cholesterol acyl transferase in the myocardial tissue. Thiobarbituric acid reactive substances decreased and activities of glutathione reductase and catalase improved in the myocardial tissue of the treated rats, indicating the free radical scavenging activity of the extract.

The effect of methanol extract of *D. gangeticum* on lipid peroxidation and antioxidants in mitochondria and tissue homogenates of normal, ischemic and ischemia-reperfused rats was studied by Kurian et al. (2008). Myocardial lipid peroxidation products in cardiac tissue homogenates and mitochondrial fractions were significantly increased during ischemia reperfusion. Antioxidant enzymes in the myocardial tissue homogenate and mitochondria showed significant decrease during ischemia reperfusion, followed by decreased activity of mitochondrial respiratory enzymes. Daily oral feeding of rats with methanol extract of *D. gangeticum* for 30 days produced a significant effect on the activity of mitochondrial and antioxidant enzymes. The extract inhibited lipid peroxidation, and also scavenged hydroxyl and superoxide radicals. The results of this study showed that *D. gangeticum* possesses the ability to scavenge free radicals generated during ischemia and ischemia reperfusion, thereby preserving the mitochondrial respiratory enzymes that offer cardioprotection. Aqueous extract of *D. gangeticum* is reported to improve the antioxidant status of the heart and

FIGURE 5.2 *D. gangeticum*. Reproduced with permission from Dr D. Suresh Baburaj, Ooty.

attenuate the degree of lipid peroxidation after ischemia reperfusion (Kurian and Paddikkala 2009). The aqueous root extract is also known to protect the mitochondrial and sarcoplasmic ATPase in the myocardium, resulting in improvement of cardiac function after ischemia reperfusion injury (Kurian and Padikkala 2010).

5.2.7 *Evolvulus alsinoides* (L.) L.

Evolvulus alsinoides is a small, hairy, procumbent, diffuse herb with small woody root stock. It bears light blue flowers. It is known by various names, such as *Viṣṇukrānta*, *Nīlapuṣpā* and *Aparājita* (Warrier et al. 2007c). Purohit et al. (1996) studied the anti-ulcer activity of *E. alsinoides* using the modified method of Shay et al. (1945). Alcohol extract of the herb was administered orally to ulcer-induced rats for five days. There was a significant decrease in free acidity and volume of gastric content of extract-treated rats when compared with controls. There was also a highly significant increase in the pH of the gastric content. The extract caused reduction in the intensity of gastric ulceration, evidenced by reduced ulcer index in the extract-treated group. The chloroform and ethyl acetate extracts of the herb possess significant anti-inflammatory activity when tested for their ability to reduce carrageenan and formalin-induced paw edema (Reddy and Rao 2013).

5.2.8 *Glycyrrhiza glabra* L.

Known as *Yaṣṭimadhu* (sweet wood) in Sanskrit, this is a tall, perennial, under-shrub. Its dried, underground stems and roots are sweet and used in Ayurveda for a variety of purposes. The licorice of commerce is the dried underground stems and roots (Figure 5.3). Its Sanskrit synonyms are *Madhūkam*, *Madhudrava*, *Madhuyaṣṭi* and *Yaṣṭimadhūkam* (Warrier et al. 2007c).

FIGURE 5.3 Roots of *G. glabra*. Reproduced with permission from Dr D. Suresh Baburaj, Ooty.

Several reports are available on the wound-healing property of *G. glabra*. Zaki et al. (2005) studied the healing effect of licorice extract on open skin wounds in adult New Zealand rabbits. Full-thickness wounds (15 × 15 mm) were made on the shaven areas of the rabbits. Creams incorporating 5%, 10% and 15% (w/w) of hydroalcohol extract were prepared and applied twice daily. Dexpanthenol ointment was used as the standard control. Healing was assessed by reduction in wound area. Results of this study proved that 10% licorice cream was a potent healing agent, showing results better than dexpanthenol cream.

Recently Zangeneh et al. (2019) investigated the wound-healing potential of ointment prepared with *G. glabra* aqueous extract. After creating the cutaneous wounds, the animals were randomly divided into four groups and treated with 3% *G. glabra* aqueous extract ointment, Eucerin ointment or 3% tetracycline ointment. At days 10, 20 and 30, *G. glabra* aqueous extract ointment could significantly decrease the level of the wound area and enhance the level of wound contraction. The content of fibrocyte, hexuronic acid and hydroxyproline was increased in comparison with the basal ointment and control groups. The study demonstrated that the aqueous extract of *G. glabra* accelerated wound-healing activity in experimental models.

Recurrent aphthous ulcers are among the most common oral mucosal diseases found in children and adults (Messier et al. 2012). These ulcers have varied etiology that includes bacterial infection by *Streptococcus sanguinis*, genetics, autoimmune causes and folic acid deficiency. Recurrent aphthous ulcers are treated by relieving pain, promoting healing and preventing secondary infection (Zunt 2001).

Investigators have recently focused their attention on the effect of licorice on pain control and reduction of the healing time of aphthous ulcers. Burgess et al. (2008) reported that Canker Melts GX patches which contain licorice extract alter the course of the aphthous ulcers by reduction of lesion duration, size and pain, thereby speeding the healing process. In a randomized, double-blind clinical trial involving 23 subjects, Martin et al. (2008) observed an improvement in ulcer size and pain using a dissolving oral patch containing licorice extract for up to 8 days, compared to the use of a placebo patch. According to the results of a recent study licorice bioadhesive can be effective in the reduction of pain, the inflammatory halo and the necrotic center of aphthous ulcers (Moghadamnia et al. 2009).

5.2.9 *Hygrophila auriculata* (K. Schum.) Heine

H. auriculata is an herbaceous, medicinal plant growing in marshy places. In Ayurveda it is known as *Kōkilākṣa*, on account of the similarity of the color of the flower to the eyes of the Indian cuckoo (Figure 5.4). It is used in the treatment of several diseases, including rheumatoid arthritis (Warrier et al. 2007c).

The first report on the successful treatment of *Vātaśōṇita*, using *Hygrophila spinosa* was published 133 years ago in the *British Medical Journal* (Jayesingha 1887). Six patients admitted to the Civil Hospital, Kurunagala, Ceylon were treated with infusion of the herb and four of them were successfully cured of their dropsy. A typical case report reads thus:

> Case vi.-Adrian, aged 30; male; civil condition, single; race, Singalese; birthplace, Myombo. This was a Singalese man, native of Myombo, but living from his younger days at Polgahawella as a cultivator; admitted into hospital on February 7th, 1887, suffering with extensive general dropsy depending on anemia. His condition on admission was the following: face pale and bloated, conjunctivae and tongue pale and bloodless, heart-sounds normal, abdomen distended with fluid, appetite fair, bowels regular, urine scanty, feet oedematous. He was put upon thymol, santonin and extract of opium, but he did not pass any anchylostoma duodenale; then he was put upon squilla and tinct. digitalis, and subsequently on iron, and finding no improvement, he was treated with small doses of liq. arsenicalis, and as there was no diminution in dropsy, he was put upon inf. asteracantha from March 20th, and continued till the end of the month. During that period he passed from 80 to 104 ounces of urine a day. He was discharged from the hospital cured of his dropsy on April 9th, 1887.
>
> **(Jayesingha 1887)**

Patra et al. (2009) investigated the anti-inflammatory and antipyretic activities of the petroleum ether, chloroform, alcoholic and aqueous extracts of the leaf of *H. spinosa*. Anti-inflammatory activity of the extracts was studied in rats, using carrageenan-induced paw edema. Antipyretic activity was evaluated on the basis of the effects on Brewer's yeast-induced pyrexia in rats. Chloroform and alcoholic extracts of leaves of *H. spinosa* produced significant anti-inflammatory and antipyretic activities in a dose-dependent manner. However, petroleum ether and aqueous extracts did not show significant anti-inflammatory and antipyretic activities. These two extracts also reduced elevated

FIGURE 5.4 Stem of *H. auriculata* with flowers. Reproduced with permission from Dr D. Suresh Baburaj, Ooty.

body temperature in rats at 200 and 400 mg/kg body weight doses throughout the observation period of 6 hours.

The herb shows significant diuretic activity, as reported by Hussain et al. (2009). The alcohol extract of *H. auriculata* at doses of 200 mg/kg caused a significant increase in the total urine volume and concentrations of Na^+, K^+ and Cl^- in the urine in the rats. This finding supports its traditional use.

5.2.10 *Ipomoea mauritiana* Jacq

I. mauritiana grows in the moist regions of tropical India. It is known as *Kṣīravidāri*, *Payasvini* and *Bhūmikūśmāṇḍa*. The tubers are galactagogue, cholagogue and demulcent (Warrier et al. 2007c). Mahajan et al. (2015) carried out an investigation to validate the folkloric claim of the aphrodisiac potential of *I. digitata*, using neem oil-induced sterile male albino rats. Experimental animals were made sterile by feeding emulsified neem oil for 15 days. Thereafter, they were administered orally with two doses of *I. digitata* root powder (250 mg/kg and 500 mg/kg body weight) suspended in water twice daily for 40 days. Changes in organ weight, sperm density, sperm motility, serum levels of testosterone, luteinizing hormone, follicle stimulating hormone and histological changes were measured to evaluate the effect of treatment. Significant increase in sperm density and sperm motility, along with increase in the weight of testes and epididymis were observed in rats made infertile with neem oil and treated with *I. digitata* root powder at both dose levels. Both the doses produced a concomitant increase in the serum levels of testosterone, luteinizing hormone and follicle-stimulating hormone. Testes of albino rats treated with *I. digitata* showed presence of spermatozoa and sperm bundle, confirming the restoration of spermatogenesis.

A major characteristic of defective human spermatozoa is the presence of large amounts of DNA damage, which is largely oxidative and is closely associated with defects in spermiogenesis. Spermiogenesis is disrupted by oxidative stress, leading to the production of defective gametes with poorly remodeled chromatin that are particularly susceptible to free radical attack. To worsen the situation, these defective cells have a tendency to undergo an unusual form of apoptosis associated with high amounts of superoxide generation by the sperm mitochondria. This leads to significant oxidative DNA damage that eventually culminates in the DNA fragmentation found in infertile patients (Aitken and Curry 2011). Studies have shown that the tuberous root of *I. digitata* possesses considerable antioxidant activity (Alagumanivasagam et al. 2010, 2011). It is plausible that the free radical scavenging activity of *I. digitata* may be responsible for its aphrodisiac properties.

5.2.11 *Moringa oleifera* Lam.

It is a middle-sized, graceful tree with corky grey bark and easily breakable branches (Figure 5.5). The root, bark, leaf and seeds are used. *Śigru*, *Śōbhāñjana*, *Śvētadāru* and *Śīghraphalō* are some of its Sanskrit names (Warrier et al. 2007d).

Fouad et al. (2019) attempted to isolate the bacteria from abscesses in camels and evaluated the antibacterial activity of *M. oleifera* extracts. Abscess in camels is one of the most important bacterial infections, causing anemia and emaciation. The disk diffusion method and minimum inhibitory concentration were used for the evaluation of the antibacterial activity of *M. oleifera* extracts against isolated bacteria from camel abscesses. The following bacteria were isolated from the abscesses: *Corynebacterium pseudotuberculosis*, *Corynebacterium ulcerans*, *Staphylococcus aureus*, *Escherichia coli*, *Klebsiella pneumoniae*, *Pseudomonas aeruginosa*, *Micrococcus* spp., *Proteus vulgaris*, *Citrobacter* spp. and *Staphylococcus epidermidis*. The ethanol extracts of *M. oleifera* showed pronounced antibacterial activity against all the tested organisms. This shows that *M. oleifera* can be used for controlling pyogenic bacteria.

M. oleifera root bark also possesses antibacterial activity, as reported by Dewangan et al. (2010). The antibacterial activity of different extracts of root bark was tested against *Staphylococcus aureus*,

FIGURE 5.5 *M. oleifera* with fruits. Reproduced with permission from Dr D. Suresh Baburaj, Ooty.

Escherichia coli, *Salmonella gallinarum* and *Pseudomonas aeruginosa*. Both Gram-positive and Gram-negative organisms showed variable sensitivity to different extracts of *M. oleifera* root bark prepared using methanol, acetone, ethyl acetate, chloroform and water. While ethyl acetate and acetone extracts showed maximum antibacterial activity, the aqueous extract had minimum activity against the test organisms.

Methanol extract of the root of *M. oleifera* was screened by Ezeamuzie et al. (1996) for anti-inflammatory effect using the rat paw edema and the rat 6-day air pouch inflammatory models. Orally administered extract inhibited carrageenan-induced rat paw edema in a dose-dependent manner. On the 6-day air pouch acute inflammation induced with carrageenan, the extract was much more potent. When delayed (chronic) inflammation was induced in the 6-day air pouch model using Freund's complete adjuvant, the extract was still effective, though less than in acute inflammation. The results suggest that the root of *M. oleifera* contains anti-inflammatory compounds that may be useful in the treatment of acute and chronic inflammatory conditions

The immunomodulatory potential of *M. oleifera* stem bark in lipopolysaccharide-induced human monocytic cell line THP-1 was evaluated by Vasanth et al. (2015). Lipopolysaccharide-induced THP-1 cell showed significant increase in pro-inflammatory cytokines (TNF-α, IL-6 and IL-1β), nitric oxide and reactive oxygen species. Pretreatment with *M. oleifera* stem bark showed marked inhibition of production of pro-inflammatory cytokines (TNF-α, IL-6 and IL-1β) in lipopolysaccharide-induced THP-1 cells. There was also significant inhibition of nitric oxide and R.O.S. generation, which is a critical step mediating immune response.

5.2.12 *Piper longum* L.

P. longum is one of the famous three acrids (*trikaṭu*), the others being *P. nigrum* and *Zingiber officinale*. It is known in Sanskrit as *Pippali*, *Māgadhi*, *Kaṇa* and *Kṛṣṇa* (Warrier et al. 2007d).

5.2.12.1 Effect on Digestive Efficiency

To study the impact of *P. longum* on gut microbiota Peterson et al. (2019) applied *in vitro* anaerobic cultivation of human fecal microbiota, followed by 16S r.R.N.A. sequencing to study the modulatory effects of the herb. *P. longum* was found to possess substantial power to modulate fecal bacterial communities to include potential prebiotic and beneficial repressive effects. Glycosyl hydrolase gene representation is strongly modulated by *P. longum*, suggesting that polysaccharide substrates present in it provide selective benefits to gut communities. The experimental data suggest that substrates present in *P. longum* may encourage beneficial alterations in gut communities, thereby altering their collective metabolism to contribute to digestive efficiency. This study validates the ayurvedic use of *P. longum* to correct "lowering of abdominal fire".

5.2.12.2 Antibacterial Activity

Singh and Rai (2013) evaluated the antibacterial activity of various solvent extracts of fruits of *P. longum* against different Gram-positive and Gram-negative bacteria using the disk diffusion method. The petroleum ether extract was resistant towards all the tested bacterial strains while the ethyl acetate extract was highly active. Among all the Gram-positive bacteria, *Staphylococcus aureus* was highly sensitive, with inhibition zone 24.33 mm in the presence of 500 mg/ml ethyl acetate extract, while in case of Gram-negative bacterial strains *Pseudomonas aeruginosa* and *Vibrio cholerae* were highly sensitive. Hexane extract was least inhibitory towards all the bacterial strains.

5.2.12.3 Anti-Inflammatory Activity

Guineensine is a dietary N-isobutylamide present in *P. longum* and shown to inhibit cellular endocannabinoid uptake. Considering the role of endocannabinoids in inflammation and pain reduction, Reynoso-Moreno et al. (2017) evaluated guineensine in mouse models of acute and inflammatory pain and endotoxemia. Significant dose-dependent anti-inflammatory effects like inhibition of inflammatory pain, inhibition of edema formation and acute analgesia were observed. Moreover, guineensine inhibited pro-inflammatory cytokine production in endotoxemia. Both hypothermia and analgesia were blocked by the CB_1 receptor inverse agonist rimonabant, but the pronounced hypolocomotion was CB1 receptor (Cannabinoid receptor type 1)-independent. A screening of 45 CNS-related receptors, ion channels and transporters revealed apparent interactions of guineensine with the dopamine transporter DAT, 5HT2A and sigma receptors. The potent pharmacological effects of guineensine might relate to the reported anti-inflammatory effects of *P. longum*.

Anti-inflammatory activity of oil isolated from *P. longum* dried fruits was studied by Kumar et al. (2009) in rats, using the carrageenan-induced right hind paw edema method. The oil inhibited carrageenan-induced rat paw edema. The anti-inflammatory activity was comparable to that of ibuprofen. The fruit decoction also showed anti-inflammatory activity against carrageenan-induced rat paw edema (Sharma and Singh 1980).

5.2.12.4 Anti-Cancer Effect

Guo et al. (2019) prepared three extracts of *P. longum* using reflux, ultrasonic and supercritical fluid extraction techniques. Active compounds were isolated by a bioassay-guided method and anti-inflammatory activity, anti-proliferation activity and cytotoxicity were evaluated. The anti-inflammatory activity and cytotoxicity of supercritical fluid extracts were stronger than those of the other two extracts. Supercritical fluid extract and piperine were found to reduce colony formation, inhibit cell migration and promote apoptosis through increasing cleaved P.A.R.P. (Poly (ADP-ribose) polymerase) and the ratio of Bax/Bcl-2.

Ovadje et al. (2014) assessed and validated the anticancer potential of ethanol extract of *P. longum* against human cancer cells (the malignant melanoma cell line G-361, human colorectal cancer cell lines HT-29 and HCT116, ovarian adenocarcinoma cell line OVCAR-3, pancreatic adenocarcinoma cell line BxPC-3 and normal-derived colon mucosa NCM460 cell line). Following treatment with the extract, cell viability was assessed using a water-soluble tetrazolium salt and apoptosis induction was observed following nuclear staining. The *in vivo* effect of the extract was studied using Balb/C mice and CD-1 nu/nu immunocompromised mice.

Results indicated that ethanol extract of *P. longum* selectively induces caspase-independent apoptosis in cancer cells, without affecting non-cancerous cells. This is achieved by targeting the mitochondria, leading to the dissipation of the mitochondrial membrane potential and increase in R.O.S. production. Release of the apoptosis-inducing factor (A.I.F.) and endonuclease G from isolated mitochondria confirms the mitochondria as a potential target of *P. longum*. *In vivo* studies indicate that oral administration is able to halt the growth of colon cancer tumors in immunocompromised mice, without associated toxicity. These results demonstrate that *P. longum* is a potentially safe and non-toxic alternative in cancer therapy (Ovadje et al. 2014).

Piperine (1-Piperoylpiperidine), the main component of *P. longum*, is an alkaloid with a long history of medicinal use. It exhibits a variety of biochemical and pharmaceutical activities, including chemopreventive activities, without significant cytotoxic effects on normal cells, at least at doses of <250 μg/ml. The term *chemoprevention* refers to natural agents with the ability to interfere with tumorigenesis and metastasis, or at least, attenuate the cancer-related symptoms (Zadorozhna et al. 2019).

Reports on the antibacterial, anti-inflammatory and anticancer activities justify the use of *P. longum* in the treatment of *plīha* (splenomegaly) and *yakṛt* (hepatomegaly). *Agnimāndya* or "lowering of abdominal fire" is a term used to describe a reduction in digestive efficiency. Dietary piperine, by favorably stimulating digestive enzymes of pancreas, enhances the digestive capacity and significantly reduces gastrointestinal food transit time (Srinivasan 2007).

5.2.13 *Plumbago zeylanica* L.

Agni, *Vahni*, *Hutāśa*, *Dahana* and *Śikhi* are the various Sanskrit names of *Citraka*, a component of numerous ayurvedic formulations (Figure 5.6). In Kerala, red-flowered *P. indica* is used as *Citraka*, whereas in North India, white-flowered *P. zeylanica* is preferred. The blue-colored *Plumbago* (*P. auriculata* Lam. syn. *P. capensis* Thunb.) is native to South Africa and is not used in Kerala as a medicinal plant (Warrier et al. 2007d).

Hemorrhoids are vascular structures in the anal canal that get swollen or inflamed. They often result in pain and swelling in the area of the anus. External hemorrhoids often result in pain and swelling in the area of the anus (Schubert et al. 2009). *P. zeylanica* possesses many properties that are helpful in the treatment of hemorrhoids.

Checker et al. (2009) studied the anti-inflammatory effects of plumbagin (5-hydroxy-2-methyl-1, 4-naphthoquinone) isolated from *P. zeylanica*. Plumbagin inhibited T-cell proliferation in response to polyclonal mitogen concanavalin A (Con A) by blocking cell cycle progression. It also suppressed expression of early and late activation markers CD69 and CD25 respectively, in activated T-cells.

FIGURE 5.6 *P. zeylanica* with flowers. Reproduced with permission from Dr D. Suresh Baburaj, Ooty.

At these immunosuppressive doses, plumbagin did not reduce the viability of lymphocytes. The inhibition of T-cell proliferation by plumbagin was accompanied by a decrease in the levels of Con A- induced IL-2, IL-4, IL-6 and IFN-gamma cytokines. Similar immunosuppressive effects of plumbagin on cytokine levels were observed *in vivo*. To characterize the mechanism of inhibitory action of plumbagin, the mitogen-induced IκBα degradation and nuclear translocation of NF-κB was studied in lymphocytes. Plumbagin completely inhibited Con A-induced IκBα degradation and NF-κB activation.

Shaikh et al. (2016) evaluated the *in vivo* and *in vitro* anti-inflammatory potential of *P. zeylanica*. The extracts prepared using water, ethanol and hexane were evaluated *in vitro* for COX-1 and 2 inhibition and antioxidant activities. Results showed that *P. zeylanica* inhibited COX-2 activity more as compared to COX-1. The ethanol extract of *P. zeylanica* demonstrated effective DPPH, OH and superoxide radical scavenging activity as well. The results obtained justify the use of *P. zeylanica* as an anti-inflammatory single herb remedy in the ayurvedic treatment of hemorrhoids.

Increased microvascular density has been reported in hemorrhoidal tissue, suggesting that neovascularization might be another important phenomenon of hemorrhoidal disease (Lohsiriwat 2012). Chung et al. (2004) reported that endoglin (CD105), which is one of the binding sites of TGF-β and is a proliferative marker for neovascularization, was expressed in more than half of hemorrhoidal tissue specimens compared to none taken from the normal anorectal mucosa. Moreover, these investigators found that microvascular density increased in hemorrhoidal tissue especially when thrombosis and stromal vascular endothelial growth factors (V.E.G.F.) were present. Han et al. (2005) also demonstrated that there was a higher expression of angiogenesis-related protein such as V.E.G.F. in hemorrhoids.

While studying the growth-inhibiting property of plumbagin derived from *P. zeylanica* on human prostate cancer PC-3M-luciferase cells, Hafeez et al. (2013) observed that plumbagin-caused inhibition of the growth and metastasis of PC-3M cells, accompanying inhibition of the expression of angiogenesis markers CD31 and VEGF.

The roots of *P. zeylanica* are recommended in Ayurveda as appetizer and digestive. It has been established that the feeding of *P. zeylanica* root stimulates the proliferation of coliform bacteria in mice. Supporting data on the above inference was adduced by using a special dilution method of obtaining bacterial counts. The data were statistically analyzed and the increase in coliform bacteria due to *P. zeylanica* feeding was shown to be significant at the 5% level. This coliform proliferation action of *P. zeylanica* is similar to that of *Mexaform*, a drug used to counteract the action of antibiotics. The authors therefore suggested that the claims that *P. zeylanica* is a digestive-stimulant and an appetizer is most probably due to its action as an intestinal flora normalizer (Iyengar and Pendse 1966).

P. zeylanica was evaluated for its wound-healing activity in rats. Wound-healing activity was studied using excision and incision wound models in rats following topical application of 10% w/v extract in saline. The herb was found to possess significant wound-healing activity (Reddy et al. 2002).

Medicines for hemorrhoids need to have the ability to control bacterial infections. Kaewbumrung and Panichayupakaranant (2014) reported that extraction and a one-step purification of the crude extract of *P. indica* using silica-gel vacuum chromatography provided a plumbagin derivative-rich *P. indica* root extract. Antibacterial activities of the standardized extract and three naphthoquinones, plumbagin, elliptinone and 3, 3'-biplumbagin, against *Propionibacterium acnes*, *Staphylococcus aureus* and *Staphylococcus epidermidis* were evaluated using the microdilution assay. The bactericidal activities of the extract against these bacteria were much stronger than those of elliptinone and 3, 3'-biplumbagin and almost equal to those of plumbagin.

5.2.14 *Saraca asoca* (Roxb.) de Wilde

S. asoca is a rainforest tree with beautiful foliage and fragrant, orange-yellow flowers. It is an erect, evergreen tree, with deep green leaves growing in dense clusters. The flowers are borne in heavy,

FIGURE 5.7 Inflorescence of *S. asoca*. Reproduced with permission from Dr D. Suresh Baburaj, Ooty.

lush bunches and turn red before wilting (Figure 5.7). It is reputed to cure dysfunctional bleeding (Warrier et al. 2008). Interestingly, in Ayurveda the bark of the tree is used only in the treatment of gynecological disorders.

S. indica is the major ingredient of the ayurvedic medicine *Aśōka ariṣṭa* indicated in menorrhagia, metrorrhagia, leucorrhea, primary amenorrhea, subfertility and menstrual disorders in general (Middelkoop and Labadie 1983). Women suffering from dysfunctional menorrhagia and menorrhagia caused by the use of an intra-uterine contraceptive device showed abnormally high levels of PGE_2 and PGFα in their endometrial tissue, as reported by Willman et al. (1976), who measured the concentrations of prostaglandins by radioimmunoassay in samples of endometrial tissue from 155 women. Lindner et al. (1980) examined the role of prostaglandins in three pivotal events of the female reproductive cycle – ovulation, luteolysis, and menstruation. They concluded that PG-synthase inhibitors can find useful applications in the management of menstrual disorders, such as functional dysmenorrhea and menorrhagia. As the bark of *S. asoca* is the main ingredient in *Aśōka ariṣṭa*, Middelkoop and Labadie (1985) tested extracts and pure compounds found in this bark for their influence on the PGH_2 synthase activity *in vitro*.

Extracts of *S. asoca* bark and pure compounds isolated from the bark were tested for properties that might inhibit the conversion of arachidonic acid by the PGH_2 synthase. They were assayed spectrophotometrically, with adrenaline as cofactor. Methanol- and ethyl acetate extracts of *S. asoca* inhibited the conversion and the observed inhibition was confirmed in an oxygraphic assay. Two procyanidin dimers from the ethyl acetate extract showed enzyme-catalyzed oxidation in the assay. The ether extract of the bark was also found to contain substances which were capable of being oxidized by the PGH_2 synthase. The holistic action of the components of the bark may explain the mode of action of *S. asoca* (Middelkoop and Labadie 1985)

Leucorrhoea is caused by bacterial or fungal infections in the mucous membranes of the vagina. Therefore, an effective medicine against it should have bactericidal and fungicidal activity or should be able, especially in the case of a fungal infection, to lower the pH of the mucous membranes (Spence and Melville 2007). Chloroform, methanol, aqueous and ethanol extracts of the stem bark of *S. indica* were investigated for their antibacterial and antifungal activities against standard strains of the bacteria *Staphylococcus aureus*, *Escherichia coli*, *Pseudomonas aeruginosa*, *Bacillus cereus*, *Klebsiella pneumoniae*, *Proteus mirabilis*, *Salmonella typhimurium*, *Streptococcus pneumoniae* and the fungi *Candida albicans* and *Cryptococcus albidus*. Methanol and aqueous extracts exhibited antimicrobial activity with MIC ranging from 0.5–2% and 1–3%

respectively. Methanol extract exhibited the strongest activity against both bacteria and fungi (Sainath et al. 2009).

It can be assumed that the mild oxytocic effect of a drug could stop the uterine bleeding by constriction of the blood-vessels in the myometrium. This effect can be assayed by measuring *in vitro* the direct uterine activity of the drug. Satyavati et al. (1970) reported marked uterine stimulating activity of *S. indica in vitro.*

5.2.15 *Strychnos potatorum* L. f.

S. potatorum is a deciduous tree. The seeds of the tree are commonly used in Ayurveda, as well as for purifying water (Figure 5.8) (Warrier et al. 2008). No clinical studies have been reported for the anti-diabetic activity of *S. potatorum*. However, a few studies indicate that the herb has the ability to regulate blood glucose levels. Dhasarathan and Theriappan (2011) investigated the anti-diabetic activity of *S. potatorum* seeds. A diabetic state was induced in Wistar albino rats by intraperitoneal injection of alloxan at a dose of 100 mg/kg of body weight. Treatment with alloxan reduced body weight and liver weight. The blood glucose level dropped by 53% with extract treatment, demonstrating the anti-diabetic potential of the plant. Serum enzymes A.S.T. and A.L.T. were increased from 24 and 18 IU/1 to 60 and 65 IU/1 respectively, whereas A.L.P. was reduced to 5 IU/1 from 14 IU/1. The total serum protein level also increased up to 5 mg/ml in extract-treated animals. The insulin level showed an increase of up to 61 µg/ml within 30 days of extract treatment compared to controls. The plant extract lowered the initial cholesterol level of 219 µg/ml to 170 µg/ml. In liver, the A.S.T., A.L.T. and A.L.P. enzymes were decreased significantly (Dhasarathan and Theriappan 2011).

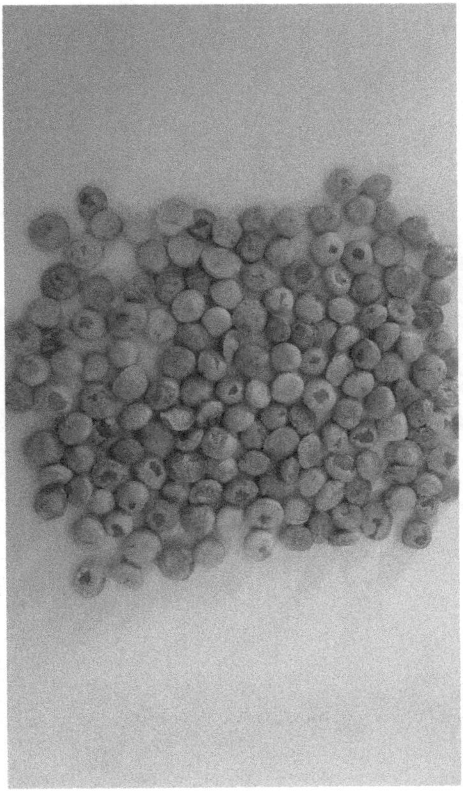

FIGURE 5.8 Seeds of *S. potatorum*. Reproduced with permission from Dr D. Suresh Baburaj, Ooty.

The anti-diabetic effect of seeds of *S. potatorum* Linn. was further evaluated in a model of diabetes mellitus using streptozotocin (Biswas et al. 2012). Changes in fasting blood sugar were monitored periodically for 12 weeks along with weekly measurement of body weight, food and water intake for 4 weeks. The anti-diabetic effects were compared with glipizide as the reference hypoglycemic drug. *S. potatorum* Linn. (100 mg/kg *p.o.*) significantly reduced fasting blood sugar, the effects being comparable with glipizide (40 mg/kg, *p.o.*), which is an established hypoglycemic drug. The herb also increased body weight in streptozotocin-induced diabetic rats. Biswas et al. (2012) remark that, as the development of diabetes by streptozotocin is related to increased generation of free radicals (Van Dyke et al. 2010), it is possible that the observed anti-diabetic effect is mediated, at least partly, through its antioxidant effect. Previous studies have demonstrated the anti-arthritic, anti-inflammatory and antioxidant activity of *S. potatorum* Linn. (Ekambaram et al. 2010), which may be attributed to the presence of antioxidants such as flavonoids and phenols (Mallikharjuna et al. 2007). The available research reports indicate that *S. potatorum* has promising anti-diabetic activity meriting further pharmacological studies.

5.2.16 *Tabernaemontana divaricata* (L.) R.Br. ex Roem. & Schultes

T. divaricata is an evergreen shrub native to India. It is now cultivated throughout South-East Asia and the warmer regions of continental Asia. The plant generally grows to a height of 5–6 feet and the large shiny leaves are deep-green in color. The waxy flowers are found in small clusters on the stem tips. The flowers appear sporadically throughout the year (Figure 5.9). The roots have a bitter taste. It is used in the treatment of eye diseases and diabetes mellitus (Warrier et al. 2008).

FIGURE 5.9 *T. divaricata*. Reproduced with permission from Dr. D Suresh Baburaj, Ooty.

The leaves are rich in alkaloids like conophylline, conophyllidine (Kam et al. 1993), voaphylline, N_1-methylvoaphylline, voaharine, pachysiphine, apparicine, (-)-mehranine, conofoline (Kam and Anuradha 1995), taberhanine, voafinine, *N*-methylvoafinine, voafinidine, voalenine and conophyllinine (Kam et al. 2003).

Fujii et al. (2009) studied the anti-diabetic effects of conophylline in rats by oral administration. Crude conophylline-containing extracts were prepared from the leaves of *T. divaricata* and administered to both normal and streptozotocin-induced diabetic Sprague Dawley rats. Conophylline was orally absorbed and showed an increase in its plasma level in normal and diabetic rats. After a single oral administration, the plasma conophylline concentration reached its maximum from 1.5 h to 3 h and gradually decreased in 24 h. The alkaloid lowered the blood glucose level and increased the plasma insulin level in the diabetic rats after daily administration for 15 days. Rats treated with conophylline at 0.11 and 0.46 mg/kg/day had fasting blood glucose levels of 411 ± 47 and 381 ± 65 mg/dl, respectively; while the glucose level of control rats was 435 ± 46 mg/dl. The fasting blood glucose levels in Goto-Kakizaki rats were also decreased by conophylline in a dose-dependent manner after repetitive administration for 42 days. (The Goto-Kakizaki rat is a model for type 2 diabetes, bred from non-diabetic Wistar rats selected from a normal population with a glucose tolerance response slightly deviating from the normal range (Goto et al. 1975)). These results suggest that the extract from conophylline-containing plants like *T. divaricata* may be useful in the treatment of type 2 diabetes mellitus.

Conophylline is also reported to increase the β-cell mass in neonatal streptozotocin-treated rats (Kodera et al. 2009). Streptozotocin (S.T.Z., 100 μg/g) was injected into neonatal rats, and then conophylline (2 μg/g) was administered to them for 1 week. In neonatal S.T.Z. rats treated with control solution, the plasma glucose concentration increased for several days, and after 8 weeks the plasma glucose concentration was very high compared to that of normal rats. The glucose response to intraperitoneal glucose tolerance test was significantly reduced in neonatal S.T.Z. rats treated with conophylline. The β-cell mass and the insulin content of the pancreas were also significantly increased in the conophylline group.

Reduction of the β-cell mass is crucial to the pathogenesis of diabetes mellitus. Therefore, the discovery of bioactives, which induce differentiation of pancreatic precursors to β- cells, could be a new approach to treat diabetes. To identify such agents, Kojima and Umezawa (2006) screened many compounds, using pancreatic AR42J cells, a model of pancreatic progenitor cells. They found that conophylline, originally extracted from leaves of *Ervatamia microphylla* and present in *T. divaricata* too, was effective in converting AR42J into pancreatic β-cells. The differentiation-inducing activity of conophylline is very similar to that of activin A. However, unlike activin A, conophylline does not induce apoptosis. Conophylline also causes differentiation of cultured pancreatic precursor cells obtained from fetal and neonatal rats (Kojima and Umezawa,2006; Sidthipong et al. 2013).

Kanthlal et al. (2014) investigated the anti-diabetic effect of aerial parts of *T. divaricate* and their ability to prevent oxidative stress in alloxan-induced diabetic rats. Two doses of methanol extract of aerial parts of *T. divaricata* (100 and 200 mg/kg, *per os*.) were tested in alloxan-diabetic rats. Administration of a 200 mg/kg dose attenuated the increased level of glucose produced by alloxan. The extract also alleviated the effect of the oxidative damage similar to the standard drug glibenclamide.

5.2.17 *Terminalia arjuna* (Roxb. ex DC) Wight & Arn.

Terminalia arjuna or the Arjun tree is usually found growing on river banks or near dry river beds. It is a tall tree usually with a buttressed trunk, having a smooth, grey bark. It bears pale yellow flowers and the fibrous woody fruit has five wings. The bark is a reputed remedy for heart disease (Kumar and Prabhakar 1987; Warrier et al. 2008).

The first study on cardiotonic and other properties of *T. arjuna* was conducted by Colabawalla (1951). She investigated the efficacy of bark powder in treating congestive cardiac failure (C.C.F.), essential hypertension and cirrhosis of the liver in 30 patients. Seven out of 17 C.C.F. cases showed marked improvement after administration of pulverized Arjuna bark. C.C.F. due to a congenital anomaly of the heart and valvular disease were also brought under control. Interestingly, four out of nine cases of C.C.F. due to chronic bronchitis were also relieved by the treatment.

T. arjuna relieved symptomatic complaints of essential hypertension like giddiness, insomnia, lassitude, headache and inability to concentrate. Pressor responses to the drug showed great variability. In two cases, the drug was found to be hypotensive and to five patients it provided symptomatic alleviation only. The author speculated that the ability of *T. arjuna* to lower blood pressure might be due to its general tonic effect. Marked improvement in the stamina and well-being of patients who responded to the treatment was observed (Colabawalla 1951).

The mechanism of cardiovascular action of *T. arjuna* was investigated by Singh et al. (1982). Aqueous solution of the alcoholic extract produced sustained dose-dependent hypotension and bradycardia in cats and dogs. The drug did not modify pressor responses to norepinephrine or splanchnic nerve stimulation, thus ruling out the blockade of autonomic ganglia, adrenergic neurone and peripheral adrenergic receptors. However, carotid occlusion response was inhibited, suggesting involvement of the central nervous system. The contention is further supported by results of bilateral stellate ganglionectomy and intracerebroventricular and intravertebral arterial administration of the drug.

As bradycardia was found to be associated with decreased blood pressure, Singh et al. (1982) reported that the effect of the drug is mediated either via the C.N.S. or the vagal and sympathetic nerves to the heart. Bilateral vagotomy blocked the lowering of heart rate following intravenous administration of the drug and bradycardia was observed when very low doses were administered centrally. These data point towards the C.N.S. as the principal site of action of *T. arjuna* in bringing forth cardiovascular responses. The hypotensive action of the drug in experimental animals is substantiated by the observation of Colabawalla (1951) that the drug was found to lower blood pressure in human subjects.

Administration of *T. arjuna* powder is fairly effective in patients with symptoms of stable angina pectoris, as evidenced by the report of Dwivedi and Agarwal (1994). Anginal frequency, blood pressure, body mass index, blood sugar, cholesterol and HDL-cholesterol were studied in 15 stable and 5 unstable angina patients before and 3 months after *T. arjuna* therapy. There was a 50% reduction in anginal episodes in the stable angina group. Treadmill test performance improved from moderate to mild changes in 5 patients, and one with mild changes became negative for ischemia. The time to the onset of angina and appearance of ST-T changes on the treadmill test was delayed significantly after *T. arjuna* treatment. Nevertheless, there was an insignificant reduction in anginal frequency in patients with unstable angina. The herb lowered systolic blood pressure and body mass index to a significant level and increased HDL-cholesterol slightly along with marginal improvement in left ventricular ejection fraction in the stable angina patients (Dwivedi and Agarwal 1994).

The effect of *T. arjuna* on angina pectoris, congestive heart failure and left ventricular mass was studied in patients of myocardial infarction with angina and or ischemic cardiomyopathy. Bark powder of *T. arjuna* was administered (500 mg 8 hourly) to ten patients of post-myocardial infarction angina and two patients of ischemic cardiomyopathy, post-operatively, for a period of three months. These patients were also on conventional treatment. Twelve patients of post-myocardial infarction angina receiving only conventional treatment served as controls. Significant reduction in anginal frequency was noted in both groups. However, only *T. arjuna*-treated patients showed significant improvement in the left ventricular ejection fraction and a reduction in left ventricular mass on echocardiography following three months of therapy. Both groups showed significant symptomatic relief in coronary heart failure. Prolonged treatment with *T. arjuna* did not produce any adverse effects (Dwivedi and Jauhari 1997).

The efficacy of *T. arjuna* in the treatment of chronic stable angina was studied by Bharani et al. (2002) in a double-blind, placebo-controlled, crossover study comparing its effect with isosorbide mononitrate. Fifty-eight males with chronic stable angina with evidence of provocable ischemia on the treadmill exercise test received *T. arjuna* bark extract (500 mg 8 hourly), isosorbide mononitrate or a matching placebo for one week each, separated by a wash-out period of at least three days in a randomized, double-blind, crossover design. *T. arjuna* therapy was associated with significant decrease in the frequency of angina and need for isosorbide dinitrate. The treadmill exercise test parameters improved significantly during therapy with *T. arjuna*. The total duration of exercise increased, maximal S.T. segment depression during the longest equivalent stages of submaximal exercise decreased, time to recovery decreased and higher double products were achieved during *T. arjuna* therapy. Similar improvements in clinical and treadmill exercise test parameters were observed with isosorbide mononitrate compared to placebo therapy. No significant differences were observed in clinical or treadmill exercise test parameters when *T. arjuna* and isosorbide mononitrate therapies were compared (Bharani et al. 2002).

Twelve patients with refractory chronic congestive heart failure received bark extract of *T. arjuna* (500 mg 8 hourly), or matching placebo for two weeks each, separated by a two week wash-out period, in a double-blind crossover design as an adjuvant to maximally tolerable conventional therapy (Phase I). Echocardiographic and other routine evaluations were carried out at baseline, and at the end of *T. arjuna* and placebo therapy. Compared to placebo, *T. arjuna* therapy was associated with an improvement in symptoms and signs of heart failure, improvement in N.Y.H.A. Class, decrease in echo-left ventricular end-diastolic and end-systolic volume indices, increase in left ventricular stroke volume index and increase in left ventricular ejection fractions. On long-term evaluation in an open design (Phase II), in which Phase I participants continued *T. arjuna* in fixed dosage (500 mg 8 hourly) in addition to flexible diuretic, vasodilator and digitalis dosage for 20–28 months on out-patient basis, patients showed continued improvement in symptoms, signs, effort tolerance and N.Y.H.A. Class, with improvement in quality of life (Bharani et al. 1995).

Kapoor et al. (2015) evaluated the cardioprotective effects of *T. arjuna* on immuno-inflammatory markers in coronary artery disease (C.A.D.) as an adjuvant therapy. One hundred and sixteen patients with stable C.A.D. were administered placebo or *T. arjuna* powder along with medications in a randomized double-blind clinical trial. To understand the specificity and efficacy of *T. arjuna*, its effect was evaluated through microarray and *in silico* analysis in a few representative samples. Data were further validated via real-time P.C.R. each at baseline, 3 months and 6 months, respectively. rIL-18 cytokine was used to induce inflammation *in vitro* to compare its effects with atorvastatin. *T. arjuna* significantly down-regulated triglycerides, V.L.D.L.-C. and immuno-inflammatory markers in stable C.A.D. subjects. Microarray and pathway analysis of a few samples from *T. arjuna* or placebo-treated groups and real-time P.C.R. validation further confirmed the observations. This study demonstrates the anti-inflammatory and immunomodulatory effects of *T. arjuna* are helpful in attenuating the ongoing inflammation and immune imbalance in medicated C.A.D. subjects.

The *in vitro* effects of *T. arjuna* ethanol bark extract on platelet function indices were demonstrated by Malik et al. (2009). Twenty patients of angiographically proven C.A.D. were included in Group I and 20 age- and sex-matched controls were included in Group II. Platelet activation was monitored by determining P-selectin (CD62P) expression, intracellular free calcium release and platelet aggregation. The ethanol extract was able to inhibit platelet aggregation significantly both in patient and control groups. Significant attenuation in calcium release and expression of CD62P was also observed with the extract. This study clearly demonstrates that the bark extract of *T. arjuna* decreases platelet activation and may possess antithrombotic properties. This effect may be brought about by desensitizing platelets to the agonist by competing with platelet receptor or by interfering with signal transduction. This study shows that *T. arjuna* can be used in the treatment of C.A.D. and related cardiovascular disorders.

5.2.18 *Terminalia belerica* (Gaertn.) Roxb.

T. belerica is a large deciduous tree common on plains and lower hills in South-East Asia. The pericarp of the fruit is one of the three ingredients of the famous ayurvedic *rasāyana*, *triphala*

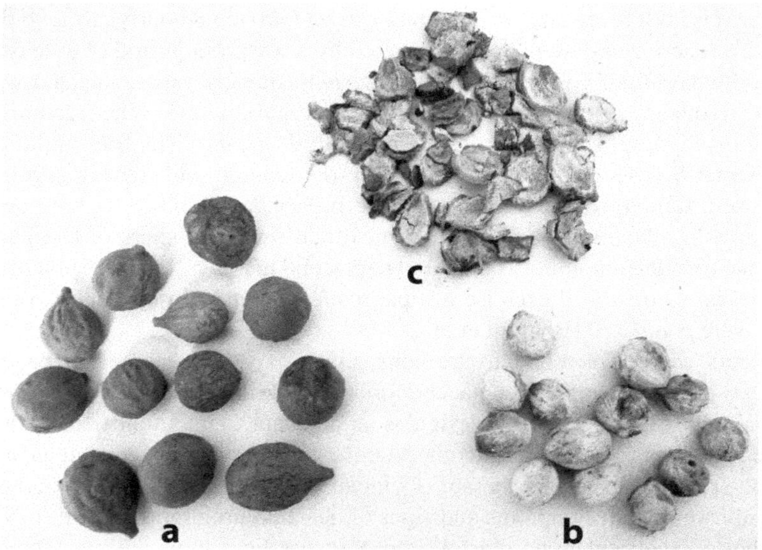

FIGURE 5.10 Fruits (a), seeds (b) and pericarp (c) of *T. belerica*. Reproduced with permission from Dr D. Suresh Baburaj, Ooty.

(Figure 5.10). Its popular Sanskrit names are *Vibhītaki* (fearless) and *Anilaghnaka* (killer of wind), alluding to its *vātā*-pacifying property (Warrier et al. 2008).

Trivedi et al. (1979) conducted a clinical study on the usefulness of the powder of pericarp of *T. belerica* in treating respiratory diseases. They reported that the powder produced antispasmodic, anti-asthmatic and antitussive effects.

Plants used in the treatment of hyperactive airway disorders usually exhibit spasmolytic effect through a combination of mechanisms (Gilani et al. 2005a, 2005b, 2008a). Therefore, Gilani et al. (2008b) investigated the pharmacological basis for the medicinal use of *T. belerica* in hyperactive respiratory disorders. They conducted the studies *in vitro* and *in vivo*. *T. belerica* crude extract caused relaxation of spontaneous contractions in isolated rabbit jejunum at 0.1–3.0 mg/mL. The extract inhibited the carbachol- (1 M) and K^+ (80 mM)-induced contractions in a pattern similar to that of dicyclomine, but different from nifedipine and atropine. The extract shifted the Ca^{++} concentration-response curves to right, like nifedipine and dicyclomine. In guinea pig ileum, the extract produced a rightward parallel shift of acetylcholine-curves, followed by a non-parallel shift at a higher concentration, with the suppression of maximum response, similar to dicyclomine, but different from nifedipine and atropine. *T. belerica* crude extract exhibited a protective effect against carbachol-mediated bronchoconstriction in rodents. In guinea pig trachea, the extract relaxed the carbachol-induced contractions, shifted carbachol-curves to the right and inhibited the contractions of K^+. An anticholinergic effect was distributed both in organic and aqueous fractions, while Ca^{++} channel-blocking activity was present in the aqueous fraction. These results indicate that the *T. belerica* fruit possesses a combination of anticholinergic and Ca^{++} antagonist effects, which supports its ayurvedic use in respiratory diseases (Gilani et al. 2008b).

5.2.19 *Tinospora cordifolia* (Willd.) Miers ex Hook. f. & Thoms

T. cordifolia is one of the most important herbs of Ayurveda. It is a large, deciduous, extensively spreading climber. The stem is the commonly used plant part. Its Sanskrit synonyms are *Amṛtā*, *Guḍūci* and *Chinnarūha*. It is a very powerful *rasāyana* herb (Warrier et al. 2008).

5.2.19.1 In Ethnomedicine

Ethnobotanical studies show that *T. cordifolia* is utilized by many indigenous people of India. For example, tribal people of Bastar district in Chhattisgarh use a paste of the roots in the treatment of leprosy (Sinha et al. 2013). The herb is used for same purpose by the people of the Kumaon region (Shah and Joshi 1971), the Dehra Dun-Siwalik region (Sharma et al. 1979) and Salsette Island near Bombay (Shah 1984; Gupta et al. 2010).

5.2.19.2 Clinical Studies

Kiṭibha kuṣḍha or psoriasis is a dreadful dermatological condition affecting 2.5% of the world population. It is a common, chronic and non-infectious skin disease characterized by well-defined, slightly raised, dry erythematous macules with silvery scales and typical extensor distribution, affecting any sex and having incidence at any time throughout the life (Menter et al. 2008). Saini (2017) carried out a clinical study on 20 patients of psoriasis. The patients were aged between 20 and 60 years. Powder of *T. cordifolia* stem (5 g) was administered twice a day in lukewarm water for 90 days. On completion of the trial, patients were examined and results evaluated on the basis of the Psoriasis Area and Severity Index (P.A.S.I.) (Fredriksson and Pettersson 1978; Bhor and Pande 2006). Other symptomatic characteristics like *dāha* (burning index) and *kaṇḍu* (itching index) were also considered. The mean P.A.S.I. score at the beginning of the trial was 24.14 and it was reduced to 12.10 giving a relief of 49.87%. This was statistically significant. There was significant improvement in *dāha* and *kaṇḍu* also.

Diabetic ulcers of the lower extremity are a major concern, preceding most of the amputations in general surgery. More than 7 million people in India suffer from diabetic foot ulcers. The patient has to spend 25–30% of his annual income on the diabetic foot (Shobhana et al. 2001). Treatment with growth factors is expensive with risk of infection transmission, and these factors may not achieve optimum wound concentration. Therefore, Purandare and Supe (2007) evaluated the role of generalized immunomodulation in diabetic ulcers using *T. cordifolia* as an adjuvant therapy. They studied its influence on parameters and determinants of healing, on bacterial eradication and on polymorphonuclear phagocytosis.

A prospective double-blind, randomized, controlled study lasting for over 18 months was conducted in 50 patients. The ulcer was classified by wound morphology and severity with Wound Severity Score (Pecoraro-Reiber system) (Reiber and Raugi 2005). Mean ulcer area, depth and perimeter were measured and swabs taken from culture. Blood was collected to assess polymorphonuclear phagocytosis. Medical therapy, glycemic control, debridement and wound care were optimized. At 4 weeks, parameters were reassessed (Purandare and Supe 2007).

Diabetic patients with foot ulcers on *T. cordifolia* therapy showed significantly better final outcome, with improvement in wound-healing. Reduced debridements and improved phagocytosis were statistically significant. The results indicate beneficial effects of immunomodulation with *T. cordifolia* for ulcer healing (Purandare and Supe 2007).

5.2.19.3 Effects on Immunity

The alcoholic and aqueous extracts of *T. cordifolia* are reported to have beneficial effects on the immune system. Nagarkatti et al. (1994) reported a significant improvement in Kupffer cell function and a trend towards normalization in rats with damaged liver. Rege et al. (1993) tested the efficacy of *T. cordifolia* in reversing immunosuppression. Thirty patients were randomly divided into two groups, matched with respect to clinical features, impairment of hepatic function and immunosuppression. Group I received conventional treatment with vitamin K, antibiotics and biliary drainage. Group II received *T. cordifolia* extract (16 mg/kg/day orally) in addition, during the period of biliary drainage. Hepatic function remained similar in the two groups after drainage. Nevertheless, the phagocytic and killing capacities of neutrophils normalized only in patients receiving *T. cordifolia*. Clinical evidence of septicemia was observed in 50% of patients in Group I as against none in Group II. Post-operative survival in Groups I and II was 40% and 92.4%

respectively. Thus the study shows that *T. cordifolia* improves surgical outcome by strengthening host defenses.

Rege et al. (1989) conducted a clinical study to determine the immune status of patients with obstructive jaundice. Screening of 16 patients for phagocytic and microbicidal activity of polymorphonuclear cells revealed a significant depression of these functions, as compared to normal values. Using rats, an animal model of cholestasis was also established. This model revealed a significant depression of activity of polymorphonuclear cells and peritoneal macrophages. These cellular abnormalities were found to precede and predispose to infection. The rats also showed an increased susceptibility to *Escherichia coli* infection. The authors made an attempt to improve this immunosuppression in these animals by treating them with aqueous extract of *T. cordifolia* (100 mg/kg for 7 days), following development of cholestasis. *T. cordifolia* extract improved the cellular immune functions. Death rate following *Escherichia coli* infection was significantly reduced. This study demonstrated that cholestasis results in immunosuppression, and therefore indicates the need for an immunomodulator in management of obstructive jaundice. *T. cordifolia* has the potential to consolidate host defense mechanisms. The immunomodulatory property of *T. cordifolia* has been confirmed by many other studies (Thatte et al. 1987; Dahanukar et al. 1988; Thatte and Dahanukar 1989; Rege et al. 1999; Manjrekar et al. 2000; Dikshit et al. 2000; Nair et al. 2004; 2006; Alsuhaibani and Khan 2017).

5.2.19.4 Antibacterial Activity

Extracts of *T. cordifolia* have remarkable antibacterial activity against several organisms like *Streptococcus mutans* (Agarwal et al. 2019), *Escherichia coli*, *Proteus vulgaris*, *Enterobacter faecalis*, *Salmonella typhi*, *Staphylococcus aureus*, *Serratia marcesenses* (Jeyachandran et al. 2003), *Salmonella typhimurium* (Alsuhaibani and Khan 2017), *Pseudomonas aeruginosa*, *Klebsiella pneumoniae*, *Escherichia coli*, *Bacillus subtilis*, *Proteus mirabilis* (Chakraborty et al. 2014), *Klebsiella pneumoniae* and *Pseudomonas* spp. (Mishra et al. 2013).

5.2.19.5 Anti-Inflammatory Activity

Wesley et al. (2008) studied the anti-inflammatory activity of alcohol extract of *T. ordifolia* on carrageenan-induced hind paw edema and cotton pellet granuloma models in male Wistar rats. Hind paw edema was produced by a subplantar injection of carrageenan and the paw volume was measured plethysmographically. A sub-acute model of cotton pellet granuloma was produced by implantation of sterile cotton in the axilla under ether anesthesia. Ethanol extract of the herb was administered orally and diclofenac sodium was used as a standard drug. The alcohol extract was able to inhibit the edema caused by the two models of inflammation.

The water extract of the stem of *T. cordifolia* that grows on *Azadirachta indica* (neem tree) significantly inhibited acute inflammatory response evoked by carrageenan administered orally and intraperitoneally. Significant inhibition of primary and secondary phases of inflammation was observed in a model of adjuvant-induced arthritis. The extract also significantly inhibited antibody formation by typhoid H antigen. A mild analgesic effect as well as potentiation of morphine analgesia was also observed (Pendse et al. 1977; Jana et al. 1999). The aqueous extract of stem was reported to produce a significant anti-inflammatory effect in both cotton pellet-induced granuloma and formalin-induced arthritis rat models (Gulati 1980; Gulati and Pandey 1982).

5.2.20 *Vitex negundo* L.

Vitex negundo is a large, erect and aromatic tree. The numerous flowers are borne in panicles. It is widely used in Ayurveda (Warrier et al. 2008). The bronchodilatory and bronchial hyperreactivity-reducing effects of alcohol extract of *V. negundo* was studied in experimental animals by Patel et al. (2010). Effect of extract was studied on toluene diisocyanate-induced asthma. Bronchodilator activity was studied on the histamine and acetylcholine aerosol-induced bronchospasm in guinea pigs and bronchial hyperreactivity was studied on broncho-alveolar lavage fluid in egg albumin-sensitized guinea pigs. The effect on the hypersensitivity reaction was studied by heterologous passive cutaneous anaphylaxis model using rats and mice. *In vitro* mast cell stabilizing activity was studied using p-Methoxy-N- methylphenethylamine as the degranulating agent.

Histopathological studies clearly showed the protective effect of extract against toluene diisocyanate-induced asthma. The extract offered significant protection against histamine and acetylcholine aerosol-induced bronchospasm in guinea pigs. A significant decrease in total leukocyte and differential leukocyte count in the broncho-alveolar lavage fluid of the egg albumin-sensitized guinea pigs was observed after administration of the extract. Treatment with the extract for three days inhibited the hypersensitivity reaction. The extract dose dependently protected the mast cell disruption induced by p-Methoxy-N- methylphenethylamine. Alcohol extract of *V. negundo* therefore offers not only bronchodilation, but also decreases bronchial hyperreactivity, thereby providing protection against asthma (Patel et al. 2010).

Considering the use of *V. negundo* in Ayurveda to treat respiratory disorders, Haq et al. (2012) carried out an investigation on its cough-relieving potential. The antitussive effect of the butanol extract of *V. negundo* on sulfur dioxide-induced cough was studied in mice. The safety profile of the extract was examined by observing acute neurotoxicity, median lethal dose and behavior. The extract dose-dependently inhibited the cough provoked by sulfur dioxide gas and exhibited maximum protection after 60 minutes of administration. At 1000 mg kg (–1), the extract caused maximum cough-suppressive effects. The LD_{50} value of *V. negundo* was found to be greater than 5000 mg kg (–1). Toxicity tests showed no signs of neural impairment and acute behavioral toxicity at antitussive doses. The extract was well tolerated at higher doses. This study demonstrated the antitussive effect of *V. negundo* devoid of toxicity.

Patel and Deshpande (2013) isolated and identified active constituents of leaves of *V. negundo* for its effects on bronchial hyperresponsiveness. Effects of aqueous sub-fraction, acetone sub-fraction, and chloroform sub-fraction on bronchial hyperresponsiveness and serum bicarbonate level were studied using egg-albumin-induced asthma in guinea pigs. The structure of the aqueous sub-fraction was determined by various spectroscopic methods. Animals pretreated with the aqueous sub-fraction showed significantly lower serum bicarbonate level, when compared to untreated sensitized animals. Animals pretreated with aqueous sub-fraction showed significantly lower eosinophils count, when compared with untreated sensitized animals. Histopathology of lungs of sensitized guinea pigs showed infiltration of inflammatory cells like eosinophils and neutrophils due to marked inflammation in lungs. On the other hand, lungs of animals pretreated with the aqueous sub-fraction showed normal airway, blood vessels, and broncho-alveolar space. The authors concluded that the aqueous sub-fraction possesses anti-eosinophilic activity and reduces bronchial hyperresponsiveness. The structure of the bioactive compound in the aqueous sub-fraction was determined as 5-hydroxy-3, 6, 7, 3', 4'-pentamethoxy flavone. The study provided a rationale for the use of *V. negundo* in the treatment of asthma and various inflammatory, allergic and immunological diseases (Patel and Deshpande 2013).

Khan et al. (2015) explored the mechanisms underlying the effectiveness of *V. negundo* in hyperactive respiratory disorders. Different doses of the crude extract of *V. negundo* were tested for *in vivo* bronchodilatory activity in anesthetized rats. The underlying mechanisms were studied in isolated guinea pig tracheal strips, suspended in organ baths at 37°C. Intravenous doses of the crude extract exhibited a dose-dependent bronchodilatory effect against carbachol-induced bronchoconstriction, similar to aminophylline.

In isolated guinea-pig tracheal strips, the extract relaxed carbachol and high K^+-induced pre-contractions, similar to papaverine. Diltiazem also relaxed both contractions with more potency against high K^+ pre-contraction. Pre-incubation of the tracheal strips with the extract potentiated the isoprenaline inhibitory concentration response curves, similar to papaverine. Results of this study suggest that crude extract of *V. negundo* possesses a combination of papaverine-like phosphodiesterase inhibitor and diltiazem-like Ca^{++} entry blocking constituents, which partially explain its bronchodilatory effect, thus validating its medicinal use in asthma (Khan et al. 2015).

5.3 ROLE OF CO-ADMINISTERED ADJUVANTS

Ayurveda recommends medicinal substances to be mixed with adjuvants like honey, sesame oil, mustard oil, clarified butter, rock-salt or jaggery. These substances enhance the action of the therapeutic agent. For example, honey is used in Ayurveda as a *yōgavāhi* or bioavailability enhancer. Ayurvedic physicians believe that it penetrates the tissue elements through porous channels, thereby strengthening and enhancing the action of all other herbs. *Āśukāri* is the property of honey that facilitates the penetration at the subtle level (Prakash and Rao 2014).

Looking from the perspective of modern pharmacology, honey seems to cause some effects on cytochrome P-450 isozymes (CYPs) in healthy patients and animals. CYPs are the major enzymes involved in drug metabolism (Guengerich 2008). Oral or intravenous administration of honey was found to decrease the plasma concentration of diltiazem in rabbits (Koumaravelou et al. 2002a). Similar results have been reported in the case of carbamazepine also (Koumaravelou et al. 2002b). Tushar et al. (2007) investigated the influence of seven days of honey administration on the activity of CYP3A4, CYP2D6 and CYP2C19 drug-metabolizing enzymes in healthy volunteers, using appropriate biomarkers and probe drugs. They observed that honey obtained from western ghats of southern India may induce CYP3A4 enzyme activity, but not CYP2D6 and CYP2C19 enzyme activities. Therefore, it is possible that the addition of honey to single herbs may alter the bioavailability of certain bioactive components.

5.4 CONCLUSION

Single herb remedies were widely used by traditional Ayurveda practitioners. However, with the advent of commercially manufactured ayurvedic medicines, this practice has nearly faded away. The major advantages of using single herb remedies are the less expensive nature of the therapy and swiftness of results. When compared to the treatment of diseases using compound ayurvedic formulations, the expense incurred with single herb therapy is minimal. However, a thorough understanding of the patient and the condition of the disease is essential, since the various applications are highly condition-specific (Gayathri et al. 1996).

Improper use of single herb remedies raises the danger of causing side-effects. Many of the cardiovascular clinical trials reported side-effects of garlic use. The most frequently reported ones were gastrointestinal symptoms such as abdominal pain, feelings of fullness, anorexia, and flatulence (Helou and Harris 2007). Another commonly used spice-cum-medicine is ginger and it is known to cause constipation. Nevertheless, as these and other single herb remedies are always administered in ayurvedic practice along with adjuvants like sesame oil, jaggery, honey, milk and clarified butter, the chances of their causing side-effects are remote (Table 5.1).

Numerous single herb remedies are recommended in Ayurveda. The medicinal applications described in this chapter represent only a fraction of what is available in Ayurveda. A survey of the literature reveals that clinical studies have been reported only for five herbs, while animal/*in vitro* studies are available for substantiating the therapeutic properties attributed to the remaining herbs (Table 5.1). Therefore, there is an obvious need to generate evidences for these claims through clinical studies. Results accrued from such studies can facilitate the application of the medical knowledge to fortification of food and beverages as well. Considering the advantages associated with single herb therapy described in Ayurveda, there is a need to study it in detail.

REFERENCES

Agarwal, S., P. H. Ramamurthy, B. Fernandes, A. Rath, and P. P. Sidhu. 2019. Assessment of antimicrobial activity of different concentrations of *Tinospora cordifolia* against *Streptococcus mutans*: An *in vitro* study. *Dental Research Journal* 16, no. 1: 24–8.

Aitken, R. J., and B. J. Curry. 2011. Redox regulation of human sperm function: From the physiological control of sperm capacitation to the etiology of infertility and DNA damage in the germ line. *Antioxidants and Redox Signaling* 14, no. 3: 367–81.

Alagumanivasagam, G., A. Kottaimuthu, D. S. Kumar, K. Suresh, and R. Manavalan. 2010. *In vivo* antioxidant and lipid peroxidation effect of methanolic extract of tuberous root of *Ipomoea digitata* (Linn.) in rat fed with high fat diet. *International Journal of Applied Biology and Pharmaceutical Technology* 1: 214–20.

Alagumanivasagam, G., A. Kottaimuthu, and R. Manavalan. 2011. Antioxidant potential of methanolic extract of tuberous root of *Ipomoea digitata* (Linn). *Asian Journal of Chemistry* 23: 1393–4.

Alsuhaibani, S., and M. A. Khan. 2017. Immune-stimulatory and therapeutic activity of *Tinospora cordifolia*: Double-edged sword against salmonellosis. *Journal of Immunology Research* 2017. doi:10.1155/2017/1787803.

Bayan, L., P. H. Koulivand, and A. Gorji. 2014. Garlic: A review of potential therapeutic effects. *Avicenna Journal of Phytomedicine* 4, no. 1: 1–14.

Beji, R. S., R. B. Mansour, I. B. Rebey et al. 2019. Does *Curcuma longa* root powder have an effect against CCl_4-induced hepatotoxicity in rats: A protective and curative approach. *Food Science and Biotechnology* 28, no. 1: 181–9.

Bharani, A., A. Ganguli, L. K. Mathur, Y. Jamra, and P. G. Raman. 2002. Efficacy of *Terminalia arjuna* in chronic stable angina: A double-blind, placebo-controlled, crossover study comparing *Terminalia arjuna* with isosorbide mononitrate. *Indian Heart Journal* 54, no. 2: 170–5.

Bharani, A., A. Ganguly, and K. D. Bhargava. 1995. Salutary effect of *Terminalia arjuna* in patients with severe refractory heart failure. *International Journal of Cardiology* 49, no. 3: 191–9.

Bhat, S. D., B. K. Ashok, R. N. Acharya, and B. Ravishankar. 2012. Anticonvulsant activity of raw and classically processed *Vacha* (*Acorus calamus* Linn.) rhizomes. *Ayu* 33, no. 1: 119–22.

Bhattacharya, I. C. 1968. Effect of Acorus (Vacha) oil on the amphetamine induced agitation, hexobarbital-sleeping time and on instrumental avoidance conditioning in rats. *Journal of Research in Indian Medicine* 2: 195–201.

Bhor, U., and S. Pande. 2006. Scoring systems in dermatology. *Indian Journal of Dermatology, Venereology and Leprology* 72, no. 4: 315–21.

Bisht, A., G. R. S. Bisht, M. Singh, R. Gupta, and V. Singh. 2011. Chemical composition and antimicrobial activity of essential oil of tubers of *Cyperus rotundus* Linn. collected from Dehradun (Uttarakhand). *International Journal of Research in Pharmaceutical and Biomedical Sciences* 2: 661–5.

Biswas, A., S. Chatterjee, R. Chowdhury et al. 2012. Antidiabetic effect of seeds of *Strychnos potatorum* Linn. in a streptozotocin-induced model of diabetes. *Acta Poloniae Pharmaceutica - Drug Research* 69, no. 5: 939–43.

Block, E. 2010. Allium botany and cultivation, ancient and modern. In *Garlic and other Alliums: The lore and the science*, 1–32. Cambridge: The Royal Society of Chemistry.

Bose, B. C., R. Vijayvargiya, A. Q. Saiti, and S. K. Sharma. 1960. Some aspects of chemical and pharmacological studies of *Acorus calamus* Linn. *Journal of the American Pharmacists Association* 49: 32–4.

Burgess, J. A., P. F. Van der Ven, M. Martin, J. Sherman, and J. Haley. 2008. Review of over-the counter treatments for aphthous ulceration and results from use of a dissolving oral patch containing Glycyrrhiza complex herbal extract. *The Journal of Contemporary Dental Practice* 9, no. 3: 88–98.

Chakraborty, B., A. Nath, H. Saikia, and M. Sengupta. 2014. Bactericidal activity of selected medicinal plants against multidrug resistant bacterial strains from clinical isolates. *Asian Pacific Journal of Tropical Medicine* 7S1: S435–41.

Chandrashekar, R., P. Adake, and S. N. Rao. 2013. Anticonvulsant activity of ethanolic extract of *Acorus calamus* rhizome in Swiss albino mice. *Journal of Scientific and Innovative Research* 2: 846–51.

Chavan, R. D., P. Shinde, K. Girkar, R. Madage, and A. Chowdhary. 2016. Assessment of anti-influenza activity and hemagglutination inhibition of *Plumbago indica* and *Allium sativum* extracts. *Pharmacognosy Research* 8, no. 2: 105–11.

Checker, R., D. Sharma, S. K. Sandur, S. Khanam, and T. B. Poduval. 2009. Anti-inflammatory effects of plumbagin are mediated by inhibition of NF-kappaB activation in lymphocytes. *International Immunopharmacology* 9, no. 7–8: 949–58.

Chung, Y. C., Y. C. Hou, and A. C. Pan. 2004. Endoglin (CD105) expression in the development of haemorrhoids. *European Journal of Clinical Investigation* 34, no. 2: 107–12.

Colabawalla, H. M. 1951. An evaluation of the cardiotonic and other properties of *Terminalia arjuna. Indian Heart Journal* 3: 205–30.

Dahanukar, S. A., U. M. Thatte, N. Pai, P. B. More, and S. M. Karandikar. 1988. Immunotherapeutic modification by *Tinospora cordifolia* of abdominal sepsis induced by caecal ligation in rats. *Indian Journal of Gastroenterology: Official Journal of the Indian Society of Gastroenterology* 7, no. 1: 21–3.

Dandiya, P. C., R. M. Baxter, G. C. Walker, and H. Cullumbine. 1959a. Studies on *Acorus calamus*. Part II. Investigation of volatile oil. *Journal of Pharmacy and Pharmacology* 11, no. 1: 163–8.

Dandiya, P. C., and H. Cullumbine. 1959. Studies on *Acorus calamus* III. Some pharmacological actions of the volatile oil. *Journal of Pharmacology and Experimental Therapeutics* 125, no. 4: 353–9.

Dandiya, P. C., H. Cullumbine, and E. A. Sellers. 1959b. Studies on *Acorus calamus* IV Investigations on mechanism of action in mice. *Journal of Pharmacology and Experimental Therapeutics* 126: 334–7.

Dandiya, P. C., and M. K. Menon. 1963. Effect of asarone and β-asarone on conditioned responses, fighting behaviour and convulsions. *British Journal of Pharmacology and Chemotherapy* 20: 436–42.

Dandiya, P. C., and J. D. Sharma. 1962. Studies on *Acorus calamus* Part V. Pharmacological actions of asarone and β-asarone on central nervous system. *Indian Journal of Medical Research* 50: 46–60.

Daswani, P. G., S. Brijesh, P. Tetali, J. Tannaz, and T. J. Birdi. 2011. Studies on the activity of *Cyperus rotundus* Linn. tubers against infectious diarrhea. *Indian Journal of Pharmacology* 43, no. 3: 340–4.

Deshpande, P. J., M. Sahu, and K. Pradeep. 1982. *Crataeva nurvala* Hook and Forst. (Varuna). The Ayurvedic drug of choice in urinary disorders. *Indian Journal of Medical Research* 76 (Suppl.): 46–53.

Dewangan, G., K. M. Koley, V. P. Vadlamudi, A. Mishra, A. Poddar, and S. D. Hirpurkar. 2010. Antibacterial activity of *Moringa oleifera* (drumstick) root bark. *Journal of Chemical and Pharmaceutical Research* 2: 424–8.

Dhalla, N. S., and I. S. Bhattacharya. 1968. Further studies on neuropharmacological actions of Acorus oil. *Archives Internationales de Pharmacodynamie et de Thérapie* 172, no. 2: 356–65.

Dhasarathan, P., and P. Theriappan. 2011. Evaluation of anti-diabetic activity of *Strychnos potatorum* in alloxan induced diabetic rats. *Journal of Medicine and Medical Sciences* 2: 670–4.

Dikshit, V., A. S. Damre, K. R. Kulkarni, A. Gokhale, and M. N. Saraf. 2000. Preliminary screening of imunocin for immunomodulatory activity. *Indian Journal of Pharmaceutical Sciences* 62: 257.

Dwivedi, S., and M. P. Agarwal. 1994. Antianginal and cardioprotective effects of *Terminalia arjuna*, an indigenous drug, in coronary artery disease. *Journal of Association of Physicians of India* 42, no. 4: 287–9.

Dwivedi, S., and R. Jauhari. 1997. Beneficial effects of *Terminalia arjuna* in coronary artery disease. *Indian Heart Journal* 49, no. 5: 507–10.

Ekambaram, S., S. S. Perumal, and V. Subramanian. 2010. Evaluation of antiarthritic activity of *Strychnos potatorum* Linn. seeds in Freund's adjuvant induced arthritic rat model. *BMC Complementary and Alternative Medicine* 10: 56. doi:10.1186/1472-6882-10-56.

El-Gohary, H. M. A. 2004. Study of essential oils of the tubers of *Cyperus rotuntdus* L and *Cyperus alopecuroides* Rottb. *Bulletin of Faculty of Pharmacy, Cairo University* 42: 157–64.

Ezeamuzie, I. C., A. W. Ambakederemo, F. O. Shode, and S. C. Ekwebelem. 1996. Anti-inflammatory effects of *Moringa oleifera* root extract. *International Journal of Pharmacognosy* 34, no. 3: 207–12.

Fouad, E. A., S. M. Azza, A. Elnaga, and M. M. Kandil. 2019. Antibacterial efficacy of *Moringa oleifera* leaf extract against pyogenic bacteria isolated from a dromedary camel (*Camelus dromedarius*) abscess. *Veterinary World* 12, no. 6: 802–8.

Fredriksson, T., and U. Pettersson. 1978. Severe psoriasis - Oral therapy with a new retinoid. *Dermatologica* 157, no. 4: 238–44.

Fujii, M., I. Takei, and K. Umezawa. 2009. Antidiabetic effect of orally administered conophylline-containing plant extract on streptozotocin-treated and Goto-Kakizaki rats. *Biomedicine and Pharmacotherapy = Biomedecine and Pharmacotherapie* 63, no. 10: 710–6.

Gayathri, M. B., M. A. Kareem, Unnikrishnan P. M. Sarneswar, and P. M. Unnikrishnan. 1996. Single drug therapy in netraroga. *Ancient Science of Life* 16, no. 2: 122–35.

Gilani, A. H., A. Khan, M. Raoof et al. 2008a. Gastrointestinal, selective airways and urinary bladder relaxant effects of *Hyoscyamus niger* are mediated through dual blockade of muscarinic receptors and Ca++ channels. *Fundamental and Clinical Pharmacology* 22, no. 1: 87–99.

Gilani, A. H., A. Khan, F. Subhan, and M. Khan. 2005a. Antispasmodic and bronchodilator activities of St. John's wort are putatively mediated through dual inhibition of calcium influx and phosphodiesterase. *Fundamental and Clinical Pharmacology* 19, no. 6: 695–705.

Gilani, A. H., A. U. Khan, T. Ali, and S. Ajmal. 2008b. Mechanisms underlying the antispasmodic and bronchodilatory properties of *Terminalia bellerica* fruit. *Journal of Ethnopharmacology* 116, no. 3: 528–38.

Gilani, A. H., A. J. Shah, M. N. Ghayur, and K. Majeed. 2005b. Pharmacological basis for the use of turmeric in gastrointestinal and respiratory disorders. *Life Sciences* 76, no. 26: 3089–105.

Goncharov, N., A. N. Orekhov, N. Voitenko, A. Ukolov, R. Jenkins, and P. Avdonin. 2016. Organosulfur compounds as nutraceuticals. In *Nutraceuticals: Efficacy, safety and toxicity*, ed. C. Gupta, 555–68. London: Academic Press.

Goncharov, N. V., P. V. Avdonin, A. D. Nadeev, I. L. Zharkikh, and R. O. Jenkins. 2015. Reactive oxygen species in pathogenesis of atherosclerosis. *Current Pharmaceutical Design* 21, no. 9: 1134–46.

Goto, Y., M. Kakizaki, and N. Masaki. 1975. Spontaneous diabetes produced by selective breeding of normal Wistar rats. *Proceedings of the Japan Academy* 51, no. 1: 80–5.

Guengerich, F. P. 2008. Cytochrome P450 and chemical toxicology. *Chemical Research in Toxicology* 21, no. 1: 70–83.

Gulati, O. D. 1980. Clinical trial of *Tinospora cordifolia* in rheumatoid arthritis. *Rheumatism* 15: 143–8.

Gulati, O. D., and D. C. Pandey. 1982. Anti-inflammatory activity of *Tinospora cordifolia. Rheumatism* 17: 76–83.

Guo, Z., J. Xu, J. Xia, Z. Wu, J. Lei, and J. Yu. 2019. Anti-inflammatory and antitumour activity of various extracts and compounds from the fruits of *Piper longum* L. *Journal of Pharmacy and Pharmacology* 71, no. 7: 1162–71.

Gupta, A., A. K. Mishra, P. Bansal et al. 2010. Antileprotic potential of ethnomedicinal herbs: A review. *Drug Invention Today* 2: 191–3.

Hafeez, B. B., W. Zhong, J. W. Fischer et al. 2013. Plumbagin, a medicinal plant (*Plumbago zeylanica*)-derived 1,4-naphthoquinone, inhibits growth and metastasis of human prostate cancer PC-3M-luciferase cells in an orthotopic xenograft mouse model. *Molecular Oncology* 7, no. 3: 428–39.

Han, W., Z. J. Wang, B. Zhao et al. 2005. Pathologic change of elastic fibers with difference of microvessel density and expression of angiogenesis-related proteins in internal hemorrhoid tissues. *Zhonghua Weichang Waike Zazhi* 8, no. 1: 56–9.

Haq, R. U., A. U. Shah, A. U. Khan et al. 2012. Antitussive and toxicological evaluation of *Vitex negundo. Natural Product Research* 26, no. 5: 484–8.

Hazra, R., and D. Guha. 2003. Effect of chronic administration of *Acorus calamus* on electrical activity and regional monoamine levels in rat brain. *Biogenic Amines* 17: 161–9.

Hazra, R., K. Ray, and D. Guha. 2007. Inhibitory role of *Acorus calamus* in ferric chloride-induced epileptogenesis in rat. *Human and Experimental Toxicology* 26, no. 12: 947–53.

Helou, L., and I. M. Harris. 2007. Garlic. In *Herbal Products: Toxicology and clinical pharmacology*, 2nd edition, ed. T. S. Tracy, and R. L. Kingston, 123–49. NJ: Humana Press Inc.

Hussain, M. S., K. F. H. N. Ahmed, and M. Z. H. Ansari. 2009. Preliminary studies on diuretic effect of *Hygrophila auriculata* (Schum) Heine in rats. *International Journal of Health Research* 2, no. 1: 59–64.

Iyengar, M. A., and G. S. Pendse. 1966. *Plumbago zeylanica* L. (Chitrak). A gastrointestinal flora normaliser. *Planta Medica* 14, no. 3: 337–51.

Jana, U., R. N. Chattopadhyay, and B. P. Shaw. 1999. Preliminary studies on anti-inflammatory activity of *Zingiber officinale Rosc., Vitex negundo* Linn. and *Tinospora cordifolia* (Willd) Miers in albino rats. *Indian Journal of Pharmacology* 31: 232–23.

Jayesingha, W. A. 1887. On *Hygrophila spinosa* (vel *Asteracantha longifolia*). *The British Medical Journal* July 16: 118–9.

Jeyachandran, R., T. F. Xavier, and S. P. Anand. 2003. Antibacterial activity of stem extracts of *Tinospora cordifolia* (Willd.) Hook. f & Thomson. *Ancient Science of Life* 23, no. 1: 40–3.

Jones, D. P. 2010. Redox sensing: Orthogonal control in cell cycle and apoptosis signalling. *Journal of Internal Medicine* 268, no. 5: 432–48.

Kaewbumrung, S., and P. Panichayupakaranant. 2014. Antibacterial activity of plumbagin derivative-rich *Plumbago indica* root extracts and chemical stability. *Natural Product Research* 28, no. 11: 835–7.

Kam, T. S., and S. Anuradha. 1995. Alkaloids from *Tabernaemontana divaricata. Phytochemistry* 40, no. 1: 313–6.

Kam, T. S., K. Y. Loh, and C. Wei. 1993. Conophylline and conophyllidine: New dimeric alkaloids from *Tabernaemontana divaricata. Journal of Natural Products* 56, no. 11: 1865–71.

Kam, T. S., H. S. Pang, and T. M. Lim. 2003. Biologically active indole and bisindole alkaloids from *Tabernaemontana divaricate. Organic and Biomolecular Chemistry* 8: 1292–7.

Kanthlal, S. K., B. Anil Kumar, J. Joseph, R. Aravind, and P. R. Frank. 2014. Amelioration of oxidative stress by *Tabernamontana divaricata* on alloxan-induced diabetic rats. *Ancient Science of Life* 33, no. 4: 222–8.

Kapoor, D., D. Trikha, R. Vijayvergiya, K. K. Parashar, D. Kaul, and V. Dhawan. 2015. Short-term adjuvant therapy with *Terminalia arjuna* attenuates ongoing inflammation and immune imbalance in patients with stable coronary artery disease: *In vitro* and *in vivo* evidence. *Journal of Cardiovascular Translational Research* 8, no. 3: 173–86.

Karamalakova, Y. D., G. D. Nikolova, T. K. Georgiev, G. Gadjeva, and A. N. Tolekova. 2019. Hepatoprotective properties of *Curcuma longa* L. extract in bleomycin-induced chronic hepatotoxicity. *Drug Discoveries and Therapeutics* 13, no. 1: 9–16.

Khan, M., A. J. Shah, and A. H. Gilani. 2015. Insight into the bronchodilator activity of *Vitex negundo. Pharmaceutical Biology* 53, no. 3: 340–4.

Khare, A. K., and M. K. Sharma. 1982. Experimental evaluation of antiepileptic activity of Acorus oil. *Journal of Scientific Research in Plant Medicine* 3: 100–3.

Kodera, T., S. Yamada, Y. Yamamoto et al. 2009. Administration of conophylline and betacellulin-δ4 increases the β-cell mass in neonatal streptozotocin-treated rats. *Endocrine Journal* 56, no. 6: 799–806.

Kojima, I., and K. Umezawa. 2006. Conophylline: A novel differentiation inducer for pancreatic β cells. *The International Journal of Biochemistry and Cell Biology* 38, no. 5–6: 923–30.

Koumaravelou, K., C. Adithan, C. Shashindran, M. Asad, and B. K. Abraham. 2002a. Influence of honey on orally and intravenously administered diltiazem kinetics in rabbits. *Indian Journal of Experimental Biology* 40, no. 10: 1164–8.

Koumaravelou, K., C. Adithan, C. Shashindran, M. Asad, and B. K. Abraham. 2002b. Effect of honey on carbamazepine kinetics in rabbits. *Indian Journal of Experimental Biology* 40, no. 5: 560–3.

Kumar, A., S. Panghal, S. S. Mallapur, M. Kumar, V. Ram, and B. K. Singh. 2009. Anti-inflammatory activity of *Piper longum* fruit oil. *Indian Journal of Pharmaceutical Sciences* 71, no. 4: 454–6.

Kumar, D. S., and Y. S. Prabhakar. 1987. On the ethnomedical significance of the Arjun tree, *Terminalia arjuna* (Roxb.) Wight & Arnot. *Journal of Ethnopharmacology* 20, no. 2: 173–90.

Kurian, G. A., and J. Paddikkala. 2009. Administration of aqueous extract of *Desmodium gangeticum* (L) root protects rat heart against ischemic reperfusion injury induced oxidative stress. *Indian Journal of Experimental Biology* 47, no. 2: 129–35.

Kurian, G. A., and J. Paddikkala. 2010. Role of mitochondrial enzymes and sarcoplasmic ATPase in cardioprotection mediated by aqueous extract of *Desmodium gangeticum* (L) DC root on ischemic reperfusion injury. *Indian Journal of Pharmaceutical Sciences* 72, no. 6: 745–52.

Kurian, G. A., S. Philip, and T. Varghese. 2005. Effect of aqueous extract of the *Desmodium gangeticum* DC root in the severity of myocardial infarction. *Journal of Ethnopharmacology* 97, no. 3: 457–61.

Kurian, G. A., N. Yagnesh, R. S. Kishan, and J. Paddikkala. 2008. Methanol extract of *Desmodium gangeticum* roots preserves mitochondrial respiratory enzymes, protecting rat heart against oxidative stress induced by reperfusion injury. *Journal of Pharmacy and Pharmacology* 60, no. 4: 523–30.

Lawal, O. A., and A. O. Oyedeji. 2009. Chemical composition of the essential oils of *Cyperus rotundus* L. from South Africa. *Molecules* 14, no. 8: 2909–17. doi:10.3390/molecules14082909.

Lindner, H. R., U. Zor, F. Kohen et al. 1980. Significance of prostaglandins in the regulation of cyclic events in the ovary and uterus. *Advances in Prostaglandin and Thromboxane Research* 8: 1371–90.

Lohsiriwat, V. 2012. Hemorrhoids: From basic pathophysiology to clinical management. *World Journal of Gastroenterology* 18, no. 17: 2009–17.

Madan, B. R., R. B. Arora, and K. Kapila. 1960. Anticonvulsant, antiveratrinic and antiarrhythmic actions of *Acorus calamus* Linn. an Indian indigenous drug. *Archives Internationales de Pharmacodynamie et de Thérapie* 124: 201–11.

Mahajan, G. K., R. T. Mahajan, and A. Y. Mahajan. 2015. Improvement of sperm density in neem oil-induced infertile male albino rats by *Ipomoea digitata* Linn. *Journal of Intercultural Ethnopharmacology* 4, no. 2: 125–8.

Malhotra, C. L., P. K. Das, and N. S. Dhalla. 1962. Investigations on the mechanism of potentiation of barbiturate hypnosis by hersaponin, Acorus oil, reserpine and chlorpromazine. *Archives Internationales de Pharmacodynamie et de Thérapie* 138: 537–47.

Malik, N., V. Dhawan, A. Bahl, and D. Kaul. 2009. Inhibitory effects of *Terminalia arjuna* on platelet activation in vitro in healthy subjects and patients with coronary artery disease. *Platelets* 20, no. 3: 183–90.

Mallikharjuna, P. B., L. N. Rajanna, Y. N. Seetharam, and G. K. Sharanabasappa. 2007. Phytochemical studies of *Strychnos potatorum* L.f.- A medicinal plant. *E-Journal of Chemistry* 4, no. 4: 510–18.

Manis, G., A. Rao, and K. S. Karanth. 1991. Neuropharmacological activity of *Acorus calamus*. *Fitoterapia* 62: 131–7.

Manjrekar, P. N., C. I. Jolly, and S. Narayanan. 2000. Comparative studies of the immunomodulatory activity of *Tinospora cordifolia* and *Tinospora sinensis*. *Fitoterapia* 71, no. 3: 254–7.

Martin, M. D., J. Sherman, J. P. van der Ven, and J. Burgess. 2008. A controlled trial of a dissolving oral patch concerning Glycyrrhiza (licorice) herbal extract for the treatment of aphthous ulcers. *General Dentistry* 56, no. 2: 206–10.

Mehrbod, P., E. Amini, and M. Tavassoti-Kheiri. 2009. Antiviral activity of garlic extract on influenza virus. *Iranian Journal of Virology* 3, no. 1: 19–23.

Menter, A., A. Gottlieb, S. R. Feldman et al. 2008. Guidelines of care for the management of psoriasis and psoriatic arthritis: Section 1. Overview of psoriasis and guidelines of care for the treatment of psoriasis with biologics. *Journal of the American Academy of Dermatology* 58: 826–50.

Messier, C., F. Epifano, S. Genovese, and D. Grenier. 2012. Licorice and its potential beneficial effects in common oro-dental diseases. *Oral Diseases* 18, no. 1: 32–9.

Middelkoop, T. B., and R. P. Labadie. 1983. Evaluation of Asoka aristha, an indigenous medicine in Sri Lanka. *Journal of Ethnopharmacology* 8, no. 3: 313–20.

Middelkoop, T. B., and R. P. Labadie. 1985. The action of *Saraca asoca* Roxb. de Wilde bark on the PGH_2 synthetase enzyme complex of the sheep vesicular gland. *Zeitscrift fuer Naturforschung* 40c: 523–6.

Mishra, A., S. Kumar, and A. K. Pandey. 2013. Scientific validation of the medicinal efficacy of *Tinospora cordifolia. The Scientific World Journal* 2013. Article ID 292934. doi:10.1155/2013/292934.

Mishra, B. 1983. *Cakradatta*. Varanasi: Chowkhamba Sanskrit Series Office.

Moghadami, M. 2017. A narrative review of influenza: A seasonal and pandemic disease. *Iranian Journal of Medical Sciences* 42, no. 1: 2–13.

Moghadamnia, A. A., M. Motallebnejad, and M. Khanian. 2009. The efficacy of the bioadhesive patches containing licorice extract in the management of recurrent aphthous stomatitis. *Phytotherapy Research* 23, no. 2: 246–50.

Mooss, N. S. 1978. *Vaidyamanōrama*, Part 2. Kottayam, Kerala: Vaidyasarathy Press (P) Ltd.

Mooss, N. S. 1979. *Vaidyamanōrama*, Part 1. Kottayam, Kerala: Vaidyasarathy Press (P) Ltd.

Munday, R. 2012. Harmful and beneficial effects of organic monosulfides, disulfides, and polysulfides in animals and humans. *Chemical Research in Toxicology* 25, no. 1: 47–60.

Murthy, K. R. S. 2017. *Śārṅgadhara Samhita*. Varanasi: Chaukhamba Orientalia.

Nagarkatti, D. S., N. N. Rege, N. K. Desai, and S. A. Dahanukar. 1994. Modulation of Kupffer cell activity by *Tinospora cordifolia* in liver damage. *Journal of Postgraduate Medicine* 40, no. 2: 65–7.

Nair, P. K., S. J. Melnick, R. Ramachandran, E. Escalon, and C. Ramachandran. 2006. Mechanism of macrophage activation by (1,4)-alpha-D-glucan isolated from *Tinospora cordifolia. International Immunopharmacology* 6, no. 12: 1815–24.

Nair, P. K., S. Rodriguez, R. Ramachandran et al. 2004. Immune stimulating properties of a novel polysaccharide from the medicinal plant *Tinospora cordifolia. International Immunopharmacology* 4, no. 13: 1645–59.

Ovadje, P., D. Ma, P. Tremblay et al. 2014. Evaluation of the efficacy and biochemical mechanism of cell death induction by *Piper longum* extract selectively in in-vitro and in-vivo models of human cancer cells. *PLoS One* 17, no. 11; 9(11):e113250. doi:10.1371/journal.pone.0113250.

Patel, J. I., and S. S. Deshpande. 2013. Anti-eosinophilic activity of various subfractions of leaves of *Vitex negundo. International Journal of Nutrition, Pharmacology, Neurological Diseases* 3, no. 2: 135–41.

Patel, M. D., M. A. Patel, and G. B. Shah. 2010. Evaluation of the effect of *Vitex negundo* leaves extract on bronchoconstriction and bronchial hyperreactivity in experimental animals. *Pharmacologyonline* 1: 644–54.

Patra, A., S. Jha, P. N. Murthy et al. 2009. Anti-inflammatory and antipyretic activities of *Hygrophila spinosa* T. Anders leaves (*Acanthaceae*). *Tropical Journal of Pharmaceutical Research* 8, no. 2: 133–7.

Pendse, V. K., A. P. Dadhich, P. N. Mathur, M. S. Bal, and B. R. Madan. 1977. Anti-inflammatory, immunosuppressive and some related pharmacological actions of the water extract of Neem Giloe (*Tinospora cordifolia*): A preliminary report. *Indian Journal of Pharmacology* 9: 221–4.

Peterson, C. T., D. A. Rodionov, S. N. Iablokov et al. 2019. Prebiotic potential of culinary spices used to support digestion and bioabsorption. *Evidence Based Complementary and Alternate Medicine* 2019: 8973704. doi:10.1155/2019/8973704.

Prabhakar, Y. S., and D. S. Kumar. 1990. The Varuna tree, *Crataeva nurvala* Buch.-Ham., a promising plant in the treatment of urinary stones: A review. *Fitoterapia* 61: 91–111.

Prakash, S., and R. R. Rao. 2014. Honey in ayurvedic medicine. In *Honey in traditional and modern medicine*, ed. L. Boukraâ, 13–20. Boca Raton: CRC Press.

Purandare, H., and A. Supe. 2007. Immunomodulatory role of *Tinospora cordifolia* as an adjuvant in surgical treatment of diabetic foot ulcers: A prospective randomized controlled study. *Indian Journal of Medical Sciences* 61, no. 6: 347–55.

Purohit, M. G., G. K. Shanthaveerappa, and S. Badami. 1996. Antiulcer and anticatatonic activity of alcoholic extract of *Evolvulus alsinoides* (Convolvulaceae). *Indian Journal of Pharmaceutical Sciences* 58: 110–2.

Reddy, P. A., and J. V. Rao. 2013. Evaluation of anti-inflammatory activity of *Evolvulus alsinoides* plant extracts. *Journal of Pharmaceutical and Scientific Innovation* 2, no. 3: 24–6.

Reddy, J. S., P. R. Rao, and M. S. Reddy. 2002. Wound healing effects of *Heliotropium indicum*, *Plumbago zeylanica* and *Acalypha indica* in rats. *Journal of Ethnopharmacology* 79, no. 2: 249–51.

Rege, N. N., R. D. Bapat, R. Koti, N. K. Desai, and S. Dahanukar. 1993. Immunotherapy with *Tinospora cordifolia*: A new lead in the management of obstructive jaundice. *Indian Journal of Gastroenterology: Official Journal of the Indian Society of Gastroenterology* 12, no. 1: 5–8.

Rege, N. N., H. M. Nazareth, R. D. Bapat, and S. A. Dahanukar. 1989. Modulation of immunosuppression in obstructive jaundice by *Tinospora cordifolia*. *Indian Journal of Medical Research* 90: 478–83.

Rege, N. N., U. M. Thatte, and S. A. Dahanukar. 1999. Adaptogenic properties of six rasayana herbs used in Ayurvedic medicine. *Phytotherapy Research: PTR* 13, no. 4: 275–91.

Reiber, G. E., and G. J. Raugi. 2005. Preventing foot ulcers and amputations in diabetes. *Lancet* 366, no. 9498: 1676–7.

Reynoso-Moreno, I., I. Najar-Guerrero, N. Escareño, M. E. Flores-Soto, J. Gertsch, and J. M. Viveros-Paredes. 2017. An endocannabinoid uptake inhibitor from black pepper exerts pronounced anti-inflammatory effects in mice. *Journal of Agricultural and Food Chemistry* 65, no. 43: 9435–42.

Sainath, R. S., J. Prathiba, and R. Malathi. 2009. Antimicrobial properties of the stem bark of *Saraca indica* (Caesalpiniaceae). *European Review for Medical and Pharmacological Sciences* 13, no. 5: 371–4.

Saini, N. K. 2017. A clinical study: Role of giloy (*Tinospora cordifolia*) in the treatment of *kitibha* with special reference to psoriasis. *International Ayurvedic Medical Journal* 5: 1147–52.

Satyavati, G. V., D. N. Prasad, S. P. Sen, and P. K. Das. 1970. Further studies on the uterine activity of S. *indica* L. *Indian Journal of Medical Research* 58, no. 7: 947–60.

Schubert, M. C., S. Sridhar, R. R. Schade, and S. D. Wexner. 2009. What every gastroenterologist needs to know about common anorectal disorders. *World Journal of Gastroenterology* 15, no. 26: 3201–9.

Shah, G. L. 1984. Some economically important plants of Salsette Island and Bombay. *Journal of Economic Taxonomy and Botany* 5: 753–65.

Shah, N. C., and M. C. Joshi. 1971. An ethnobotanical study of Kumaon region of India. *Economic Botany* 25, no. 4: 414–22.

Shaikh, R. U., M. M. Pund, and R. N. Gacche. 2016. Evaluation of anti-inflammatory activity of selected medicinal plants used in Indian traditional medication system *in vitro* as well as *in vivo*. *Journal of Traditional and Complementary Medicine* 6, no. 4: 355–61.

Sharma, A. K., and R. H. Singh. 1980. Screening of anti-inflammatory of certain indigenous drugs on carrageen induced hind paw edema in rats. *Bulletin of Medical and Ethanobotanical Research* 2: 262–4.

Sharma, J. D., P. C. Dandiya, R. M. Baxter, and S. I. Kandel. 1961. Pharmacodynamical effects of asarone and β-asarone. *Nature* 192: 1299–300.

Sharma, P. K., S. K. Dhyani, and V. Shanker. 1979. Some useful and medicinal plants of the district Dehradun and Siwalik. *Journal of Science Research in Plant Medicine* 1: 17–43.

Shay, H., S. A. Komarow, S. Fels et al. 1945. A simple method for the uniform production of gastric ulceration in the rat. *Gastroenterology* 5: 43–61.

Shobhana, R., P. R. Rao, A. Lavanya, V. Vijay, and A. Ramachandran. 2001. Footcare economics- cost burden to diabetic patients: A study from southern India. *Journal of the Association of Physicians of India* 49: 530–3.

Sidthipong, K., S. Todo, I. Takei, I. Kojima, and K. Umezawa. 2013. Screening of new bioactive metabolites for diabetes therapy. *Internal and Emergency Medicine* 8, no. Suppl. 1: 57–9.

Singh, C., and N. P. Rai. 2013. *In vitro* antibacterial activity of *Piper longum* L. fruit. *International Journal of Pharmaceutical Sciences Review and Research* 18: 89–91.

Singh, N., K. K. Kapur, S. P. Singh, K. Shanker, J. N. Sinha, and R. P. Kohli. 1982. Mechanism of cardiovascular action of *Terminalia arjuna*. *Planta Medica* 46, no. 2: 102–4.

Singh, R. G., and Usha Kapoor, S. 1991. Evaluation of antilithic properties of Varun (*Crataeva nurvala*): An indigenous drug. *Journal of Research and Education in Indian Medicine* 10: 35–9.

Sinha, M. K., D. K. Patel, and V. K. Kanungo. 2013. Medicinal plants used in the treatment of skin diseases in Central Bastar of Chhattisgarh, India. *Global Advanced Research Journal of Medicinal Plants* 2: 1–3.

Spence, D., and C. Melville. 2007. Vaginal discharge. *British Medical Journal* 335, no. 7630: 1147–51.

Srinivasan, K. 2007. Black pepper and its pungent principle-piperine: A review of diverse physiological effects. *Critical Reviews in Food Science and Nutrition* 47, no. 8: 735–48.

Thatte, U. M., S. Chhabria, S. M. Karandikar, and S. A. Dahanukar. 1987. Immunotherapeutic modification of *E. coli* induced abdominal sepsis and mortality in mice by Indian medicinal plants. *Indian Drugs* 25: 95–7.

Thatte, U. M., and S. A. Dahanukar. 1989. Immunotherapeutic modification of experimental infections by Indian medicinal plants. *Phytotherapy Research* 3, no. 2: 43–9.

Trivedi, V. P., S. Nesamany, and V. K. Sharma. 1979. A clinical study of the anti-tussive and antiasthmatic effects of Vibhitakphal Churna (*Terminalia bellerica* Roxb.) in the cases of Kasa-Swasa. *Journal of Research in Ayurveda and Siddha* 3: 1–8.

Tushar, T., T. Vinod, S. Rajan, C. Shashindran, and C. Adithan. 2007. Effect of honey on CYP3A4, CYP2D6 and CYP2C19 enzyme activity in healthy human volunteers. *Basic and Clinical Pharmacology and Toxicology* 100, no. 4: 269–72.

Uddin, S. J., K. Mondal, J. A. Shilpi, and M. T. Rahman. 2006. Antidiarrhoeal activity of *Cyperus rotundus*. *Fitoterapia* 77, no. 2: 134–6.

Upadhyaya, Y. 1975a. *Grahaṇīdōṣacikitsā* (Chapter on treatment of diarrhea). In *Aṣṭāṅgahṛdaya*, 361–7. Varanasi: Chowkhamba Sanskrit Sansthan.

Upadhyaya, Y. 1975b. *Aṣṭāṅgahṛdaya*, *Annasvarūpavijñānīyam* (Description of food). In *Aṣṭāṅgahṛdaya*. Varanasi: Chowkhamba Sanskrit Sansthan, 50–67.

Upadhyaya, Y. 1975c. *Aṣṭāṅgahṛdaya*. Varanasi: Chowkhamba Sanskrit Sansthan.

Van Dyke, K., N. Jabbour, R. Hoeldtke, C. Van Dyke, and M. Van Dyke. 2010. Oxidative/nitrosative stresses trigger type I diabetes: Preventable in streptozotocin rats and detectable in human disease. *Annals of the New York Academy of Sciences* 1203: 138–45.

Vasanth, K., G. Minakshi, K. Ilango, R. M. Kumar, A. Agrawal, and G. Dubey. 2015. *Moringa oleifera* attenuates the release of pro-inflammatory cytokines in lipopolysaccharide stimulated human monocytic cell line. *Industrial Crops and Products* 77: 44–50.

Warrier, P. K., V. P. K. Nambiar, and C. Ramankutty. 2007a. *Indian Medicinal Plants- A compendium of 500 species*, Vol. 1, 1–420. Hyderabad: Orient Longman Private Ltd.

Warrier, P. K., V. P. K. Nambiar, and C. Ramankutty. 2007b. *Indian Medicinal Plants- A compendium of 500 species*, Vol. 2, 1–416. Hyderabad: Orient Longman Private Ltd.

Warrier, P. K., V. P. K. Nambiar, and C. Ramankutty. 2007c. *Indian Medicinal Plants- A compendium of 500 species*, Vol. 3, 1–423. Hyderabad: Orient Longman Private Ltd.

Warrier, P. K., V. P. K. Nambiar, and C. Ramankutty. 2007d. *Indian Medicinal Plants- A compendium of 500 species*, Vol. 4, 1–444. Hyderabad: Orient Longman Private Ltd.

Warrier, P. K., V. P. K. Nambiar, and C. Ramankutty. 2008. *Indian Medicinal Plants- A compendium of 500 species*, Vol. 5, 1–592. Hyderabad: Orient Longman Private Ltd.

Weber, N. D., D. O. Andersen, J. A. North, B. K. Murray, L. D. Lawson, and B. G. Hughes. 1992. *In vitro* virucidal effects of *Allium sativum* (garlic) extract and compounds. *Planta Medica* 58, no. 5: 417–23.

Wesley, J. J., A. J. Christina, and N. Chidambaranathan. 2008. Effect of alcoholic extract of *Tinospora cordifolia* on acute and subacute inflammation. *Pharmacologyonline* 3: 683–7.

Willman, E. A., W. P. Collins, and S. G. Clayton. 1976. Studies in the involvement of prostaglandins in uterine symptomatology and pathology. *British Journal of Obstetrics and Gynaecology* 83, no. 5: 337–41.

Zadorozhna, M., T. Tataranni, and D. Mangieri. 2019. Piperine: Role in prevention and progression of cancer. *Molecular Biology Reports* 46, no. 5: 5617–29.

Zaki, A. A., M. H. El-Bakry, and A. A. Fahmy. 2005. Effect of licorice on wound healing in rabbits. *The Egyptian Journal of Hospital Medicine* 20: 58–65.

Zangeneh, A., M. Pooyanmehr, M. M. Zangeneh, R. Moradi, R. Rasad, and N. Kazemi. 2019. Therapeutic effects of *Glycyrrhiza glabra* aqueous extract ointment on cutaneous wound healing in Sprague Dawley male rat. *Comparative Clinical Pathology* 28, no. 5: 1507–14.

Zunt, S. L. 2001. Recurrent aphthous ulcers: Prevention and treatment. *Journal of Practical Hygiene* 10: 17–24.

6 Biological Effects of Ayurvedic Formulations

G.R. Arun Raj, Kavya Mohan, R. Anjana, Prasanna N. Rao, U. Shailaja and Deepthi Viswaroopan

CONTENTS

6.1 Introduction 136
6.2 Ariṣṭa and Āsava (Fermented Liquids) 136
6.2.1 Abhayāriṣṭa 136
6.2.2 Amṛtāriṣṭa 137
6.2.3 Āragvadhāriṣṭa 137
6.2.4 Aśōkāriṣṭa 137
6.2.5 Daśamūlāriṣṭa 137
6.2.6 Drākṣāriṣṭa 138
6.2.7 Jīrakāriṣṭa 138
6.2.8 Kanakāsava 138
6.2.9 Kumāryāsava 138
6.2.10 Pippalyāsava 138
6.2.11 Vāśāriṣṭa 138
6.3 Cūrṇa (Powders) 139
6.3.1 Avipattikara Cūrṇa 139
6.3.2 Dāḍimāṣṭaka Cūrṇa 139
6.3.3 Haridrākhaṇḍa 139
6.3.4 Puṣyānuga Cūrṇa 139
6.3.5 Rajanyādi Cūrṇa 139
6.3.6 Rāsnādi Cūrṇa 140
6.3.7 Vaiśvānara Cūrṇa 140
6.4 Ghṛta (Medicated Clarified Butter) 140
6.4.1 Aśvagandha Ghṛta 140
6.4.2 Brahmī Ghṛta 140
6.4.3 Jātyādi Ghṛta 141
6.4.4 Kalyāṇaka Ghṛta 141
6.4.5 Mahātiktaka Ghṛta 141
6.4.6 Nārasimha Rasāyana 141
6.4.7 Phala Sarpis 142
6.4.8 Sukumāra Ghṛta 142
6.4.9 Tṛphalā Ghṛta 142
6.4.10 Vidāryādi Ghṛta 142
6.5 Guṭika (Pills) 142
6.5.1 Bilvādi Guṭika 142
6.5.2 Chandraprabhā Vaṭi 143
6.5.3 Dhānvantaram Guṭika 143
6.5.4 Kaiśōra Guggulu 143

6.5.5 Kāñcanāra Guggulu ... 143
6.5.6 Mānasamitra Vaṭaka ... 143
6.6 Kvātha (Decoctions) ... 144
6.6.1 Amṛtōttaram Kvātha ... 144
6.6.2 Balāguḷūcyādi Kvātha ... 144
6.6.3 Daśamūla Kvātha ... 144
6.6.4 Daśamūla Kaṭutraya Kvātha ... 144
6.6.5 Dhānvantaram Kvātha ... 144
6.6.6 Drākṣādi Kvātha ... 145
6.6.7 Gandharvahastādi Kvātha ... 145
6.6.8 Mahāmañjiṣṭādi Kvātha ... 145
6.6.9 Mahārāsnādi Kvātha ... 145
6.6.10 Nayōpāya Kvātha ... 145
6.6.11 Paṭōla Kaṭurōhiṇyādi Kvātha ... 145
6.6.12 Punarnavādi Kvātha ... 146
6.6.13 Sahacarādi Kvātha ... 146
6.7 Avalēha (Electuaries) ... 146
6.7.1 Agastya Harītaki Rasāyana ... 146
6.7.2 Cyavanaprāśa ... 146
6.7.3 Māṇibhadra Guḍa ... 147
6.7.4 Bilvādi Lēhya ... 147
6.8 Taila (Medicated Oils) ... 147
6.8.1 Aṇu Taila ... 147
6.8.2 Balā Taila ... 148
6.8.3 Gandharvahastādi Ēraṇḍa Taila ... 148
6.8.4 Irimēdādi Taila ... 148
6.8.5 Kṣīrabalā Taila ... 148
6.8.6 Mahānārāyaṇa Taila ... 149
6.8.7 Mahāmāṣa Taila ... 149
6.9 Conclusion ... 149
References ... 150

6.1 INTRODUCTION

Ayurveda makes use of several classes of medicine in the treatment of diseases. According to *Śārṅgadhara Samhita*, composed in the 13th century A.D. and considered to be the authoritative text on ayurvedic pharmacy, expressed juices (*svarasa*), decoctions (*kvātha*), spirituous liquors (*āsava*, *ariṣṭa*), oils (*taila*), clarified butter (*ghṛta*), electuaries (*avalēha*), pastes (*lēpa*), powders (*cūrṇa*) and pills (*guḷika*) are the important classes of medicines. Instructions for preparing these dosage forms are also available in it (Murthy 2017). Ayurvedic medicines are prepared using plant parts, such as roots, bark, leaves, flowers, fruits, seeds and gums, and minerals and animal products (Parasuraman et al. 2014). These medicines bring about a wide range of effects on the body. This chapter deals with the various biological effects of a few representative ayurvedic formulations.

6.2 ARIṢṬA AND ĀSAVA (FERMENTED LIQUIDS)

6.2.1 ABHAYĀRIṢṬA

Abhayāriṣṭa is traditionally prepared by the fermentation of decoction of *Terminalia chebula*, *Vitis vinifera*, *Embelia ribes* and *Madhuca indica* (Anonymous 1978a). Intake of *Abhayāriṣṭa* before meals induces laxative action and also corrects constipation (Prajkta 2015). It possesses antioxidant activity (Lal et al. 2010) and is effective in decreasing inflammation, pain, itching and promotes vein elasticity in the perianal area (Odukoya et al. 2009). It is helpful in mild ascites and dropsy, relieves urinary obstruction, increases digestive power and is of value in enlargement of the liver and spleen and in anemia (Mutha et al. 2013). It can also be effectively used in the management of internal bleeding and piles (Shekhawat et al. 2017).

6.2.2 Amṛtāriṣṭa

Amṛtāriṣṭa is a polyherbal hydroalcoholic preparation which is used as a remedy for all types of fevers (Wadkar et al. 2016). It contains 5–8% of self-generated alcohol and also water, with the help of which its absorption takes place in the body. Its main content is *Amṛta* (*Tinospora cordifolia*), which is a natural immunity booster and has antipyretic action. *Amṛtāriṣṭa* is also effective in the management of chronic malaria (Sapra 2013), disorders of the liver and spleen, indigestion and rheumatoid arthritis (Taksale and Parulkar 2017). It also possesses antioxidant properties (Wadkar et al. 2016).

6.2.3 Āragvadhāriṣṭa

Āragvadhāriṣṭa is useful in the management of piles, foul wounds, leukoderma and intestinal parasites (Sekar and Mariappa 2008). It is specially indicated in *Kiṭibha* (psoriasis) type of *kuṣḍha* (skin disease) (Anonymous 2014). It is even effective in treating chronic skin conditions featuring lumps in the armpits or groin, such as hidradenitis suppurativa (Chavhan et al. 2017).

6.2.4 Aśōkāriṣṭa

Aśōkāriṣṭa is specially indicated in gynecological disorders including amenorrhea, metrorrhagia, menorrhagia and dysfunctional uterine bleeding (Shikha and Vidhu 2015). It contains herbs having *balya* (immune-modulator), *rasāyana* (rejuvenation), *dīpanīya* (carminative), *pācanīya* (digestive), *mēdhya* (brain tonic) and *hṛdya* (cardiotonic) properties (Vaidya 2013). It has good reducing power, and superoxide as well as free radical scavenging activity (Geeta 1995; Dushing and Laware 2012; Meena and Gaurav 2014; Katakdound 2017). It tones up uterine musculature and normalizes menstrual flow (Meena and Gaurav 2014). It is administered to enhance endometrial receptivity to avoid abortion. It is also effective in ovulation disorder (Mundewadi 2009). It nourishes the blood and the reproductive system, maintains the healthy production of female hormones, helping to regulate the menstrual cycle and ease the transition into menopause (Kumar et al. 2013). It is useful in premenstrual syndrome and menstrual discomforts. It can be effectively used in the treatment of menopause-associated symptoms and can promote the quality of life of a menopausal woman (Tomar et al. 2017). It helps in relieving backache and abdominal pain, reduces irritation, improves strength and stamina and ensures active and energetic life throughout the month (Oberoi et al. 2016).

6.2.5 Daśamūlāriṣṭa

Daśamūlāriṣṭa, prepared from *daśamūla*, is traditionally used as an analgesic as well as an anti-arthritic agent. Therefore, it is especially used to reduce inflammation. It provides strength to postpartum women and helps the uterus to regain normal size and shape. It is effective in cough, rheumatism, ovulation disorder (Nagarkar and Jagtap 2017) and polycystic ovary syndrome (Karandikar 2018). *Daśamūlāriṣṭa* has been shown to possess anti-inflammatory activity in carrageenan-induced rat paw edema models and cotton pellet-induced granuloma (Parekar et al. 2012).

Also, it is known that *Daśamūlāriṣṭa* has peripheral as well as central analgesic activity in various animal models (Bhalerao et al. 2015).

6.2.6 Drākṣāriṣṭa

Drākṣāriṣṭa is a formulation with *drākṣa* (raisins) as the chief ingredient. It is primarily indicated in digestive impairment, respiratory disorders, weakness (Alam and Gupta 2018) and sciatica (pain which may arise from compression and/or irritation of one of five spinal nerve roots which give rise to each sciatic nerve). It is also effective as an appetizer. The mechanism of action of *Drākṣāriṣṭa* is similar to that of the commonly used N.S.A.I.D.s. Hence, its traditional use in arthritis and lumbago stood the test of time, not by a mere placebo effect, but by the potent analgesic and anti-inflammatory molecules in this age-old ayurvedic medicine (Kabir et al. 2012).

6.2.7 Jīrakāriṣṭa

Jīrakāriṣṭa is used in irritable bowel syndrome, diarrhea, phthisis, postpartum fever and postpartum debility. It mainly acts on the digestive system and improves the digestive capacity. It reduces flatulence, bloating, excessive thirst, loose stools, and mucus content in the stool. It is also indicated in the management of respiratory disorders (Arun et al. 2016).

6.2.8 Kanakāsava

Kanakāsava is a traditional formulation containing *Dhatūra* (*Datura metel*), *Vāśā* (*Adhatoda vasica*), *Dhātaki* (*Woodfordia fruticosa*) and *Drākṣa* (*Vitis vinifera*) as major ingredients. It is used in the treatment of pulmonary diseases including cough and asthma (Sadhna and Abbulu 2010). It also possesses immuno-stimulating properties (Sarker et al. 2014).

6.2.9 Kumāryāsava

Kumāryāsava is an ayurvedic formulation containing *Aloe vera* as the major crude drug, along with 17 other minor ingredients. It possesses antioxidant activity (Manmode et al. 2012) and is used in the treatment of liver disorders (Kataria and Singh 1997; Khan et al. 2015), abdominal lumps, epilepsy, digestive impairment and menopause (Bhaskar et al. 2009). According to *Yōgaratnākara*, *Kumāryāsava* helps in correcting the hormonal axis of sex hormone and cures the condition of amenorrhea (Kalaiselvan et al. 2010). It is also effective in improving the quality of life physically and mentally during the phase of menstruation affected by pain and it can be administered in dysmenorrhea associated with premenstrual syndrome (Lal et al. 2017).

6.2.10 Pippalyāsava

Pippalyāsava is a very good remedy which cures quickly tuberculosis, abdominal lumps, emaciation, irritable bowel syndrome (Sreedas and Girish 2013), anemia and piles (Singh et al. 2011). *Pippalyāsava* is used as a carminative and in the treatment of dysentery and cough (Mansuri and Desai 2019).

6.2.11 Vāśāriṣṭa

Due to its unique processing, *Vāśāriṣṭa* becomes a very potent medicine and is therefore administered in a very low dose of 1 *Māṣa*, corresponding to 1 ml as per metric conversion. It acts as an expectorant and hence is effective in bronchitis and asthma (Santosh et al. 2003; Kamble et al. 2018).

6.3 CŪRṆA (POWDERS)

6.3.1 Avipattikara Cūrṇa

Avipattikara cūrṇa is a polyherbal ayurvedic formulation consisting of 14 ingredients viz., *Zingiber officinale*, *Piper nigrum*, *Piper longum*, *Terminalia chebula*, *Terminalia bellirica*, *Emblica officinalis*, *Cyperus rotundus*, *viḍa lavaṇa*, *Embelia ribes*, *Elettaria cardamomum*, *Cinnamomum tamala*, *Syzgium aromaticum*, *Operculina turpethum* and cane sugar (Chauhan et al. 2015). It is used as a remedy for hyperacidity, indigestion, anorexia, urinary retention, constipation and piles (Ram et al. 2009; Gyawali et al. 2013). Research evidence shows that *Avipattikara cūrṇa* possesses significant gastroprotective activity (Singhal 2016). An experimental study also suggests anti-ulcerogenic effect of *Avipattikara cūrṇa* (Gyawali et al. 2013). Another study reports that hydroalcoholic and methanolic extract of stem bark of *Operculina turpethum* (one of the main ingredients of *Avipattikara cūrṇa*) has enhanced ulcer preventive and protective activities when compared to ranitidine (Ignatius et al. 2013). *In vitro* study reveals that aqueous and methanol extracts of *Avipattikara cūrṇa* exhibit antioxidant properties (Kaushik et al. 2009).

6.3.2 Dāḍimāṣṭaka Cūrṇa

Dāḍimāṣṭaka cūrṇa is a polyherbal formulation containing pomegranate as the main ingredient. It is mainly used in gastrointestinal problems such as diarrhea, indigestion, dysentery and loss of appetite (Pooja and Bhatted 2015; Priya and Hegde 2017; Narang and Herswani 2018; Narang et al. 2019). *Dāḍimāṣṭaka cūrṇa* is prepared by blending the powders of the constituent herbs with sugar and traditionally it is recommended to take 3–5 g of the formulation with warm water or rice soup after the meal. However, it should be used cautiously with diabetic patients. The presence of several potent herbs makes this formulation ideal for treating digestive diseases (Narang and Herswani 2018).

6.3.3 Haridrākhaṇḍa

The main ingredient is *Haridra* (turmeric), which is a potent anti-allergic drug, recommended in various allergic conditions, including skin allergies like itching, blisters and so on (Thakkar 2006; Mahima et al. 2012). It is also a well-known immunomodulator (Bhakti et al. 2009; Ailani et al. 2019) and is effective in the management of allergic rhinitis as well as urticaria (Sason and Sharma 2016; Gupta and Mamidi 2016a). *Haridra* is proven to have antihistaminic properties. It helps in promoting the physical and mental health of the patient (Bhakti et al. 2009; Dhiman 2014). It is found to be very effective in relieving uterine fibroid and inflammatory disorders (Arbar and Verma 2016).

6.3.4 Puṣyānuga Cūrṇa

Puṣyānuga cūrna is one popular formulation described in Ayurveda. *Puṣyānuga cūrna* consists of 25 crude herbs and one mineral drug *gairika* (red ocher) (Anonymous 1978b). *Puṣyānuga cūrna* is prescribed for disorders like menorrhagia, leucorrhoea and disorders of the female genital tract (*pradara rōga*) (Bhuvaneswari and Seetarama 2017; Shailajan et al. 2017). It is also found effective in polycystic ovary syndrome (Khandelwal and Dipti Nathani 2016).

6.3.5 Rajanyādi Cūrṇa

Rajanyādi cūrna is composed of eight crude drugs. This formulation is reputed to stimulate digestion and cure all diseases of children (Panchal and Rajani 2017). It is used in *agnimāndya* (impairment of digestion), diarrhea (*atisāra*), fever (*jvara*), jaundice (*kāmala*), *pāṇḍu* (anemia) and respiratory distress (*śvāsa*) (Nigamanand et al. 2016; Panda et al. 2017)

6.3.6 Rāsnādi Cūrṇa

Rāsnādi cūrna is applied to the apex of the head and rubbed for a few seconds. This powder induces warmth in the head and keeps nasal congestion and related symptoms at bay (Janagal et al. 2017). *Śirōlēpa* (application of herbal paste on the scalp) with *Rāsnādi cūrna* is indicated in fever, diseases occurring above the clavicle, giddiness and delirium (Chandran et al. 2018).

6.3.7 Vaiśvānara Cūrṇa

Vaiśvānara cūrna is prepared by mixing appropriate quantities of rock salt, *Trachyspermum ammi*, *Carum roxburghianum*, *Zingiber officinale* and *Terminalia chebula* in a specific ratio. It is effective in the treatment of arthritis, constipation, abdominal pain, *āmavāta* (rheumatoid arthritis), *gulma* (lump in the abdomen), *hṛdrōga* (heart diseases), *śūla* (pricking pain), *pḷīha* (splenic disorder) and *granthi* (swellings). It is also recommended for use as *vēdana śamana* (analgesic), *śōtha praśamana* (anti-inflammatory) and *vātānulōmana* (facilitator of the natural movement of *vāta*, Anonymous 1978b). It helps in improving appetite and removal of abdominal gas and acts as a laxative. *Trachyspermum ammi* is a potent stimulant, antispasmodic and carminative in action, which are very helpful in digestive disorders (Bairwa et al. 2012). Similarly, antidiarrheal and antispasmodic effects of *Carum roxburghianum*, anti-emetic, anti-ulcerogenic, anticholinergic effect of *Zingiber officinale* and prokinetic effect of *Terminalia chebula* (Tamhane et al. 1997; Malhotra and Singh 2003; Khan et al. 2012) are responsible for the effectiveness of *Vaiśvānara cūrna* in gastric disorders. *In vitro* efficacy of *Vaiśvānara cūrna* as an antiurolithic agent (Ashok Kumar et al. 2013) and a laxative agent (Ashok Kumar et al. 2014) has been reported recently.

6.4 GHṚTA (MEDICATED CLARIFIED BUTTER)

6.4.1 Aśvagandha Ghṛta

Aśvagandha ghṛta helps to improve physical and mental health. It possesses properties of *rasāyana* (rejuvenator) and is traditionally claimed to enhance the weight of underweight children. It improves strength and internal circulation. *Aśvagandha ghṛta* consists of three ingredients viz. *Aśvagandha* (*Withania somnifera*), cow milk and clarified butter (*ghṛta*) (Mishra 1983). *Aśvagandha* contains withanolides which act as natural steroid compounds (Mishra 2015), whereas cow milk has proteins, vitamins and minerals. Clarified butter has a lipophilic property which favors the transport of active principles of the formulation to the target tissues. Thus *Aśvagandha* in the dosage form of *ghṛta* works better. *Aśvagandha ghṛta* has higher nutritional values and produces multidimensional effects in the body. *Aśvagandha ghṛta mātra basti* (medicated enema) is effective in knee osteoarthritis of the elderly population (Katti et al. 2016). It is also effective in the management of emaciation (*kārśyata*) (Patil et al. 2016; Upadhyay and Komal 2019).

6.4.2 Brahmī Ghṛta

Brahmī ghṛta contains ingredients like *Brahmi* (*Bacopa monneri*), *Vaca* (*Acorus calamus*), *Śaṅkhapuṣpi* (*Evolvulus alsinoides*), *Kuṣḍha* (*Saussurea lappa*) and clarified butter (Yadav and Reddy 2014). *Brahmī ghṛta* is an important formulation used for the treatment of memory disorders (Yadav et al. 2013). It is found effective in alleviating the symptoms of borderline mental retardation (Kute et al. 2017). Nasal instillation (*nasya*) of *Brahmī ghṛta* has the potency to prevent neurodegenerative disorders such as Huntington disease (Patel 2017a). *Brahmī ghṛta* is recommended for the management of *unmāda* (insanity), *apasmāra* (epilepsy) (Manu et al. 2017), *pāpavikāra* (diseases due to sinful acts) and *graha rōga* (diseases afflicted by evil spirits) (Yadav and Reddy 2012).

6.4.3 Jātyādi Ghṛta

Jātyādi ghṛta is widely used in the treatment of various types of obstinate ulcers (*duṣṭa vraṇa, nāḍī vraṇa*) (Anonymous 1978c). *Jātyādi ghṛta picu* is effective in fissure-in-ano and hence possesses a very good wound-healing activity (Rao et al. 2016; Swati et al. 2018).

6.4.4 Kalyāṇaka Ghṛta

Kalyāṇaka ghṛta is widely used in the management of a wide array of disorders. In *Suśruta Samhita*, *Kalyāṇaka ghṛta* is indicated in poisoning (Krishnapriya et al. 2018). The constituent herbs in *Kalyāṇaka ghṛta* have *viṣaghna* (anti-poison), *kuṣṭaghna* (skin disease-curing), *hṛdya* (cardioprotective) and *raktaśōdhaka* (blood-purifying) properties. This medicated clarified butter is found to be beneficial in personality disorders, insanity, cough, epilepsy, anemia, itching, consumption, delusion, diabetes mellitus, fever, infertility, poor intelligence, stammering speech and poor digestive power. It bestows strength, auspiciousness, long life, a good complexion and nourishment. However, it is more widely used in the management of psychiatric conditions (Sori et al. 2017; Sinimol 2019). It is also effective in schizophrenia (Gupta and Mamidi 2016b) and anovulation (Jain et al. 2016). *Kalyāṇaka ghṛta nasya* is effective in alcoholism (*madātyaya*) (Dhimdhime et al. 2016).

6.4.5 Mahātiktaka Ghṛta

Mahātiktaka ghṛta is indicated in *unmāda* (insanity), *ēka kuṣḍha* (psoriasis) (Kavyashree et al. 2017), menorrhagia (Vijay Lakshmi 2017), vitiligo (Khandekar et al. 2015), peptic ulcer, bleeding disorders, bleeding piles, herpes, gastritis, gout, anemia, blisters, schizophrenia, jaundice, fever and heart diseases (Reddy and Snehal 2017). It is also used for external application in non-healing wounds. This *ghṛta* is used for the purpose of oleation (*snēhakarma*) which is recommended before performing *vamana* (emesis) and *virēcana* (purgation) treatments (Anonymous 1978c).

6.4.6 Nārasimha Rasāyana

Nārasimha rasāyana is prescribed for the treatment of fatigue, hair loss, loss of vitality and emaciation in young men and women. This formulation also aids in improving fertility by acting as an aphrodisiac. It is said to be of use in reversing aging, improving immune function, and increasing sexual vigor and potency (Anonymous 1978c). Feeding mice with 50 mg/kg body weight of *Nārasimha rasāyana* consecutively for 5 days prior to, and for a month after, radiation, arrested the radiation-induced deleterious effects. Treatment with *Nārasimha rasāyana* increased the body weight and the organ weights of the recipient mice. It also decreased the levels of radiotoxic biochemical endpoints and the levels of serum and tissue lipid peroxides, serum alkaline phosphatase and glutamate pyruvate transaminase (Vayalil 2002).

The use of ionizing radiation compromises the radiosensitivity of normal tissues. A compound that can provide a selective benefit to normal cells against the damaging effects of ionizing radiation has been a long-sought goal. Nevertheless, most of the compounds investigated have shown inadequate clinical application owing to their inherent toxicity, undesirable side-effects and prohibitive cost. Ayurvedic *rasāyana* like *Nārasimha rasāyana* possesses radioprotective effects. Therefore, *Nārasimha rasāyana* is useful in improving the overall health of cancer patients (Baliga et al. 2013).

6.4.7 Phala Sarpis

Phala sarpis is one of the well-known medicines for successful conception, maintenance of pregnancy and is ideal for curing all disorders of the female genital tract (Otta and Tripaty 2002; Biala and Tiwari 2015; Oberoi et al. 2016). It is a valuable medicine known to be used in infertility for thousands of years. It is tonic, nourishing and rejuvenating (Parmar et al. 2017). *Uttaravasti* (enema) with *Phala sarpis* is highly effective in increasing the ovarian reserve and is a better alternative to hormonal therapy (Parmar et al. 2017). The fertility-promoting effect of *Phala sarpis* was studied in female albino rats with body weight, ovarian weight and ovarian estradiol as parameters. *Phala sarpis* proved to be successful in anovulatory infertility (Muralikumar and Shivasekar 2012). It is also effective in menorrhagia (Vanishree and Ramesh 2018) and cystocele (Gupta and Siddesh 2018).

6.4.8 Sukumāra Ghṛta

Sukumāra ghṛta is one of the medicinal clarified butters mentioned in *Sahasra Yōga* and *Ayurvedic Formulary of India* (Anonymous 1978c). *Sukumāra ghṛta* is an example of the dosage form *yamaka* or a combination of two fatty substances – clarified butter and castor oil – as ingredients. *Sukumāra ghṛta* is indicated in constipation, diseases of the abdomen/enlargement of the abdomen, abdominal lumps, splenic disease, abscess, edema, pain in the female genital tract, hemorrhoids, hydrocele, diseases due to *vāta dōṣa* and gout. It is also used in the oleation therapy, prior to *pañcakarma* (Vaidyan and Pillai 1985a). *Sukumāra ghṛta* facilitates parturition because of good cervical ripening and decreased labor pain, as well as shortening of the duration of the first stage of labor (Jadhav et al. 2014). *Yōni picu* (a sterile medicated cotton swab kept inside the vagina for a specific period of time) with *Sukumāra ghṛta* improved the tonicity of perineal muscles (Surendran et al. 2018).

6.4.9 Tṛphalā Ghṛta

Tṛphalā ghṛta is described by Ācārya Vāgbhṭa in the management of myopia (Dabhi et al. 2016). *Triphalā ghṛta*, which is generally used for *tarpaṇa* (eye bath), is saturated with a decoction of the *triphala* group of drugs (*Emblica officinalis*, *Terminalia belerica*, *Terminalia chebula*), and hence it contains both lipid and water-soluble constituents of *triphala*. Thus, it has lipophilic as well as hydrophilic properties. Therefore, it penetrates well through various layers of the cornea. In this way it is effective in computer vision syndrome and Stargardt's macular degeneration (Mohan 2013; Sawant et al. 2013; Sreekumar and Nair 2017).

6.4.10 Vidāryādi Ghṛta

Vidāryādi ghṛta acts as an effective nutritional remedy to overcome problems faced by adults suffering from malnutrition (Arun and Shivakumar 2016). It is also effective in the management of Graves' disease (Rao et al. 2018) and *vātaja pratiśyāya* (rhinitis) (Thakur and Jagtap 2016).

6.5 GUṬIKA (PILLS)

6.5.1 Bilvādi Guṭika

Bilvādi guṭika is widely used in the treatment of poisoning, especially in acute toxic pathological conditions. The first reference to this formulation is in *Aṣṭāṅgahṛdaya*, *uttarasthāna*, 36th chapter, where it is mentioned as *Bilvādi yōga* (a combination of *bilvādi* herbs). In *The Ayurvedic Formulary of India* it is named as *Bilvādi guṭika* and the same name is used in commerce (Anonymous 1978d; Shubha et al. 2017). *Bilvādi guṭika* is a proven drug of choice in conditions like cholera and cobra bites (Sunitha and Hussain 2018a).

6.5.2 Chandraprabhā Vaṭi

Chandraprabhā vaṭi is often recommended for having a young-looking, wrinkle-free and glowing skin and for treating several diseases such as those of skin, urinary system, respiratory system, gastrointestinal system, eye, teeth, diabetes, anemia (Doddamani et al. 2015), obesity and reproductive system. Besides, it is also claimed to possess pain-relieving and anti-inflammatory properties (Weerasekera et al. 2014). It successfully prevents diabetic neuropathy, diabetic retinopathy and may prevent essential hypertension (Sushama and Nishteswar 2012).

6.5.3 Dhānvantaram Guṭika

Dhānvantaram guṭika is used during pregnancy care to safeguard pregnancy. This formula is a contribution of the Kerala Ayurveda practice. It is an anti-inflammatory medicine, used to relieve arthritic pain. It is effective in fever, cough, wheezing, loss of appetite and uterine fibroids (Saroch and Singh 2015).

6.5.4 Kaiśōra Guggulu

Kaiśōra guggulu is one of the most famous ayurvedic formulations that is used traditionally to support healthy joints, muscles and connective tissue. It is known to have analgesic, anti-allergic, antibacterial and anti-inflammatory activities (Vaidyan and Pillai 1985b). Traditionally it is used for skin diseases and gout (Kurup 2001; Sinha et al. 2008). It is good for aggravated fibromyalgia patients suffering from muscle pain (Lather et al. 2011; Gupta et al. 2014). It acts as a skin health promoter, natural blood cleanser and supportive dietary herbal supplement in many conditions such as diabetes and skin diseases.

6.5.5 Kāñcanāra Guggulu

Kāñcanāra guggulu has anti-inflammatory and cyst-dissolving properties (Banothe et al. 2018). Because of these qualities, *Kāñcanāra guggulu* may check the changes in prostatic tissues and regulate urinary function (Han et al. 2007). This formulation is in clinical use for many centuries in the treatment of *gaṇḍamāla* (cervical lymphadenopathy), *arbuda* (tumor), *granthi* (cyst) and *kuṣḍha* (skin disease) (Dhiman 2014). *Kāñcanāra guggulu* supports the proper function of the lymphatic system, promotes the elimination of inflammatory toxins and is administered in malignant ulcers, syphilis, fistula, scrofula, sinus and benign prostatic hyperplasia (*mutraghata*) (Patel et al. 2015a).

6.5.6 Mānasamitra Vaṭaka

Mānasamitra vaṭaka is a medicine that improves memory and intellect and combats speech problems in children. This medicine contains 73 ingredients and is used in the dosage of one tablet per day (Anonymous 1978d). *Mānasamitra vaṭaka* is reported to improve memory and intellect and cure cognitive deficits (Thirunavukkarasu et al. 2012; Sanjaya Kumar et al. 2016).

Tubaki et al. (2003) explored its efficacy in patients with a generalized anxiety disorder (G.A.D.) with comorbid generalized social phobia. Seventy-two patients with G.A.D. with comorbid social phobia were randomly divided into three treatment groups. Groups 1 and 2 received *Mānasamitra vaṭaka* (100 mg twice daily for 30 days). Group 2, in addition to *Mānasamitra vaṭaka*, was treated with *Śirōdhāra* (therapy involving dripping of medicated oil over the forehead) for the first seven days. Group 3 received clonazepam 0.75 mg daily in divided dose for 30 days. The study was assessed using the Hamilton Anxiety Rating Scale, Beck Anxiety Inventory, Beck Depression Inventory, Epworth Sleepiness Scale (E.S.S.), World Health Organization Quality of Life B.R.E.F., and Clinical Global Impression Scales (improvement and efficacy). Patients of all the groups showed a significant reduction in clinical parameters. Nevertheless, improvement in E.S.S. was observed only in Group 2. The treatment outcome was comparable between the three groups. This is the first study conducted on the efficacy of *Mānasamitra vaṭaka* in anxiety disorders, and the results indicate that *Mānasamitra vaṭaka* is effective in the management of G.A.D. with comorbid generalized social phobia.

6.6 KVĀTHA (DECOCTIONS)

6.6.1 Amṛtōttaram Kvātha

Amṛtōttaram kvātha is an important ayurvedic formulation prepared using stems of *Tinospora cordifolia* (6 parts), the pericarp of *Terminalia chebula* (4 parts) and dried rhizomes of *Zingiber officinale* (2 parts) (Anonymous 1978e). It is reported to reduce uncomfortable symptoms of indigestion, acts as a mild laxative and is used in the treatment of chronic fever, as well as rheumatoid arthritis (Sulaiman et al. 2012; Bhokardankar et al. 2018).

6.6.2 Balāguḷūcyādi Kvātha

Balāguḷūcyādi kvātha is a polyherbal liquid preparation. This is a concentrated decoction of herbs which contain water-soluble active principles. *Balāguḷūcyādi kvātha* is prepared from three medicinal herbs viz. *Bala* (*Sida cordifolia*/*Sida rhombifolia*/*Sida retusa*), *Guḍūci* (*Tinospora cordifolia*) and *Dēvadāru* (*Cedrus deodara*). It is used in the treatment of rheumatoid arthritis, gout, joint inflammation and similar conditions (Khan et al. 2016).

6.6.3 Daśamūla Kvātha

Daśamūla is a group of ten drugs – *Aegle marmelos*, *Stereospermum suaveolens*, *Clerodendrum phlomidis*, *Oroxylum indicum*, *Gmelina arborea*, *Solanum nigrum*, *Solanum xanthocarpum*, *Uraria picta*, *Desmodium gangeticum*, *Tribulus terrestris* (Arun Raj et al. 2018). The official parts are roots. *Daśamūla kvātha* is indicated in conditions like disorders of the nervous system (*vātavyādhi*), including hemiplegia (*pakṣāghāta*) and low back pain (*kaṭīśūla*) (Gupta et al. 2016; Inamdar et al. 2017).

6.6.4 Daśamūla Kaṭutraya Kvātha

Daśamūla kaṭutraya kvātha is a unique combination of *daśamūla*, *trikaṭu* (fruits of *Piper longum*, *Piper nigrum* and dried rhizomes of *Zingiber officinale*) and *vāśā* (*Adhatoda vasica*). While all medicines are beneficial in the postpartum stage, *daśamūla* are especially useful in pacifying *vāta*. *Trikaṭu* is helpful in improving digestion, *vāśā* is useful in nourishing blood, and the combination has an added advantage due to its ability to reduce pain in joints. It is also effective in rheumatoid arthritis (Deshpande et al. 2017).

6.6.5 Dhānvantaram Kvātha

Dhānvantaram kvātha mentioned in *Aṣṭāṅgahṛdaya* is widely used in the management of diseases manifested due to *vāta* derangement and *vātarakta* (mostly diseases of the connective tissues, bones, joints and nervous system) (Sruthi and Sindhu 2012). It is also effective in traumatic conditions of bone and vital points. It is effectively used in pediatric diseases and postpartum care as well. This *kvātha* is also effective in fever, flatulence, mental disorders urinary obstructions and hernia (Upadhyay 1975a).

6.6.6 Drākṣādi Kvātha

Drākṣādi kvātha is a potent anthelmintic agent (Revathy 2018). It is effective in vomiting, burning sensation, fainting, fever, alcoholism, excessive thirst, jaundice and rheumatoid arthritis (Aiswarya and Dharmarajan 2015).

6.6.7 Gandharvahastādi Kvātha

Osteonecrosis or avascular necrosis is a pathologic process that results from a critical decrease of blood supply to the bone and elevated intra-osseous pressure. Osteonecrosis is the final common outcome of traumatic and non-traumatic insults, impairing blood circulation to the bone. Consequently, interruption of blood flow to the bone leads to the death of bone marrow and osteocytes, resulting in the collapse of the necrotic segment (Mont and Hungerford 1995). Avascular necrosis is clinically characterized by the gradual onset of pain in motion and relieved by rest in the affected joint with radiation down the affected limb, leading to muscle spasm (Epstein et al. 1965). *Gandharvahastādi kvātha* is effective in the management of avascular necrosis (Prasad et al. 2018).

6.6.8 Mahāmañjiṣṭādi Kvātha

It is effective in diabetes mellitus (Priyanka et al. 2017), avascular necrosis (Prasad et al. 2018), seborrheic dermatitis (Devi and Sharma 2017) and in childhood atopic dermatitis (Chethan Kumar et al. 2017).

6.6.9 Mahārāsnādi Kvātha

Mahārāsnādi kvātha is a formulation widely used in the treatment of neurological diseases. The original formula is believed to have been designed by Prajāpati (Radha et al. 2014). It is effective in the management of neurological diseases (*vāta vyādhi*) such as arthritis (Pandey and Chaudhary 2017; Padekar et al. 2018), lower back pain (Gupta et al. 2016), Bell's palsy (Akashlal et al. 2019; Sawarkar and Sawarkar 2019), painful joints, arthralgia, sciatica, frozen shoulder and gout (Tike et al. 2018) and in reducing joint pain in Dengue fever (Bhageshwary et al. 2016). It also helps in the improvement of balance in progressive degenerative cerebellar ataxias (Sriranjini et al. 2009) and is also effective in female infertility (Singh et al. 2014).

6.6.10 Nayōpāya Kvātha

Nayōpāya kvātha is effective in the management of bronchial asthma (*tamaka śvāsa*) when *dōṣās* are in morbid stage (Shyam et al. 2010) and is also prescribed for the treatment of allergic ailments such as cough (*kāsa*) (Balachandran and Devi 2007).

6.6.11 Paṭōla Kaṭurōhiṇyādi Kvātha

Paṭōla Kaṭurōhiṇyādi kvātha is a combination of six herbal ingredients viz. *Trichosanthes dioica*, *Picrorhiza kurroa*, *Pterocarpus santalinus*, *Marsdenia tenacissima*, *Tinospora cordifolia* and *Cissampelos pareira* (Upadhyay 1975b). This classical ayurvedic formulation has been reported by many practitioners to be effective in the treatment of liver disorders (Pawar et al. 2015). It is indicated in jaundice, skin disease, vomiting, fever, diseases due to poisoning, viral infections and liver diseases (Rao et al. 2015). It is widely used in the treatment of skin disease involving itching, pigmentation and burning sensations. It is useful in decreasing bad cholesterol and brings forth a potent antitoxic effect. It is used for metabolic corrections, liver dysfunctions and lowered immunity. It is effective in the management of spider poisoning (Sunitha and Hussain 2018b) and ulcers caused by snake bites (Roshni 2017; Vijayan et al. 2018). It improves digestion and relieves anorexia. It is a potent antimicrobial medicine too.

6.6.12 Punarnavādi Kvātha

The main ingredients of this formulation are *Trichosanthes dioica*, *Boerhavia diffusa*, *Azadiracta indica*, *Zingiber officinale*, *Picrorhiza kurroa*, *Tinosporac cordifolia*, *Andrographis paniculata* and *Berberis aristata* (Vaidyan and Pillai 1985c). It is slightly laxative, so it can easily clear up the channels or *srōtas*. It is effective in the management of uterine fibroids along with polycystic ovarian syndrome (Kumar et al. 2015), edema and hypothyroidism (Singh and Thakar 2018).

6.6.13 Sahacarādi Kvātha

Sahacarādi kvātha contains only three ingredients – *Zingiber officinale*, *Cedrus deodara* and *Barleria prionitis* (Vaidyan and Pillai 1985c). In this recipe, *Zingiber officinale* acts as an improver of "digestive fire" and *Cedrus deodara* pacifies *vāta*. *Sahacaram* means "walking along with". As *Barleria prionitis* has the specific property of *gativiśēṣatvam* (ability to cause movement), *Sahacarādi kvātha* can be administered in conditions like difficulty in walking or improper walking. The combination of all these drugs is therapeutically effective in lower back pain, sciatica, debility of lower limbs (Supraja and Vidyanath 2013) and chronic rheumatoid arthritis with bilateral hip involvement (Mamidi and Gupta 2015; Nimesh and Shetty 2018) It is also effective in nerve-related diseases like hemiplegia, epilepsy (Kumar et al. 2018), disc prolapse, facial palsy and paralysis (Ratha et al. 2018).

6.7 AVALĒHA (ELECTUARIES)

6.7.1 Agastya Harītaki Rasāyana

Agastya harītaki rasāyana is a formulation useful in the management of various types of respiratory disorders like cough, asthma, and hiccough, and also for piles, diarrhea, heart disease, anorexia, intermittent fever and graying of hair (Sharma 1983a). It is used as a *rasāyana* (nutrient to body and mind with adaptogenic and immuno-stimulatory properties) (Ram et al. 2013; Keshava et al. 2014). It is also proven for its immunomodulatory (Aher and Wahi 2011), cytoprotective (Dahanukar 1983), anticaries (Jagtap and Karkera 1999), antispasmodic and hypolipidemic properties (Ahirwar et al. 2003).

6.7.2 Cyavanaprāśa

Caraka was the first author to mention *Cyavanaprāśa*. He explains that it is the *rasāyana* par excellence. It is an excellent rejuvenator with the ability to delay aging and enhance general well-being. It is effective in cardiovascular diseases, cough, dyspnea, rheumatic diseases, hoarseness of voice and impotence. It promotes intellect as well (Sharma 1983b). The major ingredient of *Cyavanaprāśa* is gooseberry (*Emblica officinalis*), which is a known antioxidant and rich source of vitamin C (Parle and Bansal 2006). The primary action of *Cyavanaprāśa* is to strengthen the immune system and to support the body's natural ability to produce hemoglobin and white blood cells (Pole 2006). As a result, it is reputed as a highly admired *rasāyana*, offering deep nourishment to tissues, preserving youthfulness and promoting systemic health and well-being.

Madan et al. (2015) investigated the immune-stimulatory effects of *Cyavanaprāśa* using *in vitro* assays, evaluating the secretion of cytokines such as tumor necrosis factor-alpha (TNF-α), interleukin-1beta (IL-1β) and macrophage inflammatory protein-1-alpha (MIP-1-α) from murine bone marrow–derived dendritic cells which play a crucial role in immuno-stimulation. The effects of

Cyavanaprāśa on phagocytosis in murine macrophages (RAW264.7) and natural killer cell activity were also investigated. At non-cytotoxic concentrations, *Cyavanaprāśa* enhanced the secretion of all the three cytokines from dendritic cells. *Cyavanaprāśa* also stimulated both macrophage and natural killer cell activity *in vitro*. These findings substantiate the immunoprotective role of *Cyavanaprāśa* at the cellular level.

Chronic exposure to ultraviolet radiation induces premature skin aging. U.V. irradiation generates reactive oxygen species (R.O.S.), which play a pivotal role in skin photoaging. Takauji et al. (2016) examined the effect of *Cyavanaprāśa* on skin photoaging. Hairless mice were fed with *Cyavanaprāśa* in drinking water for three weeks, and then repeatedly exposed to ultraviolet light B (U.V.B.) irradiation to induce skin photoaging. The effects of *Cyavanaprāśa* were examined in cells cultured *in vitro*. *Cyavanaprāśa* was added to the culture medium and examined for its effect on the growth of human keratinocytes, and for the ability to eliminate R.O.S. generated by paraquat.

U.V.B. irradiation caused symptoms like rough skin, erythema and skin edema in hairless mice. But the administration of *Cyavanaprāśa* relieved these symptoms. *Cyavanaprāśa* also significantly suppressed epidermal thickening, a typical marker of skin photoaging, in mice. The authors found that *Cyavanaprāśa* enhanced the growth of human keratinocytes, and efficiently eliminated R.O.S. These findings suggest that *Cyavanaprāśa* may have beneficial effects on slowing skin photoaging (Takauji et al. 2016).

In the last 60 years *Cyavanaprāśa* has grown to industrial production and marketing in packed forms to a large number of consumers as a health food. At present *Cyavanaprāśa* has acquired a large consumer base in India and a few countries outside India. The reported clinical studies indicate that individuals who consume *Cyavanaprāśa* regularly for a definite period of time showed improvement in overall health status and immunity. Randomized controlled trials of high quality with larger sample size and longer follow-up are essential to assess evidence on the clinical use of *Cyavanaprāśa* as an immunity booster. More studies involving measurement of current biomarkers of immunity are required to establish the clinical relevance of *Cyavanaprāśa* (Narayana et al. 2017).

6.7.3 Māṇibhadra Guḍa

Māṇibhadra guḍa is a famous ayurvedic formulation described in *Aṣṭāṅgahṛdaya*. It contains *Embelia ribes*, *Emblica officinalis*, *Terminalia chebula*, *Operculina turpethum* and jaggery. It is effective in the management of worms, piles, skin diseases and fistula (Upadhyay 1975c; Patel et al. 2015b).

6.7.4 Bilvādi Lēhya

Bilvādi lēhya contains *Aegle marmelos* (root), *Cyperus rotundus* (tuber), *Cuminum cyminum* (seed), *Eletteria cardamomum* (seed), *Cinnamomum zeylanicum* (bark), *Mesua ferrea* (flower), *Zingiber officinale* (dried rhizome), *Piper nigrum* (fruit), *Piper longum* (fruit) and jaggery (Anonymous 1978f). *Bilvādi lēhya* is effective in reducing the symptoms of hyperemesis such as nausea, salivation and vomiting and helps in improving the body weight and hemoglobin level of patients, along with improvement in general condition (Math and Jana 2019).

6.8 TAILA (MEDICATED OILS)

6.8.1 Aṇu Taila

Aṇu taila is prescribed for nasal instillation (*nasya karma*) in various disease conditions. *Aṇu taila* can provide better result in sneezing (*kṣavathu*), nasal obstruction (*nāsāvarōdha*), watery nasal discharge (*tanusrāva*), retracted tympanic membrane, anosmia (*gandhahāni*), itching (*kaṇḍu*) and turbinate hypertrophy (Verma 2013). Nasal instillation of *Aṇu taila* is also reported to cure torticollis (*manyā stambha*), headache (*śiraḥśūla*), facial paralysis (*ardita*), lock-jaw (*hanusamgraha*), migraine (*ardhāvabhēdaka*) and chronic sinusitis (*duṣṭa pratiśyāya*) (Thanki et al. 2009; Bhardwaj 2012; Patil and Sawant 2012; Naveen and Kumar 2014; Dave et al. 2016; Jitesh and Amit 2016; Sharma and Soni 2017).

6.8.2 Balā Taila

Balā taila is suitable for mother and child healthcare (Sharma et al. 2017). Thus, the local application of *Balā taila* as an oil-soaked swab (*picu*) and enema (*vasti*) is a treatment of choice in diseases like uterine prolapse, abnormal uterine bleeding, cervicitis, vaginitis, pelvic inflammatory disease, urethral stricture and Lou Gehrig's disease (Patel 2017b; Patel 2018). *Balā taila* is the best-medicated oil for body massage of infants. *Balā taila* increases the strength and nutrition of tissues, thus improving the growth of the infant (Agarwal et al. 2000). Massage in infancy is helpful to prevent hypothermia and induce post-massage sleep (Raskar and Rajagopala 2015). Moderate massage pressure contributes many positive effects, including increased weight gain in pre-term infants, enhanced attentiveness, increased vagal activity, decreased cortisol levels and enhanced immune function (Butali and Arbar 2014). It also increases the strength and nutrition of tissues and is helpful in preventing hypothermia (Thakur and Baba 2017). Enema (*mātrā vasti*) with *Balā taila* is found to be effective in curing the inability to climb stairs in patients of osteoarthritis (Anurag et al. 2015). *Balā taila* application is effective on traumatic wounds, by reducing pain, minimizing the wound surface, reducing discharge, promoting epithelialization and granulation, and also avoiding hypertrophic scar formation (Tripathy et al. 2011). Massage (*abhyaṅga*) with *Balā taila* is found to be effective in the management of cerebral palsy (Ghuse et al. 2016).

6.8.3 Gandharvahastādi Ēraṇḍa Taila

Gandharvahastādi ēraṇḍa taila is known to clear the micro-channels in specific areas, facilitating proper healing and restoring physiological function (Sunil Kumar et al. 2017). Therapeutic purgation with *Gandharvahastādi ēraṇḍa taila* is effective in the treatment of rheumatoid arthritis (*āmavāta*) (Sharma et al. 2015), Meniere's disease (Rani and Madhusudan 2019), lumbar stenosis (Ratha et al. 2016), lower backache (Raghunathan et al. 2015) and sciatica (Rao et al. 2017).

6.8.4 Irimēdādi Taila

Irimēdādi taila is useful in dental problems such as stomatitis, dental caries, gingivitis, stained teeth and hyperemia of gums (Amruthesh 2008; Boloor et al. 2014; Ahlawat and Sarswat 2018; Fida et al. 2018; Patil et al. 2018; Ankita 2019). A clinical study found *Irimēdādi taila* to be equally effective as chlorhexidine, as an adjunct to mechanical plaque control in the prevention of plaque accumulation and gingivitis. The authors concluded that *Irimēdādi taila* could be an effective and safe alternative to 2% chlorhexidine gluconate mouthwash, due to its prophylactic and therapeutic benefits (Mali et al. 2016).

6.8.5 Kṣīrabalā Taila

Kṣīrabalā taila is prepared by boiling together cow milk, the paste of *Sida rhombifolia* root, the decoction of *Sida rhombifolia* root and sesame oil (Anonymous 1978g). It has rejuvenation properties and is used both topically and systemically. It is indicated in all disorders of central nervous system, facial paralysis, sciatica, hemiplegia, paraplegia, poliomyelitis and Parkinson's disease (Grampurohit et al. 2014; Nair et al. 2015; Kamble et al. 2017). Therapeutic enema with *Kṣīrabalā taila* is found to be efficacious in the whole symptom complex of painful menstruation. It also helps to prevent recurrence of dysmenorrhea (Lakshmi and Asokan 2017), cerebral palsy in children (Adinath et al. 2016) and lower backache (Tripathy et al. 2016; Singh and Satpute 2018).

6.8.6 Mahānārāyaṇa Taila

Mahānārāyaṇa taila is a rich combination of anti-arthritic ayurvedic herbs that produce no irritation on the skin and arrest further progress of chronic arthritic changes of joints, pain, stiffness, restricted movement and distortion. It restores normal joint function. It has anti-inflammatory, analgesic and anti-arthritic activities (Pawar et al. 2011; Tiwari 2015) and hence is suitable for fissure-in-ano to relieve sphincter spasm along with other symptoms, without adverse effects (Peshala et al. 2014), and in the management of sciatica (*gṛdhrasī*) (Goswami 2014), osteoarthritis and Parkinson's disease (Srivastava et al. 2015).

6.8.7 Mahāmāṣa Taila

Mahāmāṣa taila contains ingredients like *Phaseolus mungo* (*māṣa*), *daśamūla* group of drugs) and so on. It pacifies *vāta dōṣa* and gives strength to eyes through errhine therapy (*nasya*) (Renu et al. 2014). *Nasya* with *Mahāmāṣa taila* irritates the nasal mucosa, leading to an edematous response with local hyperemia, which enhances drug absorption. Since the drug administered is itself lipid in nature, there is no functional blood–brain barrier for *Mahāmāṣa taila*. During *nasya* procedure, lowering of the head and fomentation to face seems to have an impact on blood circulation to the head. The efferent vasodilator nerves which are located on the superficial surface of the face are stimulated by application of the medicinal oil and fomentation, leading to momentary hyperemia in the head region (Gupta 2017). It is effective in the management of frozen shoulder and cerebral palsy (Bagali and Prashanth 2016).

6.9 CONCLUSION

Ayurvedic medicines are designed on the basis of parameters like *rasa* (taste), *guṇa* (qualities), *vīrya* (potency), *vipāka* (post-digestive taste) and *prabhāva* (specific action) (Upadhyay 1975d). These medicines are generally prepared using herbal ingredients. When viewed through the perspective of Western medicine, the biological actions of these medicines are brought about by the numerous organic compounds belonging to classes like glycosides, saponins, tannins, alkaloids, phenols, flavonoids, terpenoids, lactones and so on (Hoffmann 2003).

Studies are required to demonstrate the therapeutic utility of ayurvedic medicines. Some recent studies on *Āmalakī rasāyana* and *Rasa sindūram* in a *Drosophila* model suggest that the effects of these formulations on fruit flies are similar to those described in classical Ayurveda (Dwivedi et al. 2012). Interestingly, both these formulations enhanced the levels of different h.n.R.N.P.s (heterogeneous nuclear RNA-binding proteins), which are involved in transcriptional and post-transcriptional regulation of gene expression, and of CBP300, a histone acetyl transferase essential for modifying chromatin for elevated gene expression (Dwivedi et al. 2012). *Āmalakī rasāyana* and *Rasa sindūram* suppressed induced apoptosis (Dwivedi et al. 2015). *Amalaki rasayana* enhanced tolerance of flies to oxidative stress, which may contribute to the increased life span attributed to *Amalaki rasayana* therapy (Dwivedi and Lakhotia 2016).

The upsurge in the popularity of Ayurveda calls for stringent quality control of ayurvedic formulations. As the majority of ayurvedic formulations are complex mixtures of various herbal ingredients, their quality control through physico-chemical parameters may not be feasible, except in a few simpler formulations (Shengule et al. 2019). Biological quality control of ayurvedic medicines, using model organisms needs to be defined (Lakhotia 2019). Thus R.&D. and quality control of ayurvedic medicines are essential to take Ayurveda through the new millennium.

REFERENCES

Adinath, L. P., B. A. Chaudhari, and P. P. Ingale. 2016. Effectiveness of ksheerabala taila matra basti in children with cerebral palsy. *Unique Journal of Ayurvedic and Herbal Medicines* 4: 8–11.

Agarwal, K. N., A. Gupta, R. Pushkarna, S. K. Bhargava, M. M. Faridi, and M. K. Prabhu. 2000. Effects of massage and use of oil on growth, blood flow and sleep pattern in infants. *Indian Journal of Medical Research* 112: 212–7.

Aher, V., and A. K. Wahi. 2011. Immunomodulatory activity of alcohol extract of *Terminalia chebula* Retz Combretaceae. *Tropical Journal of Pharmaceutical Research* 10, no. 5: 567–75.

Ahirwar, B., A. K. Singhai, and V. K. Dixit. 2003. Effect of *Terminalia chebula* fruits on lipid profiles of rats. *Journal of Natural Remedies* 3: 31–5.

Ahlawat, B., and O. Sarswat. 2018. *Arimedas* oil for oil pulling. *World Journal of Pharmaceutical and Medical Research* 4: 168–8.

Ailani, R. S., V. P. Kanani, and C. A. Dhanokar. 2019. A comparative study between *Guduchighanvati* and *Haridrakhanda* in the management of *Vataj Pratishyaya* (allergic rhinitis). *Ayurline: International Journal of Research in Indian Medicine* 3: 1–9.

Aiswarya, I. V., and P. Dharmarajan. 2015. A protocol-based approach in the management of *amavata* – A case report. *Journal of Ayurveda and Holistic Medicine* 3: 106–13.

Akashlal, M., K. Mithuna, and K. M. P. Shankar. 2019. Ayurvedic management of Bell's palsy – A single case report. *International Journal of Ayush CASE Reports* 3: 7–14.

Alam, S., and J. Gupta. 2018. Standardization of ayurvedic formulations by high performance thin layer chromatography- A review. *World Journal of Pharmacy and Pharmaceutical Sciences* 7: 530–57.

Amruthesh, S. 2008. Dentistry and Ayurveda IV: classification and management of common oral diseases. *Indian Journal of Dental Research: Official Publication of Indian Society for Dental Research* 19, no. 1: 52–61.

Ankita, A. 2019. Scientific review on popular herbs and herbal formulations of Ayurveda in dentistry. *International Journal of Ayurveda and Pharmaceutical Chemistry* 10: 227–40.

Anonymous. 1978a. *Āsava* and *Ariṣṭa*. In *The ayurvedic formulary of India, part-I, first edition*, 1–17. New Delhi: Ministry of Health and Family Planning.

Anonymous. 1978b. *Cūrṇa*. In *The ayurvedic formulary of India, part-I, First Edition*, 83–95. New Delhi: Ministry of Health and Family Planning.

Anonymous. 1978c. *Ghṛta*. In *The ayurvedic formulary of India, part-I, first edition*, 61–82. New Delhi: Ministry of Health and Family Planning.

Anonymous. 1978d. *Vaṭi and Guṭika*. In *The ayurvedic formulary of India, part-I, first edition*, 139–54. New Delhi: Ministry of Health and Family Planning.

Anonymous. 1978e. *Kvātha cūrṇa*. In *The ayurvedic formulary of India, part-I, first edition*, 41–52. New Delhi: Ministry of Health and Family Planning.

Anonymous. 1978f. *Avalēha* and *Pāka*. In *The ayurvedic formulary of India, part-I, first edition*, 23–39. New Delhi: Ministry of Health and Family Planning.

Anonymous. 1978g. *Taila*. In *The ayurvedic formulary of India, part-I, first edition*, 97–121. New Delhi: Ministry of Health and Family Planning.

Anonymous. 2014. *Kitibha-psoriasis*. New Delhi: Central Council for Research in Ayurvedic Sciences.

Anurag, V., S. Bharadwaj, S. Sharma, and S. Richa. 2015. The role of *matra Basti* with *Bala taila* in *Sandhigata Vata* w.s.r to ability to climbing stairs in patients of osteoarthritis- knee joint. *International Journal of Ayurvedic Medicine* 6: 262–6.

Arbar, A. A., and S. Verma. 2016. Clinical profile of nasal polyp in a pediatric patient: An ayurvedic approach. *Indian Journal of Health Sciences* 9, no. 3: 335–8.

Arun, K. B., U. Aswathi, Venugopal, T. S. Madhavankutty, and P. Nisha. 2016. Nutraceutical properties of cumin residue generated from ayurvedic industries using cell line models. *Journal of Food Science and Technology* 53, no. 10: 3814–24.

Arun, R., and Shivakumar. 2016. Nutraceutical effect of *Vidaryadi Ghritha* in *Karshya*. *International Ayurvedic Medical Journal* 4: 2511–7.

Arun Raj, G. R., U. Shailaja, P. N. Rao, M. P. Pujar, Mohan K. Srilakshmi, and K. Mohan. 2018. Effectiveness of Ayurveda treatment modalities in the management of spasticity in children with cerebral palsy at a tertiary care teaching hospital of southern India. *International Journal of Research in Ayurveda and Pharmacy* 9, no. 2: 96–100.

Ashok Kumar, B. S., G. Saran, R. Harshada, K. N. Bharath, A. Kumar, and K. S. Kulasekar. 2014. Evaluation of laxative activity of *Vaishvanara Churna*: An ayurvedic formulation. *Advances in Bioscience and Clinical Medicine* 2, no. 2: 64–7.

Ashok Kumar, B. S., G. Saran, R. Harshada, N. Keerthi, and S. Vandana. 2013. Evaluation of *in vitro* antiurolithiatic activity of *Vaishvanara Churna*. *Journal of Medicinal Plants Studies* 1: 142–4.

Bagali, P. H., and A. S. Prashanth. 2016. A study on frozen shoulder and its clinical management through *nasaapana* and *nasya*. *Journal of Ayurveda and Integrated Medical Sciences* 1, no. 1: 16–23.

Bairwa, R., R. S. Sodha, and B. S. Rajawat. 2012. *Trachyspermum ammi*. *Pharmacognosy Reviews* 6, no. 11: 56–60.

Balachandran, S., and R. S. Devi. 2007. Estimation of sodium benzoate in ayurvedic formulation: *Kashaya* (water decoction). *Asian Journal of Chemistry* 19: 3421–6.

Baliga, M. S., S. Meera, L. K. Vaishnav, S. Rao, and P. L. Palatty. 2013. Rasayana drugs from the ayurvedic system of medicine as possible radioprotective agents in cancer treatment. *Integrative Cancer Therapies* 12, no. 6: 455–63.

Banothe, G. D., V. Mahanta, S. K. Gupta, and T. S. Dudhamal. 2018. A clinical evaluation of *Kanchanara guggulu* and *Bala taila matra Basti* in the management of mutraghata with special reference to benign prostatic hyperplasia. *Ayu* 39, no. 2: 65–71.

Bhageshwary, J., C. Singh, P. R. Prasad, and A. Manoj. 2016. Dengue in ayurvedic perspective and its management – A review article. *International Ayurvedic Medical Journal* 4: 3385–90.

Bhakti, C., M. Rajagopala, A. K. Shah, N. Bavalatti. 2009. A clinical evaluation of *Haridra Khanda* and *Pippalyadi taila nasya* on *pratishyaya*. *Ayu* 32: 188–93.

Bhalerao, P. P., R. B. Pawade, and S. Joshi. 2015. Evaluation of analgesic activity of *Dashamoola* formulation by using experimental models of pain. *Indian Journal of Basic and Applied Medical Research* 4: 245–55.

Bhardwaj, A. 2012. Comparative therapeutic effects of various morphological forms of *nasya* (nasal route of drug delivery) in *pratishyaya* (rhinosinusitis) with reference to nasal muco-ciliary function. *Journal of Pharmaceutical and Scientific Innovation* 1: 58–64.

Bhaskar, G., A. Shariff, and S. R. B. Priyadarshini. 2009. Formulation and evaluation of topical polyherbal antiacne gels containing *Garcinia mangostana* and *Aloe vera*. *Phaarmcognosy Magazine* 5: 93–9.

Bhokardankar, P. S., Balasubramani, and M. P. Bhokardankar2018. Estimation of gallic acid in different commercial samples of *Amruthotharam kashayam* by using HPLC. *UK Journal of Pharmaceutical and Biosciences* 6: 41–5.

Bhuvaneswari, M., and D. K. Seetarama. 2017. Role of *Pushyanuga churna* in *rakta pradara*. *International Ayurvedic Medical Journal* 5: 44–67.

Biala, S., and R. Tiwari. 2015. Efficacy of *Phala Ghrita* on female infertility. *Ayushdhara* 2: 84–8.

Boloor, V. A., R. Hosadurga, A. Rao, H. Jenifer, and S. Pratap. 2014. Unconventional dentistry in India – An insight into the traditional methods. *Journal of Traditional and Complementary Medicine* 4, no. 3: 153–8.

Butali, S., and A. Arbar. 2014. A RCT of *Bala taila abhyanga* versus coconut oil massage in physiological transition of newborn. *Journal of Ayurveda and Holistic Medicine* 2: 32–9.

Chandran, S., K. S. Dinesh, B. J. Patgiri, and P. Dharmarajan. 2018. Unique contributions of Keraleeya Ayurveda in pediatric health care. *Journal of Ayurveda and Integrative Medicine* 9, no. 2: 136–42.

Chauhan, G., A. K. Mahapatra, K. A. Babar, and A. Kumar. 2015. Study on clinical efficacy of *Avipattikar choorna* and *Sutasekhar rasa* in the management of *Urdhwaga amlapitta*. *Journal of Pharmaceutical and Scientific Innovation* 4, no. 1: 11–5.

Chavhan, S. G., S. A. Naik, and L. P. Gedam. 2017. *Hidradenitis suppurativa* -A successful ayurvedic management-A case study. *World Journal of Pharmaceutical and Life Sciences* 3: 586–98.

Chethan Kumar, V. K., P. Soumya, and R. Anjana. 2017. Ayurvedic management of childhood atopic dermatitis – A case report. *Journal of Ayurvedic and Herbal Medicine* 3: 57–9.

Dabhi, D., R. Manjusha, C. R. Harisha, and V. J. Shukla. 2016. Ingredients identification and pharmaceutical analysis of *triphala ghrita* – A compound ayurvedic formulation. *World Journal of Pharmaceutical Research* 5: 606–12.

Dahanukar, S. A. 1983. Cytoprotective effect of *Terminalia chebula* and *Asparagus racemosus* on gastric mucosa. *Indian Drugs* 20: 442–5.

Dave, P. P., K. H. Bhatta, D. B. Vaghela, and K. S. Dhiman. 2016. Role of *Vyaghri haritaki avaleha* and *Anu taila nasya* in the management of *dushta pratishyaya* (chronic sinusitis). *International Journal of Ayurvedic Medicine* 7: 49–55.

Deshpande, S. V., V. S. Deshpande, and S. S. Potdar. 2017. Effect of *panchakarma* and ayurvedic treatment in postpartum rheumatoid arthritis (*amavata*): A case study. *Journal of Ayurveda and Integrative Medicine* 8, no. 1: 42–4.

Devi, P. S., and V. S. Sharma. 2017. Management of seborrhoeic dermatitis-an ayurvedic approach – A case report. *European Journal of Pharmaceutical and Medical Research* 4: 540–1.

Dhiman, K. 2014. Ayurvedic intervention in the management of uterine fibroids: A case series. *Ayu* 35, no. 3: 303–8.

Dhimdhime, R. S., K. B. Pawar, D. T. Kodape, S. R. Dhimdhime, and S. Y. Kashinath. 2016. Urine formation and its various diagnostic methods w.s.r. to Ayurveda. *Ayushdhara* 3: 811–21.

Doddamani, S. H., M. N. Shubhashree, S. K. Giri, N. Kavya, and G. Venkateshwarlu. 2015. The safety of ayurvedic herbo-mineral formulations on renal function: An observational study. *International Journal of Research in Ayurveda and Pharmacy* 6: 209–302.

Dushing, Y. A., and S. L. Laware. 2012. Antioxidant assessment of *Ashokarishta* – A fermented polyherbal ayurvedic formulation. *Journal of Pharmacy Research* 5: 3165–8.

Dwivedi, V., E. M. Anandan, R. S. Mony et al. 2012. *In vivo* effects of traditional ayurvedic formulations in *Drosophila melanogaster* model relate with therapeutic applications. *PLoS One* 7, no. 5: e37113.

Dwivedi, V., and S. C. Lakhotia. 2016. Ayurvedic *Amalaki Rasayana* promotes improved stress tolerance and thus has anti-aging effects in Drosophila melanogaster promotes improved stress tolerance and thus has antiaging effects in *Drosophila melanogaster. Journal of Biosciences* 41: 697–711.

Dwivedi, V., S. Tiwary, and S. C. Lakhotia. 2015. Suppression of induced but not developmental apoptosis in *Drosophila* by ayurvedic *Amalaki Rasayana* and *Rasa-Sindoor. Journal of Biosciences* 40, no. 2: 281–97.

Epstein, N. N., D. L. Tuffanelli, and J. H. Epstein. 1965. Avascular bone necrosis: A complication of long term corticosteroids therapy. *Archives of Dermatology* 92: 170–80.

Fida, A., S. A. Qureshi, and F. Mumtaz. 2018. Assessment of herbal preparation (*Irimedadi taila*), an adjunctive in treating plaque- caused gingivitis. Pakistan Journal of Medical and Health Sciences 12: 567–9.

Geeta, A. 1995. Treatment of *pradara* with *Ashokarishta. Journal of Research in Ayurveda and Siddha* 16: 177–80.

Ghuse, R., M. Vhora, and V. K. Kori. 2016. Clinical study on the management of cerebral palsy with *yoga Basti* (medicated enema). *Journal of Research in Traditional Medicine* 2, no. 3: 62–9.

Goswami, D. 2014. Mahavatavidhwansan rasa along with *Mahanarayan taila kati basti* in *gridhrasi. International Journal of Ayurveda and Pharma Research* 2: 108–12.

Grampurohit, P. L., N. Rao, and S. S. Harti. 2014. Effect of *anuvasana Basti* with *ksheerabala taila* in *sandhigata vata* (osteoarthritis). *Ayu* 35, no. 2: 148–51.

Gupta, A. 2017. Role of *navana nasya* with *Mahamasha taila* and *Shirobasti* with *ksheerabala taila* in the management of *ardita*: A comparative clinical study. *International Journal of Research in Ayurveda and Pharmacy* 8, no. 6: 74–8.

Gupta, K., and P. Mamidi. 2016a. Ayurvedic management of chronic idiopathic urticaria: A case report. *Journal of Pharmaceutical and Scientific Innovation* 5, no. 4: 141–3.

Gupta, K., and P. Mamidi. 2016b. Ayurvedic management of schizophrenia: Report of two cases. *International Research Journal of Pharmacy* 7, no. 9: 41–4.

Gupta, M., and S. Siddesh. 2018. Ayurvedic management of "Cystocele" (*Phalini yonivyapat*) – A case study. *World Journal of Pharmaceutical and Medical Research* 4: 258–61.

Gupta, M. K., D. S. Gaur, V. K. Singh, and N. Urmaliya. 2014. Evaluation and comparison of efficacy of *Kaishora guggulu* plus *Amrita guggulu* in the management of *utthana vatarakta* w.s.r. to gout. *Innoriginal International Journal of Sciences* 1: 22–7.

Gupta, S., V. Patil, and R. Sharma. 2016. Diagnosis and management of *Katishoola* (low back pain) in Ayurveda: A critical review. *Ayushdhara* 3: 764–9.

Gyawali, S., G. M. Khan, S. Lamichane et al. 2013. Evaluation of antisecretory and antiulcerogenic activities of *Avipattikar churna* on the peptic ulcers in experimental rats. *Journal of Clinical and Diagnostic Research: JCDR* 7, no. 6: 1135–9.

Han, H. Y., S. Shan, X. Zhang, N. L. Wang, X. P. Lu, and X. S. Yao. 2007. Downregulation of prostate specific antigen in LNCaP cells by flavonoids from the pollen of *Brassica napus* L. *Phytomedicine: International Journal of Phytotherapy and Phytopharmacology* 14, no. 5: 338–43.

Hoffmann, D. 2003. An introduction to phytochemistry. In *Medical herbalism: The science and practice of herbal medicine*, 16–40. Vermont: Healing Arts Press.

Ignatius, V., M. Narayanan, V. Subramanian, and B. M. Periyasamy. 2013. Antiulcer activity of indigenous plant *Operculina turpethum* Linn. *Evidence-Based Complementary and Alternative Medicine: ECAM* 2013. doi:10.1155/2013/272134.

Inamdar, S. N., A. S. Prashanth, and R. Kumar. 2017. Clinical evaluation of *Basti* and *nasya* in *pakshaghata* (hemiplegia). *Paryeshana International Journal of Ayurvedic Research* 1: 54–5.

Jadhav, S., K. Padmasaritha, S. Siddesh, and B. Mishra. 2014. A comparative clinical study on the efficacy of *Sukumara gritha* in *Sukhaprasava. International Journal of Ayurveda and. Pharma Research* 2: 35–41.

Jagtap, A. G., and S. G. Karkera. 1999. Potential of the aqueous extract of *Terminalia chebula* as an anticaries agent. *Journal of Ethnopharmacology* 68, no. 1–3: 299–306.

Jain, K., S. Sharma, and C. M. Jain. 2016. Clinical study of *Kalyanaka ghrit uttar basti* in *vandhyatwa*. *International Ayurvedic Medical Journal* 1: 195–201.

Janagal, B., C. Singh, R. P. Purvia, and M. Adlakha. 2017. A conceptual study of *shirodhara* in the management of *shirahshoola*. *Ayushdhara* 4: 1045–50.

Jitesh, T., and J. Amit. 2016. Management of *vataja pratishyaya* in children as per Ayurveda: A review. *European Journal of Biomedical and Pharmaceutical Sciences* 3: 134–136.

Kabir, A., M. B. Samad, N. M. D'Costa, and J. M. Hannan. 2012. Investigation of the central and peripheral analgesic and anti-inflammatory activity of *Draksharishta* an Indian ayurvedic formulation. *Journal of Basic and Clinical Pharmacy* 3, no. 4: 336–40.

Kalaiselvan, V., A. K. Shah, F. B. Patel, C. N. Shah, M. Kalaivani, and A. Rajasekaran. 2010. Quality assessment of different marketed brands of *Dashmoolaristam*, an ayurvedic formulation. *International Journal of Ayurveda Research* 1, no. 1: 10–13.

Kamble, S. S., J. Jain, and O. P. Diwedi. 2017. Ayurveda principles towards the management of Parkinson's disease (*kampavat*): A conceptual study. *European Journal of Biomedical and Pharmaceutical Sciences* 4: 301–3.

Kamble, S. S., S. Khuje, J. Jain, and O. P. Dwivedi. 2018. Medical emergency and their management: An Ayurveda perspective. *World Journal of Pharmaceutical and Medical Research* 4: 167–70.

Karandika, A. N. 2018. PCOS with infertility and its Ayurveda management – A case study. *International Journal of Ayurveda and Pharma Research* 6: 78–80.

Katakdound, S. D. 2017. Ayurvedic management of recurrent abortions due to uterine fibroid. *Ancient Science of Life* 36, no. 3: 159–62.

Kataria, M. Singh, and L. N. 1997. Hepatoprotective effect of Liv. 52 and *kumaryasava* on carbon tetrachloride induced hepatic damage in rats. *Indian Journal of Experimental Biology* 35, no. 6: 655–7.

Katti, A., K. A. Bhusane, and N. A. Murthy. 2016. Effect of *Ashwagandha ghrita matrabasti* on knee osteo arthritis of elderly population. *International Journal of Research in Ayurveda and Pharmacy* 7, no. 1: 84–9.

Kaushik, U., P. Lachake, C. S. Shreedhara, and H. N. A. Ram. 2009. *In-vitro* antioxidant activity of extracts of *Avipattikar Churna*. *Pharmacology Online* 3: 581–9.

Kavyashree, K., H. P. Savitha, and S. K. Shrilata Shetty. 2017. Psychocutaneous disorders in Ayurveda- an appraisal. *Journal of Ayurveda and Holistic Medicine* 5: 61–70.

Keshava, D. V., M. S. Baghel, C. R. Harisha, M. Goyal, and V. J. Shukla. 2014. Pharmacognostical and phytochemical evaluation of *Agastya haritaki rasayana* – A compound ayurvedic formulation. *Journal of Ayurveda and Holistic Medicine* 2: 4–10.

Khan, M., A. U. Khan, N. U. Rehman, and A. H. Gilani. 2012. Gut and airways relaxant effects of *Carum roxburghianum*. *Journal of Ethnopharmacology* 141, no. 3: 938–46.

Khan, M. A., A. Gupta, J. L. Sastry, and S. Ahmad. 2015. Hepatoprotective potential of *kumaryasava* and its concentrate against CCl_4-induced hepatic toxicity in Wistar rats. *Journal of Pharmacy and Bioallied Sciences* 7, no. 4: 297–9.

Khan, T. A., R. Mallya, and A. Gohel. 2016. Standardization of marketed ayurvedic formulation, *Balaguloochyadi kashayam*- physicochemical, microbial evaluation and ephedrine content. *Journal of Applied Pharmaceutical Science* 6: 184–9.

Khandekar, A., J. Jadhav, and S. S. Danga. 2015. Management of vitiligo: An ayurvedic perspective. *Indian Journal of Drugs in Dermatology* 1: 41–3.

Khandelwal, R., and S. Dipti Nathani. 2016. An ayurvedic approach to PCOS: A leading cause of female infertility. *International Journal of Ayurveda and Medical Sciences* 1: 77–82.

Krishnapriya, S., P. K. Sreerudran, and G. Hussain. 2018. A review on *Kalyanaka ghrita* as *vishaghna*. *Journal of Biological and Scientific Opinion* 6, no. 4: 83–5.

Kumar, B. P., K. P. Kanti, and S. Ghosh. 2015. Management of uterine fibroid along with polycystic ovarian syndrome through ayurvedic medicine: A case study. *International Ayurvedic Medical Journal* 3: 2302–6.

Kumar, P. P., M. R. K. Rao, A. A. Elizabeth, K. Prabhu, R. L. Sundaram, and S. Dinakar. 2018. GC-MS analysis of one ayurvedic medicine *Sahacharadi kashayam*. *International Journal of Pharmacy and Technology* 10: 31214–30.

Kumar, S., I. Sharma, and R. Narayan. 2017. A case report of successful ayurvedic management of facial paralysis. *World Journal of Pharmaceutical and Medical Research* 3: 201–20.

Kumar, T., Y. K. Larokar, and V. Jain. 2013. Standardization of different marketed brands of *Ashokarishta*: An ayurvedic formulation. *Journal of Scientific and Innovative Research* 6: 993–9.

Kurup, P. N. V. 2001. Ayurveda. In *Traditional Medicine in Asia* ed. R. R. Choudhury, and U. M. Rafel, 1–16. New Delhi: W.H.O.

Kute, A., N. K. Ojha, and A. Kumar. 2017. A study on improvement of IQ level in borderline mentally retarded children by the use of *Brahmi ghrita* and *Jyotishmati taial*. *International Journal of Ayurveda and Pharma Research* 5: 1–9.

Lakhotia, S. C. 2019. Need for integration of Ayurveda with modern biology and medicine. *Proceedings of Indian National Science Academy* 85: 697–703.

Lakshmi, V. 2017. Ayurvedic approach of menorrhagia: *Asrigdara*. *International Journal of Ayurveda and Pharma Research* 5: 55–9.

Lakshmi, V. S., and V. Asokan. 2017. A case report on *ksheerabala taila matrabasti* in *udavartini yonivyapat* (primary dysmenorrhea). *World Journal of Pharmaceutical and Medical Research* 3: 236–9.

Lal, P. T., S. K. Lal, and S. Garg. 2017. A clinical comparative evaluation of efficacy and safety of *Kumaryasava* and *Rajahpravartani vati* in the management of *prathmik kastartava* w.s.r. to (primary dysmenorrhoea): A prospective open label single center study. *International Journal of Ayurveda and Pharma Research* 5: 1–13.

Lal, U. R., S. M. Tripathi, S. M. Jachak, K. K. Bhutani, and I. P. Singh. 2010. Chemical changes during fermentation of *Abhayarishta* and its standardization by HPLC-DAD. *Natural Product Communications* 5, no. 4: 575–9.

Lather, A., V. Gupta, P. Bansal, M. Sahu, K. Sachdeva, and P. Ghaiye. 2011. An ayurvedic polyherbal formulation *Kaishore Guggulu*: A review. *International Journal of Pharmaceutical and Biological Archives* 2: 497–503.

Madan, A., S. Kanjilal, A. Gupta et al. 2015. Evaluation of immunostimulatory activity of *Chyawanprash* using in vitro assays. *Indian Journal of Experimental Biology* 53, no. 3: 158–63.

Mahima, A. Rahal, R. Deb et al. 2012. Immunomodulatory and therapeutic potentials of herbal, traditional/indigenous and ethnoveterinary medicines. *Pakistan Journal of Biological Sciences: PJBS* 15, no. 16: 754–74.

Malhotra, S., and A. P. Singh. 2003. Medicinal properties of ginger (*Zingiber officinale* Rosc.). *Natural Product Radiance* 2: 296–301.

Mali, G. V., A. S. Dodamani, G. N. Karibasappa, P. Vishwakarma, and V. M. Jain. 2016. Comparative evaluation of *Arimedadi* oil with 0.2% chlorhexidinegluconate in prevention of plaque and gingivitis. *Journal of Clinical Diagnosis and Research* 10, no. 7: ZC31-4. doi:10.7860/JCDR/2016/19120.8132.

Mamidi, P., and K. Gupta. 2015. Ayurvedic management of chronic rheumatoid arthritis with bilateral hip involvement: A case report. *Journal of Pharmaceutical and Scientific Innovation* 4, no. 6: 329–32.

Manmode, R., J. Manwar, M. Vohra, S. Padgilwar, and N. Bhajipale. 2012. Effect of preparation method on antioxidant activity of ayurvedic formulation *Kumaryasava*. *Journal of Homeopathy and Ayurvedic Medicine* 1: 114. doi:10.4172/2167-1206.1000114.

Mansuri, A., and S. Desai. 2019. 32 Factorial design for optimization of HPLC-UV method for quantification of gallic acid in *Lohasava* and *Pippalyasava*. *Indian Journal of Pharmaceutical Education and Research* 53, no. Suppl 2: s347–55.

Manu, P., S. K. Shetty, and H. P. Savitha. 2017. Critical review on effect of *Brahmi ghrita* in psychiatric disorders. *International Journal of Research in Ayurveda and Pharmacy* 8, no. 1: 16–8.

Math, S. G., and P. Jana. 2019. A case study on ayurvedic management of *garbhini chardi* w.s.r. to hyperemesis gravidarum. *Journal of Ayurveda and Integrated Medical Sciences* 1: 102–5.

Meena, P., and P. Gaurav. 2014. Role of ayurvedic formulations in an ovarian cyst: A case study. *International Journal of Ayurveda and Pharma Research* 2: 126–8.

Mishra, B. 1983. *Bālarōga cikitsa* (Treatment of children's diseases). In *Cakradatta*, 521–9. Varanasi: Chowkhamba Sanskrit Series Office.

Mishra, R. K. 2015. Effect of *Ashwagandha ghrita* and *ashwagandha* granules on growth w.s.r. of biochemical values. *International Journal of Ayurveda and Pharma Research* 3: 7–12.

Mohan, R. 2013. Role of *Tarpana kriya kalpa* with *Triphala ghrita* in the treatment of computer vision syndrome, a widespread eye complaint of present scenario. *Unique Journal of Ayurvedic and Herbal Medicines* 1: 1–5.

Mont, M. A., and D. S. Hungerford. 1995. Non-traumatic avascular necrosis of the femoral head. *Journal of Bone and Joint Surgery. American Volume* 77, no. 3: 459–74.

Mundewadi, A. 2009. Female infertility. *Ayurvedic Herbal Treatment* 3: 121–5.

Muralikumar, V., and M. Shivasekar. 2012. Fertility effect of ayurvedic medicine (*Phala sarpis*) in animal model. *International Journal of Research in Ayurveda and Pharmacy* 3, no. 5: 664–7.

Murthy, K. R. S. 2017. *Śārṅgadhara Samhita*. Varanasi: Chowlhamba Orientalia.

Mutha, S. S., V. S. Kasture, S. A. Gosavi, R. D. Bhalke, and S. S. Pawar. 2013. Standardization of *Abhyarishta* as per WHO guidelines. *World Journal of Pharmacy and Pharmaceutical Sciences* 3: 510–18.

Nagarkar, B., and S. Jagtap. 2017. Effect of new polyherbal formulations DF1911, DF2112 and DF2813 on CFA induced inflammation in rat model. *BMC Complementary and Alternate Medicine* 17, no. 1: 194. doi:10.1186/s12906-017-1711-6.

Nair, C. G., R. R. Geethesh, S. Honwad, and R. Mundugaru. 2015. A comparative study on the anti-inflammatory effects of *trividha paka* of *Ksheerabala taila*. *International Journal of Research in Ayurveda and Pharmacy* 6, no. 6: 692–5.

Narang, R., and I. Herswani. 2018. Ayurveda review on *Dadimashtaka churna* and its clinical importance. *Journal of Drug Delivery and Therapeutics* 8, no. 4: 80–2.

Narang, R., M. Rai, and S. Kamble. 2019. Role of *Dadimashtaka churna* in the management of *grahani dosha* w.s.r. to disease prevalence in school going children. *World Journal of Pharmaceutical and Medical Research* 5: 83–6.

Narayana, D. B. A., S. Durg, P. R. Manohar, A. Mahapatra, and A. R. Aramya. 2017. *Chyawanprash*: A review of therapeutic benefits as in authoritative texts and documented clinical literature. *Journal of Ethnopharmacology* 197: 52–60.

Naveen, D., and T. P. Kumar. 2014. Ayurvedic resolution to migraine. *Journal of Homeopathy and Ayurvedic Medicine* 3: 3. doi:10.4172/2167-1206.1000160.

Nigamanand, B., S. Chaubey, R. C. Tiwari, and H. B. Naithani. 2016. A critical review on *Gajapippali* with special reference to *Samhita Grantha*s. *International Journal of Ayurveda and Pharma Research* 4: 64–9.

Nimesh, P. K., and S. Shetty. 2018. A comparative clinical study to evaluate the effect of *Sahacharadi kwatha* and *Nagaradi kwatha* in the management of *Janusandhigata vata*. *International Ayurvedic Medical Journal* 6: 2023–9.

Oberoi, A., P. R. Lal, and P. R. Lal. 2016. Ayurvedic concepts of female fertility- A review. *International Journal of Ayurvedic and Herbal Medicine* 6: 2313–20.

Odukoya, O. A., M. O. Sofidiya, O. O. Ilori, M. O. Gbededo, J. O. Ajadotuigwe, and O. O. Olaleye. 2009. Hemorrhoid therapy with medicinal plants: Astringency and inhibition of lipid peroxidation as key factors. *International Journal of Biological Chemistry* 3, no. 3: 111–8.

Otta, S. P., and R. N. Tripaty. 2002. Clinical trial of *Phalaghrita* on female infertility. *Ancient Science of Life* 22, no. 2: 56–63.

Padekar, P. K., A. Dixit, D. R. Bairwa, K. S. Sakhitha, and V. N. Rao. 2018. A conceptual review of *panchavidha kashaya Kalpana*. *Ayurpub* 3: 1140–6.

Panchal, K. B., and A. Rajani. 2017. Probable mode of action of *Rajanyadi churna* on *grahani dosha* in children – A review article. *Pharma Science Monitor* 8: 277–84.

Panda, P., D. Dattatray, S. K. Das, and M. M. Rao. 2017. A review on pharmaceutical and therapeutical uses of *churna* (powder) in Ayurveda. *International Ayurvedic Medical Journal* 5, no. 11.

Pandey, S., and A. K. Chaudhary. 2017. A review on *Rasna saptak kwath*: An ayurvedic polyherbal formulation for arthritis. *International Journal of Research in Ayurveda and Pharmacy* 8, no. Suppl 1: 4–11.

Parasuraman, S., G. S. Thing, and S. A. Dhanaraj. 2014. Polyherbal formulation: Concept of Ayurveda. *Pharmacognosy Reviews* 8, no. 16: 73–80.

Parle, M., and N. Bansal. 2006. Traditional medicinal formulation, *Chyawanprash*-A review. *Indian Journal of Traditional Knowledge* 5: 484–8.

Parmar, M., T. Agrawal, and G. Parmar. 2017. Role of *Phalasarpi uttarbasti* in the management of low antral follicles: A case study. *International Journal of Ayurveda and Pharma Research* 5: 80–2.

Parekar, R. R., K. K. Dash, P. A. Marathe, A. A. Apte, and N. N. Rege. 2012. Evaluation of anti-inflammatory activity of root bark of *Clerodendrum phlomidis* in experimental models of inflammation. *International Journal of Applied Biology and Pharmaceutical Technology* 3: 54–60.

Patel, D. V. 2017a. Role of *Brahmi ghrita nasya* in the prevention of neurodegenerative disorder: Huntington's disease. *World Journal of Pharmacy and Pharmaceutical Sciences* 6: 1264–70.

Patel, D. V. 2017b. A literary review on the role of *Bala taila nasya* in the management of Lou Gehrig's disease. *World Journal of Pharmacy and Pharmaceutical Sciences* 6: 809–15.

Patel, J. 2018. Clinical efficacy of *Bala taila uttarbasti* in the management of urethral tricture: Case report. *International Journal of Current Research* 10: 73068–71.

Patel, J. K. K., T. S. Dudhamal, S. K. Gupta, and V. Mahanta. 2015a. Efficacy of *Kanchanara Guggulu* and *matra Basti* of *Dhanyaka gokshura ghrita* in *mootraghata* (benign prostatic hyperplasia). *Ayu* 36, no. 2: 138–44.

Patel, N. L., H. R. Jadav, C. R. Harisha, V. J. Shukla, and A. B. Thakar. 2015b. Pharmacognostical and preliminary physicochemical evaluation of *Manibhadra guda* – A polyherbal ayurvedic formulation. *Pharma Science Monitor* 6: 125–33.

Patil, A. R., R. Kulkarni, S. Uppinakudru, V. Swami, and S. A. Nithin. 2016. Effect of *Ashwagandha ghrita* on *karshya* (underweight): An explorative study. *International Journal of Research in Ayurveda and Pharmacy* 7, no. Suppl 2: 156–9.

Patil, S., S. A. Varma, G. Suragimath, K. Abbayya, S. A. Zope, and V. Kale. 2018. Evaluation of *Irimedadi taila* as an adjunctive in treating plaque-induced gingivitis. *Journal of Ayurveda and Integrative Medicine* 9, no. 1: 57–60.

Patil, Y. R., and R. S. Sawant. 2012. Study of preventive effect of *pratimarsha nasya* with special reference to *Anu tailam* (an ayurvedic preparation). *International Research Journal of Pharmacy* 3: 295–300.

Pawar, S., R. Kadam, and S. Jawale. 2015. An open randomized study of *Patola katurohinyadi kashayam* in alcoholic liver disease. *International Journal of Ayurveda and Pharma Research* 3: 15–21.

Pawar, S. D., S. N. Gaidhani, T. Anandan et al. 2011. Evaluation of anti-inflammatory, analgesic and anti-arthritic activity of *Mahanarayana tailam* in laboratory animals. *Journal of Pharmacology and Toxicology* 12: 33–42.

Peshala, K. K. V. S., M. Sahu, and L. Singh. 2014. A study of anal sphincter tone in acute fissure-i-in-ano patients treated with *Mahanarayana* oil. *International Journal of Research in Ayurveda and Pharmacy* 5, no. 3: 318–21.

Pole, S. 2006. *Cyavanaprāśa*. In *Ayurvedic medicine: The principles of traditional practice*, 296–7. Edinburgh: Churchill Livingston, Elsevier.

Pooja, B. A., and S. Bhatted. 2015. Ayurvedic management of *pravahika* – A case report. *Ayu* 36, no. 4: 410–12.

Prajkta, K. 2015. *Parikartika*: Case study. *International Journal of Research in Ayurveda and Pharmacy* 6, no. 4: 449–51.

Prasad, M. R. P., N. Ghodela, and T. S. Dudhamal. 2018. Management of avascular necrosis of head of femur in Ayurveda: A review. *International Journal of Research in Ayurveda and Pharmacy* 9, no. 3: 155–8.

Priya, S. P., and P. L. Hegde. 2017. Classical uses of *Dadima*: A review. *International Journal of Research in Ayurveda and Pharmacy* 8, no. 6: 26–30.

Priyanka Kumari M., and K. L. Meena. 2017. A conceptual study of *prameha* and its relevance in modern era. *International Journal of Ayurveda and Pharmaceutical Chemistry* 7: 349–58.

Radha, A., J. Sebastian, M. Prabhakaran, and D. S. Kumar. 2014. Observations on the quality of commercially manufactured ayurvedic decoction, *Maharasnadi Kvatha*. *Hygeia Journal for Drugs and Medicines* 6, no. 1: 74–80.

Raghunathan, P., K. U. Pillai, and A. K. Mishra. 2015. Effect of *Gandharvahastadi eranda taila* in *katigraha*. *Anveshana Ayurveda Medical Journal* 1: 352–4.

Ram, H. N. A., U. Kaushik, P. Lachake, C. S. Shreedhara, and B. Sathyanarayana. 2009. Antiulcer activity of aqueous extract of *Avipattikar churna*. *Pharmacology Online* 1: 1169–81.

Ram, T. S., B. Srinivasulu, and A. Narayana. 2013. Pragmatic usage of *Haritaki* (*Terminalia chebula* Retz): An ayurvedic perspective vis-à-vis current practice. *International Journal of Ayurveda and Pharma Research* 1: 72–82.

Rani, D. S., and B. G. Madhusudan. 2019. Ayurvedic approach to Meniere's disease. *International Ayurvedic Medical Journal* 7: 95–9.

Rao, M. M., P. H. Kumar, P. Panda, and B. Das. 2016. Comparative study of efficacy of *Jatyadi ghrita pichu* and *Yasthimadhu ghrita pichu* in the management of *Parikartika* (Fissure-In-A+no). *International Journal of Ayurveda and Pharma Research* 4: 1–9.

Rao, M. R. K., S. N. Kumar, S. Jones et al. 2015. Phytochemical and GC MS analysis of an ayurvedic formulation, *Patolakaturohinyadi kwatham*. *International Journal of Pharmaceutical Sciences Review and Research* 34: 6–12.

Rao, V. G., S. T. Kamat, and D. C. Patil. 2017. Effect of *Eranda taila prayoga* and *Dashamuladi niruha basti* in *gridhrasi*. *International Journal of Applied Ayurved Research* 3: 659–66.

Rao, V. G., P. Seethadevi, and H. S. Vatsala. 2018. Management of Graves' disease with special reference to *Bhasmaka roga* – A case report. *European Journal of Pharmaceutical and Medical Research* 5: 514–18.

Raskar, S., and S. Rajagopala. 2015. *Abhyanga* in new born baby and neonatal massage – A review. *International Journal of Ayurveda and Pharma Research* 3: 5–10.

Ratha, K. K., P. S. Aswani, D. P. Dighe, M. M. Rao, S. K. Meher, and A. K. Panda. 2018. Management of venous ulcer through Ayurveda: A case report. *Journal of Research in Ayurvedic Sciences* 2, no. 3: 202–8.

Ratha, K. K., B. L. Dhar, and H. Jayram. 2016. Management of a case of lumbar stenosis with ayurvedic intervention. *Asian Journal of Pharmaceutical and Clinical Research* 9, no. Suppl 2: 11–13.

Reddy, A. P., and A. Bankar Snehal. 2017. Management of *visarpa* with special reference to pemphigus foliaceus with *panchakarma*: A case study. *World Journal of Pharmaceutical Research* 6: 1171–81.
Renu, V. T., K. V. Mamatha, V. Hamsaveni, and B. Mishra. 2014. An insight to retinitis pigmentosa and management by ayurvedic therapies. *International Journal of Ayurveda and Pharma Research* 2: 48–53.
Revathy, C., N. L. Gowrishankar, S. Adisha et al. 2018. Phytochemical screening and anthelmintic activity: Assay of *Drakshadi kashayam*. *International Journal of Pharmacy and Pharmaceutical Research* 12: 374–84.
Roshni, K. 2017. Efficacy of ayurvedic treatment in *Vicharchika* with special reference to *Jalukavacharana*-A case study. *International Journal of Applied Ayurved Research* 3: 851–8.
Sadhna, K., and K. Abbulu. 2010. Holistic approach to management of asthma. *International Journal of Research in Ayurveda and Pharmacy* 1: 367–83.
Sanjaya Kumar, Y. R., K. G. Vasanthakumar, N. T. Selvam, and M. V. Acharya. 2016. Acute and subacute toxicity study of *Manasamitra vataka* without musk in experimental animal models. *International Journal of Pharmacy and Pharmaceutical. Research* 5: 87–99.
Santosh, M. K., D. Shaila, and I. S. Rao. 2003. Standardization of selected *asavas* and *arishtas*. *Asian Journal of Chemistry* 15: 884–90.
Sapra, U. K. 2013. Effect of ayurvedic treatment in chronic malaria. *Journal of Ayurveda and Holistic Medicine* 1: 37–8.
Sarker, M. M. R., H. Imam, M. S. Bhuiyan, and M. S. K. Choudhuri. 2014. *In vitro* assessment of *Prasarani sandhan*, a traditional polyherbal ayurvedic medicine, for immunostimulating activity in splenic cells of BALB/c mice. *International Journal of Pharmacy and Pharmaceutical Sciences* 6: 531–4.
Saroch, V., and J. Singh. 2015. A review on ayurvedic medicines for *amlapitta* (hyperacidity). *Anveshana Ayurveda Medical Journal* 1: 40–2.
Sason, R., and A. Sharma. 2016. A conceptual study of *sheetpita*, *udard* and *kotha* W.S.R to urticaria: A review. *International Journal of Ayurveda and Pharma Research* 4: 52–6.
Sawant, D. P., G. R. Parlikar, and S. V. Binorkar. 2013. Efficacy of *triphala ghrita netratarpan* in computer vision syndrome. *International Journal of Research in Ayurveda and Pharmacy* 4, no. 2: 244–8.
Sawarkar, G., and P. Sawarkar. 2019. Management of *ardita* (Bell's palsy) through Ayurveda- A case study. *International Journal of Recent Scientific Research* 10: 32040–3.
Sekar, S., and S. Mariappa. 2008. Traditionally fermented biomedicines, *arishtas* and *asavas* from Ayurveda. *Indian Journal of Traditional Knowledge* 7: 548–56.
Shailajan, S., Y. Patil, and S. Menon. 2017. Marker-based chemoprofiling of a traditional formulation: *Pushyanuga churna*. *Journal of Applied Pharmaceutical Science* 7: 239–45.
Sharma, A., and R. K. Soni. 2017. Ayurvedic treatment of allergic rhinitis-A case study. *International Journal of Ayurveda and Pharma Research* 5: 63–5.
Sharma, K. K., A. K. Srivastava, V. Singh, and G. D. Shukla. 2015. A clinical study to evaluate the efficacy of *virechana karma* and *Simhanada guggulu* in the management of *amavata*. *World Journal of Pharmacy and Pharmaceutical Sciences* 4: 1285–97.
Sharma, P., V. Nariyal, S. Sharma, and A. Sharma. 2017. Reproductive and child health care through *Bala taila*: A review on Ayurveda formulation. *International Journal of Research in Ayurveda and Pharmacy* 8, no. 3: 1–4.
Sharma, P. V. 1983a. *Kāsacikitsitam* (Treatment of cough). In *Caraka Samhita*, Vol. 2, 301–19. Varanasi: Chaukhambha Orientalia.
Sharma, P. V. 1983b. *Rasāyanāddhyāya Prathama Pāda* (First quarter of chapter on rejuvenators). In *Caraka Samhita*, Vol. 2, 3–12. Varanasi: Chaukhambha Orientalia.
Shekhawat, N. S., M. Singh, R. Gupta, and O. P. Dave. 2017. Management of internal bleeding piles by ligation and plication followed by *matra Basti* A case report. *Ayu* 38, no. 3–4: 139–43.
Shengule, S., S. Mishra, D. L. Patil, K. S. Joshi, and B. K. Patwardhan. 2019. Phytochemical characterization of Ayurvedic formulations of *Terminalia arjuna*: A potential tool for quality assurance. *Indian Journal of Traditional Knowledge* 18: 127–32.
Shikha, M., and A. Vidhu. 2015. A comparative study of prepared and marketed *Asokarista* with respect to physicochemical parameters and phytochemical markers. *International Journal of Pharmacognosy and Phytochemical Research* 7: 144–9.
Shubha, P. U., V. Sudheendra, and S. R. Honwad Ballal 2017. A review on *Bilwadi gutika*. *International Ayurvedic Medical Journal* 5: 501–7.
Shyam, P. M., A. P. Ramachandran, G. S. Acharya, and K. T. Shrilatha. 2010. Evaluation of the role of *nithya-virechana* and *Nayopayam kashaya* in *tamaka shwasa*. *Ayu* 31, no. 3: 294–9.

Singh, A., A. K. Singh, and A. Kumar. 2011. *Pippalyasava* and its indications in diseases caused by poor digestion. *Biomedical and Pharmacology Journal* 4, no. 2: 259–62.

Singh, K., and A. B. Thakar. 2018. A clinical study to evaluate the role of *Triphaladya guggulu* along with *Punarnavadi kashaya* in the management of hypothyroidism. *Ayu* 39, no. 1: 50–5.

Singh, P., R. B. Yadav, and S. Shakya. 2014. An analytical study of *Prajasthapan Mahakashaya* on *vandhyatwa* w. s. r. to female infertility. *International Journal of Ayurveda and Pharma Research* 2: 111–31.

Singh, T. N., and K. Satpute. 2018. Review of *gridhrasi* management in ayurvedic text. *Ayurlog: National Journal of Research in Ayurved Science* 6: 1–7.

Singhal, P. 2016. Effect of *virechana karma* (purgation) alone and with *Avipattikara choorna* and *Shankha bhasma* in treatment of *pittaja parinama shoola* (peptic ulcer). *Ayudrpharm International Journal of Ayurveda and Allied Sciences* 5: 128–36.

Sinha, K. R. G., V. Laxminrayana, S. V. L. N. Prasad, and S. Khanum. 2008. Standardisation of *Yograj guggulu*, an ayurvedic polyherbal formulation. *Indian Journal of Traditional Knowledge* 7: 389–96.

Sinimol, T. P. 2019. Probable mode of action of *Kalyanaka Ghrita* in unmada (insanity) based on analysis of *rasa panchaka* of ingredients – A review. *International Journal of Ayurveda and Pharma Research* 7: 19–30.

Sori, A., R. Pervaje, and B. S. Prasad. 2017. An individualized treatment protocol development in the management of *pakshaghata* (acute stage of ischemic stroke): A review. *International Journal of Research in Ayurveda and Pharmacy* 8, no. 5: 10–13.

Sreedas, E., and K. J. Girish. 2013. Clinical study on *Pippalyasava* and *Surana vataka* in *grahani roga* (irritable bowel syndrome). *International Ayurveda Medical Journal* 1: 1–6.

Sreekumar, K., and A. S. Nair. 2017. Ayurvedic management of Stargardt's macular degeneration- A case report. *International Journal of Ayurveda and Pharma Research* 5: 54–7.

Sriranjini, S. J., P. K. Pal, K. V. Devidas, and S. Ganpathy. 2009. Improvement of balance in progressive degenerative cerebellar ataxias after ayurvedic therapy: A preliminary report. *Neurology India* 57, no. 2: 166–71.

Srivastava, A. K., V. Singh, G. D. Shukla, K. K. Pandey, and S. K. Sharma. 2015. A holistic approach to hepatitis-B induced osteoarthritis by ayurvedic management: A case report. *International Journal of Green Pharmacy* 9: S100–4.

Sruthi, C. V., and A. Sindhu. 2012. A comparison of the antioxidant property of five ayurvedic formulations commonly used in the management of *vata vyadhis*. *Journal of Ayurveda and Integrative Medicine* 3, no. 1: 29–32.

Sulaiman, C. T., E. M. Anandan, and I. Balachandran. 2012. Using biochemical parameters to standardize an ayurvedic formulation. *Journal of Tropical Medicinal Plants* 13: 133–5.

Sunitha, G., and G. Hussain. 2018a. Pharmaceutical and analytical study of *Visha bilwadi gutika*. *Journal of Pharmaceutical and Scientific Innovation* 6, no. 6: 120–4.

Sunitha, G., and G. Hussain. 2018b. *Lootha visha*: A case report. *European Journal of Biomedical and Pharmaceutical Sciences* 5: 703–7.

Supraja, R., and R. Vidyanath. 2013. A case report on low back pain. *Journal of Biological and Scientific Opinion* 1, no. 3: 209–10.

Surendran, E. S., M. A. Lal, K. M. P. Sankar, and V. C. Deep. 2018. Conservative management of young age onset pelvic organ prolapse through ayurvedic management: A case report. *Journal of Research in Ayurvedic Sciences* 2, no. 4: 254–8.

Sushama, B., and K. Nishteswar. 2012. Pharmaco-therapeutic profiles of *Chandraprabha vati*- an ayurvedic herbo-mineral formulation. *International Journal of Pharmaceutical and Biological Archives* 3: 1368–75.

Swati, V., L. Gupta, S. Sharma, P. Kapil, and Sumanlata. 2018. Role of *Jatyadi ghrita* in the management of *parikartika* w.s.r to fissure-in-ano. *International Ayurvedic Medical Journal* 6: 582–8.

Takauji, Y., K. Morino, K. Miki, M. Hossain, D. Ayusawa, and M. Fujii. 2016. *Chyawanprash*, a formulation of traditional ayurvedic medicine, shows a protective effect on skin photoaging in hairless mice. *Journal of Integrative Medicine* 14, no. 6: 473–9.

Taksale, A. G., and G. D. Parulkar. 2017. *Amritarishta*: A medico review. *World Journal of Pharmaceutical Research* 6: 357–60.

Tamhane, M. D., S. P. Thorat, N. N. Rege, and S. A. Dahanukar. 1997. Effect of oral administration of *Terminalia chebula* on gastric emptying: An experimental study. *Journal of Postgraduate Medicine* 43, no. 1: 12–13.

Thakkar, S. J. 2006. TLC profile and physicochemical parameters of *Haridra khanda* – An ayurvedic formulation. *Ancient Science of Life* 25, no. 3–4:: 104–9.

Thakur, J., and A. Jagtap. 2016. Management of *vataja pratishyaya* in children as per Ayurveda: A review. *European Journal of Biomedical and Pharmaceutical Sciences* 3: 134–6.

Thakur, S. K., and R. P. Babar. 2017. Bala taila abhyangya for weight gain in lbw infant – A case study. *World Journal of Pharmaceutical and Medical Research* 3: 183–5.

Thanki, K. H., N. P. Joshi, and N. B. Shah. 2009. A comparative study of *Anu taila* and *Mashadi taila nasya* on *ardita* (facial paralysis). *Ayu* 30: 201–4.

Thirunavukkarasu, S. V., L. Upadhyay, and S. Venkataraman. 2012. Effect of *Manasamitra vatakam*, an Ayurvedic formulation, on aluminium-induced neurotoxicity in rats. *Tropical Journal of Pharmaceutical Research* 11, no. 1: 75–83.

Tike, S., S. Nilesh, and P. Baghe. 2018. Ayurvedic management of lumbar canal stenosis w.s.r. *katigata vata*: A case study. *European Journal of Biomedical and Pharmaceutical Sciences* 5: 742–5.

Tiwari, M. 2015. Ayurvedic pathophysiology of Parkinson's disease (*kampvata*). *European Journal of Biomedical and Pharmaceutical Sciences* 2: 1108–18.

Tomar, R., S. Sharma, I. Kumari et al. 2017. Clinical efficacy and safety of *Ashokarishta*, *Ashvagandha churna* and *Pravala pishti* in the management of menopausal syndrome: A prospective open-label multicenter study. *Journal of Research in Ayurvedic Sciences* 1, no. 1: 9–16.

Tripathy, R., P. Namboothiri, and S. P. Otta. 2016. Open label comparative clinical rial of *Dvipanchamooladi taila* and *Ksheerabala taila matra vasti* in the management of low backache. *International Journal of Ayurveda and Pharma Research* 4: 19–26.

Tripathy, R. N., S. P. Otta, and A. Siddram. 2011. Bala taila parisheka – A traditional approach in wound healing. *Indian Journal of Traditional Knowledge* 10: 643–50.

Tubaki, B. R., C. R. Chandrashekar, D. Sudhakar, T. N. Prabha, G. S. Lavekar, and B. M. Kutty. 2003. Clinical efficacy of *Manasamitra avtaka* (an ayurvedic medication) on generalized anxiety disorder with comorbid generalized social phobia: A randomized controlled study. *Journal of Alternative and Complementary Medicine* 18: 612–21.

Upadhyay, P. S., and S. Komal. 2019. *Ashwagandha ghrita* in *bal karshya* (childhood undernutrition). *International Journal of Health Sciences and Research* 9: 267–71.

Upadhyay, Y. 1975a. *Garbhavyāpad Śārīram (Anomalies of pregnancy).* In *Aṣṭāṅgahṛdaya*, 179–84. Varanasi: Chowkhamba Sanskrit Sansthan.

Upadhyay, Y. 1975b. *Śōdhanādi gaṇa* (Group of herbs used in elimination). In *Aṣṭāṅgahṛdaya*, 104–8. Varanasi: Chowkhamba Sanskrit Sansthan.

Upadhyay, Y. 1975c. *Kuṣḍhacikitsitam* (Treatment of skin diseases). In *Aṣṭāṅgahṛdaya*, 405–13. Varanasi: Chowkhamba Sanskrit Sansthan.

Upadhyay, Y. 1975d. *Āyuṣkāmīyam* (Desire for long life). In *Aṣṭāṅgahṛdaya*, 1–17. Varanasi: Chowkhamba Sanskrit Sansthan.

Vaidya, S. B. G. 2013. *Nighaṇṭu Ādarśa (Pūrvārdha)*, 1st edition, 484. Varanasi: Chaukhamba Vidya Bhawan.

Vaidyan, K. V. K., and A. S. G. Pillai. 1985a. *Ghṛtayōgaṅgaḷ* (Formulae of clarified butters). In *Sahasrayōgam*, 339–96. Alleppey: Vidyarambham Publishers.

Vaidyan, K. V. K., and A. S. G. Pillai. 1985b. *Vaṭakādiyōgaṅgaḷ* (Formulae of tablets). In *Sahasrayōgam*, 167–76. Alleppey: Vidyarambham Publishers.

Vaidyan, K. V. K., and A. S. G. Pillai. 1985c. *Kaṣāyayōgaṅgaḷ* (Formulae of decoctions). In *Sahasrayōgam*, 25–132. Alleppey: Vidyarambham Publishers.

Vanishree, S. K., and M. Ramesh. 2018. *Asrigdhara chikitsa* w.s.r to menorrhagia- A case study. *International Ayurvedic Medical Journal* 6: 1332–7.

Vayalil, P. K., G. Kuttan, and R. Kuttan. 2002. Protective effects of rasayanas on cyclophosphamide and radiation-induced damage. *Journal of Alternative and Complementary Medicine* 8, no. 6: 787–96.

Verma, S. 2013. Role of *Katphaladi kwatha* and *Anu taila nasya* in the management of *vataja pratishyaya* (allergic rhinitis). *International Ayurvedic Medical Journal* 1: 1–7.

Vijayan, V., P. Neethu, C. Athulya, and A. Rajesh. 2018. Post-snake bite ulcer management – A case study. *International Ayurvedic Medical Journal* 6: 2441–4.

Wadkar, K. A., M. S. Kondawar, and S. G. Lokapure. 2016. Standardization of marketed *Amritarishta-* A herbal formulation. *International Journal of Pharmacognosy* 3: 392–9.

Weerasekera, K. R., C. Wijayasiriwardhena, I. Dhammarathana, M. H. A. Tissera, and H. A. S. Ariyawansha. 2014. Establishment quality and purity of *Chandraprabha vati* using sensory characteristics, physiochemical parameters, qualitative screening and TLC fingerprinting. *International Journal of Herbal Medicine* 2: 26–29.

Yadav, K. D., and K. R. Reddy. 2014. Evaluation of *Brahmi ghrita kalpa* in ayurvedic literature. *Journal of Biological & Scientific Opinion* 2, no. 5: 305–9.

Yadav, K. D., and K. R. C. Reddy. 2012. Standardization of *Brahmi ghrita* with special reference to its pharmaceutical study. *International Journal of Ayurvedic Medicine* 3: 16–21.

Yadav, K. D., K. R. C. Reddy, and V. Kumar. 2013. Encouraging effect of *Brahmi ghrita* in amnesia. *International Journal of Green Pharmacy* 7, no. 2: 122–6.

7 Evidence Building in Ayurveda
Generating the New and Optimizing the Old Could Be Strategic

Sanjeev Rastogi, Arindam Bhattacharya and Ram Harsh Singh

CONTENTS

7.1 Introduction 161
7.2 Assembling Evidence in Ayurveda 163
7.2.1 Direct and Indirect Evidence 163
7.2.2 Yukti (Experiment) 164
7.2.3 Āpta (Documented Evidence) 165
7.3 Journeying through the Evidence in Ayurveda 165
7.3.1 Evidence Pertaining to the Etiopathogenesis and Disease Presentation 165
7.4 Optimizing the Old and Generating the New Evidence 167
7.4.1 Optimizing the Old: Re-Appropriation, Revalidation and Research Synthesis 167
7.4.2 Generating the New: Priority Areas for Clinical Research in Ayurveda 168
7.5 Areas of Priority of Clinical Research in Ayurveda 169
7.5.1 Areas Relevant to Contemporary Healthcare Needs 169
7.5.2 Areas Where Ayurveda Has Proven Strength 169
7.5.3 Taking Cues from Clinical Observations 169
7.5.4 Areas of Textual Strength in Ayurveda 169
7.5.5 Areas Important to Establishing Standards in Ayurvedic Clinical Practice 169
7.5.6 Ayurveda in Vulnerable Population 170
7.6 Research Designs Needed in Ayurveda 170
7.7 Research Designs for Ayurvedic Core Concepts of Clinical Medicine 171
7.8 Conclusion 171
References 172

7.1 INTRODUCTION

Despite their traditional prevalence and new-found popularity, traditional systems of medicine have faced allegations of being healthcare practices based on empirical, anecdotal or experiential evidence bases. Ayurveda has been no exception to this; if anything, due to its connection to living systems of religion (unlike in Chinese medicine), the allegations against Ayurveda are stronger. These allegations mostly arise from the dominant scientific stream of laboratory-based modern medicine, which occupies not only the major share in national healthcare systems but also exercises a disproportionately large influence on the minds of policy-makers. To develop Ayurveda into a dependable, reliable and reproducible system of medicine, the ayurvedic intellectual fraternity must make a concerted effort to examine the science of evidence-building in Ayurveda (Singh 2010a).

One of the features which distinguishes a traditional system of medicine from folk medicine is the presence of a systematic organization of its knowledge, a sort of higher order of learning.

All textbooks in the ayurvedic canon exhibit this property, by being organized into sections of fundamentals (*sūtra*) passing through cause and diagnostics (*nidāna*), treatment (*cikitsa*), prognosis (*indriya*) and successful intervention (*siddhi*). As an illustration, the organization of *Caraka Samhita* is presented (Figure 7.1). Since most of these texts were composed at a time when the knowledge base and practice of Ayurveda was flourishing, one may conclude that the very existence of such a higher order of learning indicates a certain level of sophistication and intelligence in the process. By having such an organization, Ayurveda has risen above the random and chance-based approach of folk medicine.

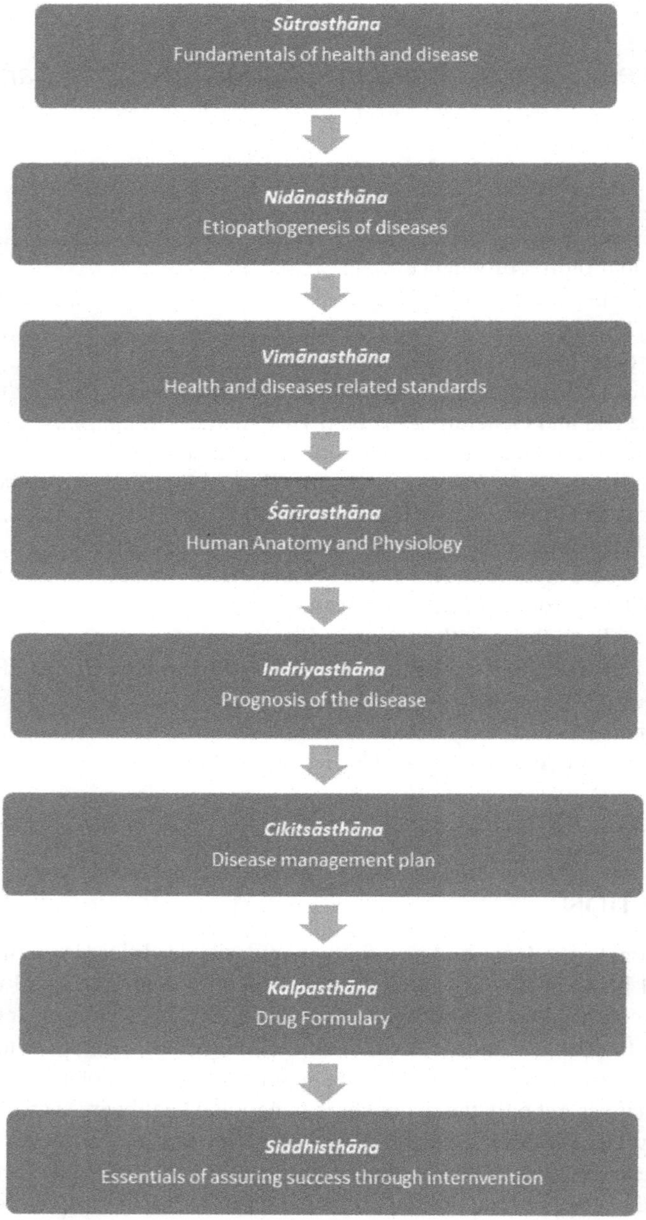

FIGURE 7.1 Organizational order of *Caraka Samhita*.

In between the hardened skeptics and the blind followers of Ayurveda, there exists a not-insignificant population of moderate practitioners and academics of Ayurveda, who have sought to demystify, decode and interpret the principles of Ayurveda using the language and logic of modern science (Rastogi and Singh 2012). This population of moderate in-betweeners may be further divided into two cohorts. In the first cohort are those who find the Ayurveda modalities useful, but do not subscribe to its fundamentals and philosophy. People do reverse pharmacology for new drug development, taking cues from ayurvedic classics. *Pañcakarma* and *kṣārasūtra* applications are used in intractable conditions. *Prakṛti* analysis is carried out for possible personalized disease management propositions. Such individuals are actually progressive and liberated members of modern science, who consider that carefully chosen techniques of patient care in Ayurveda may be provided some sort of place in modern science. The second cohort of moderates are those who sense the deep philosophical and biological connotations behind the dictums of Ayurveda and hence warn that without understanding the ayurvedic science and its philosophy, its applications may be deleterious (Valiathan 2016).

To make a distinction between modern functional biology and the ayurvedic conceptualization of system functioning, a new term "Ayurvedic biology" was even proposed by such enlightened thinkers who stressed that Ayurveda should be understood within its own frame of knowledge without getting adulterated or diluted by any uncanny scientific idea (Valiathan 2006). However, the efforts of such moderates may go waste in contemporary times when evidence-based practice is a public policy requirement for both national healthcare systems and scientific research bodies (Patwardhan 2013). Evidence-based practice has found favor with the patient community as well, since with the rising levels of education, awareness and the ease of access to information through digital technologies, patients are able to make more informed decisions when evidence-based practices are implemented in healthcare delivery. For this reason, assembling evidence has become a crucial step for the further development of Ayurveda (Patwardhan 2014).

7.2 ASSEMBLING EVIDENCE IN AYURVEDA

The practice of Ayurveda is guided by the dictum of unity and interconnectivity of microcosm and macrocosm (*yat piṇḍē tathā brahmāṇḍē*). This dictum has led to distinct and unique methods of decision-making in Ayurveda. From an early stage in its development as a healthcare system, Ayurveda has emphasized the need for meticulous observations, experimentation and the uniquely peculiar method of challenge and re-challenge to improvise the theories through debates. Ayurvedic texts, in common with the canonical texts of Chinese medicine, are much more than pharmacopeias or formularies, since these textbooks assemble the evidence arising out of these observations, experimentations and debates.

With regard to evidence generation, Ayurveda utilizes its own unique system. It strives to generate "unquestionable knowledge", i.e. evidence by distinctly distinguishing the *sata* (truth or existent) from *asata* (untruth or nonexistent). A useful aphorism in this context is the saying that the *sata* never ceases to exist, whereas *asata* never actually comes into existence, as enunciated in the *Bhagavad Gīta* (Fosse 2007; Rastogi 2010). The methods employed for acquiring true knowledge are *pratyakṣa* (direct evidence), *anumāna* (indirect evidence), *yukti* (experimental evidence) and *āpta* (documented evidence).

7.2.1 Direct and Indirect Evidence

Duly acknowledging the limiting factors related to the means of gaining knowledge, and suggesting the ways to remove them in order to acquire absolute knowledge is akin to the idea of bias and errors of current research methodology. The ayurvedic proposition of factors limiting the procurement of knowledge is actually more pervasive than the idea of systemic and random errors alone (Figure 7.2). Science, however, has made fascinating progress in countering these limiting factors

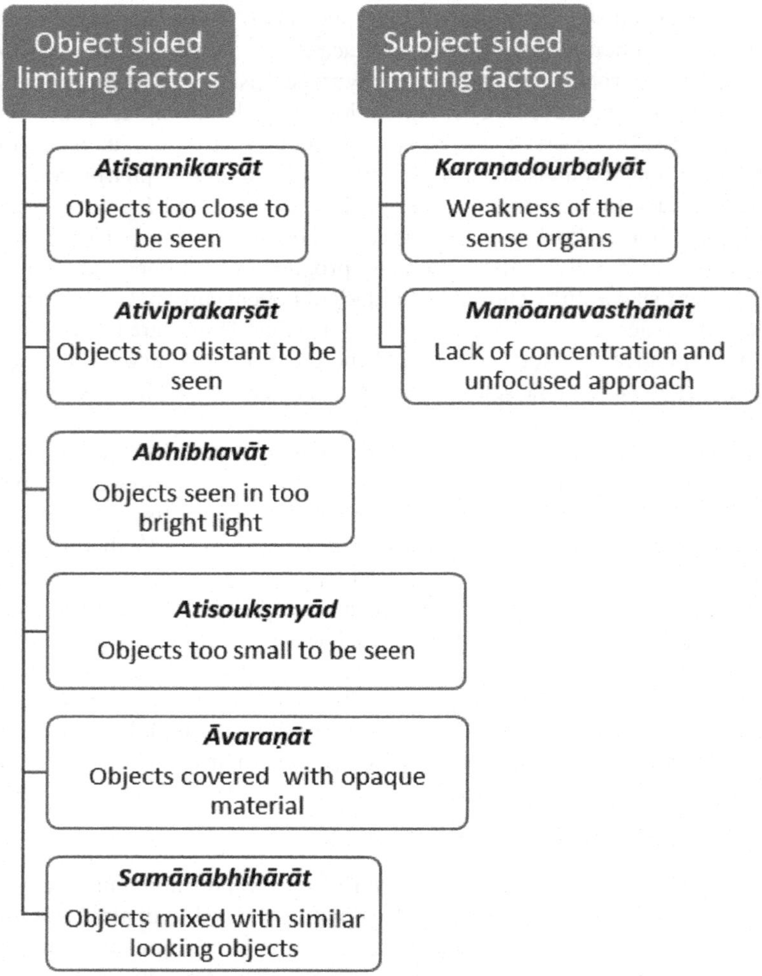

FIGURE 7.2 Factors limiting knowledge gain.

related to the direct observations by devising appropriate technologies and obtaining a fair amount of reliable knowledge. Inferences of indirect observation are principally based upon the factual observations made in the past and finding a resemblance between past events and the happenings of the present. Indirect evidence, however, finds a much wider application in contemporary science and forms the basis of almost every imaging investigation done currently.

7.2.2 Yukti (Experiment)

This is most fascinating of all research methods adopted in the past. *Yukti* begins with formulating a hypothesis about the disease in terms of its cause, presentation and prognosis and subsequently proposes a rational plan to manage it on the basis of the conceived hypothesis. *Samprāpti* is the process of pathogenesis, where disease-related offensive and individual-related defensive mechanisms interact to generate a new biological entity – *vikṛti* – which was nonexistent before this mutual interaction. The treatment proposition in Ayurveda essentially aims to dissociate the *samprāpti* (*samprāptivighaṭana*) and hence to restore the pre-morbid state (Singh 2010b). A *yukti* hypothesis is like a superiority hypothesis testing where assumed treatment plan is hypothesized to be better than other treatment options on the basis of previous experiences or textual knowledge (Wang et al. 2017).

7.2.3 Āpta (Documented Evidence)

After assembling the knowledge from all resources and checking it for all possible errors, what remains is the absolute knowledge, called *āpta* in Ayurveda. The concept of *āpta* is uniquely described in Ayurveda as being knowledge (1) devoid of *raja* (bias) and *tama* (errors), and (2) which is internally as well as externally valid in reference to time (*yēṣām trikālamamalam*). Such knowledge may be considered highest on the evidence hierarchy and is proposed as *vākyam asamśayam* (undoubted statements) in Ayurveda. *Caraka Samhita* states that the strength of knowledge gathered through perseverance, by those who are devoid of *raja* and *tama*, is true beyond time and space (Tripathi 1983a).

7.3 JOURNEYING THROUGH THE EVIDENCE IN AYURVEDA

A journey through the various disciplines of Ayurveda is an extraordinary and exemplary account of an evolutionary course taken to make it dynamic, versatile and applicable through the ages. In almost every single segment, ranging from diagnostics to treatment, drug procurement to drug processing and formulating, basic sciences and applied sciences, Ayurveda presents a glittering account of its diligent and meticulous efforts, applying the intuitive wisdom and deductive inferences generated through trial and error to reach its propositions recounted in various classics. While documenting such observations, Ayurveda seems to adhere strongly to the principles of ethics, which are very similar to the ethics of scientific writing, as known today. *Caraka Samhita* clearly states that it is the original work of Agnivēśa redacted by Caraka and subsequently replenished by Dṛḍhabala, which is the utmost example of adhering to the ethics of scientific writing (Tripathi 1983b). Similar publication honesty and ethics were followed by all subsequent treatises of Ayurveda which borrowed from the three big classics (*Bṛhatrayi* namely *Caraka Samhita*, *Suśruta Samhita* and *Aṣṭāṅgahṛdaya*) with due acknowledgment and citations of the original work.

7.3.1 Evidence Pertaining to the Etiopathogenesis and Disease Presentation

Modern scientific medicine derives its scientific base almost entirely from randomized control trials, conditioned by the clarity of its output and arrived at in a relatively short time frame (rarely more than 5–10 years). The practitioners of modern medicine allege that Ayurveda cannot and does not measure up to this standard. This is wrong, for much of the evidence in Ayurveda actually comes from what may be described as self-styled longitudinal cohort studies passing through succeeding generations with the observations carried forward from one generation of investigators to another. The time span of these studies could be at least several decades, spanning up to a few hundred years. It is only with the evidence from such extremely long time span studies that one may arrive at the etiological description (*hētu*) of various diseases in Ayurveda. We see, surprisingly, that the *hētu* are not only generic in terms of causes, leading to the gross disequilibrium in various *dōṣa*s, but are also specifically leading only to a particular set of disorders. Such an approach enabled ancient exponents of Ayurveda to unravel and discover the existence of multiple etiopathogenetic factors, despite the overall clinical manifestations being relatively similar and monotonous. Diabetes is a good example of the case. Being incriminated for disharmonic energy expenditure and intake for a large part of its history, now it is fairly understood that diabetes is a heterogeneous disease arriving from too many distinct pathways joining a common tract to make the distinct manifestations (Pietropaolo et al. 2007; Staimez et al. 2019). This is obvious to note that the treatment would not be effective unless it acts precisely at the path of its origin and this makes diabetes a difficult to cure condition with insufficient care, despite all the advancements made in its management (Rastogi

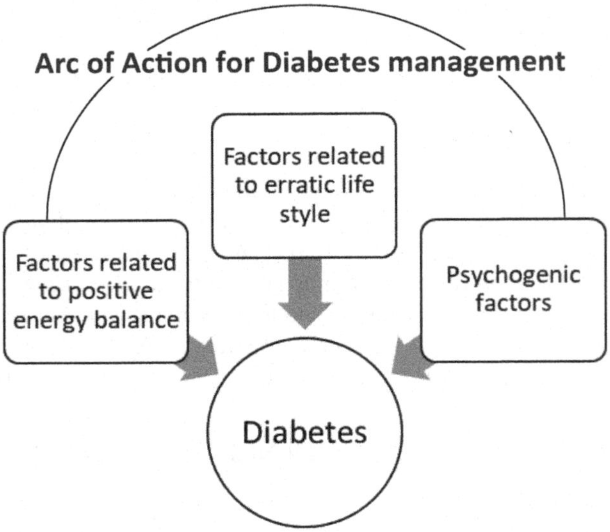

FIGURE 7.3 Heterogenic pathogenesis in diabetes and arc of action for its management.

2019). For a comprehensive management plan of diabetes, an arc of action would eventually be required, cutting across all the possible causes of the disease in a given case (Figure 7.3).

Another example of the success of the ayurvedic evidence gathering is the acknowledgment of phenotypic variations of pathology. Phenotypic variability in a disease with differential symptom dominance in different disease subpopulations is another recent realization in modern science, although it is defined exponentially in Ayurveda, in almost every single disease. A disease distinguishable into subtypes on the basis of precipitating *dōṣa* subsequently also defines the phenotypic variability of the disease by means of dominating clinical features. The diseases, therefore, can be of *vāta*, *pitta* or *kapha* subtype, or their variable combinations or can be exogenous, where *dōṣa* disharmony arises subsequent to the initiation of the primary cause. *Sandhivāta* (osteoarthritis) which is a *vāta nānātmaja* (caused by *vāta* alone) disease is another good example of the case. Keeping *vāta* alone as the major cause of the disease, it varies phenotypically owing to the primary pathology leading to *vāta* imbalance. If it is due to an excess of *kapha*, it would present *dhātvagni māndya* (hypometabolic state) as a cause and would present as obese osteoarthritis. On the other hand, if it

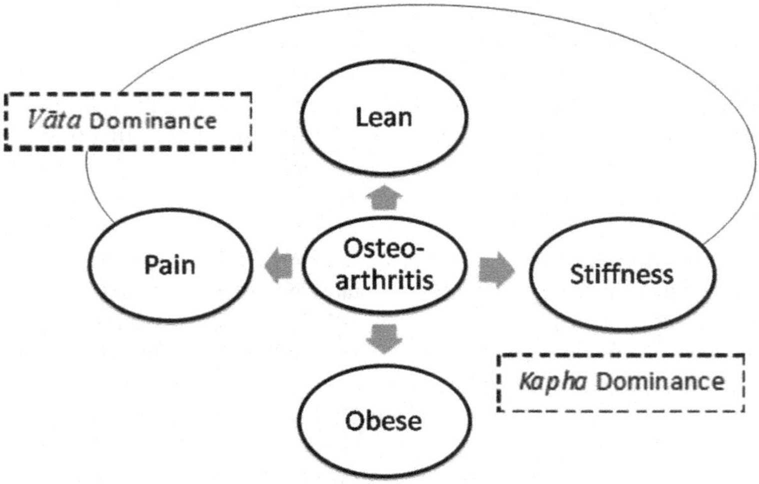

FIGURE 7.4 Phenotype variability in osteoarthritis warranting a search for its cause.

is marked with *pitta* and *vāta* predominance, it would present with *dhātu kṣaya* (degenerative state) marked by emaciation (Figure 7.4). Phenotypical variability in osteoarthritis is recently recognized in modern healthcare, and a differential treatment plan is proposed to tackle such cases originating through different routes, despite their common clinical features (Deveza et al. 2017).

7.4 OPTIMIZING THE OLD AND GENERATING THE NEW EVIDENCE

In tune with the ayurvedic classical dictum of managing diseases which call for replenishing the deficits, depleting the excess and maintaining the balance (Tripathi 1983c), it is proposed that a similar strategy could also be adopted for research in Ayurveda to generate evidence. One of the expectations that modern science has from Ayurveda is that its practitioners must produce translational benefits from books to bedside. This may prove to be particularly difficult in the case of matters like aphorisms and dictums which have been used to guide the practice of ayurvedic medicine. On the contrary, it may be argued that these dictums are actually the result of a translational effort, which has reduced complex observations in textbooks to easily remembered rules of practice. Such dictums already stand proven, since they have passed the test of time (Rastogi 2012).

Although research has been conducted in Ayurveda for the past many decades, these largely failed to give a reasonable reshaping of Ayurveda in the contemporary world and also could not help improve its clinical practice to a great extent. Ayurvedic practitioners, by and large, remain unclear about the possibility of their prescriptions to work in a given case. What has been missed by ayurvedic research done in the past decades is their contextual appropriation to link the discrete pieces of knowledge available to form a single compact and comprehensive knowledge having practical utility. Designing appropriate models to test if ayurvedic fundamentals work is also an important task. However, it seldom reached beyond its status from debate to practical designing (Patwardhan 2012).

7.4.1 Optimizing the Old: Re-Appropriation, Revalidation and Research Synthesis

Ayurvedic classical texts are full of information that is generic and are related to health maintenance, causes of diseases, diagnostic methods and treatment approaches. Ayurvedic literature is found following a temporal pattern of textual compilations, moving from the complex to the simpler narrations. Eventually the oldest texts of Ayurveda comprising the classical triad like *Caraka Samhita* and *Suśruta Samhita* are found to be the most complex, in regard to their contents, whereas latter-day texts, including their commentaries, have become much simpler for their easy comprehensibility and acceptability among *vaidya*s of average intellect. This simplification of classical texts without diluting their principles was the ancient translational approach of Ayurveda, to make it widely acceptable, by offering ease in understanding and practice. Presentation of knowledge in a compact form (*sūtra*), although facilitating its memorization and allowing the practitioner to have their contextual inference by applying their own wit (*yukti*), also proved limiting by creating a rigid grid around the knowledge and its application, which is not the case in contemporary medical learning and practice, which is more descriptive and elaborate. The difference in learning in Ayurveda and modern medicine is so large that the interpretation of knowledge and its application is highly subjective in Ayurveda, whereas it is highly objective in modern healthcare. It is for this reason that there is little space for variability in modern healthcare where the responses may be highly predictable. This is the opposite in Ayurveda, where it is highly variable, and the responses are rarely predictable. Continuous research in contemporary science is regularly incorporated in the texts allowing the readers to update them about evolving knowledge and to modify their practice in tune with the emerging knowledge. This approach of research synthesis is largely missing in Ayurveda.

Re-appropriation in this context stands for revisiting the classical knowledge of Ayurveda and finding if they seem appropriate in the contemporary context. There can be references to traditional contexts of living styles, which may not be truly appropriate in the current context. Such references may be re-appropriated. Looking at the causative factors of *vātarakta* elaborating a bumpy ride on a fast-moving animal (camel or horse) as the cause of the disease, seeing that such causes are actually not existing any more despite a sustained presence of the disease in the population argues for finding the re-appropriation of such causes in newer contexts. A motorcycle ride may be a re-appropriated cause of *vātarakta* in the modern context. Drug formulations and their dosing schedules may also require a similar contextual re-appropriation, focusing upon the drug dose as well as its scheduling. The time required for a *pañcakarma* procedure may also need a similar re-appraisal.

Revalidation studies are vital in order to revive the evidence which has been accumulated through millennia in Ayurveda. These are again referring to the cues contextual to the cause of a disease and its cure by various means and modalities of Ayurveda. Without questioning their validity in the time when they originated, it is genuine to ask if these are still valid since the application of this knowledge has spanned a gap of over five millennia. The question of revalidating Ayurveda does not seem absurd, as we see old technologies getting replaced by new ones which are better productive and less expensive. The couching technique of cataract surgery originally described by Suśruta in 500 B.C. has no place in today's ophthalmology, except for its historical value (Kansupada and Sassani 1997). The same is the case with the laparotomy method recommended by Caraka in the case of intestinal obstruction or perforation (Shastri and Chaturvedi 2009). The ancient diagnostic technique of pouring a drop of oil into the urine (*Tailabindu parīkṣa*), which had been a practical examination technique of finding the amount of bile salts in the urine and so to judge a clinical condition, has much better, accurate and reliable alternatives today, with ease and swiftness of operation (Kar et al. 2012). Ayurvedic drug formulary and pharmaceutics had largely been relying upon organoleptic techniques of drug identification and quality of finished products. There are remarkable advances made in the area of drug development and standardization in the recent past. Whether we really need today the techniques which had been described in Ayurveda many thousand years ago, is a question hard to be ignored. Revalidation is largely required in Ayurveda to understand its philosophy first, then its applications which may be the derivations of its fundamentals. A revalidation is also required to filter in Ayurveda what is still valid and what may have only historical importance. In the past century, China has been able to successfully do such a revalidation for its traditional Chinese medicine, though the process still goes on with divided opinion on the endpoints of such a process. It is India's turn now to do it for Ayurveda.

Synthesizing and amalgamating the existing research to a meaningful conclusion is yet another fascinating area having high relevance to Ayurveda. Modern biology synthesizes the research findings through meta-analysis and systematic reviews which are not of much relevance to traditional medicine. What makes more sense to Ayurveda is to see what research material is meaningful, to understand it more explicitly, whether it is done by Ayurveda scientists or not. There are a number of researches conducted across the world having great significance to Ayurveda. Such research may require screening and if needed, to be viewed in a new ayurvedic perspective, keeping the inferences obtained from the primary research as pilot observations. The amalgamation of researches conducted in Ayurveda is also needed, even though these may not be rigorous studies. A combination of a few weak studies, when put together through data pooling, may come out to be a strong study. This can have high applicability in Ayurveda.

7.4.2 Generating the New: Priority Areas for Clinical Research in Ayurveda

Compared to the clinical research in biomedicine, Ayurveda presents its own unique requirements of clinical research pertaining to its specific needs. These needs are in addition to the usual efficacy and safety-related researches needed in Ayurveda or any other system of healthcare. The following examples may be considered as the priority areas for clinical research in Ayurveda, although it should not remain limited to those listed here, and, therefore, many other novel areas not listed may subsequently be added to the list. It is further emphasized that clinical research in Ayurveda may be carried out by adopting various models using Ayurveda either as a stand-alone therapy or integrating it with modern medicine in various proportions (Box 7.1).

BOX 7.1 MODELS OF RESEARCH IN AYURVEDA

1. Stand-alone model
2. Integrative model
   a. Primary modern care followed by Ayurveda care
   b. Primary Ayurveda care followed by modern care
   c. Simultaneous care by due adoption of both systems at the same time

7.5 AREAS OF PRIORITY OF CLINICAL RESEARCH IN AYURVEDA

7.5.1 Areas Relevant to Contemporary Healthcare Needs

Areas where conventional effective care is not available, accessible or affordable can be the first priority research area in Ayurveda. This area may include the diseases of high prevalence, morbidity and mortality, having a high economic burden or requiring highly complex care not easily accessible to everyone. Ayurveda may find its research-based role here to ease the situation by extending its helping hand as a complementary, adjunct or supportive therapy along with the modern healthcare currently practiced.

7.5.2 Areas Where Ayurveda Has Proven Strength

These are certain clinical areas where Ayurveda anecdotally and conventionally is believed to have an edge over modern medicine. Hepatobiliary diseases, joint diseases, skin diseases and gastrointestinal diseases are such areas where Ayurveda is preferred by common people on the basis of the response observations in the segment. Developing these areas further through research and experimentation to maximize the benefits and to make the interventions highly specific, dependable and cost-concerned is highly desirable. The ultimate objective of research in these areas should be to develop Ayurveda as first-line therapy in specific conditions where it may dependably be considered as a primary therapy (Chauhan et al. 2015).

7.5.3 Taking Cues from Clinical Observations

While practicing Ayurveda, we find many unexpected observations which have not been seen and reported before. These observations may be related to various domains of clinical knowledge e.g. the cure of a rare disease, symptomatic relief in an intractable disease, side-effects of a relatively safe medicine or some other observations which may be of significance to other clinicians in practice. All such observations may become the subject of more detailed inquiry through systematic clinical researches.

7.5.4 Areas of Textual Strength in Ayurveda

Many areas are elaborately described and praised in Ayurveda, but are not used in contemporary practice. Such practices, which have a theoretical basis, but are not practiced, may be explored to prove the relevance and translational possibility of such descriptions. *Rasāyana*, *vājīkaraṇa*, various *dincarya* and *ṛtucarya* rituals and preventive *pañcakarma* are classical examples of this knowledge which needs contemporary reappraisal.

7.5.5 Areas Important to Establishing Standards in Ayurvedic Clinical Practice

Many fundamental questions which are important from an ayurvedic perspective are required to be taken as a priority in clinical research. These are typically related to the development of practice guidelines, cost-effectiveness analysis, determination of dose and duration of therapy on the basis of primary and secondary endpoint observations and establishment of long-term safety of ayurvedic interventions in the conditions where a prolonged therapy is required.

7.5.6 Ayurveda in Vulnerable Population

Ayurveda is still required to prove its worth in marginal populations having special healthcare needs. Ayurveda may be required to find a clear role in pediatric and geriatric healthcare needs and also need to establish its safety and utility in pregnancy and lactation. Such research work is essential to establish a firm ground for Ayurveda since it involves over half of the population requiring medical care.

7.6 RESEARCH DESIGNS NEEDED IN AYURVEDA

Although there shall be no doubt about the sanctity of a double-blind, randomized clinical trial as for being the gold standard in clinical research, it is faced with much difficulty in traditional medicine research. The complexity of ayurvedic formulations, their intake methodologies, dietary recommendations on the basis of *prakṛti* and bio-physical purificatory processes of *pañcakarma* together make a package of ayurvedic intervention which is often difficult to evaluate through a blinded placebo. A system to system comparison is therefore proposed to find out if there are any significant differences between the two approaches (Kessler and Michalsen 2012). To make such system comparisons more reliable, dummy techniques are often employed, giving an opportunity to treat both the groups with a placebo and a real intervention either of Ayurveda or of modern medicine (Furst et al. 2012). The World Health Organization has recommended a black-box approach for the clinical research in traditional medicine due to the complexity of various components employed together for treatment in such systems with almost equal weightage to all components (Fatima et al. 2017).This has been found beneficial to seeing what the whole system does rather than getting intrigued by the mechanistic details of individual components. Such mechanistic studies, however, can be of much use in improving the formula further, once the utility of a package is proved.

Besides such clinical trials, a number of other research designs are very useful in the context of Ayurveda. Observational studies that happened to be the most trusted tool of research in Ayurveda for millennia are still a tool of the first order for their capacity to cater to the primary data about epidemiology, preferences, cost and general effects. Longitudinal studies composing big cohorts are supposed to give the most reliable answers to eternal quests of the safety of ayurvedic drugs in the scientific world.

The ayurvedic approach to treating a disease is more individualized and dynamic compared to modern medicine. Ayurveda proposes that pathogenesis is a dynamic process and so has a flexible approach of treatment as per the changing impetus of the host–pathogen interactions. Depending upon their changing relationship in reference to various factors including the drug treatment, the future course of intervention may be entirely different than what has been proposed initially. This is in contrast to the fixed dose and duration regimens of modern medicine where there is little flexibility, and the same course is to be followed either till the termination of the disease or as lifelong therapy. In such cases, N of 1 trials making longitudinal observations in individual patients seems to be the best research design.

Individual case studies and series are still the best learning material in Ayurveda since they bring the individual experiences down to everyone's understanding and stimulate looking at the causes of such observations occasionally made in practice. Case-based reviews have been the new approach in various journals that allow having a deeper discussion in the case rather than mere reporting of what has been observed (Pierce et al. 2014).

7.7 RESEARCH DESIGNS FOR AYURVEDIC CORE CONCEPTS OF CLINICAL MEDICINE

Ayurvedic core concepts of clinical medicine like *prakṛti*, *tridōṣa*, *āma*, *sapta dhātu ōja* and *agni* require some specific postulates and methods for their assessment. These concepts are first required to be clearly elucidated in terms of ayurvedic biology as well as conventional biology, and subsequently methods are required to be generated for their exploration on the basis of existing knowledge. *Āma* sets a good example of the case. Being the cause of many local and systemic disorders, *āma* is incriminated as the main culprit in many disorders ranging from joint diseases to gastrointestinal diseases. Looking at the literature, it is clearly seen that the presence and absence of *āma* are differentiable by a set of distinguishing clinical features. Reduction in *āma* features in the intervention for the reduction in pathology or *samprāpti vighaṭana* as postulated in Ayurveda. In the case of joint diseases marked with the presence of *āma*-related features like swelling, stiffness and immobility, upon the adoption of non-steroidal anti-inflammatory drugs (N.S.A.I.D.s), many such symptoms get reduced to some extent transiently. This observation proposes that these N.S.A.I.D.s have *āma*-reducing properties from an ayurvedic perspective. Now, since the mechanism of action and the action pathways are well-known for most of the N.S.A.I.D.s, it is possible to relate the *āma*-reducing ayurvedic drugs with the mechanism of action utilized by the N.S.A.I.D.s. Exploration of *prakṛti* on the basis of genomic studies is another example of such exploratory work.

Agni presents another fundamental dictum of Ayurveda, which is incriminated in almost every disease. *Agni* has its physiological expressions as hypo, hyper, erratic or normal, reflective of *prakṛti* of the person. There can be pathological expressions as well, incompatible with the *prakṛti*. Genomic studies conducted recently have identified the genetic basis of the speed of metabolism in individuals and identified the subsets of the people as fast metabolizers and slow metabolizers on the basis of gene expressions (Dey and Pahwa 2014). Such metabolic variations, if expressed abnormally owing to various incriminated causes routing through diet and lifestyle, can become the progenitor of disease as is visualized in Ayurveda. Basic ayurvedic approaches to deal with a disease therefore links to the fixing of *agni* to its normalcy in reference to the *agni* of the person in the pre-morbid state. As *agni* has its expression at multiple levels from tissues to cells, this would be hard to get assessed by the colloquial terms of appetite or hunger, expressive of gastrointestinal fire (*jatharāgni*). *Dhātvagni* and *bhūtāgni*, playing at the tissue and cellular level, may play more significantly referring to systemic pathology. There are a number of biomarkers expressive of various metabolic activities in the body and their increased or decreased levels may be a clear indication of undergoing biological processes. This may present a fascinating postulate to evaluate *dhātvagni* and *bhūtāgni* through levels of biomarkers signifying various biological activities relevant to a particular tissue (Rastogi 2012; Lurie 2012). There can be a number of such novel designs of research to ayurvedic fundamentals of clinical medicine, utilizing the existing tools and by indigenizing their usage as per the needs of Ayurveda.

7.8 CONCLUSION

Modern scientific medicine has become *modern* and *scientific* by using two inter-related techniques – measurement and documentation. By being able to reduce any natural phenomena into measurable quantities (take as an illustration of the Bristol Stool scale), modern medicine has been able to "see" patterns and cause–effect relations. The emphasis on documentation has enabled modern medicine to record, disseminate, analyze, criticize and archive knowledge. Put together, measurement and documentation have given concepts like R.C.T.s which enable medicine to become more efficient.

In the case of Ayurveda, neither measurement nor documentation has played a major role in the generation, aggregation and dissemination of knowledge. There was a strong awareness of the need for having high-quality evidence, as shown by concepts like *vākyam asamśayam*. But at no point was the need felt for quantifying it to render it more measurable; documentation was incidental and mostly for the purpose of archiving and for the intergenerational transfer of knowledge. Therefore, for modern healthcare systems to insist that *prakṛti* scales be developed which may then be reliably compared between different R.C.T.s is asking a little too much. It also misses the point that the intuitional wisdom and heuristic decision making of an experienced practitioner can never be reduced to "modern scientific evidence". Nevertheless, ayurvedic practitioners have made efforts, and sooner or later, the sheer numbers of these efforts will produce some significant breakthroughs. The challenge for Ayurveda, as indeed for all traditional medicine systems, is to develop its own distinctive systems for generating and utilizing medical evidence. It cannot adopt a cut-and-paste approach from modern scientific medicine.

REFERENCES

Chauhan, A., D. K. Semwal, S. P. Mishra, and R. B. Semwal. 2015. Ayurvedic research and methodology: Present status and future strategies. *Ayu* 36, no. 4: 364–9.

Deveza, L. A., L. Melo, T. P. Yamato, K. Mills, V. Ravi, and D. J. Hunter. 2017. Knee osteoarthritis phenotypes and their relevance for outcomes: A systematic review. *Osteoarthritis and Cartilage* 25, no. 12: 1926–41.

Dey, S., and P. Pahwa. 2014. Prakriti and its associations with metabolism, chronic diseases, and genotypes: Possibilities of new born screening and a lifetime of personalized prevention. *Journal of Ayurveda and Integrative Medicine* 5, no. 1: 15–24.

Fatima, S., N. Haider, A. Alam, A. Quamri, L. Unnisa, and R. Zama. 2017. Preventive, promotive and curative aspects of dementia in complementary medicine (Unani): Through-black box design. *International Journal of Herbal Medicine* 5: 1–5.

Fosse, L. M. 2007. Chapter 2 (Theory). In *Bhagavad Gīta*, 11–16. New York: Yoga Vidya.com.

Furst, D. E., M. M. Venkatraman, M. McGann et al. 2012. Double-blind, randomized, controlled, pilot study comparing classic ayurvedic medicine, methotrexate, and their combination in rheumatoid arthritis. *Journal of Clinical Rheumatology : Practical Reports on Rheumatic and Musculoskeletal Diseases* 17, no. 4: 185–92.

Kansupada, K. B., and J. W. Sassani. 1997. *Documenta Ophthalmologica* 93, no. 1–2: 159. doi:10.1007/BF02569056.

Kar, A. C., R. Sharma, B. K. Panda, and V. P. Singh. 2012. A study on the method of *Taila bindu Pariksha* (oil drop test). *Ayu* 33, no. 3: 396–401.

Kessler, C., and A. Michalsen. 2012. The role of whole medical systems in global medicine. *Complementary Medicine Research* 19, no. 2: 65–66.

Lurie, D. 2012. Ayurveda and pharmacogenomics. *Annals of Ayurvedic Medicine* 1: 126–8.

Patwardhan, B. 2012. The quest for evidence-based Ayurveda: Lessons learned. *Current Science* 102: 1406–17.

Patwardhan, B. 2013. Time for evidence-based Ayurveda: A clarion call for action. *Journal of Ayurveda and Integrative Medicine* 4, no. 2: 63–6.

Patwardhan, B. 2014. Bridging Ayurveda with evidence-based scientific approaches in medicine. *EPMA Journal* 5, no. 1: 19. doi:10.1186/1878-5085-5-19.

Pierce, R. J., R. Falter, S. Cross, and B. Watson. 2014. Using case-based reviews to improve student exam performance. *Currents in Pharmacy Teaching and Learning* 6, no. 6: 822–5.

Pietropaolo, M., E. Barinas-Mitchell, and L. H. Kuller. 2007. The heterogeneity of diabetes: Unraveling a dispute: Is systemic inflammation related to islet autoimmunity? *Diabetes* 56, no. 5: 1189–97.

Rastogi, S. 2010. Building Bridges between Ayurveda and modern science. *International Journal of Ayurveda Research* 1, no. 1: 41–6.

Rastogi, S. 2012. Prakriti analysis in Ayurveda: Envisaging the need of better diagnostic tools. In *Evidence-based practice in complementary and alternative medicine*, ed. S. Rastogi, F. Chiappelli, M. H. Ramchandani and R. H. Singh, 99–111. Berlin: Springer.

Rastogi, S. 2019. Understanding diabetes: Uncovering the leads from Ayurveda. In *Translational Ayurveda*, ed. S. Rastogi, 123–39.Singapore: Springer Nature.

Rastogi, S., and R. H. Singh. 2012. Transforming Ayurveda: Stepping into the realm of evidence-based practice. In *Evidence-based practice in complementary and alternative medicine*, ed. S. Rastogi, F. Chiappelli, M. H. Ramchandani, and R. H. Singh, 33–49. Heidelberg: Springer.

Shastri, K. N., and G. N. Chaturvedi. 2009.*Udara cikitsā adhyāya* (Treatment of liver diseases). In *Caraka Samhita*, Part 2, 381–415. Varanasi: Chaukhambha Bharati Academy.

Singh, R. H. 2010a. Exploring issues in the development of Ayurvedic research methodology. *Journal of Ayurveda and Integrative Medicine* 1, no. 2: 91–5.

Singh, R. H. 2010b. Exploring larger evidence-base for contemporary Ayurveda. *International Journal of Ayurveda Research* 1, no. 2: 195–6.

Staimez, L. R., M. Deepa, M. K. Ali, V. Mohan, R. L. Hanson, and K. V. Narayan. 2019. Tale of two Indians: Heterogeneity in type 2 diabetes pathophysiology. *Diabetes/Metabolism Research and Reviews*: e3192. doi:10.1002/dmrr.3192.

Tripathi, B. N. 1983a. *Tisraisaṇīyam Addhyāya* (Chapter dealing with three objectives of life). In *Caraka Samhita*, Part 1, 223–52. Varanasi: Chaukhamba Surbharati.

Tripathi, B. N. 1983b. *Bhūmika* (Introduction). In *Caraka Samhita*, Part 1, 5–24. Varanasi: Chaukhamba Surbharati.

Tripathi, B. N. 1983c. *Dīrghanjīvitīya Addhyāya* (Chapter dealing with principles of longevity). In *Caraka, Samhita*, Part 1, 1–49. Varanasi: Chaukhamba Surbharati.

Valiathan, M. S. 2006. *Towards ayurvedic biology: A decadal vision Document-2006*. Bangalore: Indian Academy of Sciences.

Valiathan, M. S. 2016. Ayurvedic Biology: The first decade. *Proceedings of the Indian National Science Academy* 82, no. 1. doi:10.16943/ptinsa/2016/v82i1/48376.

Wang, B., H. Wang, X. M. Tu, and C. Feng. 2017. Comparisons of superiority, non-inferiority, and equivalence trials. *Shanghai Archives of Psychiatry* 29, no. 6: 385–8.

8 Conservation – A Strategy to Overcome Shortages of Ayurveda Herbs

S. Noorunnisa Begum and K. Ravikumar

CONTENTS

8.1 Introduction 176
8.2 Flora of India 176
8.2.1 Botanical Profile of Medicinal Plants 177
8.2.2 Distribution of Medicinal Plants in India 177
8.2.3 Biodiversity Hotspots 179
8.3 Medicinal Plants in Trade 179
8.3.1 Substitutes and Adulterants 179
8.3.2 Species Traded in High Volume 181
8.3.3 Major Supply Sources 182
8.4 Threat Assessment 182
8.5 Conservation 183
8.5.1 *In Situ* Conservation 183
8.5.2 *Ex Situ* Conservation 183
8.5.3 Medicinal Plant Conservation Areas (M.P.C.A.s) 184
8.5.4 Success Stories of Conservation 185
8.5.4.1 *Commiphora wightii* (Arn.) Bhandari 185
8.5.4.2 *In Situ* Conservation 185
8.5.4.3 *Decalepis hamiltonii* (Wight & Arn.) 186
8.5.4.4 *In Situ* Conservation 186
8.5.4.5 *Saraca asoca* (Roxb.) W. J. de Wilde 186
8.5.4.6 *In Situ* Conservation 186
8.5.5 Medicinal Plants Development Areas (M.P.D.A.s) 187
8.5.6 Botanic Gardens 188
8.5.7 Seed Banks 188
8.5.8 Cultivation 189
8.6 Measurement and Monitoring of Biodiversity 189
8.6.1 Pilot-Scale Measurements 189
8.6.2 GIS System – A Conservation Tool 190
8.7 Mapping of Plant Distribution 191
8.7.1 Global and Indian Distribution Maps 191
8.7.2 Regional Distribution Maps 191
8.7.3 Eco-Distribution Maps 191
8.8 Sustainable Harvesting 192
8.8.1 Success Story 192
8.8.1.1 *Terminalia chebula* Retz. 192
8.9 Classical Ways of Conserving Ayurvedic Herbs 193
8.10 Conclusion 193
References 194

8.1 INTRODUCTION

Over-exploitation of living resources takes place all over the world in order to meet short-term needs. But very often the process destroys exactly those resources on which the welfare of millions of people depends in the long term. This disregard for sustainable utilization in fact widens the gap between rich and poor countries of the world, as there is an obvious relation between conservation and development. According to the Food and Agriculture Organization of the United Nations (F.A.O.), closed tropical forests or rainforests are disappearing at an alarming rate. Nearly 76,000 km^2 of rainforests are being destroyed every year (Hamann 2009).

Destruction of flora has had its effect on the medicinal plant industry. Many years ago, the over-exploitation of wild-growing *Rauvolfia serpentina* Benth. ex Kurz. in India for export exhausted the supply to such a point that the government of India placed an embargo on the export of this species. This created a major problem in the United States of America, as the United States Pharmacopeia requires that *Rauvolfia serpentina* be of Indian origin when used in a crude form. Another glaring example of a plant that has been over-exploited in India for export to other Asian countries is *Coptis teeta* Wall., which has now attained the status of an endangered plant (Akerele 2009).

On being aware of the urgent need for the global conservation of plant wealth, health professionals and plant conservation specialists came together for the first time at the W.H.O./I.U.C.N./W.W.F. International Consultation on Conservation of Medicinal Plants, held in Chiang Mai, Thailand, 21–26 March 1988. They reaffirmed their commitment to the collective goal of "Health for All by the Year 2000" through the primary healthcare approach, and to the principles of conservation and sustainable development outlined in the World Conservation Strategy. The conference adopted the famous Chiang Mai Declaration, which recognized the urgent need for international cooperation and co-ordination to establish programs for the conservation of medicinal plants to ensure that adequate quantities are available for future generations. The members of the Chiang Mai International Consultation called upon all people to commit themselves to *Save the Plants that Save Lives* (Anonymous 2009a).

8.2 FLORA OF INDIA

India has the distinction of being one of the 17 mega-biodiversity countries of the world, possessing 4 out of 36 of the world's biodiversity hot spots (Myers et al. 2000; Arisdason and Lakshminarasimhan 2017; Hrdina and Romportl 2017). Among the Himalayan regions, the north-east Indian region harbors several floristically rich forest patches and a high number of endemics. It has been estimated that the north-eastern region comprises approximately 7500 species of flowering plants that constitute nearly 40% of the total floristic wealth of the country, which is about 19,400 taxa (Karthikeyan 2000). Pteridophytes are represented by 2479 species followed by 1265 bryophytes and 67 species of gymnosperms (Ravindranath et al. 2006; Chitale et al. 2014; Adhikari et al. 2015; Roy et al. 2015). A total of 4381 species and infra-specific taxa of vascular plants belonging to 1007 genera and 176 families are recorded as strict endemics to the Indian political boundary. Out of that, 4303 species and infra-specific taxa are angiosperms, 12 species are gymnosperms and 66 are pteridophytes (Singh et al. 2015).

Currently, India is experiencing a series of environmental problems like climate change, habitat modification, excessive land-use and land-cover change, environmental pollution, over-exploitation of biological resources and alien species invasion (Barik et al. 2018). Depending on the scale and intensity of these changes, several species are being lost and genetic variability within the species is being eroded. Anthropogenic factors have also played a pivotal role in controlling species distribution and have led to the extinction of numerous species, while several others have become threatened (Reaka-kudla et al. 1996).

The report of National Medicinal Plants Board (N.M.P.B.) points out that out of 6580 medicinal plant species, 1622 botanicals corresponding to 1178 plant species have been found to be traded in all of India. Only 242 species witness high volume trade (>100 M.T.) annually (Goraya and Ved 2017).

Of the medicinal plants of India, 6580 species have thus far been documented in published literature (IMPLAD 2017). In undocumented form, the knowledge of medicinal plants and their uses may exceed 10,000 species. Their fuller documentation is an important unfinished national ethnobotanical agenda that must engage with the health practices of thousands of ethnic communities that constitute the cultural fabric of India (Mukherjee 2009). As per I.M.P.L.A.D. 2017, the species of medicinal plants recorded in different systems of medicine are presented in Table 8.1.

8.2.1 Botanical Profile of Medicinal Plants

Most of the 6580 medicinal plants are higher flowering plants. One-third are trees, shrubs and climbers, another third are herbs and the rest are lower plants like algae, fungi, lichens, bryophytes, pteridophytes, gymnosperms and angiosperms. Medicinal plants of India belong to 2200 genera in 386 families. Most of them belong to the families Asteraceae, Euphorbiaceae, Lamiaceae, Fabaceae, Rubiaceae, Poaceae, Acanthaceae, Rosaceae and Apiacea. Asteraceae has the highest number of medicinal plants, with 419 species belonging to this family alone (IMPLAD 2017).

8.2.2 Distribution of Medicinal Plants in India

With a geographical area of 329 million hectares, India is located north of the equator 08°04′–37°06′N and 68°07′–97°25′E. It is bounded by the Indian Ocean in the south, the Arabian Sea in the west, the Bay of Bengal in the east and the Himalayas in the north. India is blessed with a variety of terrain and climate. The majority of India is tropical to subtropical, which means that the environmental temperature is conducive to the growth and development of vegetation. The rainfall also varies greatly from place to place. India receives rain from the monsoons originating in the Arabian Sea and the Bay of Bengal. The range of topography, temperature and rainfall are responsible for the development of a wide variety of macro- and micro-climates, resulting in the rich biological diversity of the Indian subcontinent. The country is divided into ten biogeographic zones based on biota and environmental realms (Singh and Kushwaha 2008).

Medicinal plant species occur naturally across different biogeographic zones of India (Figure 8.1). The biogeographic zones are: (i) Trans Himalaya (5.6%), (ii) Himalaya (6.4%), (iii) Desert (6.4%), (iv) Semi-arid (16.6%), (v) Western Ghats (4.4%), (vi) Deccan Peninsula (42%), (vii) Gangetic plain (10.8%), (viii) Coasts (2.5%), (ix) Northeast (5.2%) and (x) Islands (0.3%).The values in parentheses

TABLE 8.1
Cross-Tabulation Showing Medicinal Plant Species Recorded in Different Indian Medical Systems

	Ayurveda	Folk	Homoeo	Siddha	Sowa-Rigpa	Unani	Western
Ayurveda	**1539**	876	176	758	248	429	74
Folk	876	**5340**	161	873	187	332	80
Homoeo	176	161	**490**	145	69	137	102
Siddha	758	873	145	**1152**	211	337	59
Sowa-rigpa	248	187	69	211	**252**	179	23
Unani	429	332	137	337	179	**496**	63
Western	74	80	102	59	23	63	**190**

Source: I.M.P.L.A.D. (2016). Indian Medicinal Plants Database, TransDisciplinary University, Bangalore India.

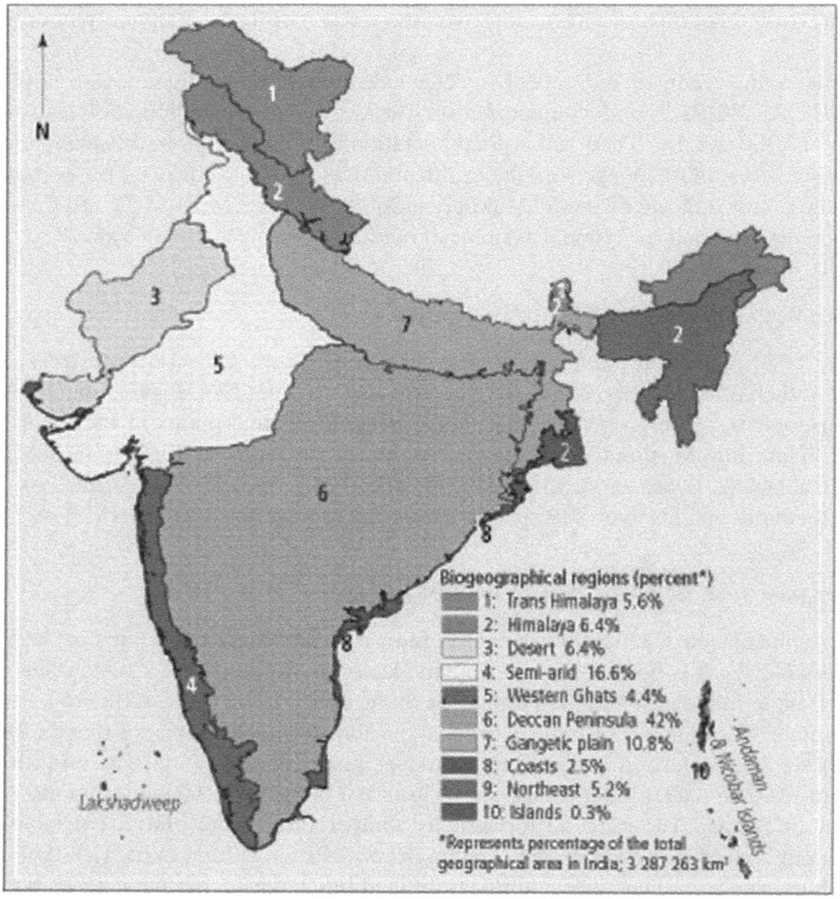

FIGURE 8.1 Biogeographic zones in India (Singh and Kushwaha 2008). Reproduced with permission from the Commonwealth Forestry Association, England (cfa@cfa-international.org).

represent the percentage of the total geographical area of the country – 32,87,263 km^2 (Rodgers and Panwar 1988). India has almost all the representative global ecological zones of south Asia viz., (i) tropical rainforest, (ii) tropical moist deciduous forest, (iii) tropical dry forest, (iv) tropical shrubland, (v) tropical desert, (vi) tropical mountain, (vii) subtropical mountain and (viii) temperate mountain (Chauhan 2007). The largest area of these biogeographic regions is covered by three ecological zones – tropical shrubland, tropical dry forest and tropical moist deciduous forest (Singh and Kushwaha 2008).

The "Trans Himalayan" zone has about 700 known medicinal plant species. The "Himalayan" biogeographic zone consists of North-West Himalaya (2A), West Himalaya (2B), Central Himalaya (2C) and East Himalaya (2D) biotic provinces. The North-West and West Himalaya (2A and 2B) regions are estimated to harbor approximately 1700 medicinal plant species. The Central and Eastern Himalayan (2C and 2D) biotic provinces put together are estimated to harbor around 1200 medicinal plant species. The "Desert" biogeographic zone consisting of Kutch (3A) and Thar (3B) harbor around 500 medicinal plant species. The "Semi-Arid" zone consisting of Punjab (4A) and Gujarat Rajware (4B) is estimated to possess around 1000 medicinal plant species. The "Western Ghats" biogeographic zone consisting of Malabar coast (5A) and Western Ghat Mountains (5B) has 2000 medicinal plant species. The "Deccan Peninsula" with Deccan Plateau South (6A), Central Plateau (6B), Eastern Plateau (6C), Chhota Nagpur (6D) and Central Highlands (6E) has the highest proportion of the country's total medicinal plant diversity, i.e. 3000 medicinal plant species. The "Gangetic Plain" zone covers Upper Gangetic Plain (7A) and Lower Gangetic Plan (7B) with 1000 medicinal plant species. The "North-East India" with two biotic provinces namely Brahmaputra Valley (8A) and Assam Hills (8B) harbor 2000 medicinal plant species. The "Islands" biogeographic zone consists of the Andaman Islands (9A), Nicobar Islands (9B) and Lakshadweep Islands (9C) harboring around 1000 medicinal plant species. Finally, the "Coasts" comprising the West Coast (10A) and the East Coast (10B) are estimated to harbor over 500 medicinal plant species (Ravikumar et al. 2005) (Figure 8.1).

8.2.3 Biodiversity Hotspots

There are 36 terrestrial biodiversity hotspots in the world. Out of these global hotspots, India has four, namely Eastern Himalayas, Western Ghats (and Sri Lanka), North-East India and Andaman Islands (Indo-Burma) and Nicobar Island (Sundaland). The Indo-Burma hotspot is the largest among them with an area of 2.37 million km^2. However, Western Ghats–Sri Lanka is the smallest one with an area of 0.19 million km^2. These hotspots support a unique biodiversity and are highly species-rich areas with a high rate of endemism. The Himalaya hotspot alone includes all of the world's mountain peaks higher than 8000 m, and several of the world's deepest river gorges. The Indo-Burma hotspot is the richest one in biodiversity and includes most of the north-eastern India except Arunachal Pradesh and parts of Assam. The Western Ghats–Sri Lanka hotspot stretches from Gujarat to Kanyakumari and 400 km farther up in Sri Lanka. It has some of the last remaining rainforests and associated high biodiversity. In spite of its higher human population density, the Western Ghats–Sri Lanka hotspot possesses the highest concentration of endemic species. Sundaland lies in South-East Asia and includes Thailand, Singapore, Indonesia, Brunei and Malaysia. The Nicobar Islands represent India (Singh and Kushwaha 2008). The biodiversity hotspots in North-East India and Western Ghats are the repositories of a large number of wild medicinal plant species and even the biogeographic zone "Indian Desert" harbors around 500 medicinal plant species. There is a need for appropriate conservation action and resource augmentation efforts in different parts of the country for specific species which occur in different biogeographical regions (Figure 8.2).

As per the I.M.P.L.A.D. 2017 of the 6580 medicinal plants recorded in different Indian medical systems, the Ayurveda medical system reports 1539 medicinal plant species spread across 190 families. The top ten families are shown in Figure 8.3.

8.3 MEDICINAL PLANTS IN TRADE

In order to have a sharper focus on conservation of plant species which are under sizable commercial exploitation, there is a need to analyze the pattern and quantum of consumption of medicinal plant materials by the herbal sector. According to a published report of the N.M.P.B. (Goraya and Ved 2017), out of 6580 medicinal plant species traditionally used by Indian communities, only 1622 botanicals corresponding to 1178 plant species are found to be traded in all of India. Of these 42% are herbs, 27% trees and 31% are shrubs and climbers. Only 242 species witness high volume trade (>100 M.T.) annually. Diverse parts of plants (leaf, flowers, fruit, seed, bark, root, resin, gum) serve as medicinal raw drugs. Nearly 53% of the medicinal plant species are subject to destructive methods of harvest, as the medicinal parts harvested include underground parts, wood, bark and the whole plant. It is observed that 85% of the traded species and 70% of the demand are even today met from wild sources.

8.3.1 Substitutes and Adulterants

Adulteration and substitution of herbal drugs is a burning problem in the herbal industry, and it has evolved to be a major threat to research on commercial natural products. The deforestation and

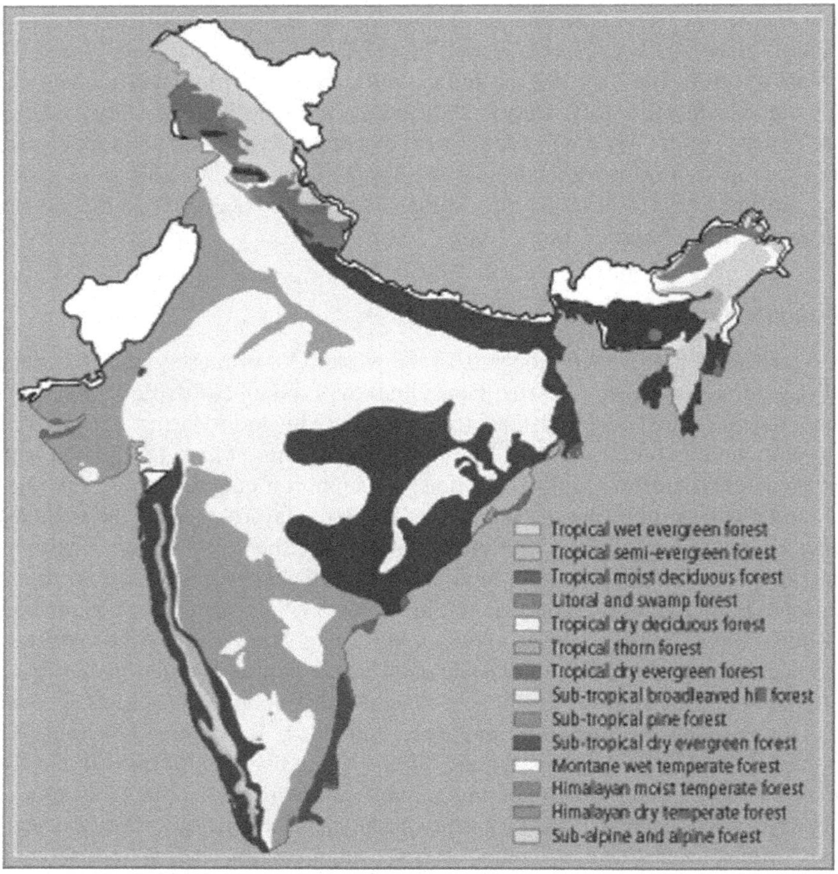

FIGURE 8.2 Forest types of India (Singh and Kushwaha 2008). Reproduced with permission from the Commonwealth Forestry Association, England (cfa@cfa-international.org).

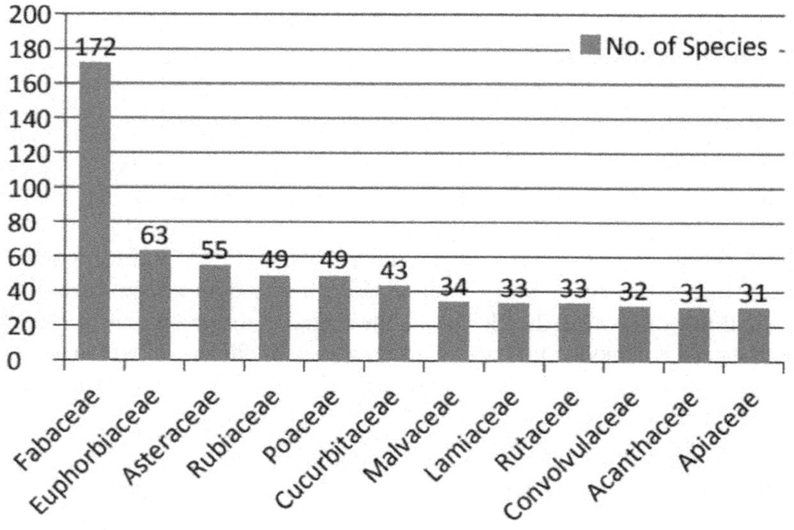

FIGURE 8.3 Top ten botanical families.

TABLE 8.2
Important Medicinal Plants with Their Substitutes and Known Adulterants

Sl.No.	Highly Traded Medicinal Plant	Trade Name	Part Traded	Substituents/Known Adulterants
1	*Abies spectabilis*	Talispatra, Talisa	Leaf Needle	*Abies pindrow* Royle, *A. densa* Griff., *Rhododendron anthopogon* D. Don and *Taxus wallichiana* Zucc.
2	*Aconitum heterophyllum*	Atiss, Ativisa	Root (Tuber)	*Aconitum balfourii* Stapf., *A. deinorrhizum* Stapf., *A. chasmanthum* Stapf *ex* Holmes and *A. falconeri* Stapf
3	*Centella asiatica*	Brahmibooti, Vallaarai	Leaves and Whole plant	*Bacopa monnieri* (L.) Wettst., Bacopa floribunda *Bacopa floribunda* (R.Br.) Wettst
4	*Berberis aristata*	Daruhaldi, Daruharidra	Root, Stem and Fruits	*B. lyceum* Royle, *B. asiatica* Roxb., *B. chitria* Lindl., *B. tinctoria* Lesch and *B. umbellata* Wall.
5	*Cinnamomum cassia*	Dalchini, Tejpatta	Bark (Stem)	*Cinnamomum* ssp.
6	*Cinnamomum sulphuratum*	Dalchini	Bark (Stem)	*Cinnamomum cassia* (Nees & T. Nees) J. Presl, *C. zeylanica* Blume, *C. malabathrum* (Lam.) J. Presl., *C. verum* J. Presl and *C. wightii* Meisn.
7	*Cinnamomum tamala*	Tejpatta Tvakapatra	Bark (Stem), Leaves	*Cinnamomum zeylanica* Blume and *C. malabathrum* (Lam.) J. Presl.
8	*Terminalia chebula*	Harda, Haritaki	Fruits	*Terminalia pallida* Brandis, *Terminalia citrina* (Gaertn.) Roxb. ex Flem.
9	*Boerhavia diffusa* L.	Punarnava, Mukaraai	Root and Whole plant	*Boerhavia rependa* Willd. and *Trianthema portulacastrum* L.
10	*Hemidesmus indicus*	Anatmool, Sveta sariva	Roots	*Cryptolepis buchananii* Roem. & Schult., *Ichnocarpus frutescens* R. Br. and *Decalepis hamiltonii* Wight & Arn.
11	*Mesua ferrea*	Nagekesar	Flowers (Stamens)	*Mammea suriga* Buch.-Ham. *ex* Roxb., *Cinnamomum* spp., and *Dillenia pentagyna* Roxb.
12	*Phyllanthus amarus*	Bhumiamla	Whole plant	*P. urinaria* L., *P. reticulatus* Poir., *P. virgatus* G. Forst., *P. debilis* Klein ex Willd., *P. fraternus* Webster, *P. maderaspatensis* L.
13	*Piper longum*	Pipal/Thippili	Piper, Pippali	*Piper retrofractum* Vahl, *P. betle* L. and *P. peepuloides* Roxb.
14	*Uraria picta*	Prsniparni	Roots	*Uraria lagopodioides*

Source: Ved and Goraya 2008

extinction of many species and incorrect identification of many plants have resulted in adulteration and substitution of raw drugs. Table 8.2 lists selected medicinal plants with their possible adulterants and substitutes.

8.3.2 Species Traded in High Volume

Of the total 242 top traded medicinal plants, 225 are included in the list of ayurvedic medicinal plants. These species are sourced from the wild, cultivated or imported, primarily for use as herbal raw drugs. For example, *makoi* (*Solanum nigrum* Linn.) is found naturally growing in habitats outside forests and as agriculture weeds, and it is this wild-grown population that forms the major source of its supply to the end-users. This species has, however, been recently brought under cultivation also primarily to meet the part supply of its fruits. Cultivation of *Atees* (*Aconitum heterophyllum* Wall. ex Royle.), a red-listed Himalayan species, has been recently initiated, even as major supplies of this herb continue to be met from wild collections (Goraya and Ved 2017).

8.3.3 Major Supply Sources

Analysis of the top traded species in the Ayurveda medical system reveals that the major supply source of 11 of these species is imports and that 49 are largely sourced from cultivation. The major source of supply of the remaining species is wild collections from forests (108 species). Further analysis of the 108 species reveals that 33 species are obtained from Himalayan forests and the remaining 75 species are from tropical forests.

8.4 THREAT ASSESSMENT

Local health traditions (L.H.T.s) cannot be revitalized without ensuring the health of the medicinal plant resource base. Given that the funds, human resources and efforts available are limited, it is very much needed to prioritize and assess the threat status of medicinal plants, in order to focus the conservation action. To accomplish the prioritization of medicinal plants within a reasonable time and cost, the Conservation Breeding Specialist Group (C.B.S.G.) of the Species Survival Commission has developed a rapid assessment methodology called C.A.M.P. (Conservation Assessment and Management Plan) (Anonymous 2020a). C.A.M.P. workshops provide strategic guidance for the application of intensive management and information collection techniques to the threatened plants. They also provide a comprehensive means of testing the applicability of the I.U.C.N. criteria to the threatened taxa. The Foundation for Revitalization of Local Health Traditions (F.R.L.H.T.) and the University of Trans-disciplinary Health Sciences and Technology (T.D.U.) anchored C.A.M.P. workshops in 19 states across the country and assessed 354 medicinal plants with I.U.C.N. threat status. Of these 47 species of endemic medicinal plants were appended into the I.U.C.N. Database (Anonymous 2020b). The details of the C.A.M.P. workshops are provided in Table 8.3.

To illustrate species above near-threatened status, analysis of 1539 medicinal plant species showed 24 medicinal plant species to be critically endangered, 26 species endangered and

TABLE 8.3
Summary of the C.A.M.P.s Held in 19 States of the Country by F.R.L.H.T.

Sl. No.	State	No. of Red-Listed Species with Assessed Conservation Status	Year and Location of C.A.M.P. Workshop (1995–2017)
1	Andhra Pradesh	47	2001 at Hyderabad
2	Arunachal Pradesh	44	2003 at Guwahati
3	Assam	16	2003 at Guwahati
4	Chhattisgarh	47	2003 at Bhopal
5	Himachal Pradesh	62	1998 at Kullu, 2003 at Shimla
6	Jammu and Kashmir	62	1998 at Kullu, 2003 at Shimla
7	Karnataka	81	1995, 1996, 1997, 1999 all at Bangalore
8	Kerala	85	1995, 1996, 1997, 1999 all at Bangalore
9	Madhya Pradesh	50	2003, 2006 both at Bhopal
10	Maharashtra	35	2001 at Pune
11	Meghalaya	25	2003 at Guwahati
12	Nagaland	28	2015 at Dimapur
13	Orissa	40	2007 at Bhuvaneshwar
14	Rajasthan	38	2007 at Jaipur
15	Sikkim	24	2003 at Guwahati
16	Tamil Nadu	80	1995, 1996, 1997, 1999 all at Bangalore
17	Tripura	21	2016 at Agartala
18	Uttaranchal	60	2003 at Shimla
19	West Bengal	43	2007 at Kolkata

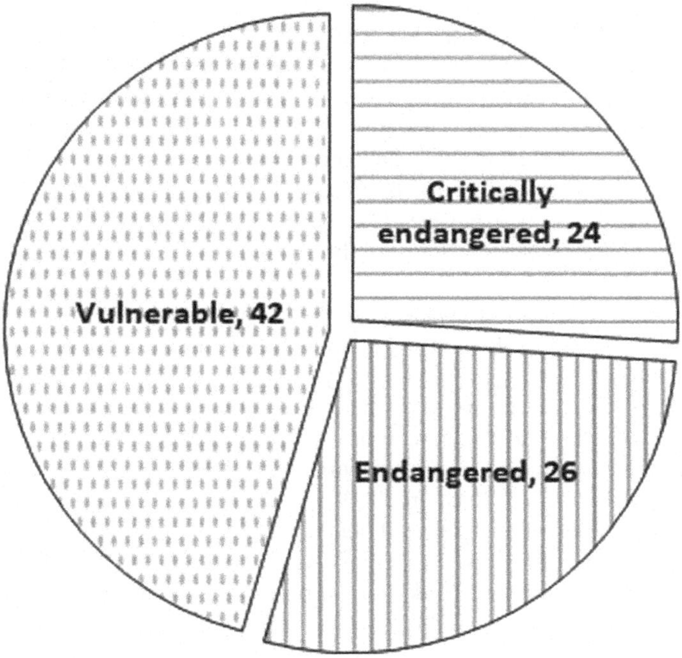

FIGURE 8.4 Threatened medicinal plant species.

42 species as vulnerable (see Figure 8.4). Selected traded medicinal plants of conservation concern are: *Aconitum heterophyllum*, *Boswellia serrata* Roxb., *Chlorophytum borivilianum* Linn., *Cochlospermum religiosum* (Linn.) Alston., *Gymnema sylvestre* R. Br., *Oroxylum indicum* Vent., *Pterocarpus marsupium* Roxb., *Rauvolfia serpentina*, *Taxus wallichiana* Zucc., *Sterculia urens* Roxb. and *Stereospermum. chelonoides* (Linn. f.) DC.

8.5 CONSERVATION

8.5.1 *In Situ* Conservation

Most medicinal plants are endemic species, and their medicinal properties are mainly because of the presence of secondary metabolites that respond to stimuli in natural environments, and may not be expressed under culture conditions (Figueiredo et al. 2009; Coley et al. 2003). *In situ* conservation is the process of conserving the living species, especially the wild and endangered species in their natural habitats and environment. *In situ* conservation of biodiversity includes biosphere reserves and national parks (Hamilton and Hamilton 2006). *In situ* conservation of whole communities allows us to protect indigenous plants and maintain natural communities, along with their intricate network of relationships (Gepts 2006).

8.5.2 *Ex Situ* Conservation

Ex situ conservation is the conservation of biological diversity outside their natural habitats in locations that imitate their natural habitats. This involves the conservation of genetic resources, as well as wild and cultivated species. The approach draws on a diverse array of techniques and facilities, including seed banks *in vitro* plant tissue and microbial culture collections, artificial propagation of plants, with possible reintroduction into the wild and botanic gardens (Maunder et al. 2004).

Conservation of plant genetic resources can be achieved *in situ* as well as *ex situ*. Both cultivated and domesticated plant species are also maintained in their natural habitats as well as in field conditions (Cruz-Cruz 2013; Langhu and Deb 2014). Due to habitat destruction and transformation of the natural environment, several species have been lost from the ecosystems. Therefore, in situ methods alone are insufficient for conserving the threatened species. Under these circumstances, *ex situ* conservation is a viable alternative for preventing the extinction of threatened species. In some cases, it is the only viable strategy to conserve certain species. *In situ* and *ex situ* methods are complementary and not mutually exclusive. The selection of appropriate strategy should be based on a number of criteria including the status of the species and feasibility of applying the chosen methods (Engelmann 2012).

Ex situ conservation is not always sharply separated from *in situ* conservation, but it is an effective complement to it, especially for those over-exploited and endangered medicinal plants with slow growth, low abundance and high susceptibility to replanting diseases (Hamilton 2004; Havens et al. 2006; Yu et al. 2010). *Ex situ* conservation aims to cultivate and naturalize threatened species to ensure their continued survival and sometimes to produce large quantities of planting material used in the creation of drugs, and it is often an immediate action taken to sustain medicinal plant resources (Swarts and Dixon 2009; Pulliam 2000).

Many species of previously wild medicinal plants can not only retain high potency when grown in gardens far away from the habitats where they naturally occur but can have their reproductive materials selected and stored in seed banks for future replanting (Hamilton 2004). For example, F.R.L.H.T. established several *ex situ* conservation sites to complement *in situ* conservation. *Ex situ* conservation was undertaken to improve the livelihood of indigenous people and enhance the use of plants through the establishment of Medicinal Plants Conservation Parks (M.P.C.P.s). It comprises nurseries, the establishment of living collections of a limited number of specimens of the medicinal plants and promotion of kitchen herbal gardens and home herbal gardens (Singh et. al. 2008).

8.5.3 Medicinal Plant Conservation Areas (M.P.C.A.s)

M.P.C.A.s are managed as "hands off" areas with only the following interventions, wherever required – fire management, soil and moisture conservation and weed management/encouraging native vegetation. On-field research, collection of germplasm for research and multiplication and right of way and water to the local communities are also allowed. All harvesting operations, thus, stand suspended in the M.P.C.A.s. M.P.C.A.s are sites with known medicinal plant richness, based on literature and local interaction and are less disturbed, but easily accessible. They are relatively free from local rights and livelihood issues. They form compact manageable units and encompass different forest or vegetation types and altitude ranges.

During 1993–2019 TDU guided the Ministry of Environment, Forests and Climate Change; State Forest Departments and N.M.P.B. technically to establish the world's largest network of 210 *in situ* conservation areas for conserving wild gene pools of medicinal botanicals. These conservation sites are called M.P.C.A.s (Medicinal Plants Conservation Areas). Most M.P.C.A.s are of the average size of around 200 hectares. The 210 M.P.C.A.s are located across 21 States spread in the North, South, East and West of India. Around 90 M.P.C.A.s were designed and created by N.M.P.B. during 2005–2015 and the rest by State Forest Departments under the technical guidance of F.R.L.H.T.-T.D.U. This is the largest *in situ* conservation program in the entire tropical world.

Of the 210 M.P.C.A.s in India, the first 34 M.P.C.A.s were established during 1993–1997 with the support of D.A.N.I.D.A. (Danish International Development Agency) in southern Indian states of Karnataka, Kerala and Tamil Nadu. These M.P.C.A.s were created in forest areas with rich floristic diversity. Later, species-specific M.P.C.A.s were established to capture the viable wild population of threatened medicinal plants. For example, Kollur M.P.C.A. was established in the Udupi district of

Karnataka in collaboration with the State Forest Department of Karnataka. This M.P.C.A. is located close to the famous Mookambika temple and is spread over 300 hectares. Along with *Saraca asoca*, more than 20 species of threatened red-listed plants and around 200 species of other wild medicinal plant species occurring in the M.P.C.A. are also being conserved.

The M.P.C.A. established at Kollur, for long-term *in situ* conservation of the wild gene pool of *Saraca asoca* has been an important highlight of the pioneering medicinal plant conservation program initiated by F.R.L.H.T. in southern India. Similar efforts, for other threatened or highly traded medicinal plants of southern India, have resulted in the establishment of the Anappadi M.P.C.A. (Kerala) to conserve *Utleria salicifolia* Bedd. ex Hook.f., Kulamavu M.P.C.A. (Kerala) for *Coscinium fenestratum* (Goetgh.) Colebr. and Nambikoil M.P.C.A., near the Kalakad Mundanthurai Tiger Reserve (K.M.T.R.) for *Decalepis arayalpathra* (J. Joseph & V. Chandras) Venter. In total, 74 M.P.C.A.s were established under the Country Cooperation Framework I project, the Country Cooperation Framework II project and the Global Environment Facility project and 12 M.P.C.A.s were established with the support of N.M.P.B. In order to conserve the important threatened, endemic and highly traded medicinal plants there is a need to establish a nationwide network of M.P.C.A.s across different forest types. More than one M.P.C.A. will have to be established across the range of distribution of the species, to capture the viable population and conserve the gene pool.

8.5.4 Success Stories of Conservation

Wild medicinal plant species face various types of threats such as loss of habitat due to fragmentation and exotic plants, over-exploitation, human interference, disease, predation, landslides and trade. Understanding the bottlenecks in the reproductive biology of these species, followed by an informed *in situ* conservation action, can contribute to their recovery and long-term conservation. Medicinal species can be successfully protected under the *in situ* conservation program in reserve forest areas where they occur naturally. Three such examples are presented here.

8.5.4.1 *Commiphora wightii* (Arn.) Bhandari

Commiphora wightii is one of the high-value medicinal plants and an important endangered medicinal plant of dry regions of India. The oleoresin of this plant is extensively used in the Ayurveda, Siddha and Unani systems of medicine. It is widely used to control cholesterol and obesity. Unscientific tapping methods, in order to increase yield of oleo-gum resin to meet the increasing market demand have caused mortality of plants, leading to the near-extinction of this species.

Its common names are *Guggul* (Hindi), *Guggulu* (Sanskrit), *Kukkulu* and *Mahiṣākṣi* (Tamil). It is critically endangered (Rajasthan) according to A2c, d; A4c, d ver. 3.1 of the I.U.C.N. red list categories and criteria (Reddy et al. 2012). In India, its wild populations occur in the arid, rocky regions of Rajasthan and Gujarat, along with a very limited presence in Maharashtra and Madhya Pradesh. Annual domestic consumption of guggul gum by herbal industries was estimated at 1000–2000 M.T. for the years 2014–2015 (Goraya and Ved 2017). The major portion of this was reportedly imported from Pakistan. Oleo-gum resin obtained from other species of *Commiphora* (*Balsamodendron*) are the adulterants.

8.5.4.2 *In Situ* Conservation

Under the N.M.P.B.-funded project, Gujarat State Forest Department established M.P.C.A.s at Tharavada and Mathal. Both of these M.P.C.A.s harbor sizeable wild populations of this species. Rajasthan State Forest Department has established two M.P.C.A.s in dry, deciduous and thorn forests at Barkochra M.P.C.A., Ajmer Forest Division and Gajroop Sagar M.P.C.A. in Jaisalmer District under the Government of India–U.N.D.P. project coordinated by F.R.L.H.T.

N.M.P.B. has supported projects through Department of Environment & Forest, Gujarat State Government during the financial years 2007–2008 and 2010–2011 with the objective of establishing Medicinal Plants Conservation Development Areas (M.P.C.D.A.s). Six M.P.C.D.A.s, namely Mangvana, Gugliayna, Tharvada, Ler, Mathal and Kurboi Nabahoi in Kachchh circle (Gujarat) were established to cover an area of about 1200 hectares. Many activities such as base line survey, inventory and mapping work, establishment of M.P.C.D.A.s for guggal, raising of seed production areas for getting seeds, awareness generation among forest staff, farmers and local community through various training programmes were initiated. The wild population of guggul was augmented to be used in future for the raw drug, gene bank and research and development work. Apart from these, various activities were also initiated for raising of nurseries, area development, soil moisture conservation and fencing of the area. During the project period eight training programmes were organized for farmers and forest staff for mass scale cultivation of guggul in Gujarat State.

8.5.4.3 *Decalepis hamiltonii* (Wight & Arn.)

Decalepis hamiltonii is the only species without a cousin in the botanical world. Over-exploitation and destructive harvesting of the roots are the major threats to the plant, causing population reduction. Its vernacular names are *Śariba*, *Śvēta śāriva* (Sanskrit), *Māgāḷi bēru* (Kannada), *Nannāri* (Malayalam), *Mākāḷi kiḷaṅgu* (Tamil) and *Māreḍu geḍḍalu* (Telugu). It is endangered under A2c, d of ver. 3.1 of the I.U.C.N. red list categories and criteria (Mishra et al. 2017). It is endemic to the Deccan region and occurs in the hilly tracts of Eastern and Western Ghats of Andhra Pradesh, Karnataka and Tamil Nadu. It is sparsely seen in Kerala. Annual domestic consumption of *Śvēta śāriva* by herbal industries was estimated at 200 M.T. for the year 2014–2015 (Goraya and Ved 2017). It is extensively used for the preparation of pickles in Andhra Pradesh, Karnataka and Tamil Nadu. Most of the herbal industries use this species as a substitute for *Hemidesmus indicus* (L.) R. Br. ex Schult.

8.5.4.4 *In Situ* Conservation

A sizable wild population is being conserved in Savanadurga M.P.C.A. in Bangalore rural district and Devarayanadurga M.P.C.A. in Tumkur district by Karnataka Forest Department.

8.5.4.5 *Saraca asoca* (Roxb.) W. J. de Wilde

Saraca asoca is one of the most sacred plants of Hindus and a boon for women to treat disorders related to menstruation and fertility. Nevertheless, in view of the high quantum of use of the bark of this tree species by the Indian herbal industry (>2000 M.T. per year), it seems highly improbable that any significant proportion of such large quantity of bark could be obtained from this species, which exists only as an avenue tree and is not seen with a sizeable wild or planted population in India.

Its vernacular names are *Aśōka* (Hindi), *Aśōka*, *Hēmapuṣpa* (Sanskrit), *Aśōka mara* (Kannada), *Aśōkam* (Malayalam), *Aśōka maram* (Tamil) and *Aśōkamu* (Telugu). This herb is endangered (Karnataka) A1c, d ver. 2.3 and critically endangered (Orissa) A2c, d ver. 3.1 of the I.U.C.N. red list categories and criteria (Patwardhan et al. 2014). Globally the species occurs wild only in the Indian subcontinent and in Sri Lanka. Within India its wild population is found in Karnataka, Orissa, Goa and sporadically in Andhra Pradesh, Kerala, Tamil Nadu, Meghalaya and Mizoram. Annual domestic consumption of *Aśōka chāl* (bark) by herbal industry was estimated at 2000 M.T. for the year 2014–2015 (Goraya and Ved 2017). It is, however, reportedly substituted/adulterated with plant materials obtained from *Humboldtia vahliana* Wight (Caesalpiniaceae) (N. Sasidharan, K.F.R.I., personal communication), *Shorea robusta* Gaertn. (Dipterocarpaceae) and *Mallotus nudiflorus* (L.) Kulju & Welzen (Euphorbiaceae) (Noorunnisa Begum et al. 2014).

8.5.4.6 *In Situ* Conservation

Karnataka State Forest Department in collaboration with F.R.L.H.T. has established Kollur M.P.C.A. in Udupi which harbors wild populations of this species.

8.5.5 Medicinal Plants Development Areas (M.P.D.A.s)

During implementation of participatory conservation projects, it is often realized that in addition to intangible benefits through implementation of M.P.C.A. programs like soil and water conservation, climate improvement and ethical values, people's participation in medicinal plants conservation efforts will be limited if direct material benefits are not adequately provided to participating village communities. Therefore, when the D.A.N.I.D.A. project *Strengthening the Medicinal Plants Resource Base in Southern India in the Context of Primary Health Care* (1993 to 2004) was initiated, it was decided to take degraded forest patches near M.P.C.A.s available under Joint Forest Management guidelines as medicinal plants development areas (M.P.D.A.s) (Singh et. al. 2008).

The objective of the M.P.D.A. program was to develop a model for conservation, development and sustainable use of plant resource under participatory forest management. Therefore, this program was aimed at providing economic benefits to communities in conservation and management of the medicinal plant resources in M.P.D.A.s. A total of 12 M.P.D.A.s were established. Four were in Karnataka and eight in Tamil Nadu. The area of each M.P.D.A. ranged from 10 to 50 ha. As degraded forest areas were not available near most of the M.P.C.A.s, the M.P.D.A.s were established in areas at some distance from the M.P.C.A.s. In such cases the local communities participating in the M.P.D.A. program were different from those participating in the M.P.C.A. program. The idea behind undertaking the M.P.D.A. program was that this approach would be useful for adoption under the Joint Forest Management program being implemented on a large scale throughout the country (Singh et. al. 2008).

Four M.P.D.A. models were attempted. In one, degraded forest areas were taken up for eco-restoration through planting of trees and herbs of medicinal value. In the second, natural bushy vegetation was cleared and the cleared area planted with a variety of medicinal plants. In the third, essential oil-bearing plants already growing in the area were encouraged to grow and enrichment planting was carried out. The material was harvested to distill oil for sale. The fourth model was tried in Tamil Nadu toward the end of project period under which areas already under Joint Forest Management were considered for augmentation and simultaneous harvesting of medicinal plants to benefit the participating village communities (Singh et. al. 2008).

The first model was found to be emphasizing eco-restoration rather than medicinal plant development and was too expensive to be replicable. The expected benefits to village communities were not at par with the expenditure incurred. The second model suffered from several drawbacks. Clearing of natural vegetation removed the medicinal plants already growing in that area. Clearing reduced biodiversity and increased soil erosion on slopes. Clearing and planting also increased expenses and reduced net benefits to the local communities. The third model was the most successful of all the four models tried.

After the project was over, the sustainability aspect of the third M.P.D.A. attracted much attention. The Medicinal Plants Development Agency was formed in place of the M.P.D.A. management committee and a new M.o.U. was signed between D.A.N.I.D.A. and Tamil Nadu Forest Department in August 2007 to jointly manage and sustain the M.P.D.A. Several factors contributed to the success of the third M.P.D.A. First of all, the village community participating in the M.P.D.A. already had the necessary skills in raising aromatic plants and their distillation. The members having become unemployed after the closure of the Cinchona Department were badly in need of employment which was provided with the establishment of the M.P.D.A. A coherent village community facilitated formation of village-level institutions for the M.P.D.A. and subsequent activities. The approach of converting the M.P.D.A. produce into high-value products like essential oils and processed herbs for marketing also contributed to the success. Abundant availability of dry *Eucalyptus* leaves from nearby *Eucalyptus* plantations free of cost for extraction of *Eucalyptus* oil was in fact a boon. Support and guidance provided by Tamil Nadu Forest Department and the non-governmental organization H.O.P.E. (Health of People and Environment) was also an important factor (Singh et. al. 2008).

The approach adopted in the fourth model demonstrated that by utilizing the strengths of Joint Forest Management, there was good scope for medicinal plant resource augmentation. This approach also effectively involved women's self-help groups, micro-credit groups, local healers, non-governmental organizations and educational institutions. This approach showed great potential, but as this model was tried toward the end of project period, no definite conclusions could be drawn before the end of the project

The M.P.D.A. program aimed at compensating village communities participating in protection and management of M.P.C.A.s with some income generation from the development and harvesting of medicinal plants. The program also developed a workable model of M.P.D.A. for Joint Forest Management areas protected and managed with the involvement of local communities. This is a model that can be applied throughout the country because it ensures people's participation in conservation and development of medicinal plants and substantially contributes to the welfare of the participating communities (Singh et. al. 2008).

8.5.6 Botanic Gardens

Botanic gardens conserve and propagate rare species and genetic diversity. They play an important role in *ex situ* conservation (Havens et al. 2006), and they can maintain the ecosystems to enhance the survival of rare and endangered plant species (Huang et al. 2002). Although living collections generally consist of only a few individuals of each species and so are of limited use in terms of genetic conservation (Yuan et al. 2010), botanic gardens have multiple unique features. They involve a wide variety of plant species grown together under common conditions, and often contain taxonomically and ecologically diverse flora (Primack and Miller-Rushing 2009). Botanic gardens can play a further role in medicinal plant conservation through the development of propagation and cultivation protocols, as well as undertaking programs of domestication and variety breeding (Maunder et al. 2001).

Threatened plant species are conserved at field germplasm banks such as I.C.A.R.-N.B.P.G.R. National Gene Bank and institutional botanical gardens such as National Botanical Research Institute (N.B.R.I.) Botanic Garden, Lucknow and Jawaharlal Nehru Tropical Botanical Garden, Thiruvananthapuram, Kerala. F.R.L.H.T.-T.D.U., Bangalore has established a unique ethnomedicinal garden with more than 1500 native medicinal plants supported with a nursery. It conserves some of the important threatened and traded medicinal plant species (Anonymous 2020c). The Arya Vaidya Sala Herb Garden, by Arya Vaidya Sala, Kottakkal in Kerala has a demonstration garden set up in an eight-acre plot at Kottakkal and a live collection of 700 scientifically identified medicinal plants (Anonymous 2020d).

8.5.7 Seed Banks

Seed banks offer a better way of storing the genetic diversity of many medicinal plants *ex situ* than through botanic gardens, and are recommended to help preserve the biological and genetic diversity of wild plant species (Li and Pritchard 2009; Schoen and Brown 2001). The most noteworthy seed bank is the Millennium Seed Bank Project at the Royal Botanic Gardens in Kew, Britain (Schoen and Brown 2001). It is the largest *ex situ* conservation program in the world, presently involving 96 countries and territories. Where possible, seeds are collected and conserved in the country of origin with duplicates being sent to this seed bank for storage. Unique taxonomic diversity exists amongst the collections which represent 365 families, 5813 genera, 36,975 species and 39,669 taxa conserved (Liu et al. 2018).

Seed banks allow relatively rapid access to plant samples for the evaluation of their properties, providing helpful information for conserving the remaining natural populations (Li and Pritchard 2009; Schoen and Brown 2001). The challenging tasks of seed banking are how to reintroduce the plant species back into the wild and how to actively assist in the restoration of wild populations (Li and Pritchard 2009). An example is a community seed bank project aimed at identifying important traditional seed varieties and orienting the agricultural community toward conserving and cultivating them (Anonymous 2020e).

8.5.8 Cultivation

Although wild-harvested resources of medicinal plants are widely considered more efficacious than those that are cultivated, domestic cultivation is a widely used and generally accepted practice (Gepts 2006; Joshi and Joshi 2014; Leung and Wong 2010). Cultivation provides the opportunity to use new techniques to solve problems such as toxic components, pesticide contamination, low contents of active ingredients and the misidentification of botanicals encountered in the production of medicinal plants (Raina et al. 2011).

Cultivation under controlled growth conditions can improve the yields of active compounds, which are almost invariably secondary metabolites, ensuring production stability. Cultivation practices are designed to provide optimal levels of water, nutrients, optional additives and environmental factors including temperature, light and humidity to obtain improved yields of target products (Liu et al. 2011; Wong et al. 2014). Moreover, increased cultivation decreases the harvest volume of medicinal plants, benefits the recovery of their wild resources and decreases their prices to a more reasonable range (Hamilton 2004; Larsen and Olsen 2007; Schippmann et al. 2005).

An example of cultivation is provided by Aryavaidya Sala Kottakkal of Kerala. Over 200 acres of medicinal plant estates are being maintained at Mannarghat, Kottapupram, Thrikkakara and Kottakkal, where large-scale cultivation of rare plant species is organized. These estates also support scientific activities by providing trial cultivation and maintenance of field gene banks (Anonymous 2020d).

N.M.P.B. was established by the Government of India to encourage the cultivation of medicinal plants and their sustainable management, so that dependence on forests for the collection of herbs can be reduced. The primary function of the N.M.P.B. is to develop a proper mechanism for co-ordination between various ministries, departments, organizations and implementation of support policies and programs for conservation, cultivation, trade and export of medicinal plants. N.M.P.B. promotes cultivation of medicinal plants and offers support through a centrally sponsored scheme of the National Mission on Medicinal Plants (N.M.M.P.) since 2008. This support is now continuing under the National A.Y.U.S.H. Mission (N.A.M.), a flagship program launched by the Ministry of A.Y.U.S.H., Government of India during the XII Plan period. The program is being implemented in the country through state government-designated agencies. To meet the increasing demand for medicinal plants, N.M.B.P. focusses on *in situ* and *ex situ* conservation. It also promotes research and development and capacity-building through training and raises awareness through promotional activities like the creation of home and school herbal gardens. The Board also supports programs for quality assurance and standardization, by developing Good Agricultural and Collection Practices (G.A.C.P.s) (Anonymous 2020f).

8.6 MEASUREMENT AND MONITORING OF BIODIVERSITY

8.6.1 Pilot-Scale Measurements

Pilot-scale measurements of biodiversity were carried out in the past using different indices like species richness, the Simpson index and the Shannon-Wiener index (Singh 2002). Among them, species richness is the simplest measure of species diversity. This is done by counting the number of species in a plot or a community. The Simpson and Shannon-Wiener indices are measures that account for richness and allocation of individuals among species. Data on both number of species and number of individuals of each species are needed to calculate these indices (Singh and Kushwaha 2008).

8.6.2 GIS System – A Conservation Tool

Large area coverage of satellite imagery is a new tool for biodiversity assessment (Fuller et al. 1998, Kushwaha et al. 2000, Nagendra and Gadgil 1999). Vegetation-type maps generated from satellite imagery are the first important inputs for a two-stage biodiversity inventory at landscape level (Roy and Tomar 2000; Behera et al. 2006).

The forestry and ecology division of Indian Institute of Remote Sensing, Dehradun, developed a methodology for rapid assessment of biodiversity, encompassing large natural vegetation areas in India using a three-pronged approach (Singh and Kushwaha 2008). The technique makes use of satellite imagery to generate homogeneous vegetation strata and landscape analysis. Landscape parameters like fragmentation, patchiness (Romme 1982), porosity (Forman and Godron 1986), interspersion (Lyon 1983), juxtaposition (Lyon 1983) and proximity of the vegetation patch to biotic disturbance features, such as roads, railways and settlements, are then considered to derive the disturbance index. This is followed by field assessment by the Shannon-Wiener index of diversity in different vegetation strata and evaluation of the vegetation community for its uniqueness and determination of its biodiversity value following Belal and Springuel (1996).

The approach takes into account the terrain complexity, which plays an important role in biodiversity development. The final output, the biological richness is calculated as a function of disturbance index, terrain complexity, biodiversity value, species richness and ecosystem uniqueness. The non-spatial field data are converted into spatial data in the geographic information system (G.I.S.) domain, by assigning values ranging from 1 to10. The resultant output is scaled to four classes like very high richness, high richness, medium richness and low richness, representing plant richness across the district, state or region. Sampling of biodiversity saves considerable time and cost otherwise needed for such an inventory using ground-based methods alone. A Windows-based software module of the ARC/INFO, BioCAP was developed to aid the landscape analysis (Anonymous 2002; Singh and Kushwaha 2008).

This G.I.S. methodology, developed in 1998, was field tested extensively for biological richness assessment in North-East India (262,179 km^2), western Himalayas (339,575 km^2), Western Ghats (260,962 km^2) and the Andaman-Nicobar Islands (8249 km^2) between 1999 and 2001. It provides an enormous quantity of maps and tabular data. During phases 1 and 2 of this project, more than 10,000 plots were sampled and a detailed species database was created (Anonymous 2002; Singh and Kushwaha 2008).

Application of G.I.S. technology in predictive mapping was recently attempted by Bhandari et al. (2020). The population of *Rhododendron arboreum* Sm. is shrinking in the middle Himalayas due to low seed viability, poor regeneration, habitat fragmentation, habitat distortion and species invasion. The authors attempted to predict *R. arboreum* distribution in Uttarakhand using the MaxEnt model. The MaxEnt software is particularly popular in species distribution and environmental niche modeling, with over 1000 applications published since 2006 (Phillips et al. 2006; Merow et al. 2013).

A total of 1077 geospatial data was recorded and 300 well-distributed geo-coordinates were used to predict and estimate the distribution. The remaining data were used to validate the model. The MaxEnt model furnished an A.U.C. curve with an accurate and significant value of 0.886 $\pm$ 0.023. Bioclimatic variables like temperature seasonality, annual temperature range, altitude, annual precipitation and precipitation seasonality significantly contributed to the prediction of the distribution using the Jackknife test. In the total geographical area of 617.48 km^2 under *R. arboreum* distribution as shown by Landsat 8-generated map, 167.48 km^2 were found to be very dense, 320.75 km^2 were moderately dense and 129.25 km^2 were open. This study shows that satellite-based mapping and model-based prediction of plant species are of great importance to conservation biologists and foresters for species conservation, management and sustainable utilization (Bhandari et al. 2020).

8.7 MAPPING OF PLANT DISTRIBUTION

Understanding the natural distribution and eco-climatic limits of the fast-disappearing taxa can help conservation biologists to formulate strategies for their conservation (Ganeshaiah and Uma Shaankar 1998). With the help of data on distribution patterns and availability of the species, the causes for their reduction in number and rarity can be discovered. Systematic mapping of the occurrence of the species can also help identify regions where conservation has to be initiated. Such maps provide information on the extent of protection required and how effectively it could be carried out (Ved et al. 1998).

The process of developing the maps began with prioritization of wild medicinal plants of southern India, based on data related to their trade. This included volume, value and plant parts or products in trade. Endemism and reported rarity were also considered. A short list of nearly 300 prioritized species was finalized and data on them were collected from the published literature, herbarium records and M.P.C.A. data (Ved et al. 1998).

The survey of India has 1:1 million scale digital maps of state boundaries, rail and road networks, rivers and lakes of the country. Based on such maps, distribution maps were generated using MapInfo Professional v1.0 software. The presence of the species was indicated in the distribution maps, using black, red and blue flags. Black flags represented the herbarium specimens collected from the 22 herbaria visited in connection with the study. Red flags indicated collections from M.P.C.A. network. Locations mentioned in the literature were indicated by blue flags. Using these databases, three types of maps were developed for 60 species of medicinal plants (Ved et al. 1998).

8.7.1 Global and Indian Distribution Maps

Country-level occurrence of a species was depicted in the world map, included as inset in the Indian distribution map. State-level presence of the species was shown on the Indian map. In the case of endemic species with very restricted distribution, like *Janakia arayalpathra* J. Joseph & V. Chandrasekaran and *Pterocarpus santalinus* Linn. f., the districts of occurrence were marked on the Indian map. Indian distribution maps were based on data compiled from published floras of the respective states. The maps thus generated serve as ready-reckoner for information on the distribution of particular plant species in the country and neighboring countries (Ved et al. 1998).

8.7.2 Regional Distribution Maps

Maps were prepared for southern India, covering Kerala, Karnataka and Tamil Nadu. They show the district-level occurrence of the species, based on literature survey and recorded collections in herbaria. From a survey of the species mapped, it is observed that most of them occur in the Western Ghats biogeographic zone, which is a major evergreen and deciduous forest region of India, with great biological diversity (Ved et al. 1998).

8.7.3 Eco-Distribution Maps

These maps were generated by superimposing layers of altitude range and rainfall range over the geographical distribution of the species. Such superimposing of altitude and rainfall layers on the distribution maps provides valuable information on the ecological factors that influence the natural occurrence of the species in question (Ved et al. 1998).

Eco-distribution maps are attempted for mapping distribution, occurrence and population of critical species of conservation concern, like *Saraca asoca* and *Coscinium fenestratum*. This is not a one-time effort but a continuous upgradation process, based on research and botanical field experience. Incorporating information on precise geographical locations from literature, herbaria and field, correlated with ecological parameters like altitude range, rainfall range and soil type provides an understanding of the pattern of natural distribution of the species. F.R.L.H.T. has created such eco-distribution maps for 450 rare, endemic and threatened species. For educating the society on medicinal plants, 2000 geo-maps for threatened and traded species, prioritized from different bio-geographical regions of India have been created and disseminated as the Digital Atlas of Medicinal Plants, under the Center of Excellence program of Ministry of Environment, Forest and Climate Change. This activity has widened the distribution database of the Indian Medicinal Plant Database of F.R.L.H.T.-T.D.U.

8.8 SUSTAINABLE HARVESTING

Destructive collection practices are one of the major factors influencing the depletion of plant resources in the wild. Lack of awareness of good collection practices, growing industrial demand for the wild resources, weak guidelines and monitoring mechanisms for wild resource collection and management, competition among the local collectors, non-availability of better prices or incentives for primary collectors and an insufficient policy environment are some of the reasons for destructive collection. Unscientific collections from the wild had led to the threat of species extinction and inflicted severe genetic impoverishment among the wild populations. Sustainable harvesting can improve the livelihoods of people by ensuring a continued supply of biomass through supplementary income and employment (Leaman 2006).

The simplest definition of sustainable harvesting is "the use of plant resources at the levels of harvesting in such a way that the plants are able to continue to supply indefinitely" (Wong et al. 2001). It places an emphasis on maintenance of the population of the species in the wild, irrespective of high demand from across the globe. It is important to conserve the populations of many commercially exploited species in the wild, which face the threat of extinction on the grounds of cultural, ecological and commercial pressures.

Good collection practices or sustainable collection practices are the method of extraction of non-timber forest produce (N.T.F.P.)/medicinal plants or wild resources from the forest areas without causing any damage to their reproduction, while also avoiding damages to their associates. Applying sustainable collection practices in the wild is significant in conservation of resources and fulfilling the needs of forest-dependent communities and other stakeholders, who directly or indirectly benefit from the harvesting of N.T.F.P.s including medicinal plants (Anonymous 2009b).

8.8.1 Success Story

8.8.1.1 *Terminalia chebula* Retz.

Terminalia chebula, commonly known as myrobalan, is a deciduous tree, fruits of which have high medicinal value. Apart from this, the fruits of chebula are used in the tanning industry (Auad et al. 2020). In India, more than 10,000 M.T./year is traded. Due to the high demand for these fruits in the herbal as well as tanning industries, the fruits are collected following destructive harvesting methods such as lopping the branches and plucking immature fruits. This results in increased mortality of the trees and decreased regeneration, thereby lowering the fruit yield year after year. Regeneration in *Terminalia chebula* is also difficult, as the percentage of seed germination is very low. As it is a tree species, fruiting starts at the age of 10–15 years (Bawa 1974).

A pilot project was undertaken in the state of Karnataka to study the quantity of fruits available for collection, development of sustainable harvesting technique and marketing of the collected fruits (Anonymous 2012). The study area was mapped at the onset of the study for understanding the resource availability. Later traditional collection practices were documented. By merging traditional collection practices with modern scientific techniques, sustainable harvesting techniques were developed for collection of the fruits. The sustainable harvesting techniques emphasized collection of 80% mature or fallen fruits. During the process of drying, it was observed that the fruits were turning black. This deteriorates the quality of the fruits and the collector gets lower price in the market. Training sessions were organized to orient the collectors on sustainable and efficient post-harvesting techniques. The intervention made to dry the fruits on a rock or concrete platform helped to retain the golden colour of the fruits, thereby ensuring their quality.

A market survey was undertaken to determine marketing and pricing of the fruits. Before the project intervention, local traders were purchasing at Rs. 5–6 for a kilogram (kg) of fruits. After implementation of sustainable harvesting and post- harvesting techniques such as drying on the rock and grading the fruits based on size and colour, the fruits were directly marketed to an herbal industry in Kerala through the Forest Development Agency, which is a government enterprise, for collection and marketing of N.T.F.P. and medicinal plants. After deducting the expenditure on transportation and other administrative costs, each collector received Rs. 10.35 per kg of fruit. This study demonstrated >75 % price appreciation by adopting sustainable harvesting and post-harvesting methods (Deepa et al. 2019).

8.9 CLASSICAL WAYS OF CONSERVING AYURVEDIC HERBS

Pratinidhi Dravya (substitutes for herbal drugs) are mentioned in ayurvedic classical texts when the original drug is not available. In Ayurveda there are many drugs combined in a single formulation. Some of them are difficult to procure and some species disappeared with advancement of civilization and industrialization. This resulted in scarcity and poor availability of ingredients for the formulations, causing reduction in their efficacy. Hence ancient seers of Ayurveda advised some drugs which can be used when drugs with similar properties are not available. Such drugs are known as *Pratinidhi Dravya* or substitute drugs. This concept is based on ayurvedic principles. Post-16th-century Ayurveda texts and lexicons give specific examples of possible substitutes. One such text is *Bhāvaprakāśa* composed by *Bhāvamiśra* of Varanasi (Pandey 1961). Understanding the logic behind the ayurvedic concept of drug substitution can lead to new methods of identifying legitimate drug alternatives, and help solve the industry's problem of crude drug shortages.

An example is root of *Piper chaba* Hunter non-Blume., commonly known as *Cavya*. It is used in many formulations. But as an alternative, roots of *Asparagus adscendens* Roxb. and roots of *Piper longum* Linn. are used. In the case of *Amḷavētas* (*Garcinia pedunculata* Roxb. or *Rheum emodi* Wall. ex Meissn.) whole plant and leaves of *Rumex vesicarius* Linn. can be used. In texts, *Nāgakēsar* is correlated with the flowers and stamens of *Mesua ferrea* Linn. The alternative herbs recommended are the flowers and stamens of *Nelumbo nucifera* Gaertn. or *Nelumbium speciosum* Willd. Most excitingly, a north-west Himalayan threatened plant *ativisha*, *Aconitum heterophyllum* whose tubers are used, can be replaced by tubers of a common grass *Cyperus rotundus* Linn. (Vidhate et al. 2015).

Some other examples follow, with their substitutes in parentheses: *Viḍaṅga* (*Embelia ribes* Burm. f., *Embelia tsjeriam-cottam* A. DC., *Maesa indica* (Roxb.) A. DC., *Myrsine africana* Linn.), *Raktacandana* (*Pterocarpus santalinus*, *Daphniphyllum himalayense* (Benth.) Müll. Arg.), *Guggulu* (*Commiphora wightii*, *Commiphora berryi* (Arn.) Engl.), *Mañjiṣṭa* (*Rubia sikkimensis* Kurz., *Rubia cordifolia* Linn.), *Dāruharidra* (*Berberis aristata* DC., *Berberis asiatica*, *Berberis chitria*, *Berberis lycium*, *Berberis tinctoria*, *Coscinium fenestratum*, *Mahonia leschenaultia* Wall. Ex. Wight & Arn., *Mahonia napalensis* DC).

8.10 CONCLUSION

Ayurveda makes use of a vast array of medicinal plants. When used in combination, these herbs can cure many diseases and also help in their prevention. These herbs grow in the several biogeographic zones of India. However, habitat degradation, habitat fragmentation, developmental activities and encroachment on forest land for agriculture, urbanization and industrial expansion have put burdens on the natural distribution of these medicinal species. Therefore, there is an urgent need to conserve these species, if Ayurveda is to survive.

The conservation and sustainable use of medicinal plants have been studied extensively (Larsen and Olsen 2007; Uprety et al. 2012). Various recommendations have been made regarding their conservation, including the establishment of systems for species, and the need for coordinated conservation practices based on both *in situ* and *ex situ* strategies (Hamilton 2004). Sustainable use of wild resources can be an effective conservation alternative for medicinal plants with increasingly limited supplies (Chen et al. 2016).

The diversity of plants and animals contained in these special, protected areas is greater than in the radically altered habitats which surround them. Most of the plants in these forests have medicinal value and people frequently enter these areas to harvest such herbs for use. In some countries, such use is punishable by fines and or imprisonment. However, according to I.U.C.N., a better approach is required, one which will enable people to benefit in a sustainable manner from the natural resources that are presently conserved in those areas. I.U.C.N.'s approach to the problem of linking conservation with sustainable development is embodied in the *World Conservation Strategy*, which defines conservation as "The management of human use of the biosphere so that it may yield the greatest sustainable benefit to present generations while maintaining its potential to meet the needs and aspirations of future generations" (McNeely and Thorsell 2009).

The scientific execution of a contemporary, world-class medicinal plant conservation program needs four kinds of prior information. Firstly, knowledge about the medicinal species in high volumes of all-India trade and species that are largely sourced from wild forest habitats is required. This information is available today in the Trans-Disciplinary University from the work of the last two decades. Secondly, it is necessary to analyze species that are endemic, assessed to be under different degrees of threat as per I.U.C.N. criteria and traded in high volume. This information is partially available today, while more analytical and ground work needs to be carried out.

The third requirement is ready access to a database on the occurrence of the medicinal flora in all the 29 states and seven union territories in India. This information is partially available and needs to be enriched. Fourthly, it is essential to have reliable information on the natural geographical distribution of highly traded, endemic and threatened medicinal plant species. This information is largely available but incomplete, in the geographical distribution databases established over the last two decades by the Trans-Disciplinary University.

It is based on the above four kinds of information at state level that forest managers and policy makers can design an M.P.C.A. program focused on highly traded, endemic and threatened medicinal taxa that are state-specific. Despite the existence of various sets of recommendations for the conservation and sustainable use of medicinal plants, only a small portion of these have gained adequate protection through conventional conservation in natural reserves or botanic gardens.

In a nutshell, it may be said that the main reasons behind the existing threats to medicinal plants are loss and degradation of habitat, illegal trade, over-exploitation, over-grazing, human settlements, climatic disasters and avalanches. The need of the hour is to conserve the threatened species by bridging the gap. Review of trade regulations and their implementation, development of cultivation packages, sustainable collection practices, habitat management, conducting of surveys and periodic monitoring across known ranges may prove beneficial to effective conservation of medicinal flora of India.

REFERENCES

Adhikari, D., R. Tiwary, and S. K. Barik. 2015. Modelling hotspots for invasive alien plants in India. *PLoS One* 10, no. 7: e0134665.

Akerele, O. 2009. Medicinal plants: Policies and priorities. In *The conservation of medicinal plants*, ed. O. Akerele, V. Heywood, and H. Synge, 3–11. Cambridge: Cambridge University Press.

Anonymous. 2002. *Biodiversity characterisation at landscape level in North-East India using satellite remote sensing and geographic information system*, 1–296. Dehradun: Indian Institute of Remote Sensing.

Anonymous. 2009a. The Chiang Mai declaration. In *The conservation of medicinal plants*, ed. O. Akerele, V. Heywood, and H. Synge. Cambridge: Cambridge University Press.

Anonymous. 2009b. *Guidelines on good field collection practices for Indian medicinal plants*, 1–34. New Delhi: NMPB-AYUSH.

Anonymous. 2012. *Sustainable harvest, augmentation and marketing of selected medicinal plants raw material (collected & cultivated) through Forest Development Agency (FDA) structure of forest department*, Final technical report of the project. FRLHT. Bengaluru: NMPB.

Anonymous. 2020a. The species survival commission. https://www.iucn.org/species/about/species-survival-commission. (accessed April 8, 2020).

Anonymous. 2020b. The IUCN red list of threatened species. Version 2017-3. http://www.iucnredlist.org (accessed April 14, 2020).

Anonymous. 2020c. Ethno medicinal garden. http://envis.frlht.org/amruthvana/gardenshop.php (accessed April 4, 2020).

Anonymous. 2020d. Herbal gardens. https://www.kottakkal.shop/pages/herbal-gardens (accessed April 4, 2020).

Anonymous. 2020e. The CIKS seed bank project. http://ciks.org/old-site/seedbanks.htm (accessed April 4, 2020).

Anonymous. 2020f. Introduction. https://nmpb.nic.in/content/introduction (accessed 9 April, 2020).

Arisdason, W., and P. Lakshminarasimhan. 2017. Status of plant diversity in India: An overview, Central National Herbarium, Botanical Survey of India, Howrah. http://www.bsienvis.nic.in/Database/Status of Plant Diversity in India 17566.asps (accessed April 4, 2020).

Auad, P., F. Spier, and M. Gutterres. 2020. Vegetable tannin composition and its association with the leather effect. *Chemical Engineering Communications* 207, no. 5: 722–32.

Barik, S. K., O. N. Tiwari, D. Adhikari, P. P. Singh, R. Tiwary, and S. Barua. 2018. Geographic distribution pattern of threatened plants of India and steps taken for their conservation. *Current Science* 114, no. 3: 470–503.

Bawa, K. S. 1974. Breeding systems of tree species of a low land tropical community. *Evolution; International Journal of Organic Evolution* 28, no. 1: 85–92.

Behera, M. D., S. P. S. Kushwaha, and P. S. Roy. 2006. Rapid assessment of biological richness in a part of eastern Himalaya: An integrated three-tier approach. *Forest Ecology and Management* 207, no. 3: 363–84.

Belal, A. E., and I. Springuel. 1996. Economic value of plant diversity in arid environments. *Nature & Resources* 22: 33–9.

Bhandari, M. S., R. K. Meena, R. Shankhwar et al. 2020. Prediction mapping through MaxEnt modeling paves the way for the conservation of *Rhododendron arboreum* in Uttarakhand Himalayas. *Journal of Indian Society of Remote Sensing* 48, no. 3: 411–22.

Chauhan, R. N. 2007. The forest resource base. In *Global forest resources: Geographical approach*, 15–36. Jaipur: Oxford Book Company.

Chen, S. H., H. Yu, H. M. Luo, Q. Wu, C. F. Li, and A. Steinmetz. 2016. Conservation and sustainable use of medicinal plants: Problems, progress, and prospects. *Chinese Medicine* 11: 37. doi:10.1186/s13020-016-0108-7.

Chitale, V. S., M. D. Behera, and P. S. Roy. 2014. Future of endemic flora of biodiversity hotspots in India. *PLoS One* 9, no. 12: e115264.

Coley, P. D., M. V. Heller, R. Aizprua et al. 2003. Using ecological criteria to design plant collection strategies for drug discovery. *Frontiers in Ecology and the Environment* 1, no. 8: 421–8.

Cruz-Cruz, C. A., M. T. Gonzalez-Arnao, and F. Engelmann. 2013. Biotechnology and conservation of plant biodiversity. *Resources* 2, no. 2: 73–95.

Deepa, G. B., A. S. Mark, and R. J. Rao. 2019. Sustainable harvesting of NTFPs and medicinal plants: A participatory process. In *Prospects in conservation of medicinal plants*, ed. A. V. Raghu, M. Amruth, K. V. Muhammed Kunhi, V. P. S. Raveendran, and S. Viswanath, 43–56. Peechi: KSCSTE-KFRI.

Engelmann, F. 2012. Germplasm collection, storage and conservation. In *Plant biotechnology and agriculture*, ed. A. Altman, and P. M. Hasegawa, 255–68. Oxford: Academic Press.

Figueiredo, M. S. L., and C. E. V. Grelle. 2009. Predicting global abundance of a threatened species from its occurrence: Implications for conservation planning. *Diversity and Distributions* 15, no. 1: 117–21.

Forman, R. T. T., and M. Godron. 1986. *Landscape Ecology*, 83–120. New York: John Wiley & Sons.

Fuller, R. M., G. B. Groom, S. Mulish et al. 1998. The integration of field survey and remote sensing for biodiversity assessment: A case study in the tropical forests and wetlands of Sango Bay, Uganda. *Biological Conservation* 86, no. 3: 379–91.

Ganeshaiah, K. N., and R. Uma Shaanker. 1998. Contours of conservation - A national agenda for mapping biodiversity. *Current Science* 75: 292–8.

Gepts, P. 2006. Plant genetic resources conservation and utilization: The accomplishments and future of a societal insurance policy. *Crop Science* 46, no. 5: 2278–92.

Goraya, G. S., and D. K. Ved. 2017. *Medicinal plants in India: An assessment of their demand and supply.* Dehradun: NMPB, AYUSH, ICFR & E.

Hamann, O. 2009. The joint IUCN-WWF plants conservation programme and its interest in medicinal plants. In *The conservation of medicinal plants*, ed. O. Akerele, V. Heywood, and H. Synge, 13–22. Cambridge: Cambridge University Press.

Hamilton, A., and P. Hamilton. 2006. Approaches to *in situ* conservation. In *Plant conservation: An ecosystem approach*, 189–214. London: Earthscan.

Hamilton, A. C. 2004. Medicinal plants, conservation and livelihoods. *Biodiversity and Conservation* 13, no. 8: 1477–517.

Havens, K., P. Vitt, M. Maunder, E. O. Guerrant, and K. Dixon. 2006. *Ex situ* plant conservation and beyond. *BioScience* 56, no. 6: 525–31.

Hrdina, A., and D. Romportl. 2017. Evaluating global biodiversity hotspots – Very rich and even more endangered. *Journal of Landscape Ecology* 10, no. 1. doi:10.1515/jlecol-2017-0013.

Huang, H., X. Han, L. Kang, P. Raven, P. W. Jackson, and Y. Chen. 2002. Conserving native plants in China. *Science* 297, no. 5583: 935.

IMPLAD. 2017. *Database on Indian medicinal plants*. Bangalore: FRLHT.

Joshi, B., C., and R. K. Joshi. 2014. The role of medicinal plants in livelihood improvement in Uttarakhand. *International Journal of Herbal Medicine* 1: 55–8.

Karthikeyan, S. 2000. A statistical analysis of flowering plants of India. In *Flora of India introductory volume, part-II*, ed. Singh, N. P., D. K. Singh, P. K. Hajra, and B. D. Sharma, 201–17. New Delhi: B.S.I.

Kushwaha, S. P. S., M. D. Behera, and P. S. Roy. 2000. Biodiversity characterization using remote sensing and GIS. In *Proceedings of the symposium on biodiversity in India*, Darjeeling, India (8–11 November), 320–5.

Langhu, T., and C. R. Deb. 2014. Studies on the reproductive biology and seed biology of *Aconitum nagarum* Stapf: A threatened medicinal plant of North East India. *Journal of Research in Biology* 4: 1465–74.

Larsen, H. O., and C. S. Olsen. 2007. Unsustainable collection and unfair trade? Uncovering and assessing assumptions regarding Central Himalayan medicinal plant conservation. *Biodiversity and Conservation* 16, no. 6: 1679–97.

Leaman, D. J. 2006. Sustainable wild collection of medicinal and aromatic plants. In *Medicinal and aromatic plants*, ed. R. J. Bogers, L. E. Craker, and D. Lange, 97–107. Heidelberg: Springer Verlag.

Leung, K. W., and A. S. Wong. 2010. Pharmacology of ginsenosides: A literature review. *Chinese Medicine* 5: 20.

Li, D. Z., and H. W. Pritchard. 2009. The science and economics of *ex situ* plant conservation. *Trends in Plant Science* 14, no. 11: 614–21.

Liu, C., H. Yu, and S. L. Chen. 2011. Framework for sustainable use of medicinal plants in China. *Zhi Wu Fen Lei yu ZI Yuan Xue Bao* 33: 65–8.

Liu, U., E. Breman, T. A. Cossu, and S. Kenney. 2018. The conservation value of germplasm stored at the Millennium Seed Bank, Royal Botanic Gardens, Kew, UK. *Biodiversity and Conservation* 27, no. 6: 1347–86.

Lyon, J. G. 1983. Landsat-derived land cover classification for locating potential kestrel nesting habitat. *Photogrammetric Engineering and Remote Sensing* 49: 245–50.

Maunder, M., K. Havens, E. O. Jr. Guerrant, and D. A. Falk. 2004. Ex situ methods: A vital but underused set of conservation resources. In *Ex situ plant conservation: Supporting species survival in the wild*, ed. E. O. Guerrant Jr., K. Havens, and M. Maunder, 3–20. Washington, DC: Island Press.

Maunder, M., S. Higgens, and A. Culham. 2001. The effectiveness of botanic garden collections in supporting plant conservation: A European case study. *Biodiversity and Conservation* 10, no. 3: 383–401.

McNeely, J. A., and J. W. Thorsell. 2009. Enhancing the role of protected areas in conserving medicinal plants. In *The conservation of medicinal plants*, ed. O. Akerele, V. Heywood, and H. Synge, 199–210. Cambridge: Cambridge University Press.

Merow, C., M. J. Smith, and J. A., Jr. Silander. 2013. A practical guide to MaxEnt for modeling species' distributions: What it does, and why inputs and settings matter. *Ecography* 36, no. 10: 1058–69.

Mishra, P., A. Kumar, G. Sivaraman et al. 2017. Character-based DNA barcoding for authentication and conservation of IUCN Red listed threatened species of genus *Decalepis* (Apocynaceae). *Nature Scientific Reports* 7, no. 1: 14910. doi:10.1038/s41598-017-14887-8.

Mukherjee, T. 2009. Medicinal plants: Need for protection. In *Medicinal plants: Utilisation and conservation*, ed. P. C. Trivedi, 391–404. Jaipur: Aavishkar Publishers, Distributors.

Myers, N., R. A. Mittermeier, C. G. Mittermeier, G. A. B. da Fonseca, and J. Kent. 2000. Biodiversity hotspots for conservation priorities. *Nature* 403, no. 6772: 853–8.

Nagendra, H., and M. Gadgil. 1999. Satellite imagery as a tool for monitoring species diversity: An assessment. *Journal of Applied Ecology* 36, no. 3: 388–97.

Noorunnisa Begum, S., K. Ravikumar, and D. K. Ved. 2014. 'Asoka' – An important medicinal plant, its market scenario and conservation measures in India. *Current Science* 107: 26–8.

Pandey, H. P. 1961. *Bhāvaprakāśa*. Varanasi: Chowkhamba Sanskrit Series office.

Patwardhan, A., M. Pimputkar, M. Mhaskar, P. Agarwal, N. Barve, and R. Vasudeva. 2014. Status of *Saraca asoca*: An endangered medicinal species of conservation concern from Northern Western Ghats biodiversity hotspot. *Biodiversity Watch* January-March: 90–102.

Phillips, S. J., R. P. Anderson, and R. E. Schapire. 2006. Maximum entropy modeling of species geographic distributions. *Ecological Modelling* 190, no. 3–4: 231–59.

Primack, R. B., and A. J. Miller-Rushing. 2009. The role of botanical gardens in climate change research. *New Phytologist* 182, no. 2: 303–13.

Pulliam, H. R. 2000. On the relationship between niche and distribution. *Ecology Letters* 3, no. 4: 349–61.

Raina, R., R. Chand, and Y. P. Sharma. 2011. Conservation strategies of some important medicinal plants. *International Journal of Medicinal and Aromatic Plants* 1: 342–7.

Ravikumar, K., D. K. Ved, R. Vijaya Sankar, and N. M. Ganesh Babu. 2005. Medicinal plants diversity in India: Current status and *in situ* conservation in southern India - A case study. In *Medicinal plants research – An industry oriented perspective*, ed. R. Baskar, A. Malarvizhi, and S. Doraiswamy, 30–46. Coimbatore: G.R. Damodaran College of Science.

Ravindranath, N. H., N. V. Joshi, R. Sukumar, and A. Saxena. 2006. Impact of climate change on forests in India. *Current Science* 90: 354–61.

Reaka-Kudla, M. L., D. E. Wilson, and E. O. Wilson, ed. 1996. *Biodiversity II: Understanding and protecting our biological resources*. Washington, DC: Joseph Henry Press.

Reddy, C. S., S. L. Meena, P. H. Krishna, P. D. Charan, and K. C. Sharma. 2012. Conservation threat assessment of *Commiphora wightii* (Arn.) Bhandari- an economically important species. *Taiwania* 57: 288–93.

Rodgers, W. A., and H. S. Panwar. 1988. *Planning a wildlife protected area network in India*, 2 volumes. Project FO: IND/82/003. Dehradun, India: FAO. cited by Singh and. Kushwaha, 2008.

Romme, W. H. 1982. Fire and landscape diversity in subalpine forests of Yellowstone National Park. *Ecological Monographs* 52, no. 2: 199–211.

Roy, P. S., A. Roy, P. K. Joshi et al. 2015. Development of decadal (1985–1995–2005) land use and land cover database for India. *Remote Sensing* 7, no. 3: 2401–30.

Roy, P. S., and S. Tomar. 2000. Biodiversity characterization at landscape level using geospatial modelling technique. *Biological Conservation* 95, no. 1: 95–109.

Schippmann, U., D. J. Leaman, A. B. Cunningham, and S. Walter. 2005. Impact of cultivation and collection on the conservation of medicinal plants: Global trends and issues. *Acta Horticulturae* 676, no. 676: 31–44.

Schoen, D. J., and A. H. D. Brown. 2001. The conservation of wild plant species in seed banks. *BioScience* 51, no. 11: 960–6.

Singh, J. S. 2002. The biodiversity crisis: A multifaceted review. *Current Science* 82: 638–47.

Singh, J. S., and S. P. S. Kushwaha. 2008. Forest biodiversity and its conservation in India. *International Forestry Review* 10, no. 2: 292–304.

Singh, P., K. Karthigeyan, P. Lakshminarasimhan, and S. S. Dash. 2015. *Endemic vascular plants of India.* Kolkata: Botanical Survey of India.

Singh, R. V., P. Singh, L. A. Hansen, and L. Graudal. 2008. *Medicinal plants, their conservation, use and production in Southern India*, 1–57. Copenhagen: Forest & Landscape.

Swarts, N. D., and K. W. Dixon. 2009. Terrestrial orchid conservation in the age of extinction. *Annals of Botany* 104, no. 3: 543–56.

Uprety, Y., H. Asselin, A. Dhakal, and N. Julien. 2012.Traditional use of medicinal plants in the boreal forest of Canada: Review and perspectives. *Journal of Ethnobiology and Ethnomedicine* 8: 1–14.

Ved, D. K., V. Barve, N. Begum, and R. Latha. 1998. Eco-distribution mapping of the priority medicinal plants of southern India. *Current Science* 75: 205–8.

Ved, D. K., and G. S. Goraya. 2008. *Demand and supply of medicinal plants in India*. Dehra Dun and, Bangalore: Bishen Singh Mahendra Pal Singh & FRLHT.

Vidhate, S., M. S. Deogade, and P. Khobragade. 2015. *Abhava Pratinidhi Dravya* (substitutes for herbal drugs) acting on digestive system - A brief review. *Joinsysmed* 3: 155–60.

Yu, H., C. X. Xie, J. Y. Song, Y. Q. Zhou, and S. L. Chen. 2010. TCMGISII based prediction of medicinal plant distribution for conservation planning: A case study of *Rheum tanguticum*. *Chinese Medicine* 5: 31. doi:10.1186/1749-8546-5-31.

Yuan, Q. J., Z. Y. Zhang, J. A. Hu, L. P. Guo, A. J. Shao, and L. Q. Huang. 2010. Impacts of recent cultivation on genetic diversity pattern of a medicinal plant, *Scutellaria baicalensis* (Lamiaceae). *BMC Genetics* 11: 29. doi:10.1186/1471-2156-11-29.

Wong, K. L., R. N. Wong, L. Zhang et al. 2014. Bioactive proteins and peptides isolated from Chinese medicines with pharmaceutical potential. *Chinese Medicine* 9: 19. doi:10.1186/1749-8546-9-19.

9 Lessons to Be Learnt from Ayurveda

Nutraceuticals and Cosmeceuticals from Ayurveda Herbs

Prachi Garodia, Sosmitha Girisa, Varsha Rana, Ajaikumar B. Kunnumakkara and Bharat B. Aggarwal

CONTENTS

9.1 Introduction 200
9.2 History of Medicine and Ayurveda 201
9.3 Nutraceuticals 202
- 9.3.1 Nutraceuticals in Ayurveda 203
- 9.3.2 Chronic Diseases 203
- 9.3.3 Role of Nutraceuticals in Chronic Diseases 203
- 9.3.4 Nutraceuticals in Diabetes 203
- 9.3.5 Nutraceuticals in Cardiovascular Diseases 204
- 9.3.6 Nutraceuticals in Neurodegenerative Diseases 205
- 9.3.7 Nutraceuticals in Inflammatory Diseases 205
- 9.3.8 Nutraceuticals in Arthritis 206
- 9.3.9 Nutraceuticals in Obesity 207
- 9.3.10 Nutraceuticals in Cancer 207
- 9.3.11 Nutraceuticals in Chronic Respiratory Diseases 208

9.4 Ayurveda in Chronic Diseases 209
- 9.4.1 Ayurveda in Diabetes 210
- 9.4.2 Ayurveda in Chronic Respiratory Diseases 211
- 9.4.3 Ayurveda in Cancer 211
- 9.4.4 Ayurveda in Cardiovascular Disease 211
- 9.4.5 Ayurveda in Arthritis 211
- 9.4.6 Ayurveda in Digestive Diseases 212

9.5 Cosmeceuticals 212
- 9.5.1 History of Cosmeceuticals 212
- 9.5.2 Cosmeceuticals in Ayurveda 213
- 9.5.3 Important Cosmeceuticals 213

9.6 Conclusion 214
References 215

9.1 INTRODUCTION

Natural products have always remained as the basis for treating various diseases (Patwardhan et al. 2005). Medicinal plants from nature are recognized as potential candidates due to their drug-like effective properties (Bernhoft 2010). Ayurveda, the traditional Indian medicine, forms one of the most ancient, yet living, traditions of healing (Patwardhan et al. 2005). Nature has granted enormous varieties of medicinal plants in India, due to which the country is regarded as the medicinal garden of the world (Kumar et al. 2015). The use of plants for their medicinal values has been well described in the Indian *vēdas* to cure several diseases, and this gave rise to the widely accepted traditional medical system. With regard to this, India has several traditional medical systems such as Ayurveda, Siddha and Unani (Kumar et al. 2015).

Ayurveda is one of the ancient healthcare systems, the existence of which can be traced back to about 5000 years. According to the ancient literature on Ayurveda, it was practiced during the vedic period in India. Besides India, this medicinal system thrived as complementary medicine in other parts of the world. All therapies in Ayurveda aim to provide complete health benefits in terms of the physical, mental and spiritual state so that people can attain the real goal of life, i.e. self-fulfillment. The precise meaning of Ayurveda is the "sacred knowledge of life and longevity" and this system makes use of about 900 plants. The use of herbal treatment is the most popular form of this traditional medicine system (Sarker and Nahar 2007). According to the report of World Health Organization, those reliant on herbal sources as the complementary system has been estimated to be about 70–80% of the world population (Wise 2013). The demand for herbal medicine, pharmaceuticals, health products, nutraceuticals, food supplements and herbal and natural cosmetics is growing worldwide. In the 21st century, natural products represent more than 50% of the drugs that are used clinically and they are derived from plants, animals, microbes and fungi (Shakya 2016).

The main reasons for public interest in relying on complementary medicinal systems are the side-effects associated with synthetic drugs, high cost, drug resistance, lack of treatment for various chronic diseases and other emerging diseases (Humber 2002). Undeniably, ayurvedic treatment is more effective in most chronic diseases as compared to allopathy, but, as the majority of the population depends on modern medicine due to its fast relief, this has made Ayurveda less popular. However, in recent years, the awareness of people related to the toxicity of allopathic drugs and the high cost of the health system have led to searching for alternatives (Chauhan et al. 2015). Though treatment with Ayurveda is highly efficient, the proper mechanism of action, pharmacology and pharmacokinetics of most of the important drugs are yet to be explored, and this has made Ayurveda lag behind due to deficiency of scientific evidence and poor research. In addition to this, there is a gap in the exchange of information globally (Chauhan et al. 2015; Jaiswal and Williams 2017).

According to the National Medicinal Plant Board (N.M.P.B.), the Indian herbal industry aims to increase the trend to 80–90 billion rupees by the year 2020. Nonetheless, India has focused on popularizing the traditional medicine system of A.Y.U.S.H., which stands for Ayurveda, Yoga, Unani, Siddha and Homeopathy, in the health system media through its worldwide network that targets remedies to cure diseases and manage quality of life (Shakya 2016). Therefore, at present, there is once again a revival of interest in natural resources and there has been a great improvement in recent years in understanding Indian herbs (Sarvesh Kumar et al. 2013).

Most of the anti-cancer and anti-infection drugs that were approved from 1981 to 2002 have their origin in natural sources – 60% and 75%, respectively (Gupta et al. 2005). The promising compounds and molecules that originated from ayurvedic medicine include *Rauwolfia* alkaloids against hypertension, *Holarrhena* alkaloids against amoebiasis, guggulsterones used as hypolipidemic agents, *Mucuna pruriens* against Parkinson's disease, curcumin for inflammation, withanolides, steroidal lactones and glycosides as immunomodulators (Patwardhan 2000; Patwardhan and Mashelkar. 2009).

9.2 HISTORY OF MEDICINE AND AYURVEDA

The origin of medicine dates back to the age of life itself, and the development of this system depends on various features and forms, contents that are decided by the civilization and the environment of origin. With the advancement of human civilization, there is a change in medical science due to the appearance of newer diseases. Ayurveda grew in India on the basis of a logical foundation and has survived as a separate discipline from ancient times to the present day, based on fundamental human factors and intrinsic causes (Narayanaswamy 1981). The history of medicine reveals most of the ancient discoveries to be a result of serendipity or folklore approaches that were involved with poisonous bases, and not actually from traditional medicines (Harvey 2008). The Greek physician Hippocrates who is known as the father of medicine stated, "let food be thy medicine and medicine be thy food". Healing systems existed since the time of development of human civilization, and the present form of the modern medicine system steadily settled throughout the years through scientific observations. However, the basis of it is derived from tradition (Patwardhan 2000; Patwardhan et al. 2004). In traditional medical systems, plants formed the renewable source of food and medicine, and in this, India remained the pioneer (Chaudhary and Singh 2012). Ayurveda originated and was established between 2500 and 500 B.C. in India (Mishra et al. 2001a; Pandey et al. 2013).

Based on ayurvedic knowledge, turmeric paste wrap was employed to treat common eye infections, dress wounds and treat acne and several skin diseases, and roasted turmeric is an ingredient used as an anti-dysenteric for children (Hatcher et al. 2008). Ayurveda endured and thrived for ages to the present times with vast information on nature-based medicine, the relationship of the human body with nature and the universe and fundamentals that coordinate and affect all living beings. The ancient wisdom of Ayurveda is slowly being explored, which is why this rich knowledge could give rise to new opportunities to discover herbal drugs (Jaiswal and Williams 2017). Ayurveda represents a complete system towards health and personalized medicine that consists of ethical, physical, philosophical, psychological and spiritual health (Semwal et al. 2015). The fundamentals of Ayurveda are known by its bases founded by the schools of Hindu philosophy – *Nyāya*, *Vaiśēṣika*, *Sāmkhya*, *Yoga*, *Mīmāmsa* and *Vēdānta* (Ninivaggi 2008; Jaiswal and Williams 2017). The *Vaiśēṣika* group addressed the interpretations and observations that were intended to understand the pathological condition of the patient, whereas the *Nyāya* school spread its wisdom on the concept of having complete knowledge of the condition of patient and disease before the initiation of treatment (Jaiswal and Williams 2017). Ayurveda is believed to have originated even before the establishment of these schools, from the Hindu deity, Brahma, the creator of the universe (Mukherjee and Houghton 2009). Brahma passed on this complete wisdom to the sages and from them, the medical knowledge was passed on to their disciples and then to the common people through their writings and oral tradition. The knowledge disseminated in the form of quatrains, called *ślōka*s, incorporating the wisdom of the four *Vēdas* – *Ṛgvēda*, *Yajurvēda*, *Sāmavēda* and *Atharvavēda* (Jaiswal and Williams 2017).

When Indologists express their views about Ayurveda writings, they refer to the earliest collections of Caraka, Suśruta and Vāgbhaṭa, the Great Triad of Indian medicine (Glazier 2000). Of these, *Suśruta Samhita* forms the landmark in Ayurveda. Though the text focused on surgery, it has also described 395 medicinal plants, 57 drugs from animal origin and 64 minerals or metals used as therapeutic agents. In ancient time, besides India, these ayurvedic writings were valued in other countries and were being translated into Greek (300 B.C.E.), Tibetan, Chinese (300 C.E.), Persian and Arabic (700 C.E.) (Pan et al. 2014).

Ayurveda basically consists of eight branches, one of which is *rasāyana tantra*. The term *rasāyana* is derived from the words *rasa*- (primordial tissue or plasma) and *ayana*- (path). Thus, the term *rasāyana* means the path that *rasa* takes (Puri 2002). *Rasāyana tantra* or rejuvenation therapy is one of the eight specialized and vital branches of the *Aṣṭāṅga* Ayurveda healing system (*Aṣṭa* = eight, *aṅga* = limbs or parts) which emphasizes nutrition dynamics, rejuvenation of the body, mind

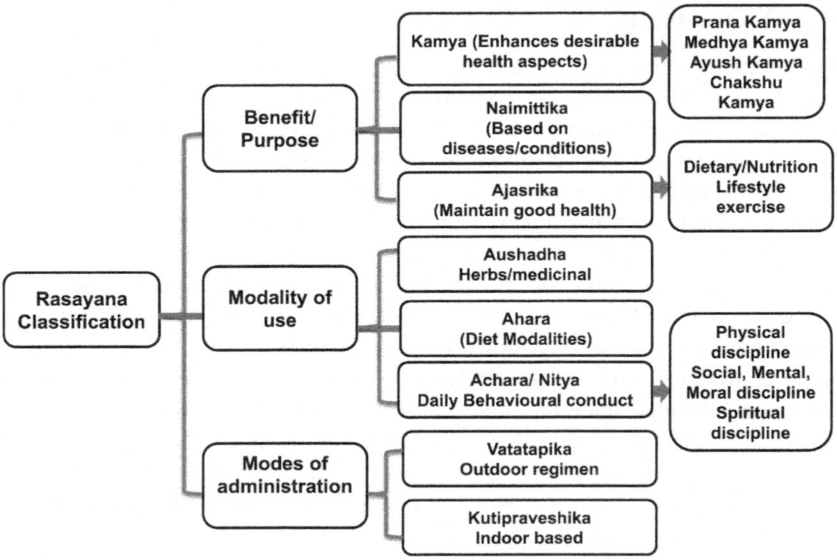

FIGURE 9.1 Classification of *rasāyana*.

and emotions and promotion of immunity and longevity (Vāgbhaṭa 2005). *Rasāyana* is defined as a means to attain the optimal quality of body tissues which can enhance mental ability, memory, intellect, complexion, immunity, longevity and youthfulness. *Rasāyana* therapy can promote a healthy state of an individual and also reverse and manage a disease condition, providing better immunity. This form of therapy is not restricted to the senile group and can be applied from pediatrics to geriatrics, and the ideal group age ranges between 16 years to 90 years. The therapies and various herbal combinations in *Rasāyana* are tailored to specifications that are required for an individual, suitable for the age, body type, status of body and expected effect (Puri 2002) (Figure 9.1).

Rasāyana tantra signifies the basic methodology of Ayurveda to prevent and cure diseases by following the procedures of *Rasāyana cikitsa*. The treatment method for aging (*Jaracikitsa*) is synonymous with *Rasāyana cikitsa*, which not only aims to specify a definite knowledge system but also acts as a procedure to enhance the efficiency of body tissues and their function (Venugopalan and Venkatasubramanian 2017). The ideal first phase to any effective *rasāyana* therapy is purification of the body and channels through a series of body purification techniques known as *pañcakarma*, which is a well-known rejuvenation and detoxification method that consists of three series of steps such as pre-treatment (*Pūrvakarma*), the primary treatment (*Pradhānakarma*) and the post-treatment (*Pascātkarma*) (Puri 2002).

9.3 NUTRACEUTICALS

The term *nutraceutical* is well-defined as "a food or a part of food that provides medical or health benefits including the prevention and treatment of a disease". However, the term does not represent the perfect terminology in the market and may refer not only to a supplement in the diet, but also to its role in the prevention and treatment of a disease or a disorder. The term *nutraceutical* was defined by Stephen DeFelice from the combination of the terms *nutrition* and *pharmaceutical* in 1989 (DeFelice 1995; Baragi et al. 2008). Herbal nutraceuticals are powerful tools for maintaining health and they act against nutritionally induced acute and chronic diseases, by promoting optimal health, longevity and quality of life (Khan et al. 2014).

9.3.1 Nutraceuticals in Ayurveda

Medicinal plants contain several secondary metabolites that have medicinal properties against diseases (Padmavathi 2013). In ayurvedic pharmaceutics there are some dosage forms like *ghṛta*, *taila*, *avalēha*, *āsavāriṣṭa* and *kṣīrapāka* that can be equated with nutraceuticals (Khamkar and Dubewar 2019). One of the techniques of Ayurveda, *Ājasṛk Rasāyana*, applies to the use of food products that can be consumed on a daily basis. These have the ability to improve the status of health by offering protection against stress factors. The nutraceuticals that can be commonly employed on a daily basis include *Cyavanprāśa* for overall health and prevention of respiratory disease, *Phala Ghṛta* for reproductive health, *Brahma Rasāyana* to protect from mental stress, *Arjuna Kṣīrapāka* and *Rasōna Kṣīrapāka* for cardioprotective functions and *Śatāvarī Ghṛta* in women's health (Rani and Sharma 2005). Thus, herbal formulations that can protect general health have gained worldwide importance on account of their diverse benefits (Kumar and Sharma 2020).

9.3.2 Chronic Diseases

The term *chronic condition* is a state of physical or mental health that can last for more than a year with functional restrictions or the condition requiring ongoing management (Basu et al. 2016). It affects health and quality of life globally, leading to death and disability (Raghupathi and Raghupathi 2018). Chronic diseases which are also known as non-communicable diseases include a range of disorders such as arthritis, Alzheimer's disease, Parkinson's disease, cancer, cardiovascular disease and diabetes (Kunnumakkara et al. 2018a; Khwairakpam et al. 2018). The general risk factors that are related to these chronic diseases include unhealthy lifestyles with physical inactivity, poor routine in diet, exposure of the body to radiation, stress, excessive consumption of tobacco or alcohol and infection to pathogenic microorganisms (Kunnumakkara et al. 2018a). Three essential components such as research, performance measurement and quality improvement are required for continued management of chronic diseases (Davis et al. 2000).

9.3.3 Role of Nutraceuticals in Chronic Diseases

Nutraceuticals may vary from dietary supplements, specific nutrients, herbal products, processed food such as cereals and beverages containing bioactive components (Prakash et al. 2012; Khamkar and Dubewar 2019). In terms of chemical origin, nutraceuticals can be isoprenoids, phenolic compounds, carbohydrate derivatives, fatty acids, structural lipids, amino acids, microbes and minerals (Prakash and Sharma 2014). On the basis of natural sources, they are termed nutrients, herbals, dietary supplements, dietary fiber and so on, out of which dietary supplements (with 19.5% annually) and natural/herbal products (with 11.6% annually) are the most rapidly growing segments in the sector. Nutraceuticals have an overall global market estimate of US$117 billion (Chauhan et al. 2013). The properties of nutraceuticals are important in terms of their effectiveness against several chronic ailments like cardiovascular disease, diabetes, metabolic syndrome, obesity, cancer and neurodegenerative disease (Chauhan et al. 2013; Islam et al. 2020).

9.3.4 Nutraceuticals in Diabetes

Diabetes mellitus is one of the most important metabolic disorders. The major cause of mortality and morbidity of the disease is due to its associated complications that are related to increased glucose levels (Prabhakar et al. 2014). It is characterized by abnormally high blood glucose levels which may be either due to insufficient insulin production or lowered effectiveness of insulin. The general forms of this disease are diabetes type 1, representing 5%, which is an autoimmune disorder and type 2 diabetes, representing 95%, which is linked with obesity. One rare type is gestational diabetes which occurs during pregnancy (Rajasekaran et al. 2008). The disease can give rise to many short-term complications like hypoglycemia, ketoacidosis or non-ketonic hyperosmolar coma, if it is not properly managed. It can also give rise to long-term problems that include cardiovascular disease, diabetic nephropathy or chronic renal failure, retinal damage that can cause blindness and severe nerve damage (Baldi et al. 2013).

In a clinical study, human subjects were provided with the composition of nutraceuticals that contained epicatechin added to an equivalent quantity of gymnemic acid and others in smaller quantities, such as *Cinnamomum tamala*, *Syzygium cumini*, *Azardichta indica*, *Trigonella foenum graceum*, *Tinospora cordifolia* and *Ficus racemose* for four and 12 months. The subjects were administered one to two grams of composition thrice daily before meals. The treatment caused regeneration of the pancreatic cells resulting in insulin production (Dhaliwal 1999). Another study evaluated the outcome of aqueous extract of *Pterocarpus marsupium* bark that decreased the levels of blood sugar from 2 hours after the extract was orally administered in alloxan-induced diabetic rats. A similar effect was observed from the alcohol extract of *Trigonella foenum-graecum* seeds and the leaves of *Ocimum sanctum* both in normal and alloxan-induced diabetic rats. The extracts also showed its effect on glucose disposition in hyperglycemic rats fed with glucose (Vats et al. 2002). Further, various extracts of *Momordica charantia* reduced fasting blood glucose by 48% with no association of nephrotoxicity and hepatotoxicity complications, suggesting it as a safe alternative in managing glucose levels (Virdi et al. 2003). Similar results were observed in another clinical study which reported that daily dose of 1 to 6 grams of cinnamon powder resulted in decreasing about 20% of fasting glucose and improving the levels of serum lipid by reducing triglyceride, L.D.L. cholesterol and total cholesterol levels (Khan et al. 2003). The ethanol extract of the rhizomes of wild ginger (*Zingiber zerumbet*) was also found to have the property of reducing blood sugar levels in a diabetes-induced rat model (Muhtadi et al. 2018).

9.3.5 Nutraceuticals in Cardiovascular Diseases

Cardiovascular diseases are the leading causes of death (Di Girolamo et al. 2020). This disease involves a number of disorders that affect the heart and blood vessels. Nearly one out of three suffers from this disease, with many complications that can result in ischemic heart attack or stroke (Park et al. 2020). Cardiovascular diseases can be mainly grouped into ischemic heart disease and cerebrovascular disease, which account for nearly 2 million and 1 million mortality rates respectively (Di Girolamo et al. 2020). Many psychological factors have been known to increase the risk of cardiovascular diseases (Larkin and Chantler 2020). Undeniably, this disease commonly affects adults above 60 years of age and factors like obesity, hyperlipidemia, diabetes mellitus, hypertension, metabolic syndrome and lifestyle risk factors that include unhealthy diet routine, smoking and less physical activity are associated with this disease (Sosnowska et al. 2017).

It has been reported that greater consumption of plant sterols could lower the total serum L.D.L., non-H.D.L. cholesterol and carotid intima-media thickness in humans (Wang et al. 2012). Additionally, grape extracts rich in polyphenol at a 700 mg dose on healthy subjects with risk of cardiovascular disease, reduced total plasma cholesterol and L.D.L.-C. concentrations (Yubero et al. 2013). The dietary intervention of grape nutraceutical containing 8 mg of resveratrol markedly improved the inflammatory condition and fibrinolytic status in patients, who were under primary prevention of cardiovascular disease. The outcome was due to suppression of h.s.-C.R.P. expression after a one-year regimen that could be related to decreased level of the pro-inflammatory cytokine T.N.F.-α and thrombogenic P.A.I.-1 (Tomé-Carneiro et al. 2012).

In humans, berberine has the property of lowering lipids and the insulin-resistance status, as observed in numerous randomized clinical trials. Moreover, berberine possesses the ability to decrease endothelial inflammation and improve the vascular health condition in patients affected by cardiovascular diseases (Cicero and Baggioni 2016). Fenugreek has been showed to lower blood glucose and cholesterol levels in animals. It decreases plasma cholesterol in chickens and rabbits, reducing risk for heart disease. Polyphenols from fenugreek seed extract protect from ethanol-induced damage in human liver cells and the effect is comparable to hepatoprotective drug silymarin (Thomas et al. 2011).

9.3.6 Nutraceuticals in Neurodegenerative Diseases

Neurodegenerative diseases represent advanced disorders of the neurons that could cause destruction or abolish the function of the neurons (Kannappan et al. 2011). Neurodegenerative diseases like Alzheimer's disease and Parkinson's disease are known to have general features in cellular and molecular mechanisms that include protein aggregation and inclusion body formation (Ross and Poirier 2004). A wide range of these disorders are characterized by damage in neurons which might be a result of toxic and aggregate-prone proteins (Taylor et al. 2002). They pose a great threat to human health and have increasingly affected the elderly population in recent years (Heemels 2016).

These diseases are well-known for their signs related to memory and cognition, affecting the ability of the person to move, speak and even breathe appropriately (Wyss-Coray 2016). The main feature of neurodegenerative diseases like Alzheimer's disease is brain inflammation, characterized by increased activation of microglia and astrocytes that increases during aging (Sawikr et al. 2017). Several nutraceuticals were shown to be neuroprotective in experimental models and serve as an alternative to other synthetic drugs like L-Dopa that are identified to cause many adverse side-effects (Hang et al. 2016). Nutraceuticals affect several neurodegenerative diseases through modulation of various signaling pathways. A nutraceutical like curcumin was shown to inhibit fAbeta (fAβ) formation from fresh Abeta (Aβ) and then destabilize the preformed fAβ *in vitro* where the fAβ destabilized by N.D.G.A. show less toxicity than the intact fAβ (Ono et al. 2004). The binding of Aβ to a neurotrophin receptor (p75N.T.R.) could activate N.F.-κB (nuclear factor kappa-B), and induce brain cell death, and curcumin could inhibit this process of N.F.-κB activation to prevent the Aβ-induced cell death in a human neuroblastoma cell line, which signifies a possible option for Alzheimer's disease treatment (Kannappan et al. 2011).

In an animal model study in rats, resveratrol was observed to inhibit the reuptake of noradrenaline and 5-HT with dose-dependent increase in hippocampal serotonin. Further, it reduced neuronal inflammation and cell death in oxidation-induced neuronal toxicity animal models. It also protected the slices of organotypic hippocampus from the hydroperoxide damage (Mecocci et al. 2014). The other possible mechanisms of neuroprotection by resveratrol are related to its antioxidant properties and the ability to modulate Aβ processing and upregulation of sirtuin 1 expression (Pocernich et al. 2011; Mecocci et al. 2014).

9.3.7 Nutraceuticals in Inflammatory Diseases

The term *inflammation* means "to set on fire". It is one of the natural responses of the body to the harmful pathogens that occurs as a form of acute and chronic type of inflammation (Kunnumakkara et al. 2018a; Gupta et al. 2018). The acute response forms a part of the innate immunity system through the persisting immune cells for a short duration. However, the chronic inflammation response can give rise to various chronic diseases which include arthritis, cancer, diabetes, cardiovascular diseases and so on, which might be due to deregulation of signaling pathways like S.T.A.T.3 (Signal transducer and activator of transcription 3) and N.F.-κB (Kunnumakkara et al. 2018a). Inflammation can be characterized by swelling, pain, redness, heat and the body's response to these irritations or injuries (Delfan et al. 2014). The production of reactive oxygen species induces oxidative stress and also promotes inflammatory processes through the stimulation of transcription factors like N.F.-κB and activator protein 1 (A.P.1), resulting in production of T.N.F.α (Boots et al. 2008).

When polyphenols from concentrated apple extract with the dominant presence of chlorogenic acid are administered to rats afflicted with colitis induced by acetic acid, it could downregulate iNOS expression and conversely upregulate the levels of copper and zinc superoxide dismutase

(CuZnSOD) (Pastrelo et al. 2017). Another group of compounds, anthocyanins prevented the inflammation in the colitis-affected mice model and reduced the levels of pro-inflammatory cytokines such as I.F.N.-γ, T.N.F. and I.L.-6 with the outcome of improved colon condition. In the case of ulcerative colitis-affected rats, the anthocyanins decreased the expression of I.F.N.-γ R.2 and part of the I.F.N.-γ receptor signal-transduction in colonic tissue, which resulted in inhibition of the I.F.N.-γ-induced inflammatory activity in the colon and also elevated the concentrations of the I.L.-22 and I.L.-10 tissue-protective cytokines in serum (Ghattamaneni et al. 2018). In D.S.S.-induced I.B.D. rats, a dietary complementation of quercetin along with fish oil containing eicosapentaenoic acid and docosahexaenoic acid improved the status of inflammation in the intestine (Camuesco et al. 2006]. Aloe and its components aloin, and aloe-gel have activity against inflammation, as they can reduce the T.N.F. and I.L.-1β expression levels in colon mucosa and decrease the plasma levels of L.T.B.4 and T.N.F. which can decrease the inflammation in the intestine of D.S.S.-induced I.B.D. mice (Park et al. 2011). Further, curcumin was shown to reduce the colonic inflammation by inhibiting pro-inflammatory pathways such as N.F.-κB and M.A.P.K. pathways in the multidrug resistance minus gene (Mdr1a–/–) I.B.D. mice (Nones et al. 2009). To add to this, the ellagic acid present in pomegranate was found to reduce the intestinal inflammation in D.S.S.-induced acute and chronic I.B.D. mice models which results in improved condition through COX2 and inducible nitric oxide synthase (iNOS) downregulation, thus inhibiting the signaling pathways like N.F.-κB, signal transducer and activator of transcription 3 (S.T.A.T.3) and p38 M.A.P.K. (Marin et al. 2013).

Curcumin demonstrated anti-inflammatory properties through arachidonic acid cascades by suppressing N.F.-kB and immunosuppressive properties through inhibition of I.L.-2 and T.N.F.-α levels. *Boswellia serrata* is another ayurvedic herb to cure the inflammatory state. It acts similarly through the arachidonic acid cascade and inhibition of leukotrienes resulting in reduction of pro-inflammatory cytokines. Another nutraceutical, apigenin, which forms the main constituent of wheatgrass inhibits the production of I.L.-1β, I.L.-8 and T.N.F. via inactivated N.F.-κB (Parian et al. 2015). Quercetin resulted in inhibition of L.P.S.-induced T.N.F.α and I.L.8 levels in macrophages and lung cells respectively which might be due to the interaction between the oxidative stress and inflammation (Boots et al. 2008).

Additionally, in one clinical study, 40 patients (23 males, 17 females) suffering from bronchial asthma, and in the age group of 18–75 years, were treated with 300 mg gum resin of *Boswellia* thrice a week for 6 weeks. The treatment resulted in improved prognosis of 70% of the patients with relief from signs and symptoms of bronchial asthma and minimization of the attacks (Kunnumakkara et al. 2018b).

9.3.8 Nutraceuticals in Arthritis

The word *arthritis* is derived from the Greek words *arthron* meaning joint and *-itis* meaning inflammation. It is a disorder of the joints characterized by chronic inflammation in one or several joints resulting in pain and disability. Arthritis includes more than 100 forms such as osteoarthritis, psoriatic arthritis, rheumatoid arthritis and related autoimmune diseases (Daily et al. 2016). It arises due to inflammation of the joints and tissues that surround the joints and other connective tissues (Kunnumakkara et al. 2018b). Among all these, osteoarthritis followed by rheumatoid arthritis are the most common ones.

Many factors can lead to the development of osteoarthritis. They are aging, excessive exercise, obesity, immune disorders, genetic predisposition, poor nutrition, injury and infection (Gupta et al. 2016). Osteoarthritis is the most common form of arthritis and remains one of the few chronic diseases related to aging, with fewer treatment options (Felson 2009). The most important feature of osteoarthritis is the advanced destruction of articular cartilage that results in impaired joint motion and severe pain, leading to the person's disability (Ameye and Chee 2006). The condition of osteoarthritis is not only limited to articular cartilage. It also affects the entire portion of the joint, including the subchondral bone, menisci, ligaments, periarticular muscle, capsule, adjacent connective tissue and the synovial membrane, giving rise to pain, swelling, deformity and instability (Sanghi et al. 2009; Ziskoven et al. 2010). Though osteoarthritis occurs in many joints, the knee, hip, hand and facet joints are mostly affected (Ziskoven et al. 2010). Osteoarthritis is one of the chronic diseases that forms an example for the pathology where the treatment could be addressed by proper nutrition (Ameye and Chee 2006).

Several nutraceuticals such as fish oil, black cumin extracts, fenugreek, licorice, coriander and dates were evaluated for their potential use in treatment of arthritis. Modified inflammatory biomarkers like erythrocyte sedimentation rate, C-reactive protein, seromucoids, fibrinogen, tumor necrosis factor-α (T.N.F.-α), prostaglandin E_2, oxidative stress (malondialdehyde) and antioxidant status (total antioxidant capacity) were measured for evaluation. It was observed that these nutraceuticals can be used against inflammation and oxidation for alternative management of arthritis (Al-Okbi 2014). Intraperitoneal injection of curcuminoid extracts inhibited acute and chronic inflammation of joints, preventing induction of arthritis. Curcumin was reported to downregulate inflammatory enzymes such as that of nitric oxide synthase (iNOS), COX 2 and arachidonate5-lipoxygenase (ALOX5); pro-inflammatory cytokines TNF-α; interleukins such as I.L.-1, I.L.-2, I.L.-6, I.L.-8, I.L.-12; and chemokines through NF-κB inactivation (Alam et al. 2014). Gallic acid treatment in R.A. F.L.S. cells significantly decreased the expression of pro-inflammatory cytokines, chemokines, COX2 and matrix metalloproteinase 9, increased apoptotic protein caspase 3 and also regulated the expression of Bcl2, Bax, p53 and Akt proteins (Yoon et al. 2013). Genistein could suppress the growth of fibroblast-like synoviocytes in rheumatoid arthritis by stimulating I.L.-1β and T.N.F.-α cytokines and E.G.F. growth factor (Zhang et al. 2012).

9.3.9 Nutraceuticals in Obesity

Obesity can be understood as a disease that is related to imbalance of energy, which can increase the risk for other chronic diseases like cardiovascular disease, type 2 diabetes, hypertension, dyslipidemia and some cancers. It is characterized by increased fat mass in the adipose tissue that serves as a storage center for consumed food (Singh et al. 2019). It is a multifaceted disorder with serious social and psychological extents, affecting people of all ages and socioeconomic groups (Chintale et al. 2013). In cases of obesity, like in most other chronic diseases, inflammation plays a major role and these states later lead to the development of another pro-inflammatory disease, atherosclerosis (Aggarwal. 2010). The main cause of obesity is increased high-fat availability and energy-dense foods. Thus, obesity is globally prevalent and nutrition and exercise play a role in its prevention and treatment (Nasri et al. 2014). The mixture of the dietary supplements of nutraceuticals fenugreek, glucomannan, *Gymnema sylvestre*, and vitamin C significantly decreased body weight and promoted fat loss in obese individuals (Chintale et al. 2013). The supplementation of conjugated linoleic acid that is found in flax seeds, nut oils, fish and eggs reduced the fat mass of obese individuals (Kasbia 2005). Supplementation of conjugated linoleic acid might also have a role towards insulin resistance which occurs in conjunction with obesity, and additionally, supplementation of conjugated linoleic acid to a high-fat diet of rodents has been shown to prevent the onset of obesity-induced muscle insulin resistance (Kasbia 2005).

9.3.10 Nutraceuticals in Cancer

Despite the advancement in cancer treatment modes such as surgery, radiation and chemotherapy, cancer still remains one of the most lethal diseases globally (Padmavathi et al. 2015; Banik et al. 2018; Ranaware et al. 2018; Sailo et al. 2018). Many natural products are used in herbal medicine, food supplements and as cooking spices. They are well-investigated in various cancer experimental settings in both *in vitro* and *in vivo* models (Tripathi et al. 2005). Similarly, nutritional modulations could also prove to be useful in treating cancer patients. Examples are foods that contain comparatively low simple carbohydrates and reasonable quantities of high-quality protein, fibers and fats (omega-3 fatty acid). Additionally, supplementation of certain nutraceuticals, micronutrients and functional foods have a role in reducing the risk of cancer development or inhibit cancer progression and also reduce the toxicity that is associated with conventional cancer chemotherapy and radiation (Tripathi et al. 2005).

The common nutraceuticals used for this purpose include curcumin, capsaicin, genistein, flavopiridol, sanguinarine, resveratrol and tocotrienol (Gupta et al. 2019). Polyphenolic and flavonoid contents from fruits and vegetables are also proven for their anti-cancer ability to downregulate the expression of various genes, proteins and signaling pathways that are involved in tumor growth and progression (Khwairakpam et al. 2018). Phytochemicals like triterpenoids, flavonoids and retinoids with their derivatives are multi-targeted agents with the properties of anti-angiogenesis and anti-inflammation mostly targeting A.M.P.K. and other metabolic pathways like the m.T.O.R axis (Albini et al. 2019). Curcumin, the golden nutraceutical, was shown to suppress the expression of inflammatory cytokines like I.L.-6, I.L.-8, and granulocyte macrophage colony stimulating factor and T.N.F.-α and I.K.K.β kinase in the saliva of head and neck cancer patients (Kunnumakkara et al. 2017).

The natural polyphenol calebin A, a novel derivative of turmeric, blocked the activation and nuclear translocation of p65-N.F.-B by T.N.F.-β (Buhrmann et al. 2019). The extract of gum resin from *Boswellia serrata* and its various analogues have shown cytostatic and apoptotic activity against glioma cells. Furthermore, one of its analogues in combination with curcumin showed anti-tumor effects in various experimental settings in both *in vitro* and *in vivo* conditions by regulating cancer-related miR.N.A.s like miR-34a and miR-27a in colon cancer (Roy et al. 2019). Curcumin induces cell cycle arrest and initiates apoptosis and also alters many targets such as T.N.F.-α, N.F.-κB and COX-2. The plant polyphenol butein could decrease the proliferation, cytotoxicity, migration and invasion by modulating various gene products such as COX-2, survivin, and M.M.P.-9. Similar modulations of these cancer hallmarks in O.S.C.C. cells through regulation of N.F.-κB and its regulated gene products were demonstrated (Bordoloi et al. 2019).

Resveratrol administered to breast cancer patients for 12 weeks showed a marked decline in the methylation of four genes which can cause an increase in the threat for breast cancer. In the case of prostate cancer, the effect of resveratrol in combination with curcumin encapsulated in the same liposome increased their anti-cancer effect (Ghani et al. 2019). The compound piceatannol was also shown to induce apoptosis in leukemic cells through the induced expression of F.A.S. and M.A.P.K. (mitogen-activated protein kinase)-mediated c-Jun and A.T.F.-2 (activating transcription factor 2) pathways. It also activated caspases and downregulated B-cell lymphoma (Bcl2) (Liu and Chang 2010).

Further studies found the regulation of N.F.-κB by the compounds that were involved in inflammatory and other signaling cascades in cancer cells (Banik et al. 2020). Honokiol prevents the progression of different cancers through the decreased expression of anti-apoptotic proteins, affects the mitochondrial-dependent pathways and elevates apoptosis through the modulation of Ca+2 channels and pro-apoptotic proteins. It affects metastatic activity by inhibiting M.M.P.s, P.I.3K./Akt/mT.O.R., epithelial to mesenchymal transition, N.F.-κB, S.T.A.T.3, and Wnt signaling pathways (Banik et al. 2019). The compound zerumbone could cause a decrease in cancer hallmarks such as survival, proliferation, angiogenesis, invasion and metastasis through the modulation of Akt, N.F.-κB, and I.L.-6/J.A.K.2/S.T.A.T.3 pathways (Girisa et al. 2019).

9.3.11 Nutraceuticals in Chronic Respiratory Diseases

Chronic respiratory diseases are associated with a high economic burden, affecting the quality of life of patients (Sleurs et al. 2019). Mainly asthma and chronic obstructive pulmonary disease (C.O.P.D.), with their associated mortality, contribute to the public burden (Roth-Walter et al. 2019). Oxidative stress is strongly involved in increasing the conditions of asthma, C.O.P.D., infections and lung cancer (Dua et al. 2019). In previous studies, the use of various nutraceuticals in terms of nutrients or bioactive compounds from plants or microbes with potential health-improving effects has been widely explored (Rivellese et al. 2019). Epigallocatechin gallate showed its potential in inhibiting the chemo-attractants and regulated the inflammatory responses in several fibrotic diseases. Further, the *in vivo* studies with the compound proved to be beneficial in improving the condition of lung injuries in cigarette smoke-exposed rats (Hwang and Ho 2018).

9.4 AYURVEDA IN CHRONIC DISEASES

When applied to all chronic diseases, Ayurveda can yield reliable and effective results, as its theory is based on three *dōṣās* – *vāta*, *pitta* and *kapha*, collectively called *tridōṣa* (Hankey 2010). The *tridōṣa* work in conjunction to maintain homeostasis in an individual, starting from fertilization and continuing throughout the lifetime. Each *dōṣā* has distinct properties and functions, as, for example, *vāta* contributes to expression of shape, signaling, movement, cell division, excretion of wastes, cognition and also regulates the activities of *kapha* and *pitta*. *Kapha* is associated with anabolism, growth, maintenance of structure, storage and stability of the body, while *pitta* is primarily responsible for metabolism, thermal regulation, homeostasis of energy, pigmentation, vision and host surveillance (Prasher et al. 2008). Health is seen as a state of balance of these three *dōṣā*s and their imbalance can lead to disease conditions which occur in the series of *ṣaḍkriyākāla*, which are the six stages of disease progression (Elder 2004; Hankey 2010). Ayurveda also highlights the importance of maintaining good digestion or *agni*. Weak *agni* could result in building up of digestive toxins known as *āma* which can later accumulate in the tissues or *dhātu*s, forming the basis for intensified *dōṣā*s, leading to the expression of disease (Elder 2004).

Ayurveda advises expansion of consciousness as the key factor required to achieve optimal health. Treatment of diseases is greatly personalized and depends on the patient's psychophysiological constitution. Different diets and lifestyles are recommended in each season and common spices, herbs, herbal mixtures and *rasāyana* preparations (for rejuvenation, longevity and anti-aging) are used in treatment (Sharma et al. 2007). These diets and lifestyles are based on the disturbed *dōṣā*s and *prakṛti* (physical and mental structure of a person) (Mishra et al. 2001b) (Figure 9.2).

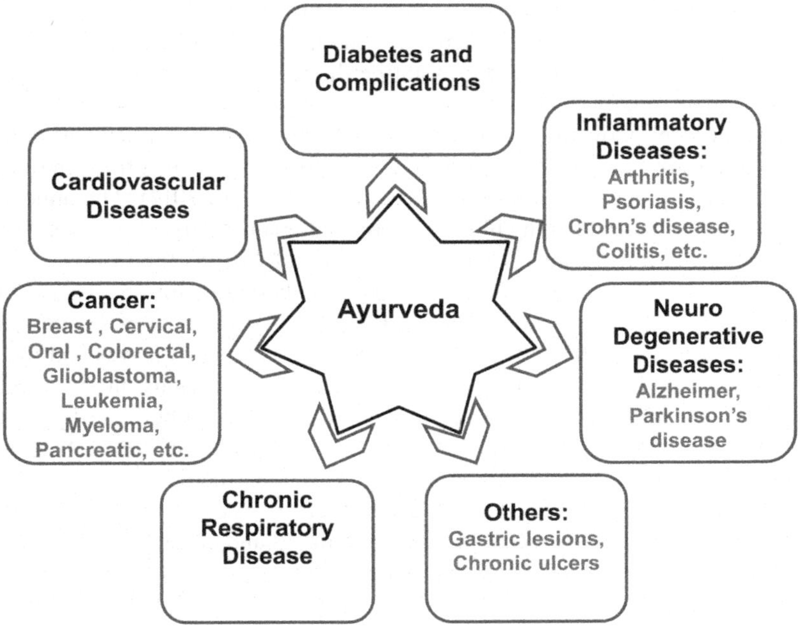

FIGURE 9.2 Ayurveda in different chronic diseases.

9.4.1 Ayurveda in Diabetes

In Ayurveda, diabetes mellitus is known as *madhumēha* (Patel et al. 2012). According to ayurvedic classification, 20 different types of *pramēha*, or polyuria exist. The term *madhumēha*, or "honeyed urine" is the diagnostic term that is used in Ayurveda representing diabetes mellitus. It refers to a subdivision of *pramēha*. This disease is *kapha*-dominant in the early stages, with the association of polyuria and involvement of fat (*mēdōdhātu*) (Elder 2004). Later, with the progression of *pramēha*, *pitta* and *vāta dōṣās* become worse and the disease becomes difficult to treat. The cause of this disease includes familial predisposition, reduced physical activity and physical and mental stress. Diet plays an important etiologic part and especially irregular meals and abundant sweet, sour and salty food items, all of which vitiate *vāta dōṣa* (Elder 2004). Various herbs are used as anti-diabetic agents in ayurvedic *materia medica* and most of these herbs have been investigated both in animal and human models (Nadkarni 1954).

The herb *Gymnema sylvestre* has gained importance, as chewing its leaves can reduce the sensation of sweet taste due to which it is known as *gurmār* or "sugar-destroyer" (Power 1904). An extract of *Gymnema sylvestre* leaves (400 mg/day) was administered to 22 type 2 diabetic patients for 18–20 months, in addition to conventional oral drugs. The patients showed a significant decrease in blood glucose, glycosylated hemoglobin and glycosylated plasma proteins. The conventional drugs could also be withdrawn. Five patients on the extract treatment were able to maintain their glucose homeostasis and discontinued the conventional drugs, indicating regeneration of beta cells in them, as evidenced by increased insulin levels (Baskaran et al. 1990).

Oral administration of aqueous extract of *Ficus religiosa* bark (25, 50 and 100 mg/kg) in diabetic mice models significantly reduced the blood glucose levels, signifying its anti-diabetic activity. There was also an increase in the levels of serum insulin and glycogen content in the liver and skeletal muscle of S.T.Z.-induced diabetic rats, whereas the levels of serum triglyceride and total cholesterol were reduced. The extract was also found to have a role in decreasing lipid-peroxidation activity in the pancreatic cells of S.T.Z.-induced diabetic rats (Pandit et al. 2010).

The extract of aloe gum effectively increased glucose tolerance in both normal and diabetic rats. The treatment also induced a hypoglycemic effect in alloxan-diabetic mice. Further, the extracts from *Mangifera indica*, when administered along with glucose or given an hour before the glucose load, showed hypoglycemic activity which might be the result of decreased intestinal absorption of glucose (Rajesham et al. 2012). The leaf extract of *Azadirachta indica* (Neem) (400 mg/kg body weight) was administered daily for 30 days to diabetic rats. The treatment resulted in positive effects on parameters like glucose tolerance, fasting blood glucose, insulin signaling molecules, serum lipid profile and glucose oxidation in skeletal (Gupta et al. 2017). The aqueous extract of garlic (10 ml/kg/day) decreased fasting blood glucose and serum triglyceride levels in sucrose- fed rabbits. Additionally, administration of S-methyl cysteine sulphoxide, the sulfur-containing amino acid from *Allium cepa* (200 mg/kg) to alloxan-induced diabetic rats for 45 days could control the glucose levels in blood and also normalize the activities of liver enzymes involved in glucose metabolism (Rajesham et al. 2012).The extracts from fruit pulp, seed, leaves and the whole plant of *Momordica charantia* have shown hypoglycemic activity in various animal models. Similar result was observed for the ethanol extract from the plant possessing both antihyperglycemic and hypoglycemic activities through inhibition of liver enzymes (Jia et al. 2017). Aqueous extract of *Ocimum sanctum* leaves significantly decreased sugar levels in the blood in both the normal and the alloxan-induced diabetic rats (Dwivedi and Daspaul 2013). The roots of *Salacia reticulata* have been used extensively in ayurvedic medicine to treat diabetes and obesity. Extracts of the plant are also used as food supplements in countries like Japan and the United States to prevent obesity and diabetes (Li et al. 2008). The phenolic components of the heartwood of *Pterocarpus marsupium* could also markedly lower glucose levels in S.T.Z.-induced diabetic rat models (Manickam et al. 1997).

9.4.2 Ayurveda in Chronic Respiratory Diseases

The leaves, bark, flowers, and roots of *Albizia lebbeck* are used in the treatment of cough, cold, asthma, bronchitis, bronchial catarrh and tuberculosis. The plant parts are mostly used in the forms of extract or powder (Gulati et al. 2016). The aqueous extract from the bark of *Albizia lebbeck* showed anti-asthmatic, anti-allergic and anti-anaphylactic properties in experimental studies. β-boswellic acid was also shown to inhibit the production of pro-inflammatory cytokines such as T.N.F.-α, I.L.-1, I.L.-2, I.L.-6, I.L.-12; and I.F.N.-γ through the suppression of N.F.-κB activation suggesting the effect of *Boswellia serrata* to control the inflammation and contraction of airway smooth muscle in asthmatic disease by inhibiting the enzymes for generation of pro-inflammatory mediators and bronchoconstrictors (Gulati et al. 2016).

9.4.3 Ayurveda in Cancer

Caraka Samhita and *Suśruta Samhita* describe cancer as inflammatory or non-inflammatory swelling. Cancerous conditions are mentioned as *granthi* (minor neoplasm) or *arbuda* (major neoplasm). Cancerous conditions in Ayurveda are sub-divided into *vātaja*, *pittaja*, *kaphaja* and *tridōṣaja* entities (Jain et al. 2010a). Herbs with anti-cancer properties such as *Piper longum*, *Tinospora cordifolia* and *Semecarpus anacardium* are used in Ayurveda (Poornima and Efferth 2016).

The actives from *Tinospora cordifolia* could enhance the host immune system by increasing immunoglobulins and white blood cell counts and stimulating stem cell proliferation, resulting in reduced solid tumors (Matthew and Kuttan 1999). *Basella rubra*, *Musa paradisiaca*, conch shell ash, *Elaeocarpus tuberculatus*, sulfur, potassium carbonate, *Embelia ribes* and *Zingiber officinale* are employed to treat *arbuda* neoplasms (Dash and Kashyap 1987). *Pittaja arbuda* tumors are cured with the leaves of *Ficus glomerata*, *Tectona grandis* and *Elephantopus scaber*, frequently followed by applying with honey, mixture of the fine pastes of *Aglaia roxburghiana*, *Xanthium strumarium*, *Caesalpinia sappan*, *Terminalia arjuna* and *Symplocos racemosa* (Dash and Kashyap 1987). An aqueous extract of *Phyllanthus amarus* administered in rats increased the life span of the animals and normalized the activity of γ-glutamyl transpeptidase (Rajeshkumar and Kuttan 2000).

9.4.4 Ayurveda in Cardiovascular Disease

Ayurvedic texts describe two complications of *āmavāta* – *Hṛdayaviśuddhi* and *Hṛdgraha*. The former represents the unclean state of the heart and the latter an impaired condition (Mithra 2019). The treatment for heart disease mentions the use of *Arjuna* for the first time by Vṛnda, and later by Cakrapāṇi Datta and Bhāvamiśra. All of them recommend the use of different bark powder formulations of *Arjuna* (Seth et al. 2013). The bark of *Terminalia arjuna* tree is extensively used in ayurvedic therapy for several cardiovascular diseases (Mahmood et al. 2010). The bark is demonstrated to have cardioprotective properties that range from positive inotropic, hypolipidemic, coronary vasodilatory and antioxidant effects to the initiation of stress protein in the heart of various animal models (Maulik and Katiyar 2010). *Berberis aristata*, the major source of alkaloid berberine, has also been an essential component of Ayurveda and traditional Chinese medicine for over 2000 years, due to its outstanding effects on cardiovascular disease (Feng et al. 2019).

9.4.5 Ayurveda in Arthritis

Ayurvedic management of *āmavāta* (rheumatoid arthritis) first requires that the *āma* which is excess of collagen materials, immune bodies or crystal-like uric acid be removed by reducing food intake to near or total fasting (*kṣudnigraha*). The diet should consist of small quantities of double-boiled rice and soggy lentils for the first five days along with small amounts of vegetables for the following five days, then followed by one *rōṭṭi* (flat bread) for five more days. This diet plan is recommended for a minimum of one month (Mishra et al. 2001b). Recent study with 394 chronic rheumatic patients reported relief with the ayurvedic system (Chopra 2000). In *Caraka Samhita*, *ēraṇḍa taila* (castor oil) and guggul gum have been recommended as potent anti-arthritic medicines. Other *rasāyana* formulations include *Withania somnifera*, *Daśamūla*, guggul, *Mahāyōgarāja Guggul*, *Yōgarāja Guggul* and *Triphalā Cūrṇa*. The preparations of guggul contain minerals such as gold silver, copper, iron, mica, mercury, sulfur, zinc and lead, which are useful against arthritis (Chopra et al. 2010). The administration of root extract of *Withania somnifera* in monosodium urate crystal-induced inflammation, exhibited suppressive activity in rats by reducing the expression and proliferation of inflammatory response genes, without gastric damage (Rasool and Varalakshmi 2006).

9.4.6 Ayurveda in Digestive Diseases

The simultaneous administration of alcohol extract of fenugreek seeds for 60 days inhibited the leakage of enzymatic activities of aspartate transaminase, alkaline phosphatase and alanine transaminase into serum and lipid peroxidation in liver and brain with ethanol toxicity (Srinivasan 2006). Hydroxycitric acid, the major compound in *Garcinia cambogia*, is identified to block the synthesis of lipids and fatty acids. It can also prevent the activity of enzyme A.T.P.-citrate lyase that decreases the production of acetyl-CoA, the important substance in fat and carbohydrate metabolism, leading to low L.D.L. and triglycerides formation (Meena et al. 2009). Oral administration of the extract of *Emblica officinalis* at doses of around 500 mg/kg inhibited gastric lesions in rats. The extract was found to protect against ethanol-induced stomach wall mucus depletion and reduce sulfhydryl concentration. The study suggested *Emblica officinalis* extract as antisecretory, antiulcer and cytoprotective (Al-Rehaily et al. 2002). The lipid components from fruits and vegetables provide mucosal cytoprotection through increased prostaglandins production, glycogen and glycoproteins like mucin. In many *in vivo* studies fruits and vegetables like *Allium sativum*, *Abelmoschus esculentus*, *Citrus limon*, *Citrus aurantium* and *Emblica officinalis* could increase prostaglandin E2 in gastric ulcer models, resulting in cytoprotection of gastric mucosa (Harsha et al. 2017).

9.5 COSMECEUTICALS

Cosmeceuticals are a group of cosmetic products that are known to have biologically active components with medicinal or drug-like benefits when applied. Ayurveda literature mentions more than 700 herbs, minerals and fats that can be used to maintain and increase health and skin beauty. The term *cosmeceutical* was coined by Albert Kligman in the national scientific meeting of the Society of Cosmetic Chemists. The word refers to advanced products that are able to improve skin status and beautify the skin, but which are not regarded as drugs or cosmetics (Anunciato and da Rocha Filho 2012). Cosmetics are substances that are required to be rubbed, poured, sprinkled or sprayed, or introduced or applied to the human body to cleanse, beautify, promote attractiveness or alter the appearance. These products can be used in skincare, hair care, anti-aging, and fragrances, satisfying the consumers' needs (Yadav and Chaudhary 2015).

Nutricosmetics is a new term in the field of cosmetology that describes the intake of food or oral supplements that can produce an appearance benefit. They are also known as "beauty pills", "beauty from within", and even "oral cosmetics". These products are formulated by the combination of both cosmeceuticals and nutraceuticals (Anunciato and da Rocha Filho 2012).

9.5.1 History of Cosmeceuticals

The first archaeological evidence of the use of cosmetics was found in Egypt in 3500 B.C. The origin of the word *cosmetics* is Greek, which defines cosmetics as products that have biologically active substances having similar properties to drugs or medicine. The beauty of hair and skin basically depends on the individual's job, routine, health, climatic situation and their maintenance (Saraf et al. 2014). Ayurvedic cosmetics have been used and practiced for many centuries in India and have few side-effects. The investigation of many herbal components with modern scientific techniques has led to identifying the bioactive components in Indian herbs that can bestow functional and other benefits such as anti-dandruff and deodorant (Sarvesh Kumar et al. 2013).

9.5.2 Cosmeceuticals in Ayurveda

Modern cosmeceuticals are mostly dependent on the basics of anti-aging properties such as *vayasthāpana* (age defying), *sandhānīya* (cell regenerating), *varṇya* (brightening skin-glow), *vraṇarōpaṇa* (wound-healing), *tvacya* (nurturing), *śōthahara* (anti-inflammatory), *tvacāgnivardhani* (strengthening skin metabolism) and *tvagrasāyana* (retarding aging) (Datta and Paramesh 2010). Herbal extracts are basically added to cosmetic preparations, as they have antioxidant properties. The anti-aging treatment includes therapies such as *ūrjaskara* (promotive) and *vyādhihara* (curative). Ayurveda considers healthy skin as a reflection of the general state of an individual's health and recommends the use of numerous skincare treatments that are required in different phases of life. Ayurvedic herbs purify the skin and eliminate the vitiated *tridōṣa* from the body, as they are mainly involved in skin disorders and other related diseases. Several herbs are mentioned in Ayurveda as having the ability to achieve healthy skin and glowing complexion (Sarvesh Kumar et al. 2013). Such plants are formulated and consumed in the form of capsules, tablets and other dosage forms (Padmavathi 2013). At present, cosmetics are considered to be one of the essential commodities of life. Human preferences have revived the use of natural products and due to these reasons, in recent years there has been an increased interest in research on Indian herbs (Sarvesh Kumar et al. 2013).

9.5.3 Important Cosmeceuticals

The origin of ayurvedic cosmeceuticals dates back to Indus Valley Civilization (Datta and Paramesh 2010). Ayurvedic cosmetology has its source in nature, and many vegetables, fruits and herbs are used to create a wide range of products with properties of skin rejuvenation (Jain et al. 2010b). The use of cosmetics was not only focused on beautification, but also in achieving longevity, which is termed as *Āyus and Ārōgyam* in Sanskrit. There is evidence for the use of various cosmetics by both genders in ancient India and these uses were dependent on season (*ṛtu*) and daily routine (*dina-carya*) (Datta and Paramesh 2010). Ancient Sanskrit literary works like *Abhijñāna Śākuntaḷam* and *Mēgha Sandēśam* of Kāḷidāsa and other mythological epics refer to cosmetics as *tilakam*, *kajjaḷi*, *lepa* and *agaru* (*Aquilaria agallocha*) that were used to beautify the body and to create beauty spots on the chin and cheeks (Shilpa et al. 2002). Cosmeceuticals improve skin function by stimulating collagen growth, offering protection from harmful free radicals. They also maintain good keratin structure, maintaining healthy skin (Patil et al. 2017). These cosmetics are used to maintain and improve the general appearance of the face and other body parts. They are available in the form of creams, lotions, powders, moisturizers, shampoos, hair oils and nail polishes (Pareek et al. 2012).

The Drugs and Cosmetics Act, 1940 defines cosmetics by their intended use. They can be rubbed, poured, sprinkled or sprayed, introduced or applied to human body for beautifying, cleansing, promoting attractiveness and also to alter the appearance. This range of products includes skin moisturizers, perfumes, lipsticks, fingernail polishes, cosmetics for eye makeup, facial makeup preparations, shampoos, hair colors, toothpastes and deodorants, including the components of cosmetic products. These products form the bridge between personal care products and pharmaceuticals, to provide special benefits for medicinal and cosmetic use (Dureja et al. 2005).

The different ayurvedic cosmetics come under the following groups: Cosmetics to enhance facial skin appearance, cosmetics for hair growth and care, cosmetics used in skin treatment of acne, pimples and other skin ailments, shampoos, soaps, powders and perfumes and miscellaneous products (Shilpa et al. 2002).

Skincare products are part of daily care, with their extensive properties to preserve and protect skin. The exposure of skin to U.V. radiation can cause damage due to generation of free radicals and breakage of collagen fibers. Skincare products counter inflammation, wrinkle formation and skin hyperpigmentation. Several herbal extracts prevent aging and enhance skin appearance (Preetha and Karthika 2009).

Haircare, color and style have important roles in physical appearance (Semalty et al. 2011). Procyanidins derived from apples, grape seeds and barley, isoflavones, extracts of *Centella asiatica*, plant-based 5-alpha reductase inhibitors, essential oils, vitamin C and amino acids such as taurine and L-carnitine are beneficial for haircare (Rogers 2014). The use of shampoo is the most common way of cosmetic hair treatment. The formulae of such products are revised according to variations in hair quality, habits of haircare and specific problems associated with scalp conditions (Trüeb 2001).

Medicinal plants that are used as moisturizers, skin tonics and anti-aging agents include *Aloe vera*, *Curcuma longa*, *Glycyrrhiza glabra*, *Ocimum sanctum*, *Rosa damascene*, *Rubia cordifolia* and *Triticum sativum*. Tubers of *Cyperus rotundus* and bark of *Moringa oleifera* are used in suntanning formulations. *Mesua ferrea* acts as strong astringent and *Terminalia chebula* functions as an antibacterial, antifungal and antiseptic (Shilpa et al. 2002).

Dental care is given importance in Ayurveda. Ingredients for dental care products include *Azadirachta indica* (against toothache, bacterial infection, dental caries), *Acacia arabica* (to treat swelling and bleeding gums), *Barleria prionitis* (in toothache, bleeding gums and loose teeth), *Mimusops elengi* (as an astringent, to keep gums healthy), *Pimpinella anisum* (antiseptic, mouth freshener), *Salvadora persica* (as a potent antimicrobial), *Syzygium aromaticum* (as a local anesthetic in toothache) and *Symplocos racemosus* (to strengthen gums and teeth) (Shilpa et al. 2002).

Dermatological preparations employ medicinal plants such as *Allium sativum* as an antifungal and antiseptic; *Alpinia galanga* as an antibacterial; *Azadirachta indica* and *Nigella sativa* as potent antibacterials; *Celastrus paniculata* in wound-healing and eczema; *Pongamia glabra* in herpes and scabies; and *Psoralea corylifolia* in vitiligo, leprosy, psoriasis and inflammation (Shilpa et al. 2002).

Medicinal plants for hair care products include *Acacia concinna* as a natural detergent and anti-dandruff; *Aloe vera* as a cleanser and revitalizer; *Azadirachta indica* in reducing hair loss and dandruff; *Bacopa monnieri* as a hair tonic and hair growth enhancer; *Cedrus deodara* in dandruff control; *Centella asiatica* in darkening of hair; *Eclipta alba* in prevention of hair graying and alopecia; *Emblica officinalis* in toning and dandruff control, protection and alopecia; *Hibiscus rosa-sinensis* as a natural hair dye, prevention of hair loss and dandruff control; *Hedychium spicatum* in promoting hair growth; *Lawsonia alba* as a natural hair dye, anti-dandruff and conditioner; *Sapindus trifoliatus* as a natural detergent and cleanser; *Triticum sativum* in providing nourishment, lubrication and luster; *Terminalia bellerica* in the prevention of hair graying and *Sesamum indicum* to promote hair growth and darkening (Shilpa et al. 2002).

The pods and nuts of the herb *shikakai* (*Acacia concinna*) are widely used in shampoos. The pods have neutral pH in aqueous suspension and are used to make mild detergent rich in saponins. Similarly, *Sapindus trifoliatus* (soapnuts) contains saponins that are used as a foaming agent that takes the place of soap in ayurvedic tradition (Jain et al. 2010b). *Āmla* (*Emblica officinalis*) with its rich content of vitamin C has importance in rejuvenation of both body and hair. As advised by Caraka, the regular intake of *āmla* can delay the aging process (Murari et al. 2013). Boswellic acid obtained from *Boswellia serrata* inhibits enzymes like 5-lipoxygenase, responsible for inflammation and skin damage (Preetha and Karthika 2009). The clear gel of *Aloe vera* has the ability to heal wounds, ulcers and burns by forming a coating over areas that are affected, and helps in healing. Besides providing protection against skin infections and reducing wrinkles, *Aloe vera* finds its place in shampoo that improves dry and brittle hair (Jain et al. 2010b).

9.6 CONCLUSION

Phytochemicals from various herbs, vegetables and fruits are reported to reduce different kinds of chronic diseases, including cancers and age-related pathological disorders (Kumar et al. 2014). From the time of human civilization, herbs have been considered as powerful tools to treat different diseases, and various cultures have produced their own mixtures of herbs to deal with common health problems. Sometimes these herbal mixtures are superior and more effective with a better safety profile and lower cost, as compared to their synthetic counterparts (Meena et al. 2009). Ayurveda describes the use of medicinal herbs to cure diseases. The use of ayurvedic plants have gained attention because they are easily available, biodegradable, cost-effective and have fewer side-effects. Despite the advancements in techniques to treat chronic diseases, mainstream medicine is associated with various side-effects. Therefore, adapting ayurvedic knowledge can be beneficial to treat chronic diseases without side-effects. Nutraceuticals derived from ayurvedic herbs can also offer similar benefits. Such nutraceuticals can prevent chronic diseases, improve health conditions and increase longevity.

Medicinal plants from Ayurveda are also important sources of cosmeceuticals that offer benefits to skin health and personal care. Cosmeceuticals improve the texture and function of skin. The important cosmeceuticals are derived from pomegranate, dates, grapes, turmeric, aloe, red sandalwood and so on. Therefore, Ayurveda offers the opportunity of developing nutraceuticals and cosmeceuticals that can offer protection from chronic diseases and enhance beauty.

REFERENCES

Aggarwal, B. B. 2010. Targeting inflammation-induced obesity and metabolic diseases by curcumin and other nutraceuticals. *Annual Review of Nutrition* 21: 173–99.

Alam, M. N., M. M. Rahman, and M. I. Ibrahim Khalil. 2014. Nutraceuticals in arthritis management: A contemporary prospect of dietary phytochemicals. *The Open Nutraceuticals Journal* 7: 21–7.

Albini, A., B. Bassani, D. Baci et al. 2019. Nutraceuticals and "repurposed" drugs of phytochemical origin in prevention and interception of chronic degenerative diseases and cancer. *Current Medicinal Chemistry* 26, no. 6: 973–87.

Al-Okbi, S. Y. 2014. Nutraceuticals of anti-inflammatory activity as complementary therapy for rheumatoid arthritis. *Toxicology and Industrial Health* 30, no. 8: 738–49.

Al-Rehaily, A. J., T. A. Al-Howiriny, M. O. Al-Sohaibani, and S. Rafatullah. 2002. Gastroprotective effects of 'Amla' *Emblica officinalis* on *in vivo* test models in rats. *Phytomedicine* 9, no. 6: 515–522.

Ameye, L. G., and W. S. Chee. 2006. Osteoarthritis and nutrition. From nutraceuticals to functional foods: A systematic review of the scientific evidence. *Arthritis Research and Therapy* 8, no. 4: R127.

Anunciato, T. P., and P. A. da Rocha Filho. 2012. Carotenoids and polyphenols in nutricosmetics, nutraceuticals, and cosmeceuticals. *Journal of Cosmetic Dermatology* 11, no. 1: 51–4.

Baldi, A., N. Choudhary, and S. Kumar. 2013. Nutraceuticals as therapeutic agents for holistic treatment of diabetes. *International Journal of Green Pharmacy* 7, no. 4: 278–87.

Banik, K., C. Harsha, D. Bordoloi, et al. 2018. Therapeutic potential of gambogic acid, a caged xanthone, to target cancer. *Cancer Letters* 416: 75–86.

Banik, K., A. M. Ranaware, V. Deshpande et al. 2019. Honokiol for cancer therapeutics: A traditional medicine that can modulate multiple oncogenic targets. *Pharmacological Research* 144: 192–209.

Banik, K., A. M. Ranaware, C. Harsha et al. 2020. Piceatannol: A natural stilbene for the prevention and treatment of cancer. *Pharmacological Research* 153:104635. doi:10.1016/j.phrs.2020.104635.

Baragi, P. C., B. J. Patgiri, and P. K. Prajapati. 2008. Nutraceuticals in Ayurveda with special reference to Avaleha Kalpana. *Ancient Science of Life* 28, no. 2: 29–32.

Baskaran, K., K. B. Ahamath, K. R. Shanmugasundaram, and E. R. Shanmugasundaram. 1990. Antidiabetic effect of a leaf extract from *Gymnema sylvestre* in non-insulin-dependent diabetes mellitus patients. *Journal of Ethnopharmacology* 30, no. 3: 295–300.

Basu, J., R. Avila, and R. Ricciardi. 2016. Hospital readmission rates in U. S. States: Are readmissions higher where more patients with multiple chronic conditions cluster? *Health Services Research Journal* 51, no. 3: 1135–51.

Bernhoft, A. 2010. A brief review on bioactive compounds in plants, In: Bioactive compounds in plants – Benefits and risks for man and animals, Proceedings from a symposium held at The Norwegian Academy of Science and Letters, Oslo, 13 – 14 November 2008. *The Norwegian Academy of Science and Letters*, 11–17.

Boots, A. W., G. R. Haenen, and A. Bast. 2008. Health effects of quercetin: From antioxidant to nutraceutical. *European Journal of Pharmacology* 585, no. 2–3: 325–37.

Bordoloi, D., J. Monisha, N. K. Roy et al. 2019. An investigation on the therapeutic potential of butein, a tetrahydroxychalcone against human oral squamous cell carcinoma. *Asian Pacific Journal of Cancer Prevention: A.P.J.C.P.* 20, no. 11: 3437–46.

Buhrmann, C., B. Popper, A. B. Kunnumakkara, B. B. Aggarwal, and M. Shakibaei. 2019. Evidence that calebin A, a component of *Curcuma longa* suppresses NF-B mediated proliferation, invasion and metastasis of human colorectal cancer induced by TNF-β (lymphotoxin). *Nutrients* 11, no. 12: pii: E2904. doi:10.3390/nu11122904.

Camuesco, D., M. Comalada, A. Concha et al. 2006. Intestinal anti-inflammatory activity of combined quercitrin and dietary olive oil supplemented with fish oil, rich in EPA and DHA (n-3) polyunsaturated fatty acids, in rats with DSS-induced colitis. *Clinical Nutrition* 25, no. 3: 466–76.

Chaudhary, A., and N. Singh. 2012. Intellectual property rights and patents in perspective of Ayurveda. *Ayu* 33, no. 1: 20–6.

Chauhan, A., D. K. Semwal, S. P. Mishra, and R. B. Semwal. 2015. Ayurvedic research and methodology: Present status and future strategies. *Ayu* 36, no. 4: 364–9.

Chauhan, B., G. Kumar, N. Kalam, and S. H. Ansari. 2013. Current concepts and prospects of herbal nutraceutical: A review. *Journal of Advanced Pharmaceutical Technology and Research* 4, no. 1: 4–8.

Chintale, A. G., V. S. Kadam, R. S. Sakhare, G. O. Birajdar, and D. N. Nalwad. 2013. Role of nutraceuticals in various diseases: A comprehensive review. *International Journal of Research in Pharmacy and Chemistry* 3: 290–9.

Chopra, A. 2000. Ayurvedic medicine and arthritis. *Rheumatic Disease Clinics of North America* 26, no. 1: 133–44.

Chopra, A., M. Saluja, and G. Tillu. 2010. Ayurveda-modern medicine interface: A critical appraisal of studies of ayurvedic medicines to treat osteoarthritis and rheumatoid arthritis. *Journal of Ayurveda and Integrative Medicine* 1, no. 3: 190–8.

Cicero, A. F., and A. Baggioni. 2016. Berberine and its role in chronic disease. *Advances in Experimental Medicine and Biology* 928: 27–45.

Daily, J. W., M. Yang, and S. Park. 2016. Efficacy of turmeric extracts and curcumin for alleviating the symptoms of joint arthritis: A systematic review and meta-analysis of randomized clinical trials. *Journal of Medicinal Food* 19, no. 8: 717–29.

Dash, B., and L. Kashyap. 1987. Diagnosis and treatment of Galaganda, Gandamala, Apaci, granthi and arbuda. In *Diagnosis and treatment of diseases in Ayurveda*, ed. B. Dash, and L. Kashyap, 437–66. New Delhi: Concept Publishing Company.

Datta, H. S., and R. Paramesh. 2010. Trends in aging and skin care: Ayurvedic concepts. *Journal of Ayurveda and Integrative Medicine* 1, no. 2: 110–3.

Davis, R. M., E. G. Wagner, and T. Groves. 2000. Advances in managing chronic disease. Research, performance measurement, and quality improvement are key. *British Medical Journal* 320, no. 7234: 525–6.

DeFelice, S. L. 1995. The nutraceutical revolution: Its impact on food industry R&D. *Trends in Food Science and Technology* 6, no. 2: 59–61.

Delfan, B., M. Bahmani, H. Hassanzadazar, K. Saki, and M. Rafieian-Kopaei. 2014. Identification of medicinal plants affecting on headaches and migraines in Lorestan Province, West of Iran. *Asian Pacific Journal of Tropical Medicine* 7S1: S376–9.

Dhaliwal, K. S. 1999. Method and composition for treatment of diabetes. *U.S. Patent 5,886,029*, issued March 23,1999.

Di Girolamo, C., W. J. Nusselder, M. Bopp et al. 2020. Progress in reducing inequalities in cardiovascular disease mortality in Europe. *Heart* 106, no. 1: 40–9.

Dua, K., V. Malyla, G. Singhvi et al. 2019. Increasing complexity and interactions of oxidative stress in chronic respiratory diseases: An emerging need for novel drug delivery systems. *Chemico-Biological Interactions* 299: 168–78.

Dureja, H., D. Kaushik, M. Gupta, V. Kumar, and V. Lather. 2005. Cosmeceuticals: An emerging concept. *Indian Journal of Pharmacology* 37, no. 3: 155–9.

Dwivedi, C., and S. Daspaul. 2013. Antidiabetic herbal drugs and polyherbal formulation used for diabetes: A review. *Journal of Phytopharmacology* 2: 44–51.

Elder, C. 2004. Ayurveda for diabetes mellitus: A review of the biomedical literature. *Alternative Therapies in Health and Medicine* 10, no. 1: 44–50.

Felson, D. T. 2009. Developments in the clinical understanding of osteoarthritis. *Arthritis Research and Therapy* 11, no. 1: 203. doi:10.1186/ar2531.

Feng, X., A. Sureda, S. Jafari et al. 2019. Berberine in cardiovascular and metabolic diseases: From mechanisms to therapeutics. *Theranostics* 9, no. 7: 1923–51.

Ghani, U., S. S. H. Bukhari, S. Ullah et al. 2019. A review on nutraceuticals as therapeutic agents. *International Journal of Biological Sciences* 15: 326–40.

Ghattamaneni, N. K. R., S. K. Panchal, and L. Brown. 2018. Nutraceuticals in rodent models as potential treatments for human inflammatory bowel disease. *Pharmacological Research* 132: 99–107.

Girisa, S., B. Shabnam, J. Monisha et al. 2019. Potential of zerumbone as an anti-cancer agent. *Molecules* 24, no. 4: pii: E734. doi:10.3390/molecules24040734.

Glazier, A. 2000. A landmark in the history of Ayurveda. *The Lancet* 356, no. 9235: 1119.

Gulati, K., N. Rai, S. Chaudhary, and A. Ray. 2016. Nutraceuticals in respiratory disorders. In Nutraceuticals*: Efficacy, safety and toxicity*, 75–86. ed. R.C. Gupta. New York: Academic Press.

Gupta, R., B. Gabrielsen, and S. M. Ferguson. 2005. Nature's medicines: Traditional knowledge and intellectual property management. Case studies from the National Institutes of Health (NIH), USA. *Current Drug Discovery Technologies* 2, no. 4: 203–19.

Gupta, R. C. 2016. Nutraceuticals in arthritis. In Nutraceuticals*: Efficacy, safety and toxicity*, 161–76. ed. R.C. Gupta. New York: Academic Press.

Gupta, S. C., A. B. Kunnumakkara, S. Aggarwal, and B. B. Aggarwal. 2018. Inflammation, a double-edge sword for cancer and other age-related diseases. *Frontiers in Immunology* 9: 2160. doi:10.3389/fimmu.2018.02160.

Gupta, S. C., S. Prasad, A. K. Tyagi, A. B. Kunnumakkara, and B. B. Aggarwal. 2017. Neem (*Azadirachta indica*): An Indian traditional panacea with modern molecular basis. *Phytomedicine: International Journal of Phytotherapy and Phytopharmacology* 34: 14–20.

Gupta, S. C., A. Sharma, S. Mishra, and N. Awasthee. 2019. Nutraceuticals for the prevention and cure of cancer. In *Nutraceuticals in veterinary medicine*, ed. R. C. Gupta, A. Srivastava, and R. Lall, 603–10. Cham: Springer.

Hang, L., A. H. Basil, and K. L. Lim. 2016. Nutraceuticals in Parkinson's disease. *NeuroMolecular Medicine* 18, no. 3: 306–21.

Hankey, A. 2010. Ayurveda and the battle against chronic disease: An opportunity for Ayurveda to go mainstream? *Journal of Ayurveda and Integrative Medicine* 1, no. 1: 9–12.

Harsha, C., K. Banik, D. Bordoloi, and A. B. Kunnumakkara. 2017. Antiulcer properties of fruits and vegetables: A mechanism- based perspective. *Food and Chemical Toxicology: An International Journal Published for the British Industrial Biological Research Association* 108, no. A: 104–19.

Harvey, A. L. 2008. Natural products in drug discovery. *Drug Discovery Today* 13, no. 19–20: 894–901.

Hatcher, H., R. Planalp, J. Cho, F. M. Torti, and S. V. Torti. 2008. Curcumin: From ancient medicine to current clinical trials. *Cellular and Molecular Life Sciences: CMLS* 65, no. 11: 1631–52.

Heemels, M. T. 2016. Neurodegenerative diseases. *Nature* 539, no. 7628: 179.

Humber, J. M. 2002. The role of complementary and alternative medicine: Accommodating pluralism. *JAMA* 288: 1655–6.

Hwang, Y. Y., and Y. S. Ho. 2018. Nutraceutical support for respiratory diseases. *Food Science and Human Wellness* 7, no. 3: 205–8.

Islam, J., H. Shirakawa, and Y. Kabir. 2020. Emerging roles of nutraceuticals from selected fermented foods in lifestyle-related disease prevention. In *Herbal medicine in India*, ed. S. Sen, and R. Chakraborty, 479–88. Singapore: Springer.

Jain, A., S. Dubey, A. Gupta, P. Kannojia, and V. Tomar. 2010b. Potential of herbs as cosmeceuticals. *International Journal of Research in Ayurveda and Pharmacy* 1: 71–7.

Jain, R., S. Kosta, and A. Tiwari. 2010a. Ayurveda and cancer. *Pharmacognosy Research* 2, no. 6: 393–4.

Jaiswal, Y. S., and L. L. Williams. 2017. A glimpse of Ayurveda - The forgotten history and principles of Indian traditional medicine. *Journal of Traditional and Complementary Medicine* 7, no. 1: 50–3.

Jia, S., M. Shen, F. Zhang, and J. Xie. 2017. Recent advances in *Momordica charantia*: Functional components and biological activities. *International Journal of Molecular Sciences* 18, no. 12: pii: E2555. doi:10.3390/ijms18122555.

Kannappan, R., S. C. Gupta, J. H. Kim, S. Reuter, and B. B. Aggarwal. 2011. Neuroprotection by spice-derived nutraceuticals: You are what you eat! *Molecular Neurobiology* 44, no. 2: 142–59.

Kasbia, G. S. 2005. Functional foods and nutraceuticals in the management of obesity. *Nutrition and Food Science* 35, no. 5: 344–52.

Khamkar, S. C., and A. Dubewar. 2019. Nutraceuticals in ayurvedic Sneha Kalpana with special reference to Ghrita. *Journal of Ayurveda and Integrated Medical Sciences* 4: 289–94.

Khan, A., M. Safdar, M. M. Ali Khan, K. N. Khattak, and R. A. Anderson. 2003. Cinnamon improves glucose and lipids of people with type 2 diabetes. *Diabetes Care* 26, no. 12: 3215–8.

Khan, R. A., G. O. Elhassan, and K. A. Qureshi. 2014. Nutraceuticals in the treatment and prevention of diseases-an overview. *The Pharmaceutical Innovation* 3: 47–50.

Khwairakpam, A. D., D. Bordoloi, K. K. Thakur et al. 2018. Possible use of *Punica granatum* (pomegranate) in cancer therapy. *Pharmacological Research* 133: 53–64.

Kumar, H., I. S. Kim, S. V. More, B. W. Kim, and D. K. Choi. 2014. Natural product-derived pharmacological modulators of Nrf2/ARE pathway for chronic diseases. *Natural Product Reports* 31, no. 1: 109–39.

Kumar, N., Z. A. Wani, and S. Dhyani. 2015. Ethnobotanical study of the plants used by the local people of Gulmarg and its allied areas, Jammu & Kashmir, India. *International Journal of Current Research in Bioscience and Plant Biology* 2: 16–23.

Kumar, P., and A. Sharma. 2020. Analysis of international patent applications for inventions like traditional herbal medicines. In *Handbook of research on emerging trends and technologies in Library and information science*, ed. A. Kaushik, A. Kumar, and P. Biswas, 346–56. PA: IGI Global.

Kumar, S., P. Satadru, S. K. Maurya, and D. Kumar. 2013. Skin care in Ayurveda: A literary review. *International Research Journal of Pharmacy* 4, no. 3: 1–3.

Kunnumakkara, A. B., K. Banik, D. Bordoloi et al. 2018b. Googling the guggul (*Commiphora* and *Boswellia*) for prevention of chronic diseases. *Frontiers in Pharmacology* 9: 686. doi:10.3389/fphar.2018.00686.

Kunnumakkara, A. B., D. Bordoloi, G. Padmavathi et al. 2017. Curcumin, the golden nutraceutical: Multitargeting for multiple chronic diseases. *British Journal of Pharmacology* 174, no. 11: 1325–48.

Kunnumakkara, A. B., B. L. Sailo, K. Banik et al. 2018a. Chronic diseases, inflammation, and spices: How are they linked? *Journal of Translational Medicine* 16, no. 1: 14. doi:10.1186/s12967-018-1381-2.

Larkin, K. T., and P. D. Chantler. 2020. Stress, depression, and cardiovascular disease. In *Cardiovascular implications of stress and depression*, ed. P. D. Chantler, and K. T. Larkin, 1–12. London: Academic Press.

Li, Y., T. H. Huang, and J. Yamahara. 2008. Salacia root, a unique Ayurvedic medicine, meets multiple targets in diabetes and obesity. *Life Sciences* 82, no. 21–22: 1045–9.

Liu, W. H., and L. S. Chang. 2010. Piceatannol induces Fas and FASL up-regulation in human leukemia U937 cells via Ca2+/p38alpha MAPK-mediated activation of c-Jun and ATF-2 pathways. *The International Journal of Biochemistry and Cell Biology* 42, no. 9: 1498–506.

Mahmood, Z. A., M. Sualeh, S. B. Mahmood, and M. A. Karim. 2010. Herbal treatment for cardiovascular disease the evidence based therapy. *Pakistan Journal of Pharmaceutical Sciences* 23, no. 1: 119–24.

Manickam, M., M. Ramanathan, M. A. Jahromi, J. P. Chansouria, and A. B. Ray. 1997. Antihyperglycemic activity of phenolics from *Pterocarpus marsupium. Journal of Natural Products* 60, no. 6: 609–10.

Marin, M., R. M. Giner, J. L. Ríos, and M. C. Recio. 2013. Intestinal anti-inflammatory activity of ellagic acid in the acute and chronic dextrane sulfate sodium models of mice colitis. *Journal of Ethnopharmacology* 150, no. 3: 925–34.

Matthew, S., and G. Kuttan. 1999. Immunomodulatory and anti-tumour activities of *Tinospora cordifolia. Fitoterapia* 70, no. 1: 35–43.

Maulik, S. K., and C. K. Katiyar. 2010. *Terminalia arjuna* in cardiovascular diseases: Making the transition from traditional to modern medicine in India. *Current Pharmaceutical Biotechnology* 11, no. 8: 855–60.

Mecocci, P., C. Tinarelli, R. J. Schulz, and M. C. Polidori. 2014. Nutraceuticals in cognitive impairment and Alzheimer's disease. *Frontiers in Pharmacology* 5: 147. doi:10.3389/fphar.2014.00147.

Meena, A. K., P. Bansal, and S. Kumar. 2009. Plants-herbal wealth as a potential source of ayurvedic drugs. *Asian Journal of Traditional Medicines* 4: 152–70.

Mishra, L., B. B. Singh, and S. Dagenais. 2001a. Ayurveda: A historical perspective and principles of the traditional healthcare system in India. *Alternative Therapies in Health and Medicine* 7, no. 2: 36–42.

Mishra, L., B. B. Singh, and S. Dagenais. 2001b. Healthcare and disease management in Ayurveda. *Alternative Therapies in Health and Medicine* 7, no. 2: 44–50.

Mithra, M. P. 2019. Ayurvedic approach to treat Hridroga (valvular heart disease): A case report. *Journal of Ayurveda and Integrative Medicine.* pii: S0975-9476(18)30302-4. doi:10.1016/j.jaim.2018.08.002.

Muhtadi, M., Y. N. Annissa, A. Suhendi, and L. M. Sutrisna. 2018. Hypoglycemic effect of *Zingiber zerumbet* ethanolic extracts and *Channa striata* powder in alloxan-induced diabetic rats. *Journal of Nutraceuticals and Herbal Medicine* 1: 9–15.

Mukherjee, P., and P. Houghton. 2009. The worldwide phenomenon of increased use of herbal products: Opportunity and threats. In *Evaluation of herbal medicinal products*, ed. P. Houghton, and P. Mukherjee, 1–12. London: The Pharmaceutical Press.

Murari, K., B. K. Bharti, and S. K. Bharti. 2013. Indian plants in medicine -Green economics. *Trends in Biosciences* 6: 864–5.

Nadkarni, A. K. 1954. *Indian materia medica.* Bombay: Popular Prashkan. 1954.

Narayanaswamy, V. 1981. Origin and development of Ayurveda: (a brief history). *Ancient Science of Life* 1, no. 1: 1–7.

Nasri, H., Baradaran, A., Shirzad, H. and Rafieian-Kopaei, M. 2014. New concepts in nutraceuticals as alternative for pharmaceuticals. *International Journal of Preventive Medicine* 5: 1487–99.

Ninivaggi, F. J. 2008. *Ayurveda: A comprehensive guide to traditional Indian medicine for the West.* MD: Rowman and Littlefield Publisher, Inc.

Nones, K., Y. E. Dommels, S. Martell et al. 2009. The effects of dietary curcumin and rutin on colonic inflammation and gene expression in multidrug resistance gene-deficient (mdr1a–/–) mice, a model of inflammatory bowel diseases. *British Journal of Nutrition* 101, no. 2: 169–81.

Ono, K., K. Hasegawa, H. Naiki, and M. Yamada. 2004. Curcumin has potent anti-amyloidogenic effects for Alzheimer's beta-amyloid fibrils *in vitro*. *Journal of Neuroscience Research* 75, no. 6: 742–50.

Padmavathi, G., S. R. Rathnakaram, J. Monisha, D. Bordoloi, N. K. Roy, and A. B. Kunnumakkara. 2015. Potential of butein, a tetrahydroxychalcone to obliterate cancer. *Phytomedicine* 22, no. 13: 1163–71.

Padmavathi, M. 2013. Chronic disease management with nutraceuticals. *International Journal of Pharmaceutical Science Invention* 2: 1–11.

Pan, S. Y., G. Litscher, S. H. Gao et al. 2014. Historical perspective of traditional indigenous medical practices: The current renaissance and conservation of herbal resources. *Evidence-Based Complementary and Alternative Medicine* 2014: 525340. doi:10.1155/2014/525340.

Pandey, M. M., S. Rastogi, and A. K. Rawat. 2013. Indian traditional ayurvedic system of medicine and nutritional supplementation. *Evidence- Based Complementary and Alternative Medicine* 2013:376327. doi:10.1155/2013/376327.

Pandit, R., A. Phadke, and A. Jagtap. 2010. Antidiabetic effect of *Ficus religiosa* extract in streptozotocin-induced diabetic rats. *Journal of Ethnopharmacology* 128, no. 2: 462–66.

Pareek, A., V. Jain, Y. Ratan, S. Sharma, P. K. Jain, and V. Dave. 2012. Mushrooming of herbals in new emerging markets of cosmeceuticals. *International Journal of Advanced Research in Pharmaceutical and Biosciences* 2: 473–80.

Parian, A. M., B. N. Limketkai, N. D. Shah, and G. E. Mullin. 2015. Nutraceutical supplements for inflammatory bowel disease. *Nutrition in Clinical Practice: Official Publication of the American Society for Parenteral and Enteral Nutrition* 30, no. 4: 551–8.

Park, J. H., D. Dehaini, J. Zhou, M. Holay, R. H. Fang, and L. Zhang. 2020. Biomimetic nanoparticle technology for cardiovascular disease detection and treatment. *Nanoscale Horizons* 5, no. 1: 25–42.

Park, M. Y., H. J. Kwon, and M. K. Sung. 2011. Dietary aloin, aloesin, or aloe-gel exerts anti-inflammatory activity in a rat colitis model. *Life Sciences* 88, no. 11–12: 486–92.

Pastrelo, M. M., C. C. Dias Ribeiro, J. W. Duarte et al. 2017. Effect of concentrated apple extract on experimental colitis induced by acetic acid. *International Journal of Molecular and Cellular Medicine* 6, no. 1: 38–49.

Patel, P., P. Harde, J. Pillai, N. Darji, and B. Patel. 2012. Antidiabetic herbal drugs a review. *Pharmacophore* 3: 18–29.

Patil, A. S., A. V. Patil, A. H. Patil, T. A. Patil, M. Bhurat, and S. Barhate. 2017. A review on: Standardization of herb in new era of cosmeceuticals: Herbal cosmetics. *World Journal of Pharmaceutical Research* 6: 303–20.

Patwardhan, B. 2000. Ayurveda: The designer medicine. *Indian Drugs* 37: 213–27.

Patwardhan, B., and R. A. Mashelkar. 2009. Traditional medicine-inspired approaches to drug discovery: Can Ayurveda show the way forward? *Drug Discovery Today* 14, no. 15–16: 804–11.

Patwardhan, B., A. D. B. Vaidya, and M. Chorghade. 2004. Ayurveda and natural products drug discovery. *Current Science* 86: 789–99.

Patwardhan, B., D. Warude, P. Pushpangadan, and N. Bhatt. 2005. Ayurveda and traditional Chinese medicine: A comparative overview. *Evidence-Based Complementary and Alternative Medicine* 2, no. 4: 465–73.

Pocernich, C. B., M. L. Lange, R. Sultana, and D. A. Butterfield. 2011. Nutritional approaches to modulate oxidative stress in Alzheimer's disease. *Current Alzheimer Research* 8, no. 5: 452–69.

Poornima, P., and T. Efferth. 2016. Ayurveda for cancer treatment. *Medicinal and Aromatic Plants* 5, no. 5: 5. doi:10.4172/2167-0412.1000e178.

Power, F. B. 1904. Chemical examination of *Gymnema sylvestre*. *Pharmacologic Journal* 19: 234–46.

Prabhakar, P. K., A. Kumar, and M. Doble. 2014. Combination therapy: A new strategy to manage diabetes and its complications. *Phytomedicine: International Journal of Phytotherapy and Phytopharmacology* 21, no. 2: 123–30.

Prakash, D., C. Gupta, and G. Sharma. 2012. Importance of phytochemicals in nutraceuticals. *Journal of Chinese Medicine Research and Development* 1: 70–8.

Prakash, D., and G. Sharma. 2014. *Phytochemicals of nutraceutical importance*. MA: CABI.
Prasher, B., S. Negi, S. Aggarwal et al. 2008. Whole genome expression and biochemical correlates of extreme constitutional types defined in Ayurveda. *Journal of Translational Medicine* 6: 48. doi:10.1186/1479-5876-6-48.
Preetha, P., and K. Karthika. 2009. Cosmeceuticals–an evolution. *International Journal of ChemTech Research* 1: 1217–23.
Puri, H. S. 2002. *Rasayana: Ayurvedic herbs for longevity and rejuvenation*, 331–2. Boca Raton: CRC Press.
Raghupathi, W., and V. Raghupathi. 2018. An empirical study of chronic diseases in the United States: A visual analytics approach. *International Journal of Environmental Research and Public Health* 15, no. 3: pii: E431. doi:10.3390/ijerph15030431.
Rajasekaran, A., G. Sivagnanam, and R. Xavier. 2008. Nutraceuticals as therapeutic agents: A review. *Research Journal of Pharmacy and Technology* 1: 328–40.
Rajesham, V. V., A. Ravindernath, and D. V. R. N. Bikshapathi. 2012. A review on medicinal plants and herbal drug formulations used in diabetes mellitus. *Indo-American Journal of Pharmaceutical Research* 2: 1200–12.
Rajeshkumar, N. V., and R. Kuttan. 2000. *Phyllanthus amarus* extract administration increases the life span of rats with hepatocellular carcinoma. *Journal of Ethnopharmacology* 73, no. 1–2: 215–19.
Ranaware, A. M., K. Banik, V. Deshpande et al. 2018. Magnolol: A neolignan from the Magnolia family for the prevention and treatment of cancer. *International Journal of Molecular Sciences* 19, no. 8: pii: E2362. doi:10.3390/ijms19082362.
Rani, Y., and N. K. Sharma. 2005. Nutraceuticals: Ayurveda perspective. In *Aromatic Plants-Volume 6: Traditional Medicine and Nutraceuticals*, 131–6. In *III WOCMAP Congress on Medicinal and Aromatic Plants*. doi:10.17660/ActaHortic.2005.680.192003.
Rasool, M., and P. Varalakshmi. 2006. Suppressive effect of *Withania somnifera* root powder on experimental gouty arthritis: An *in vivo* and *in vitro* study. *Chemico-Biological Interactions* 164, no. 3: 174–80.
Rivellese, A. A., P. Ciciola, G. Costabile, C. Vetrani, and M. Vitale. 2019. The possible role of nutraceuticals in the prevention of cardiovascular disease. *High Blood Pressure and Cardiovascular Prevention: The Official Journal of the Italian Society of Hypertension* 26, no. 2: 101–11.
Rogers, N. E. 2014. Cosmeceuticals for hair loss and hair care. In *Cosmeceuticals and cosmetic practice*, ed. P. K. Farris, 234–44. West Sussex: John Wiley & Sons Ltd.
Ross, C. A., and M. A. Poirier. 2004. Protein aggregation and neurodegenerative disease. *Nature Medicine* 10: S10–7.
Roth-Walter, F., I. M. Adcock, C. Benito-Villalvilla et al. 2019. Comparing BioLogicals and small molecule drug therapies for chronic respiratory diseases: An EAACI task force on immunopharmacology position paper. *Allergy* 74, no. 3: 432–48.
Roy, N. K., D. Parama, K. Banik et al. 2019. An update on pharmacological potential of boswellic acids against chronic diseases. *International Journal of Molecular Sciences* 20, no. 17: pii: E4101. doi:10.3390/ijms20174101.
Sailo, B. L., K. Banik, G. Padmavathi, M. Javadi, D. Bordoloi, and A. B. Kunnumakkara. 2018. Tocotrienols: The promising analogues of vitamin E for cancer therapeutics. *Pharmacological Research* 130: 259–72.
Sanghi, D., S. Avasthi, R. N. Srivastava, and A. Singh. 2009. Nutritional factors and osteoarthritis: A review article. *Internet Journal of Medical Update* 4, no. 1: 42–53.
Saraf, S., M. Jharaniya, A. Gupta, V. Jain, and S. Saraf. 2014. Herbal hair cosmetics: Advancements and recent findings. *World Journal of Pharmaceutical Research* 3: 3278–94.
Sarker, S. D., and L. Nahar. 2007. Natural product chemistry. In *Chemistry for pharmacy students: General, organic and natural product chemistry*, 283–359. West Sussex: John Wiley and Sons Ltd.
Sarvesh Kumar, S. Palbag, S. K. Maurya, and D. Kumar. 2013. Skin care In Ayurveda: A literary review. *International Research Journal of Pharmacy 2013* 4, no. 3: 1–3.
Sawikr, Y., N. S. Yarla, I. Peluso, M. A. Kamal, G. Aliev, and A. Bishayee. 2017. Neuroinflammation in Alzheimer's disease: The preventive and therapeutic potential of polyphenolic nutraceuticals. *Advances in Protein Chemistry and Structural Biology* 108: 33–57.
Semalty, A., M. Semalty, G. P. Joshi, and M. S. M. Rawat. 2011. Techniques for the discovery and evaluation of drugs against alopecia. *Expert Opinion on Drug Discovery* 6, no. 3: 309–21.
Semwal, R. B., D. K. Semwal, I. Vermaak, and A. Alvaro Viljoen. 2015. A comprehensive scientific overview of *Garcinia cambogia*. *Fitoterapia* 102: 134–48.
Seth, S., P. Dua, and S. K. Maulik. 2013. Potential benefits of *Terminalia arjuna* in cardiovascular disease. *Journal of Preventive Cardiology* 3: 428–32.

Shakya, A. K. 2016. Medicinal plants: Future source of new drugs. *International Journal of Herbal Medicine* 4: 59–64.

Sharma, H., H. M. Chandola, G. Singh, and G. Basisht. 2007. Utilization of Ayurveda in health care: An approach for prevention, health promotion, and treatment of disease. Part 1--Ayurveda, the science of life. *Journal of Alternative and Complementary Medicine* 13, no. 9: 1011–9.

Shilpa, G., B. K. Sevatkar, and R. Sharma. 2002. Cosmetology in Ayurveda – A review. *International Ayurvedic Medical Journal* 2: 138–42.

Singh, Y. P., S. Girisa, K. Banik et al. 2019. Potential application of zerumbone in the prevention and therapy of chronic human diseases. *Journal of Functional Foods* 53: 248–58.

Sleurs, K., S. F. Seys, J. Bousquet et al. 2019. Mobile health tools for the management of chronic respiratory diseases. *Allergy* 74, no. 7: 1292–306.

Sosnowska, B., P. Penson, and M. Banach. 2017. The role of nutraceuticals in the prevention of cardiovascular disease. *Cardiovascular Diagnosis and Therapy* 7, no. Suppl 1: S21–31.

Srinivasan, K. 2006. Fenugreek (*Trigonella foenum-Graecum*): A review of health beneficial physiological effects. *Food Reviews International* 22, no. 2: 203–24.

Taylor, J. P., J. Hardy, and K. H. Fischbeck. 2002. Toxic proteins in neurodegenerative disease. *Science* 296, no. 5575: 1991–5.

Thomas, J. E., M. Bandara, D. Driedger, and E. L. Lee. 2011. Fenugreek in western Canada. *The American Journal of Plant Science and Biotechnology* 5: 32–44.

Tomé-Carneiro, J., M. Gonzálvez, M. Larrosa et al. 2012. One-year consumption of a grape nutraceutical containing resveratrol improves the inflammatory and fibrinolytic status of patients in primary prevention of cardiovascular disease. *American Journal of Cardiology* 110, no. 3: 356–63.

Tripathi, Y. B., P. Tripathi, Behram H. Arjmandi. 2005. Nutraceuticals and cancer management. *Frontiers in Bioscience: A Journal and Virtual Library* 10: 1607–18.

Trüeb, R. M., and Swiss Trichology Study Group. 2001. The value of hair cosmetics and pharmaceuticals. *Dermatology* 202, no. 4: 275–82.

Vāgbhaṭa, Ācārya. 2005. Aṣṭāṅga Hṛdaya with Sarvāṅga Sundari commentary of Aruṇadatta and Āyurvēda Rasāyana commentary of Hēmādri, 217. ed. Harishastri Paradkara Vaidya. Varanasi: Chaukhambha Orientalia.

Vats, V., J. K. Grover, and S. S. Rathi. 2002. Evaluation of anti-hyperglycemic and hypoglycemic effect of *Trigonella foenum-Graecum* Linn, *Ocimum sanctum* Linn and Pterocarpus marsupium Linn in normal and alloxanized diabetic rats and *Pterocarpus marsupium* Linn. in normal and alloxanized diabetic rats. *Journal of Ethnopharmacology* 79, no. 1: 95–100.

Venugopalan, S. N., and P. Venkatasubramanian. 2017. Understanding the concepts rasayana in Ayurveda biology. *Journal of Natural and Ayurvedic Medicine* 1 no. 2: 000112.

Virdi, J., S. Sivakami, S. Shahani, A. C. Suthar, M. M. Banavalikar, and M. K. Biyani. 2003. Antihyperglycemic effects of three extracts from *Momordica charantia*. *Journal of Ethnopharmacology* 88, no. 1: 107–11.

Wang, P., Y. M. Chen, L. P. He et al. 2012. Association of natural intake of dietary plant sterols with carotid intima-media thickness and blood lipids in Chinese adults: A cross-section study. *PLoS One* 7, no. 3: e32736. doi:10.1371/journal.pone.0032736.

Wise, J. 2013. Herbal products are often contaminated, study finds. *British Medical Journal* 347: f6138. doi:10.1136/bmj.f6138.

Wyss-Coray, T. 2016. Ageing, neurodegeneration and brain rejuvenation. *Nature* 539, no. 7628: 180–6.

Yadav, K. D., and A. K. Chaudhary. 2015. Cosmeceutical assets of ancient and contemporary ayurvedic astuteness. *International Journal of Green Pharmacy* 9, no. 4: S1–6.

Yoon, C. H., S. J. Chung, S. W. Lee, Y. B. Park, S. K. Lee, and M. C. Park. 2013. Gallic acid, a natural polyphenolic acid, induces apoptosis and inhibits pro-inflammatory gene expressions in rheumatoid arthritis fibroblast-like synoviocytes. *Joint Bone Spine* 80, no. 3: 274–9.

Yubero, N., M. Sanz-Buenhombre, A. Guadarrama et al. 2013. LDL cholesterol-lowering effects of grape extract used as a dietary supplement on healthy volunteers. *International Journal of Food Sciences and Nutrition* 64, no. 4: 400–6.

Zhang, Y., J. Dong, P. He et al. 2012. Genistein inhibit cytokines or growth factor-induced proliferation and transformation phenotype in fibroblast-like synoviocytes of rheumatoid arthritis. *Inflammation* 35, no. 1: 377–87.

Ziskoven, C., M. Jäger, C. Zilkens, W. Bloch, K. Brixius, and R. Krauspe. 2010. Oxidative stress in secondary osteoarthritis: From cartilage destruction to clinical presentation? *Orthopedic Reviews (Pavia)* 23, no. 2: e23. doi:10.4081/or.2010.e23.

10 Ayurveda in the West

Atreya Smith

CONTENTS

10.1 Introduction 223
10.2 The Turning Point 225
10.3 The Fundamental Problems 226
10.3.1 Lack of Support from the Indian Government 226
10.3.2 Insistence of Indian Government on Basing Western Courses on the B.A.M.S. Syllabus 228
10.3.3 Western Ayurveda Is Driven by Money and Economic Factors vs Educational Priorities 229
10.3.4 Each Western Country Has Different Rules concerning T.M./C.A.M. 234
10.3.4.1 Definition of What a Practitioner Can Legally Do Therapeutically 234
10.3.4.2 Definition of Which Therapeutic Procedures Are Legal 234
10.3.4.3 Definition of Herbs vs Medicine 235
10.3.4.4 Regulatory Bodies: Self-Regulating vs Government-Regulated 237
10.4 Conclusion 238
References 238

10.1 INTRODUCTION

Ayurvedic medicine, while very ancient, is a newcomer to Europe and the Americas. The Western public at large had never heard of Ayurveda until 1989 when Dr Deepak Chopra published his bestselling book *Quantum Healing* (Chopra 1990). Today most of the Western population still has little understanding of ayurvedic medicine in spite of the World Health Organization (W.H.O.) recognizing the role of traditional medicine in healthcare over 40 years ago (Chaudhary and Singh 2011). Ayurveda was first presented to the Western public as a form of "wellness" and not as a medical system. "Ayurvedic Wellness" was primarily packaged with meditation or *hatha* yoga groups from the early 1970s. Even though schools for Ayurveda appeared in North America in the mid-1980s the focus remained on wellness and lifestyle vs medicine. Europe and South America followed suit, with schools for Ayurveda appearing in early and mid-1990s (Table 10.1).

The approach of "Wellness Ayurveda" is to determine the *Prakṛti* or constitution of the person and then give dietary and lifestyle guidelines accordingly. This approach to Ayurveda does not recognize the disease or pathologies of the individual concerned. One of the main reasons why "Wellness Ayurveda" was presented in the West instead of ayurvedic medicine was to avoid legal conflicts with the existing medical and insurance industries. By presenting "Wellness Ayurveda" as a means of disease prevention and lifestyle therapies, it was possible to avoid legal conflicts and government regulations, both at local and federal levels.

As "Wellness Ayurveda" was usually packaged with, or associated with, meditation and *hatha* yoga, it also became marginalized in the West as another "New Age" healthcare fad. In this context

TABLE 10.1
Schools Currently Teaching Ayurveda in the West

Country	Address	URL
Argentina	Fundacion Salud de Ayurved Prema, Buenos Aires, Argentina	https://www.medicinaayurveda.org/
Austria	Ayurveda Academy of Yoga in Daily Life, Vienna	https://www.ayurvedaacademy.org/
Brazil	Instituto de Cultura Hindu Naradeva Shala, Rua Coriolano, 169/171, São Paulo, Brazil	http://www.naradeva.com.br/
	Arjun Das, Av. Cisne 160, 302 Nova Lima MG, 34018-010, Brazil	www.arjundas.com
Bulgaria	Ayurveda Center, Sofia	https://ayurvedacenter.bg/
Canada	Pacific Rim College, Victoria BC	https://www.pacificrimcollege.com/faculties-programs/course-descriptions/area/area-ayurveda/
	Centre for Ayurveda & Indian Systems of Healing, Toronto	https://caishayurveda.org/
Chile	Tulsi Ayurveda, Santiago	https://www.tulsichile.cl
	Escuela Latinoamericana de Ayurveda, Santiago	https://www.ayurvida.cl
France	Institut Français de Yoga et Ayurvéda (IFYA), Nîmes	http://ifya.fr/
	Europe Ayurveda Academy, Bours	https://www.europeayurvedaacademy.org/
Germany	Euroved, Köln	https://www.euroved.com/en/
	Europäische Akademie für Ayurveda, Birstein	https://www.ayurveda-akademie.org/
Hungary	Hungarian Ayurveda Medical Foundation, Budapest	http://www.ayurveda.hu/index_eng.html
Italy	Ayurvedic Point, Milan	https://www.ayurvedicpoint.it/
New Zealand	Wellpark College of Natural Therapies, Auckland	http://www.wellpark.co.nz
Russia	Novosibirsk	https://vedatng.com/english/ayurveda-course-2018
Switzerland	European Institute of Vedic Studies (E.I.V.S.), Montreux	http://www.atreya.com/
	Swiss Ayurvedic Medical Academy (S.A.M.A.), Vevey	http://www.samayurveda.com/
UK	Ayurveda Institute, Surrey	https://www.ayurvedainstitute.co.uk/
	College of Ayurveda, Milton Keynes	https://ayurvedacollege.org/
USA		
California	American University of Complementary Medicine	http://www.aucm.org/program.asp?ProgId=11
	California College of Ayurveda	www.ayurvedacollege.com
	Dancing Shiva	https://www.dancingshiva.com/education-circle
	Kerala Ayurveda Academy	https://www.keralaayurveda.us/courses/
	Mount Madonna College of Ayurveda	https://www.mountmadonnainstitute.org/academics/college-of-ayurveda
	Narayana Ayurveda and Yoga Academy	http://www.sandiegocollegeofayurveda.com/programs
Colorado	Alandia Ayurveda Gurukula	https://www.alandiashram.org/study-ayurveda/
Florida	Florida Vedic College	www.ayurvedichealers.com
Iowa	Maharishi University of Management	https://www.mum.edu/ba-in-ayurveda-wellness
Massachusetts	Kripalu School of Ayurveda	www.kripalu.org
New York	The National Institute of Ayurvedic Medicine	http://panchakarma.com/the-national-institute-of-ayurvedic-medicine-new-york-usa-p-342.html
New Mexico	The Ayurvedic Institute	https://www.ayurveda.com/
	American Institute of Vedic Studies	https://www.vedanet.com/
Pennsylvania	Ojas – Ayurveda Wellness Center	www.ojas.us
National	National Ayurveda Medical Association	https://www.ayurvedanama.org/
	Ayurveda Schools and Universities in the U.S.A.	https://study.com/ayurveda_schools.html

"New Age" has a negative or non-serious connotation. As the *Encyclopedia Britannica* defines the term,

> New Age movement: movement that spread through the occult and metaphysical religious communities in the 1970s and 1980s. It looked forward to a 'New Age' of love and light and offered a foretaste of the coming era through personal transformation and healing.
>
> **(Anonymous 2019a)**

The unfortunate effect of this association was that the general public viewed Ayurveda as something that was basically used by unstable people who were unhappy in life, or unable to fit into a "normal" role in society. Indeed, Ayurveda was vulgarized and reduced to a simplistic wellness system of judging individuals and placing them in one of three boxes; *Vāta*, *Pitta* or *Kapha*. According to one of these three possibilities, diet, lifestyle and psychological profiles were determined. In essence, the first exposure of the West to Ayurveda in the 1970s and 1980s was disastrous.

10.2 THE TURNING POINT

The turning point for Ayurveda in the West was the publication by a mainstream publishing company of Dr Deepak Chopra's *Quantum Healing* (Chopra 1990). This book brought ayurvedic medicine to the conventional public and, more importantly, the Western medical community. The fact that Dr Chopra was an endocrinologist and chief of staff at the New England Memorial Hospital in Massachusetts was not something easily dismissed. Hence, this was the turning point for Ayurveda in the West, as the scientific and medical communities began to understand that Ayurveda was a traditional medical system and not a passing "wellness" healthcare fad, or a "New Age" health craze.

Dr Deepak Chopra was himself trained by Indian ayurvedic doctors associated with Maharishi Mahesh Yogi and the Maharishi's promotion of ancient Indian sciences (Vedic sciences). In the 1980s Maharishi Mahesh Yogi started "Maharishi Ayur-Veda" and this later became known as "Maharishi Vedic Approach to Health" (M.V.A.H.). It is mainly due to the efforts of Maharishi Mahesh Yogi that ayurvedic medicine vs Wellness began to appear in the Western countries. (Anonymous 2019b).

However, the problem of Wellness Ayurveda did not just disappear and it remains an issue today throughout the West. It is only in the last five to ten years that we see documentary films or media coverage of ayurvedic medicine vs "Wellness Ayurveda". The upsurge of ayurvedic massage being offered in Wellness Spas throughout the world is also a contributing factor. There is still a stigma attached to Ayurveda as something that is ancient, yet unscientific. This problem is often compounded by the fact that Indian ayurvedic doctors and researchers try to prove the validity of Ayurveda by using allopathic means which are not adapted to functional forms of medicine.

All of the three ancient traditional forms of medicine that are still in use today, traditional Chinese medicine (T.C.M.), Unani-Tibb and Ayurveda are functional systems of medicine. These traditional systems of medicine view the human organism via systems and functions vs modern medicine which tends to view the body primarily by form and structure. Hence, the terms anatomy and physiology in modern medicine, whereas in traditional medicine it is more correct to state physiology and anatomy. Trying to fit a functional form of individualized medicine to a double-blind placebo study is fundamentally erroneous. This is because ayurvedic medicine treats the individual and not the symptoms of the disease or the structure in which the disease is located. The same disease can be treated in many different ways depending on the cause of the disorder and who is experiencing the disorder.

10.3 THE FUNDAMENTAL PROBLEMS

This fundamental problem of "wellness" vs "medicine" leads us to the difficulties of practicing Ayurveda in Western countries. If an objective view of the current situation is taken, it is possible to distinguish four fundamental problems for Ayurveda in Western countries. Each of these problems concerns practicing and propagating Ayurveda in Western countries. They are as follows:

1. Lack of support from the Indian government
2. Insistence of the Indian government on basing Western courses on the B.A.M.S. syllabus
3. Western Ayurveda is driven by money and economic factors vs educational priorities
4. Each Western country has different rules concerning traditional medicine or complementary and alternative medicine (T.M./C.A.M.)

Every one of these problems has a wide set of issues associated with it, and needs to be looked at in detail.

10.3.1 Lack of Support from the Indian Government

The first point is that the Indian government never actively supported Ayurveda in the West until very recently, unlike the Chinese government which insisted that Western countries accept T.C.M. if they wanted to trade with China. This has given T.C.M. a 40 to 50-year advantage over Ayurveda in the West. The main issue that has come from this situation is a total lack of coherent presentations of ayurvedic medicine to the West. As stated previously Ayurveda was presented as a wellness system and not a medical system. It is very difficult to change public opinion when the general view is 40 years in the making.

The Indian government's Ministry of A.Y.U.S.H. that was established in 2014 to ensure the development and propagation of Indian systems of medicine and healthcare is basically a disaster. Westerners who have tried to work with A.Y.U.S.H. from South America, North America and Europe have found the same problems – rotating ministers who are only interested in personal advancement. Meetings are held, agreements are made and then the minister changes. The total ineffectiveness of A.Y.U.S.H. to actually do anything useful, practical or even political in Western countries is well-known, and most educational institutions in the West gave up trying to work with them many years ago. Only people who wish to push their own political and economic agendas are still working with A.Y.U.S.H. at this time, as the third point explains.

Using T.C.M. as an example is very important because this traditional system of medicine is present in all Western countries and even integrated into many medical university systems to some extent or the other. T.C.M. was never presented as a wellness system; it was presented as a traditional medical system complete in its own right. Hence, we find T.C.M. schools and medicines available in all European, North American and South American countries. The author was in Chile, an advanced economy in South America, in 2006 and could find a large assortment of T.C.M. medicines and not even one ayurvedic product.

Ayurvedic medicine vs "Ayurvedic Wellness" is not a small issue. We can see how this fundamental problem has reached deeply into all levels of Western society, including the W.H.O. For example, the first books published in the West translated the Sanskrit anatomy term *Raktadhātu* as "blood". This is incorrect as *Rakta* means "red" and *Raktadhātu* means hemoglobin and refers to the heat-carrying red blood cells. Blood, as we know, is roughly half a fluid called blood plasma which is very different from hemoglobin, even though they complement each other. In Sanskrit the blood plasma is called *Raktadhātu* and is only one aspect of this *dhātu* (tissue). Together these two substances, *Raktadhātu* and *Rasadhātu*, mix together to form "blood" which circulates in vessels that are called *Srōtāmsi*. Therefore, the Western concept of blood directly relates to *Raktavāhasrōta* (blood vessels) and not *Raktadhātu*, as blood is composed of both *Rasa* and *Rakta* together. This difference is critical in clinical treatments. Nevertheless, we see on page 17 of the W.H.O. document *Benchmarks for Training in Ayurveda, Rakta* translated as "blood" (Anonymous 2010).

These kinds of errors actually prevent Ayurveda from being used effectively as a medical system. Western practitioners of Ayurveda even go so far as publishing papers on the treatment of skin disorders through the treatment of *Raktadhātu*. Here is a flagrant lack of education, as the *Suśruta Samhita* clearly states in *Śārīirasthāna*, Chapter 5, verse 5, that there are seven layers of skin (*tvak*), the first being part of *Rasadhātu* and the other six being *Māmsadhātu* (muscle tissue) (Bhishagratna 1911). Nowhere is *Raktadhātu* mentioned as being part of the skin structure. Obviously, if the preceding explanation of ayurvedic anatomy and physiology is understood, then the role of *Rakta* mixing with *Rasa* to form "blood" denotes a disease pathway into the skin from *Rakta*. Understanding which level of the skin is being affected, epidermis (*avabhāsini*) or structure such as *stratum, basale*, or fibers (*avabhāsini, lōhīta, śvēta, tāmra, vēdhini, rōhiṇi, māmsadhāra*) is essential for curing skin disorders and not just reducing or suppressing symptoms. Yet nowhere is this information being disseminated in the West. On the contrary Indian botanicals are sold as miracle cures in the West for skin disorders because they target *Raktadhātu* or "blood" (Anonymous 2019c).

Another instance of how ingrained the "Ayurvedic Wellness" approach is in the Western mind is in the presentation of psychology. The *Caraka Samhita* gives many very clear indications on mental disorders and how to treat them, beginning in *Śārīirasthāna* chapter 1, verse 57 where *Caraka* clearly states that "mental disorders are caused by *Rajas* and *Tamas Guṇa*" (Sharma 1981a). In spite of this very clear statement Western practitioners of Ayurveda base all psychological analyses and treatment on the *Dōṣa – Vāta, Pitta* and *Kapha*, notwithstanding that the same verse 57 clearly states that the three *Dōṣa* cause pathology in the body, not the mind. This error is due to a total ignorance of classical teaching such as the *Sāmkhya Darśana*. Ayurveda uses several of the *Ṣaḍdarśana* (six views of reality), but it is the *Sāmkhya* vision which clearly indicates that the mind (*manas*) manifests before the *Tanmātra* where the three *Dōṣa* come into existence, as they fulfill their role as managers of the *Pañcabhūta* (five primordial elements or states of matter) (Sinha 1915). Therefore, it is logical to note the role of *Rajas* and *Tamas* as pathological factors in mental disturbances, as they are the attributes of *Prakṛti* (latent matter) and manifest prior to the mind (*manas*). The three *Dōṣa* can affect the human psychology, but it is not through the *Dōṣa* that Ayurveda understands or treats psychological disturbances as proposed in Western "Wellness Ayurveda". In the West it is common for people to be categorized psychologically as a *Vāta* type, *Pitta* type or *Kapha* type which translates (incorrectly) as "unreliable", "angry" and "lazy". Like many misunderstandings, the main issue is not having correct information. See *Sūtrasthāna*, chapter 8 in the *Caraka Samhita* for a clear presentation of psychological pathology and its treatment in ayurvedic medicine (Sharma 1981b).

Similarly, ayurvedic medicines are not used symptomatically in India. They are used according to an accurate diagnosis and adapted to the individual needs of the patient according to the systems, channels or tissues implicated. Without proper training in the whole ayurvedic system, the use of botanicals becomes symptomatic and mechanical. It is well known in ayurvedic medicine that botanical preparations do not work well unless the patients change their diet and lifestyle (called *Dinacarya*) to support the treatment. Thus, promoting ayurvedic pharmacology without *Dinacarya* is ineffective. It only worsens by trying to use botanicals symptomatically for which they are far inferior to chemical medications. The power of Ayurveda lies not in one aspect, but rather in the whole system which uses a variety of methods to support the complexity of the human being and restore correct hemostasis.

Yet, another example of this issue is massage therapies. Massage has always been an integral part of external ayurvedic therapies. It is used in a number of different ways as per the *Caraka Samhita*, usually considered to be the oldest text on Ayurveda. Massage can be used to lighten or purify the body (*Laṅghana*), or to strengthen the weak or elderly (*Bṛmhaṇa*). To achieve these results, many different types of substances are used in massage, varying from dry powder to warm oil decoctions. In spite of this, Wellness Ayurveda is almost exclusively associated with oil massage and oil applications. Most of these forms of massage are regional methods of massage and are not medical in nature. This adds to the confusion in the West.

The obvious example is the state of Kerala, which has been developing the sector of ayurvedic tourism since the early 1990s. Kerala has long been known for massage. In fact, when the author lived in India from 1987 to 1994, Kerala was *only* known for massage. The state of Kerala is very powerful politically in India at this time due to their economic success in promoting ayurvedic Wellness. In 2018 alone Kerala grossed over 1.3 billion US dollars in foreign tourism, a billion of which was related to ayurvedic spas and Wellness (Anonymous 2019d).

Most of the so-called ayurvedic massage being taught and practiced in the West is what is called *Kēraḷīya* massage, or the regional massage from Kerala (Ranade and Rawat 2008). All of the different regions in India have their own methods and traditions of massage, and many Western Wellness centers use massage methods from Tamil Nadu. Italy is one such example. However, these traditional methods are not the same therapeutic methods used in the ayurvedic classics such as the *Caraka Samhita*. For example, in the *Caraka Samhita* the use of oil on the body is always followed by the application of heat (see *Sūtrasthāna*, chapter 13, verse 99 (Sharma 1981c). Yet, most Western practitioners of ayurvedic massage do not give heat applications after oil massage, which can actually cause pathology due to non-assimilated oil. Ayurvedic massage uses many different substances in order to treat the cause of pathology. They are quite diverse and vary according to the needs of the patient. The correct use of ayurvedic massage means that the physician gives the assistant (or nurse) a different protocol for each patient, as each patient has a different pathology, cause of that pathology and, thus, a different form of external therapy.

Taking into account the economic and political power of promoting ayurvedic tourism (e.g., Wellness) the Indian government has little motivation to meddle with a billion-dollar industry. This is obviously contributing to the confusion of ayurvedic medicine vs Wellness. The direct result is the total rejection from the Western scientific community of ayurvedic medicine as a whole, due to the spa and wellness association developed and promoted by the state of Kerala, amongst others, in India.

Once a problem is identified, it is possible to solve it through effort. Thus, the Indian government, including the Department of A.Y.U.S.H., can with great effort, correct these misunderstandings and errors through political volition and education. The primary manner in which to correct any error is through correct education and information. Erroneous information needs to be noted and corrected. Standard education material and a united effort towards Western regulatory bodies needs to be adopted on a political front. Much has already been done on the level of pharmacology and botanicals by the Central Council for Research in Ayurveda and the Government of India publications. The same approach needs to be made with ayurvedic physiology and anatomy, standardized translations that can be used by Western regulatory bodies and schools. As the next point reveals, this is not an easy task.

10.3.2 Insistence of Indian Government on Basing Western Courses on the B.A.M.S. Syllabus

The second point is that the Bachelor of Ayurvedic Medicine and Surgery (B.A.M.S.) syllabus is over 70 years old and was never a good syllabus to begin with. Now it is a virtual calamity. The B.A.M.S. is the basic medical degree in Ayurveda which consists of five and a half years of university study with one year of internship. The B.A.M.S. syllabus is basically one-third allopathic and two-thirds Ayurveda. Leaving the ayurvedic portion of the syllabus aside, the one-third allopathic part of the syllabus is so out-of-date as to be dangerous, because it is 70 years old. It is no wonder that Indian citizens consider ayurvedic doctors to be second-class practitioners of medicine. This is 100% due to the poor state of the B.A.M.S. syllabus and its continuous support by the Indian government. There have been a number of attempts to revise and update this syllabus, and so far, every effort has failed (Anonymous 2019e). If this problem stayed in India it could be ignored. However,

A.Y.U.S.H. and other Indian educational institutions push the B.A.M.S. syllabus on the West, which is totally ridiculous for the following reasons:

a) It is a medical syllabus, and ayurvedic medicine is not a legal medical system in the West
b) It is an illogical syllabus mixing Ayurveda and allopathy, which is totally inappropriate in Western countries to the point that it is illegal in most countries to have medical knowledge taught outside of a medical university
c) Ayurveda is still mainly Wellness in the West, and the B.A.M.S. syllabus is not adapted to the needs of the educational institutions in the West teaching Ayurveda
d) A.Y.U.S.H. and others who push the B.A.M.S. syllabus also push the classical medicines of Ayurveda. These medicines are not appropriate for the West; many are illegal in Western countries or impossible to procure

Fundamentally the main reason why the B.A.M.S. syllabus cannot be used in the West is because it is a medical syllabus. Western medicine is quick to discount Ayurveda as "wellness" and "non-scientific", but as soon as the B.A.M.S. program is presented for education it is immediately called a medical program. The fact that ayurvedic medicine is not a legal medical system outside of India, Nepal and Sri Lanka immediately prevents the use of the B.A.M.S. syllabus as a basis for education. It is not possible to study an illegal medical system in a certified school.

There are many other reasons why the B.A.M.S. syllabus is inappropriate for most Western countries. Certainly, there is no other medical university system in the world today using a 70-year-old syllabus to teach medical doctors. Additionally, trying to export the ayurvedic pharmacopeia also is impractical. Common herbal ingredients such as *Sida cordifolia*, Linn. and *Withania somnifera*, Dunal are illegal in European countries due to their high level of biochemical activity (Anonymous 2019f). Another aspect of exporting the ayurvedic pharmacopeia is that this additional demand on natural resources creates a shortage of medicinal plants in India. The price of *Withania somnifera* has increased over ten times since 1993 (e.g., RS 40 per kg to RS 400 per kg).

Another reason why the B.A.M.S. syllabus is inappropriate can be seen in the syllabus. After anatomy and physiology, the students learn pharmacology (*Dravyaguṇa*) and advanced methods of drug preparations. After learning these subjects the students learn pathology and diagnosis. This means a medical student is learning about drugs and their fabrication before they have learnt how pathology develops and how to diagnose disease. This is unprecedented in Western medical schools (Anonymous 2019g).

10.3.3 Western Ayurveda Is Driven by Money and Economic Factors vs Educational Priorities

The third point is that Ayurveda in the West is driven by money and economic factors vs educational priorities. Almost all education facilities are privately owned and run, with very few exceptions, such as the Bachelor of Arts diploma in the U.K. This means that the courses being taught on Ayurveda vary radically from school to school, as there is no government standardization of the syllabus. The first question is not about money. It is about having private institutions determine the level and quality of Ayurveda that is being taught in the West. By their very nature private institutions must give priority to economic factors, instead of educational quality and content.

Typical commercial schools of Ayurveda need to push a high volume of low-quality programs on students in order to make money from teaching. In most cases private education is not a lucrative business. There are two main exceptions: 1) Elite schools that cater to the rich, and 2) volume. This is why the large schools in Europe and U.S.A. tend to flood the market with programs every weekend, often diverging into other Indian subjects such as *Vāstu Śāstra* (a part of ancient Indian architecture) in order to make money.

When education is concerned with ayurvedic medicine, volume is not ideal for the students. Having a steady comprehensive syllabus is best for learning medical sciences. One of the problems with focusing on the volume of courses to make money is that of having qualified teachers. For many years in the early 2000s the largest school in Germany would offer a weekend course and then ask a student of that course to teach the same course the following month because they did not have enough teachers. Hence, when economic factors dictate the syllabus and structure of a school, the first thing to suffer is the quality of the teaching, which further adds to the vulgarization of ayurvedic medicine.

Another aspect of volume is typically observed in the largest school in the U.S.A. where many well-known teachers are brought in to teach various aspects of Ayurveda. However, there is no core teacher to bring the different subjects and teachings together in a coherent manner. The result of this approach is a number of big-name teachers who are interesting teachers in their own right, nevertheless, leaving the students with an incoherent understanding of Ayurveda. A typical scenario with all teachers (especially Indian) is that they start their course with the basics, as they do not know how much the student has been taught. Obviously after the tenth teacher giving the same basic information the students get frustrated. The fact is that the experts teaching in these schools are not coordinated to form a coherent syllabus. That is mainly due to the focus on income, and not on training students to actually practice ayurvedic medicine.

This leads us to a primary issue. Is the goal of an institution to train students to actually practice ayurvedic medicine? Or is the goal to teach students everything *about* Ayurveda? Examples of this are various B.A. and M.A. programs in Europe and the U.S.A. which are not science-based. A Bachelor of Arts or a Masters of Arts in Ayurveda are not focused on training practitioners or doctors. They are training people to know many things "about" Ayurveda as study in literature or the arts. Students are usually disappointed after completing three years of full-time study in Ayurveda, receiving a B.A. and realizing that they do not have a clue about how to diagnose and treat a patient in clinical practice. In spite of this obvious difference in a science diploma vs a diploma in the Arts, many students are led by these schools to believe they will be able to actually practice ayurvedic medicine. The reality is that many schools in the West do not want to train practitioners, due to the legal aspects of practicing Ayurveda in their countries. Many countries are highly litigious, like the U.K. and the U.S.A. Training practitioners of ayurvedic medicine opens legal issues for the schools. For example, are the teaching staff certified doctors? One school in the U.S.A. even goes so far as to have students sign a paper on graduation that they will not practice Ayurveda in the same state as the school.

The next issue is standardization of the Ayurveda syllabus. Schools which focus on volume will avoid entering into difficult and profound aspects of Ayurveda, such as diagnosis, as they are not economically feasible. A smaller, non-volume-orientated school may focus on difficult subjects because they understand that the student will need this knowledge when working with patients. One of the main subjects to suffer in ayurvedic medicine is that of pathology. Very few schools enter into this subject in depth, and yet students who do not have this training are unable to practice effectively. The author's personal observation on the lack of training is in ayurvedic anatomy and physiology. Many students join our advanced courses in pharmacology and are unable to follow normal treatment protocols due to their lack of knowledge of anatomy and physiology according to Ayurveda. If this information is not memorized, then diagnosing pathology of *Dōṣa* is impossible. This is due to the lack of standardization of the ayurvedic syllabus in the West.

The other aspect of private schools is that the larger volume-orientated institutions push their own agendas – for economic reasons – with A.Y.U.S.H. and local governments. There is no Ayurveda professor or team of professors working on a standardized ayurvedic curriculum in the West. It is being done by a few political people whose goal is making money. This leads to a total degradation of Ayurveda medicine as a whole, because the priority is not Ayurveda – it is having an economically successful school with laws that support this. As we know in modern science, there is a problem when the people paying for studies are the same ones benefitting from the results of the studies. A conflict of interest is a major problem in "Western Ayurveda". The Indian government, ministries and A.Y.U.S.H. are very impressed with a European school which began to gross over a million Euros a year in income over several years – the same school which used students from one weekend to teach the next weekend. Because of this situation these kinds of volume educational institutional schools are deciding the future of Ayurveda in the West.

Of course, the same problem exists in India. The state of Kerala virtually controls the Indian government's stance on Ayurveda at this time. This is purely due to the growth of ayurvedic tourism which is a billion-dollar industry in Kerala alone.

Many institutions in the West are trying to use the *Benchmarks for Training in Ayurveda.* document and guidelines from the W.H.O. (Anonymous 2010). No doubt this is a step in the right direction for a standardized Ayurveda syllabus in the West. On a closer look at this document, one comes up with a number of educational contradictions. For example, the students are supposed to have one hundred hours of Sanskrit study in conjunction with Ayurveda studies in order to allow study of the classical texts. All well and good, until we read the "expectations" of this W.H.O. document which states "students should be able to read and translate Sanskrit *sūtrā*s after one hundred hours of study".

If we examine any European language, for example, French, we find that students have roughly 2400 hours of French after 12 years of school. After that they go to university and can specialize in French language and receive a B.A. in French requiring another several thousand hours of study. Now, at this point, a student is able to study to become a translator. Having any kind of expectation that 100 hours of study in any language allows a student to be able to translate is absurd. Sanskrit is a complicated language and even experts in Sanskrit can make horrible translations of Ayurveda texts. An example of this is seen in the translation of the *Caraka Samhita* into French by Jean Papin, a French Sanskrit scholar (Papin 2013). His translation manages to reverse the meaning of virtually every critical *sūtrā*; a veritable disaster. He translates verse 57 from chapter one in the *Sūtrasthāna* section as "the *Dōṣa*s *Vāta*, *Pitta* and *Kapha* are the cause of health in the body", when the *sūtrā* clearly states that "*Vāta*, *Pitta* and *Kapha* are the cause of pathology in the body". Therefore, if a reputed Sanskrit scholar cannot correctly translate verses after thousands of hours of study, how can the W.H.O. *Benchmarks*, on page 9, expect 100 hours of study to:

> translate from Sanskrit to English, and vice versa. Upon completion of this subject, students are expected to be able to read, write and understand the *śḷōkā*s in the various texts of Ayurveda (*Caraka Samhita*, etc.), translate them as required, understand, apply and interpret them scientifically.
>
> **(Anonymous 2010)**

The document *Benchmarks for Training in Ayurveda* (Anonymous 2010) has several other problems. It has too many inconsistencies and unrealistic goals. For example, on page 8 the Level 1 practitioner, under the category of "Technical Skills", is expected to:

- Independently acquire technical knowledge about diseases not necessarily covered by the program
- Create a database of clinical experience and research and communicate this information to other practitioners and the public
- Analyze the merits and demerits of contemporary healthcare systems

These points are certainly admirable goals for the development of Ayurveda outside of India. The whole issue of developing a benchmark, or guidelines, for education in Ayurveda is first standardization, followed by staff who are able to teach the agreed-upon syllabus. In the guidelines on page 8, the newly trained practitioner is expected to "acquire technical knowledge about diseases not necessarily covered by the program". This is absurd, as the reality is that there are not enough qualified teachers to teach the basic program. Where will a student find the expert to give additional technical knowledge?

The author's impression of reading the above points as a practicing *Vaidya* (traditional doctor) is that the average ayurvedic practitioner would have enormous difficulties in finding the time to create a database of all their clinical cases. In the 32 years that the author has been practicing Ayurveda he has never been able to create a database of his patients. The author has extensive written records dating back to twenty-five years, correlating to his move to Europe from India in 1994. However, he lacks the time and skill to create a database of clinical records. How could the W.H.O. expect new practitioners, who are developing their skills and struggling financially, to be able to find the time to create, enter and maintain databases?

The author has been training ayurvedic practitioners in Europe since 1998. He wonders how a newly trained practitioner would be able to "analyze the merits and demerits of contemporary health-care systems". This would require many years of study in various "health-care systems" in order to effectively be able to have an objective view of such system. This is an analysis that would be best done by a team of professional educators, and not newly trained practitioners.

Lastly, how many practicing M.D. general practitioners are fulfilling the above requirements? Is this a double standard: That newly trained practitioners of traditional medicine are required to carry out tasks that are highly time-consuming, when most medical doctors are not able to comment on other healthcare systems themselves?

Then under the category of "Communication skills" on page 8 the practitioner is expected to:

- Disseminate clinical observations and findings to other professionals in accordance with ethical principles
- Provide appropriate case history and diagnostic information when referring patients to related specialists

Again, these are admirable goals for the development of Ayurveda outside of India. However, in order to disseminate clinical findings, there would need to be the establishment of specialized publications or groups, who could coordinate and disseminate such information. The actual reality is that there is nothing like this in the West at this time. There have been a few attempts to create "Ayurvedic journals", notably in the U.S.A., Germany and the U.K. Unfortunately, these kinds of publications are filled with dubious or incorrect information. One such publication in 2006 published monographs on medicinal plants written by students of various schools in the U.S.A. The persons listed on editorial board (all famous people) never checked the content being published. I was asked to review one such monograph and found it poorly referenced, wrongly classified and incorrect in a number of statements, such as "this is a famous plant used extensively in Ayurveda". Noting that the medicinal plant in question was mainly used in folklore medicine rather than in Ayurveda, brought the ire of the editor, as she then had to change the layout of the magazine, not to mention the author who was shocked that someone actually corrected his technical writing.

The second point on referrals to "related specialists" would tend to indicate purification treatments or *Pañcakarma* (five methods of purification). For example, in Europe, there is not one Ayurveda clinic that follows the guidelines of *Pañcakarma* as per the *Caraka Samhita*. Using *Śōdhana* (reducing) therapies is a clinical procedure that requires special knowledge and training and is often a cornerstone of specialized clinical treatments in India. Hence, at the moment of writing, European practitioners of Ayurveda need to refer patients to Indian clinics for "related specialists" in *Pañcakarma,* as none exist in Europe itself, noting that there are many "*Pañcakarma* clinics" in Europe, notably Germany and Austria, who do palliative (*Śamana*) therapies rather than reducing (*Śōdhana*) therapies.

One of the main reasons why *Pañcakarma* therapies are not used as per *Caraka Samhita* in Europe is that they are labor-intensive. This fact makes the cost of offering these therapies to the European public far beyond the average person's budget. Additionally, the health insurance of most countries does not cover these kinds of procedures. *Pañcakarma* also requires a large quantity of botanicals to be therapeutically effective. According to Dr Sunil V. Joshi, director of the Vinayak *Pañcakarma* Chikitsalaya in Nagpur, India they use roughly 10 kilos of medicinal plants per patient, per week in *Pañcakarma*. This includes the fabrication of medicinal oils, pastes, enemas (*Basti*) as well as other preparations that are used in both preparation (*Pūrvakarma*) and administration of the primary therapies (*Pradhānakarma*). Dr Joshi is the author of the acclaimed book *Ayurveda and Panchakarma* (Joshi 1997). Between the cost of raw materials, the labor needed to fabricate the medicines and the labor needed to apply the therapies to the patient, the cost is too high to follow classical *Pañcakarma* guidelines.

To further explain this problem, the author can share his experience of trying to set up a *Pañcakarma* clinic in Europe in 2011. In collaboration with Dr Sunil V. Joshi, an attempt was made at costing of all required materials and of two Western therapists to carry out the procedures. In order pay the rent of the clinic, the two Western therapists (trained by Dr Joshi), the material, the medicinal plants, the oils, room and board for the patient and, finally, pay of the doctor, the cost would need to be around €5000 ($5535) per week. As *Pañcakarma* therapies as per *Caraka Samhita* require a minimum of three to four weeks, this would mean a cost of €15,000 to €20,000 ($22,140) per patient. Needless to say, the project never went beyond the business plan, as this would only cater to an elite five-star clientele.

Another main reason why *Pañcakarma* is not performed in the West is due to legal issues which are developed in point number 4 below. Basically, any therapy that is invasive (e.g., entering the body) can only be done by an allopathic M.D. Unfortunately, medical doctors in the West must follow the protocols of their health department, effectively preventing them from carrying out these kinds of ayurvedic procedures.

Thus, clinics in Europe, South America and North America offer a misrepresentation of "*pañcakarma*" in a reduced form which is closer to *Kēraḷīya* methods of *Pūrvakarma* (preparatory procedures). Prof. R.H. Singh details the differences in these methods very clearly in his book *Pañca Karma Therapy* (Singh 1992). Therapeutically this means the patients receive *Śamana* (palliative) treatments instead of *Śōdhana* (reducing or purifying) therapies. Naturally when people in the West use the wrong terms to describe the therapies being offered, this further reduces the credibility of ayurvedic medicine in the West.

Furthermore, the document *Benchmarks for Training in Ayurveda* states under the category of "Research and information-management skills" on page 8, the Level 1 practitioner is expected to:

- Understand and acquire new knowledge from ayurvedic clinical research
- Remain informed about advances in ayurvedic knowledge and apply that knowledge appropriately in clinical practice

Without a professional network of practicing ayurvedic doctors/practitioners in the West it is difficult to both:

1. Acquire new knowledge
2. Remain informed of advances in research and its application in clinical practice

Again, these are admirable goals, ones that ayurvedic practitioners and schools should strive to attain. I wonder how many M.D. general practitioners would be able to carry out these goals without peer-reviewed publications, alumni networks, and government/educational support systems. Asking newly trained ayurvedic practitioners, or newly formed private schools, to create these structures is unfair and unrealistic. With the standardization and implementation of an ayurvedic educational syllabus, Western governments also need to organize and support the creation of peer-reviewed research journals and research grants. Without this support, the ability of the individual practitioner is limited.

From an educational point of view a standardized Ayurveda syllabus for the West that is *not* based on the B.A.M.S. Indian diploma is critical for the proper growth of ayurvedic medicine. This syllabus should be developed by people in education, not those people running private schools. A panel of ayurvedic professors with Western educators would need to make a syllabus that is a compromise between Indian medicine and what is legally allowed in Western countries. At this time, most governments in the West are using a combination of the W.H.O. *Benchmarks for Training in Ayurveda* and the B.A.M.S. syllabus. Both have many shortcomings and are not adapted to the legal systems in these countries. As stated earlier, ayurvedic medicine is not a legal medical system in the West. Hence, using the current Indian medical syllabus is illogical, even if it were not out of date. This leads us to the next issue of each country having its own laws concerning traditional medicine.

10.3.4 Each Western Country Has Different Rules concerning T.M./C.A.M.

The fourth point is that every Western country has different rules concerning Non-Conventional Medicine (N.C.M.). Ayurveda is an N.C.M. outside of India, Nepal and Sri Lanka. For some reason, the Indian government and A.Y.U.S.H. try to push a standard syllabus on different countries regardless of how much their own laws on N.C.M. vary. For example, it is illegal in many countries for ayurvedic practitioners to give their patients ayurvedic medicine, as the pharmaceutical industry controls the distribution of all forms of medicine. The majority of countries in Europe and the Americas forbid Indian-made medicines, yet A.Y.U.S.H. and the B.A.M.S. syllabus only allow an educational curriculum using the medicines that are largely banned in the West (Anonymous 2019f). Obviously, the result is that Ayurveda is never certified as a valid form of Non-Conventional Medicine (N.C.M.).

There are several important points in this category:

1. Definition of what a practitioner can legally do therapeutically
2. Definition of which therapeutic procedures are legal
3. Definition of herbs vs medicine
4. Regulatory bodies – self-regulating vs government-regulated

Before addressing each of the above points it is important to note that each country in Europe, South and North America has its own laws governing health and healthcare. Therefore, the following discussion can only be general, as each country needs to be analyzed individually concerning legalities.

10.3.4.1 Definition of What a Practitioner Can Legally Do Therapeutically

In most Western countries, practitioners are not allowed to diagnose diseases and prescribe medicines. Diagnosis is the domain of medical doctors (allopathic), and only they are able to prescribe medications. This is governed by the medical authorities, who develop and give doctors the medical protocols. This means that ayurvedic doctors or practitioners are not allowed to use any medical terms when diagnosing their patients. They can use general terms, but even using Sanskrit medical terms can be viewed as medical diagnosis and therefore illegal. Additionally, practitioners and therapists are usually not able to give the patient medications (herbs, etc.). Some exceptions exist to this rule, for example, in Great Britain and a few states in the United States. In the majority of the United States, medicines can only be bought in a pharmacy. In general, the pharmaceutical lobby controls the distribution of medicines and does not allow medical doctors to give medicines. For example, this is also true in France, Spain, Portugal, Holland and Germany.

10.3.4.2 Definition of Which Therapeutic Procedures Are Legal

In many Western countries any therapeutic procedure that is invasive, e.g., enters the body of the patient, can only be done by a medical doctor. In most Western countries this would include *Nasya* (nasal), *Virēcana* (purgation) and *Vamana* (therapeutic vomiting), procedures which are the cornerstones of *Pañcakarma*. It can even go so far as to include self-administered advice given by the practitioner to the patient. For example, telling the patient to avoid eating (fasting) can be construed as an invasive therapy. Giving herbal prescriptions can also be interpreted as an invasive therapy, if they are given internally.

In some countries, like France, Spain and much of the U.S.A., a practitioner is not allowed to tell the patient to return for a consultation. This is considered medical advice and practitioners and therapists are not medical doctors. Asking the patient to return means that the practitioner has made a diagnosis and is following up a treatment.

For example, the following is a list of therapies that are generally considered illegal in the United States (Anonymous 2017):

- Practitioners may not diagnose medical disease. A practitioner cannot act in the capacity of a licensed physician and provide a diagnosis of a disease using medical terminology
- Practitioners cannot interfere with the prescriptions or recommendations made by a medical doctor. A practitioner who tells a patient not to take their medications is considered to be practicing medicine without a license
- Practitioners cannot invade the body or perform any other procedure that penetrates the skin or any orifice of the body

10.3.4.3 Definition of Herbs vs Medicine

This is a huge gray area that varies from country to country. In many Western countries herbs and herbal preparations are considered "food supplements" and not medicines. In other places there are lists of herbs that are considered medicines vs food supplements. As these lists change from country to country, what may be a supplement in one location will be considered a medicine in the next country. This is the case in Europe where, at the time of writing, no standard list of permitted herbal supplements has been universally accepted, although there is currently a list being proposed for the E.U. member states.

One of the main problems with this approach is that all botanical preparations are medicines and can be misused by lay people. Unfortunately, this happens quite often, resulting in valuable medicinal herbs being outlawed. The problem is not the herb in question, but rather incorrect legislation for botanical medicines. Classifying herbs as "food supplements" allows freedom of use by the public, but without any formal training on how to use them. Most end-users of herbal preparations choose herbs based on the recommendations of the salesperson in the store where they buy them. Obviously, the salesperson has had a few hours of training in the symptomatic use of herbs from a biochemical point of view.

On the other side of this problem, herbs are treated like chemical medicines that cause many secondary effects and are dangerous. Herbs are medicines and should be used correctly, but are far less dangerous than chemical drugs and do not cause secondary effects when used correctly. Grouping them into the same category as chemical drugs is also an error. One aspect of this is that it prevents all practitioners from prescribing herbs to patients when they are grouped with chemical medicines. Herbs are not the same as chemical medicines.

One common misconception is that herbs work like modern medicines in that they target specific places in the body. Medicinal plants, or herbs, have a long history as both medicine and promotors of health in human history. There is little doubt that – when used correctly – herbs can help people recover their health. This being said, it is important to understand that plants do not work in the same way as chemical medicines. Herbs work on homeostasis and metabolism generally. When the whole plant (e.g., the part of the plant traditionally used therapeutically) is used, the therapeutic action is general and less specific. This is exactly why using the whole plant is safe, because its action is general and working more on homeostasis than a specific place in the body.

There is a trend to change the way herbs work, by making concentrated extracts of the plant and either adding it back into the plant or marketing the extract itself as a product. People wrongly think that the extract is the same as the original herb and is as safe as using the whole plant. This is simply not true. In September 2016 I had the good fortune to meet and hear a lecture by Dr Rama Jayasundar, who has a Ph.D. from Cambridge in nuclear science. She is using nuclear magnetic resonance (N.M.R.) spectroscopy to understand medicinal plants as per Ayurveda. She has found that the whole plant in water solutions corresponds perfectly with traditional ayurvedic classification of taste, action, etc. However, when plant extracts were used the classification changed. When alcohol extracts were tested, they also changed. In fact, she found through repeated experimentation that only water-based preparations of the whole plant (e.g., part of the plant used traditionally for a therapeutic result) matched the classical texts and classifications of the herbs. Dr Jayasundar is also an Ayurveda physician and works for the Indian government in research. She is the only person to date with a doctorate in both nuclear physics and Ayurveda.

Traditional healing systems like T.C.M. or ayurvedic medicine have always understood that herbs work in a broad, nonspecific manner. They also have developed many low-tech pharmacological methods to make extracts and increase the potency of herbs. In spite of having the knowledge and skill to do this, the primary way traditional systems of herbal medicine increased potency was by combining whole plants in formulae. The ancients found this to be a safer, more balanced method of administering herbs. This in turn means that they accepted that herbs should be combined with other herbs in order to target specific locations in the body.

Another misconception is that herbs treat specific diseases. This misunderstanding is much like the preceding, in that Western man is conditioned to think of disease as a fixed set of symptoms. This implies that the absence of symptoms indicates health. Herbs rarely, if ever, treat specific diseases because a fixed set of symptoms (e.g., disease) is the result of some underlying malfunction of homeostasis or metabolism. In other words, the "disease" is simply what we can see. The cause of the symptoms is often due to a number of factors. Herbs do work very well to help restore health by working on *homeostasis* (i.e., the tendency of the body to seek and maintain a condition of balance or equilibrium within its internal environment, even when faced with external changes).

According to Ayurveda the managers of homeostasis are the cause of disease – the three *Dōṣa*, or *Vāta*, *Pitta* and *Kapha* (Sharma 1981a). When these managers do their job correctly, the body stays healthy. When they function poorly, or are disrupted in their work, then the homeostasis of the body is disrupted and disease results. Hence, traditional systems of healthcare tended to focus on the underlying causes of disorders rather than on the symptoms. This is the focus of functional medicine as previously mentioned. The current obsession with symptoms is a very recent development in medicine and the history of human healthcare. The unfortunate problem is that people want to use herbs according to the modern trend of medicine and herbs simply do not work well on a set group of symptoms, or "disease". That is to say that herbs do not work well as symptomatic medicines. This is why modern studies rarely show medicinal plants as being effective treatments for diseases. Herbs work best when the cause of the disorder is addressed, not the apparent fixed symptoms.

Another misconception common today concerns dosage and the adage that "more is better". This is especially a problem in the United States, where Americans see excess as normal. Most herbs work better in lower doses taken over longer periods of time. Plants are safest when used in lower doses and they also work better to regulate or correct the homeostasis when given in low doses. Lower doses are easier to digest and assimilate than higher doses. An old dictum in Ayurveda says, "It is not what you eat, it is what you can digest that gives health".

Still another misunderstanding is that herbs, or herbal preparations, will work well without changing diet and lifestyle. This in itself is the reason many treatments with herbs fail. It is related to the first two misconceptions that "herbs treat specific locations" and that "herbs treat disease". As noted before, herbs work generally in the body to correct underlying disturbances. Whenever possible it is best to treat these disturbances before they manifest as diseases with fixed symptoms.

As Benjamin Franklin so aptly put it, “an ounce of prevention is worth a pound of cure”. So, using herbs to treat the underlying state, or foundation, of the body is the traditional approach of using herbs. Still further, all traditional forms of healthcare emphasize diet and lifestyle as the main source of health. Sadly, the symptomatic approach of the “magic bullet to cure every disease” actively denounces any relation of diet and lifestyle to health. This was (and is) needed to remove the possibility that an individual could actually prevent or cure a disorder by his own effort to eat and live right. As symptomatic medicine is economics-based, it was/is important to remove the very idea that patients can improve their health with diet, lifestyle and self-effort.

The fact is that herbs and herbal preparations are not chemicals. They are more like concentrated food. However, they are not strong enough to overcome a poor diet, or a lifestyle that is not suited to the individual. In order for herbs to work well they need support. The patient needs to make an effort to live in a balanced manner and needs to eat real food that can be digested. When herbal treatments are supported with diet and lifestyle, the results are often incredible. Likewise, when diet and lifestyle are not changed, herbs usually fail to make any significant change therapeutically.

10.3.4.4 Regulatory Bodies: Self-Regulating vs Government-Regulated

Self-regulating vs government-regulated is a very large topic and beyond the scope of this chapter. Nevertheless, this presentation would not be complete without mention of this very controversial subject. Self-regulating Ayurveda Wellness is the current situation in almost all Western countries. Because Ayurveda is not recognized as a medical system, it is it not under government regulation in most countries.

The main disadvantage of self-regulating Ayurveda is that regulation will be done by the largest schools who tend to favor volume teaching of low quality. Smaller schools tend to have less political power, or even interest in regulations, as they tend to focus on the content and quality. The U.S.A. is a perfect example of self-regulating the profession of ayurvedic healthcare practitioners.

In order for governments to regulate Ayurveda, it would need to be considered a medical system and hence have a medical university system which teaches a standard syllabus like in India, Nepal and Sri Lanka. This is not the situation in the West, and so Ayurveda basically remains outside of government control.

One of the exceptions to the above state of affairs is the country of Switzerland. Since 2015 Ayurveda has been part of the health system of Switzerland under two newly developed government-recognized professions with certified national diplomas:

- One is the Naturopathic Practitioner in Ayurveda-Medicine under the aegis of the Swiss regulatory body of Alternative Medicine
- The second is the Complementary Therapist in Ayurveda-Therapy under the aegis of the Swiss regulatory body of Complementary Therapy

These two new Ayurveda professions are designed according to W.H.O., A.Y.U.S.H. and the Swiss regulatory directives. As an additional support the Swiss parliament has voted for revision of the law on therapeutic products (e.g., herbal) with the scope to ease the access to medicines of traditional medical systems, including the Asian remedies from Ayurveda and Chinese medicine.

These two professions can be studied now in officially government-accredited schools. Training in Ayurveda-Medicine and Ayurveda-Therapy are both meant for persons without prior training in healthcare. The health insurance companies reimburse ayurvedic treatments and remedies. Naturopathic practitioners in Ayurveda-Medicine have permission to prescribe and hand out ayurvedic medicines. Under the revised Swiss law these remedies are recognized as therapeutic products and no longer as food supplements or cosmetics (Anonymous 2019h). In essence the new Swiss model of legalizing ayurvedic medicine and providing educational guidelines and regulation is a first for the West. This is a huge step in the right direction and Switzerland is now the most advanced country in the West concerning the recognition and integration of ayurvedic medicine into their health and insurance systems. Additionally, the creation of a national diploma issued by the Federal government is currently the best diploma in Ayurveda outside of India, Nepal or Sri Lanka. Switzerland shows us the best aspect of having government regulations and support for Ayurveda as a stand-alone medical system and not grouped together with all other C.A.M. or T.M. methods such as Germany does. It shows that with public initiative the government can support and develop ayurvedic medicine in the West.

10.4 CONCLUSION

These are only some of the main points that represent the difficulties of presenting ayurvedic medicine to Western countries. It does not help that there are Indians in these countries who are not ayurvedic doctors teaching and promoting Ayurveda. There is one so-called ayurvedic doctor from the state of Maharashtra, who was arrested in Mumbai for practicing medicine without a license. This person has a very successful clinic where many Western disciples of a famous ashram are taken for ayurvedic treatments. The same person who was arrested for practicing Ayurveda without a license in India teaches "pulse diagnosis" throughout the Western countries. According to him, he learned the method from a yogi high in the Himalayas. The fact that there are no references to this kind of diagnosis that he teaches in ayurvedic literature or in oral tradition does not seem to faze Westerners who pay a fortune to be cheated by a charlatan. These problems will never change unless the Indian government takes the necessary steps to address the above points and to enforce its own laws on the practice of ayurvedic medicine.

The primary solution to the majority of the problems presented in this chapter is education. The development and presentation of a medical Ayurveda syllabus for the West is possible for the Indian government. If the correct energy, time and money are devoted to this goal it could be done in a fairly short period of time. Once created, it simply needs to be presented to the W.H.O. internationally and to individual governments on a per demand basis. This can be done by individual countries in the West, for example, as Switzerland has done. However, it would be better if the Indian government could standardize the educational content and material which would in turn standardize Ayurveda throughout the world. Ignorance can only be overcome with knowledge.

REFERENCES

Anonymous. 2010. *Benchmarks for training in Ayurveda*, 1–48. Geneva: WHO.

Anonymous. 2017. Being an Ayurvedic doctor in the United States. https://www.ayurvedanama.org/articles/2018/1/22/0c6acjnw5161xwvua1ewquOntf448v (accessed December 27, 2019).

Anonymous. 2019a. New age movement. https://www.britannica.com/topic/New-Age-movement (accessed October 12, 2019).

Anonymous. 2019b. Maharishi Mahesh yogi. https://en.wikipedia.org/wiki/Maharishi_Mahesh_Yogi (accessed October 12, 2019).

Anonymous. 2019c. Soothe your skin guide. https://www.banyanbotanicals.com/info/ayurvedic-living/living-ayurveda/health-guides/soothe-your-skin-guide/ (accessed October 12, 2019).

Anonymous. 2019d. Kerala records 6% rise in tourist arrivals. https://economictimes.indiatimes.com/industry/services/travel/kerala-records-6-rise-in-tourist-arrivals-despite-floods-and-nipah-virus-scare/articleshow/67995390.cms?from=mdr (accessed October 12, 2019).

Anonymous. 2019e. BAMS syllabus. https://www.ccimindia.org/ayurveda-syllabus.php (accessed October 12, 2019).

Anonymous. 2019f. European Union ban on Ayurvedic medicines. https://www.ncbi.nlm.nih.gov/pmc/articles/PMC3131768/ (accessed October 12, 2019).

Anonymous. 2019g. Syllabus for post - Graduate course in ayurveda. https://www.ccimindia.org/first_year_syllabus.php (accessed October 12, 2019).

Anonymous. 2019h. Franz Rutz. https://vedacenter.ch/ayurveda/ (accessed October 12, 2019).

Bhishagratna, K. L. 1911. Śarīrasaṅkhyā Vyākaraṇam (Anatomy of the human body). In *An English translation of the Sushruta Samhita*, 156–72. Calcutta: Published by the author.

Chaudhary, A., and N. Singh. 2011. Contribution of world health organization in the global acceptance of Ayurveda. *Journal of Ayurveda and Integrative Medicine* 2, no. 4: 179–86.

Chopra, D. 1990. *Quantum healing: Exploring the frontiers of mind/body medicine*. New York: Bantam Books, 1–263.

Joshi, S. V. 1997. *Ayurveda and Panchakarma: The science of healing and rejuvenation*, 1–309.WI: Lotus Press.

Papin, J. 2013. *Caraka Samhita- Traité Fondamental de la Médecine Ayurvédique*. Paris: Almora Editions, 1–256.

Ranade, S., and R. Rawat. 2008. Kerala massage. In *Ayurvedic massage therapy*, Vol. 53–56. WI: Lotus Press.

Sharma, P. V. 1981a. *Katithapuruṣīyam Śārīram* (Types of the personal self). In *Caraka Samhita*, Vol. 1, 397–411. Varanasi: Chowkhamba Orientslia.

Sharma, P. V. 1981b. *Indriyōpakramaṇīya Addhyāya* (Chapter on introductory description of sense organs). In *Caraka Samhita*, Vol. 1, 54–62. Varanasi: Chowkhamba Orientalia.

Sharma, P. V. 1981c. *Snēhāddhyāya* (Chapter on oleation). In *Caraka Samhita*, Vol. 1, 85–94. Varanasi: Chowkhamba Orientalia.

Singh, R. H. 1992. *Pañca Karma Therapy: Ancient classical concepts, traditional practices and recent advances*, 1–304. Varanasi: Chaukhamba Sanskrit Series Office.

Sinha, N. 1915. Preface. In *The Samkhya philosophy*, i–xv. Calcutta: Bhuvaneswari Asrama.

11 Ayurveda Renaissance – *Quo Vadis*?

D. Suresh Kumar

CONTENTS

11.1 Introduction 242
11.2 Renewed Interest in Ayurveda 242
11.3 Theoretical Constructs of Ayurveda 243
11.3.1 Tridōṣa (The Three Vitiators) 243
11.3.2 Dhātu (Tissue Elements) 243
11.3.3 Mathematical Modeling of Tridōṣa 244
11.3.4 Prakṛti (Constitutional Types) 246
11.3.4.1 Understanding Prakṛti 246
11.3.4.2 Gene Polymorphism of Drug Metabolizing Enzyme 247
11.3.4.3 Whole Genome Expression and Biochemical Correlates 247
11.3.4.4 Genetic Basis of Prakṛti 248
11.3.4.5 Prakṛti and Variations in Platelet Aggregation 248
11.3.4.6 Cardiovascular Responses and Prakṛti 249
11.3.4.7 Autonomic Responses and Prakrti 249
11.3.4.8 Isotonic Aerobic Exercise and Prakrti 249
11.3.4.9 DNA Methylation Analysis of Prakrti Groups 250
11.3.4.10 Immunophenotyping of Prakrti Groups 250
11.3.4.11 Association of Prakṛti with Diseases 250
11.3.4.12 Prakrti in Therapeutics 251
11.3.4.13 Uniform Method for Assessment of Prakṛti 252
11.3.5 Taste and Drug Action 253
11.3.5.1 Taste Response as Sensory Expression of Pharmacological Activity 253
11.3.5.2 Role of Taste Preferences in Diagnosis 255
11.4 Molecular Basis of Ayurvedic Medicines 255
11.5 Problems That Warrant Urgent Attention 256
11.5.1 Botanical Identity of Herbs 257
11.5.2 Identification of Formulae with Lesser Ingredients 258
11.5.3 Discovering New Dosage Forms 258
11.5.4 Substitution of Plant Parts 259
11.5.5 Pharmaceutics of Ayurvedic Medicines 260
11.5.6 Research on Posology 261
11.5.7 Research on Rasaśāstra 261
11.5.8 Objective Way of Teaching Ayurveda 262
11.5.9 Diagnosis of Diseases 262
11.5.10 Clinical Research on Ayurvedic Medicines 264
11.6 Conclusion 265
References 266

11.1 INTRODUCTION

Ayurveda has been practiced in India and many Asian countries since time immemorial. The names of the celestial physicians, Aśvinikumāra (*Dasra* and *Nāsatya*) appear in the documents excavated from Boghaz Koyi in Cappadocia region of present-day Turkey (Keswani 1974). Similarly, the discovery of the Ayurveda text *Nāvanītakam* in a ruined Buddhist monastery near Kuchar in Chinese Turkestan indicates the popularity of Ayurveda in that faraway land (Pandey and Pandey 1988). In India this medical system occupied a lofty position as a result of royal patronage. But a series of foreign invasions and the subsequent socio-economic problems caused grievous injury to it.

Western medicine was introduced into India by the Portuguese in the 16th century. The Portuguese commander Alfonso de Albuquerque conquered Goa in 1510 and founded *Hospital Real* (The Royal Hospital). It was later handed over to the Jesuits in 1591. The Jesuits managed this institution exceptionally well and introduced a rudimentary form of medical training, with Capriano Valadares as its chief. *Hospital Real* was converted in 1842 into School of Medicine and Surgery (Keswani 1974). Although it was the Portuguese who introduced Western medicine into India, it was largely the British who later established and consolidated both its practice and study in the subcontinent. Starting with small trading posts in the 17th century, the British ultimately occupied the entire Indian subcontinent. The history of Western medicine in India is almost exclusively that of the development of medicine during the British rule (1857–1947) (Chakravorty 2008).

Up to the 19th century, the impact of Western medicine on India was relatively small. It was mostly confined to the larger cities like Calcutta, Bombay and Madras, the enclaves of the white community and the army. Nevertheless, several factors, including advances in medical science and sanitary practice, growing Indian involvement and the rise of the women's medical movement, enabled Western medicine to partly free itself from its old enclavism. By 1937 when the Indian National Congress first formed governments in the provinces, it was apparent that the traditional systems like Ayurveda and Unani would enjoy the modest benefits of a minority status while the lion's share of state support went to Western-style public health and medical practice. The Bhore Committee (1943) headed by Sir Joseph William Bhore called for more doctors, nurses, midwives, dispensaries and hospitals, to bring India closer to the level of health care in the West. This Beveridge-style blueprint was put into practice in independent India. As a result Western medicine became the dominant medical system of the country (Arnold 1996).

11.2 RENEWED INTEREST IN AYURVEDA

Interest in Ayurveda started growing all over the world in the late 1970s due to two reasons. Firstly, a major shift in global healthcare management policy was instrumental in renewing interest in all forms of herbal medicine, including Ayurveda. To encourage national and international efforts to develop and implement primary healthcare throughout the world, the World Health Organization (W.H.O.) convened the International Conference on Primary Health Care (6–12 September 1978) at Alma Ata, in the former Soviet republic of Kazakhstan (Hall and Taylor 2003). This conference adopted the famous Alma Ata Declaration, which called on member nations to formulate national policies, strategies and plans to launch and sustain primary healthcare. The member states were especially encouraged to mobilize their own national resources (Anonymous 1978a). The Western world was thus encouraged to study in depth the various traditional medical systems of the world. Ayurveda was an important one among them, having a sound theoretical basis.

The second reason for reviving interest in Ayurveda was the enthusiasm generated by several spiritual organizations such as the Chinmaya Mission, International Sivananda Yoga Vedanta Center, International Society for Krishna Consciousness (ISKCON), Isha Foundation, Maharishi Foundation, Osho Foundation and The Art of Living International Center. Imparting knowledge in Ayurveda is one of their major missions, and this has helped to popularize Ayurveda in the West (Kumar 2016). In Germany, Austria and Switzerland Ayurveda is one of the fastest-growing complementary and alternative medicine methods (Kessler et al. 2013).

Coinciding with burgeoning worldwide interest in Ayurveda, important decisions were taken in India. Aimed at improving the traditional Indian systems of medicine, Government of India established in 1995 the Department of Indian System of Medicine and Homeopathy (I.S.M. & H.). In November 2003 this department was renamed the Department of A.Y.U.S.H., to further the development of education and research in Ayurveda, yoga, naturopathy, Unani, Siddha medicine and homoeopathy. The Ministry of A.Y.U.S.H. was formed on 9 November 2014 for better integration of A.Y.U.S.H. systems into healthcare in the country (Anonymous 2017).

11.3 THEORETICAL CONSTRUCTS OF AYURVEDA

There are many traditional medical systems in the world. While many of them are collections of empirical knowledge, very few of them are based on a firm theoretical foundation. Ayurveda is the foremost among them and the only other medical system that can stand at par with it is Chinese medicine. Ayurvedic theory and practice are exclusively based on the doctrine of *tridōṣa* derived from the six schools of Indian philosophy, namely, *nyāya*, *vaiśēṣika*, *sāmkhya*, *yōga*, *mīmāmsa* and *vēdānta*. These schools include an atheistic and generally materialistic tradition, centrally concerned with questions of logic. However, this tradition has generally committed itself to the view that the aim of philosophy is *mōkṣa* or liberation from the cycle of birth and death (Dasgupta 1997).

11.3.1 Tridōṣa (The Three Vitiators)

According to Ayurveda, the body is made up of *pañcabhūta* or the five primordial elements *pṛdhvi* ("earth"), *ap* ("water"), *tējas* ("fire"), *vāyu* ("air") and *ākāśa* ("sky"). The ability of the *pañcabhūta* to modulate life processes under the influence of a driving force (*ātma*) is denoted by the collective term *tridōṣa*, consisting of *vāta*, *pitta* and *kapha*. In other words, *tridōṣa* is a three-dimensional view of metabolism. *Vāta* represents *vāyu* and *ākāśa bhūta*, *pitta* represents *tējas bhūta* and *kapha* represents *pṛdhvi* and *ap bhūta*. The body is said to be healthy when the *tridōṣa* exist in steady state. Disease originates as a result of destabilization of the *tridōṣa*, which can be either of a "high" or a "low" nature. These pathological states of the *tridōṣa* can be identified by characteristic signs or symptoms (Upadhyaya 1975a). Due to dietary and behavioral indiscretions and effects of season, the *tridōṣa* get destabilized and if left untreated, they progress to produce well-defined disease entities like *jvara* (fever), *atisāra* (dysentery). *vātarōga* (neurological diseases), *kuṣḍha* (skin diseases) and so on (Upadhyaya 1975b).

11.3.2 Dhātu (Tissue Elements)

The human body is made up of seven tissue elements (*dhātu*) (*rasa* = tissue fluid, *rakta* = blood. *māmsa* = muscle, *mēdas* = adipose tissue, *asthi* = bone, *majja* = bone marrow and *śukra* = reproductive element) and their waste products known as *mala* (*purīṣa* = feces, *mūtra* = urine and *svēda* = sweat). The essential products of digestion are collectively called *rasa* which transforms sequentially into *raktam*, *māmsam*, *mēdas*, *asthi*, *majja* and *śukra*. The end product of this *dhātu* cycle is known as *ōjas* which is said to circulate in the body imparting strength and vitality. *Ōjakṣaya* or diminution of *ōjas* follows the disruption of the *dhātu* cycle and varying states of illness originate therefrom. While taking part in the *dhātu* cycle, each *dhātu* gives out its characteristic waste product (*mala*). *Dhātu* and *mala* also exist in "high" and "low" states (Upadhyaya 1975a).

A simple model diagram illustrating the ayurvedic concept of the human body is provided in Figure 11.1. The communication between the *tridōṣa*, *dhātu* and *mala* is believed to be two-way, as shown in the figure. When *vāta*, *pitta* and *kapha* are in an undisturbed steady state, the other two conceptual compartments will also be in an undisturbed steady state, and the body is said to be in perfect health.

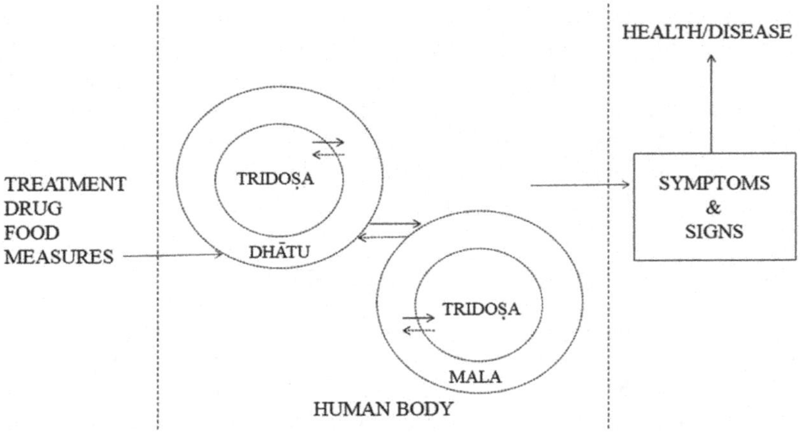

FIGURE 11.1 Relationship of *tridōṣa*, *dhātu* and *mala* among themselves as well as with symptoms and therapy. Reproduced with kind courtesy of National Institute of Indian Medical Heritage (CCRAS), Hyderabad.

11.3.3 Mathematical Modeling of Tridōṣa

Diseases and drugs are studied in Western medicine in compartmentalized models. Mathematical modeling is increasingly being used in drug design, drug delivery and medical management (Bellomo and Preziosi 1995; Chambers 2000; Siepmann and Siepmann 2008; Wang et al. 2013). Prabhakar and Kumar (1993) made a novel attempt to offer a simple model named the V:P:K code for quantification of diseases, in terms of ayurvedic principles. They transformed all the signs and symptoms of disease in terms of "high" (1), "low" (–1) or "no change" (0) of *v*, *p* and *k* states of *tridōṣa*, *dhātu* and *mala* by making basic assumptions that changes in these (*v*, *p* and *k*) will be responsible for each symptom or sign. This exercise furnished the V:P:K code of symptoms and signs of each disease (*v*, *p* and *k* represent the respective sub-components in *tridōṣa*, *dhatu* and *mala*. V, P, K are the sum total of all sub-components of *v*, *p*, *k*).

For example, an increase of *vāta* causes the appearance of symptoms or signs like leaning, black color, increased body movements, "feeling of pulsations", interest in warmth, insomnia, constipation, flatulence, borborygmi, inability to concentrate, fear and anxiety. The V:P:K code of each of these symptoms or signs will be 1:0:0. Similarly, symptoms or signs like anorexia, nausea, disinterest in talking or work, decreased libido and indigestion appear when *vāta* decreases. The V:P:K code of each symptom or sign will be –1:0:0.

As any disease is a result of a combination of symptoms and signs, variations of *v*, *p* and *k* at all levels linearly add to one another and exert their resultant influence. Thus the V:P:K code of a disease state can be derived from the V:P:K codes of constituent symptoms and signs by the linear summation principle. A calculation protocol to derive the V:P:K code of a disease is illustrated using *kapha*-dominant heart disease (*kapha hṛdrōga*) as an example (Table 11.1).

Ayurvedic nosological data on 63 important *vātaja* (*vāta*-dominant), *pittaja* (*pitta*-dominant) and *kaphaja* (*kapha*-dominant) diseases were considered to verify the model. The V:P:K codes of these diseases were derived and their validity was tested through a correlation study using regression analysis with the least square method. An inverse relationship was found to exist between *vāta* and *kapha* in the *kaphaja* disease group (Figure 11.2). This finding is in agreement with evidence from ayurvedic theory. The V:P:K code is a novel finding that can be used for generating computer-based disease maps which can be used in experimental and clinical medicine. Although there are several reported studies of mathematical modeling in Chinese medicine (Ding and Wan 2008; Ming et al. 2010; Zhao et al. 2010; Han and Huang 2012; Kim et al. 2014) this is the first report of a mathematical model in Ayurveda.

TABLE 11.1
Step-Wise Derivation of V:P:K Code of *Kapha hṛdrōga* (Prabhakar and Kumar 1993)

	V:P: K CODE													
	Tridōṣa			*Dhātu*							*Mala*			
Symptom/Sign	*Vāta*	*Pitta*	*Kapha*	*Rasa*	*Rakta*	*Māmsa*	*Mēdas*	*Asthi*	*Majja*	*Śukra*	*Purīṣam*	*Mūtram*	*Svēda*	Linear Summation
Heaviness in precordial area	−1:0:0		0:0:1											1:0:1
Heaviness of head			0:0:1											0:0:1
Cough	−1:0:0		0:0:1	0:0:1		0:0:1	0:0:1							−1:0:4
Lowering of digestive efficiency	−1:0:0	0:−1:0	0:0:1	0:0:1	0:1:0	0:0:1	0:0:1							−1:0:4
Nausea	−1:0:0		0:0:1	0:0:1		0:0:1	0:0:1							−1:0:4
Excessive sleepiness	−1:0:0		0:0:1	0:0:1		0:0:1	0:0:1							−1:0:4
Adynamia	−1:0:0		0:0:1	0:0:1		0:0:1	0:0:1							−1:0:4
Fever		0:1:0												0:1:0
Anorexia	−1:0:0	0:−1:0		0:0:1										−1:−1:1
Stiffness of body		0:−1:0		0:0:−1										0:−1:−1
Sweetness in mouth			0:0:1											0:0:1
Linear summation of *dōṣa*	−7:0:0	0:−2:0	0:0:8	0:0:5	0:1:0	0:0:5	0:0:5							−7:−1:23
V:P:K code of *Kapha hṛdrōga*														−7:−1:23

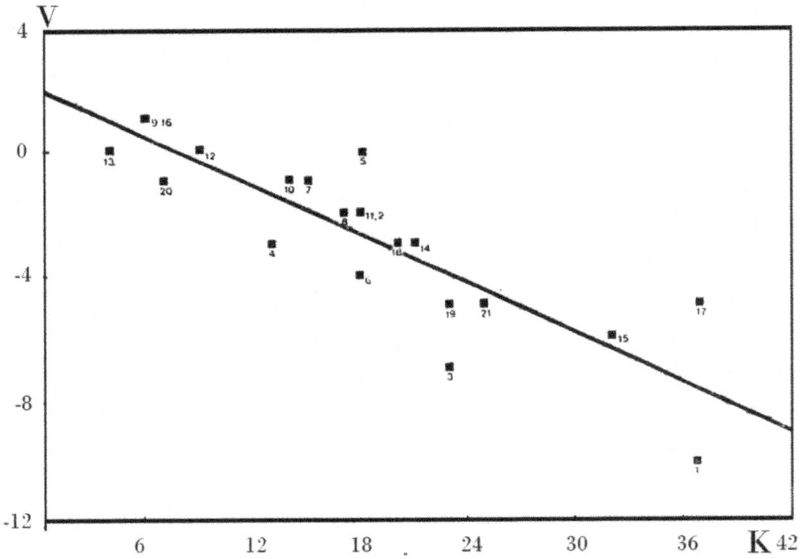

FIGURE 11.2 Graphical representation of the inverse relationship between estimates of *vāta* and *kapha* in *kaphaja* group of diseases. Reproduced with kind courtesy of National Institute of Indian Medical Heritage (CCRAS), Hyderabad.

11.3.4 Prakṛti (Constitutional Types)

Though *tridōṣa* is ubiquitous, it exists in individuals in differing magnitudes, in the state of health. This is called *prakṛti*, which is literally translated as "nature". When the *prakṛti* of an individual is challenged by dietary or behavioral indiscretions and exteroceptive factors, *vikṛti* or "distortion" appears, resulting in disease. Based on the season in which conception occurs, dietary and behavioral patterns of the pregnant mother, dominance of *tridōṣa* in the mother's reproductive tract and genetic factors, individuals are born with a characteristic physiological constitution (*prakṛti*). As there can be innumerable combinations of *tridōṣa*, there can be innumerable *prakṛti.* However, as it is humanly impossible to comprehend an endless array of *prakṛti*, Ayurveda advises that a physician need to consider only seven *prakṛti*, namely, *vāta prakṛti, pitta prakṛti, kapha prakṛti, vāta-pitta prakṛti, vāta-kapha prakṛti, pitta-kapha prakṛti* and *vāta-pitta-kapha* (homologous) *prakṛti.* These seven types are discernible by typical physical, behavioral and psychological characteristics. As individuals belonging to each *prakṛti* are prone to diseases arising out of the destabilization of the corresponding component (s) of *tridōṣa*, food, measures and medicinal substances should be selected to promote their steady state and maintain health (Upadhyaya 1975c).

11.3.4.1 Understanding Prakṛti

The first empirical test providing preliminary statistical support to the concept of *prakṛti* was carried out by Joshi (2004). Using regression modeling with a sample of 117 healthy subjects (ages 18–70), she created an algorithmic heuristic based on physical and psychological characteristics traditionally used by ayurvedic physicians to determine *prakṛti.* This heuristic was compared with the qualitative assessment of *prakrti* for each of the subject and found a 75% convergence with significance at the $p = 0.05$ level. This study was flawed in that it did not specify gender variation within the sample set, and the diagnostic criteria used by the ayurvedic physicians were not specified. Nevertheless, this test is considered to be one of the pioneering studies on the theoretical constructs of Ayurveda.

The human leukocyte antigen (H.L.A.) system is a gene complex encoding the major histocompatibility complex (M.H.C.) proteins in human beings. These proteins found on the cell surface are responsible for the regulation of the immune system. H.L.A. genes are highly polymorphic and they have many different alleles (mutants of a gene found at the same place on a chromosome), allowing greater efficiency of the adaptive immune system (Svejgaard 1978; Bodmer 1997; Jin and Wang 2003).

Patwardhan et al. (2005) made an attempt to correlate characteristics of *prakṛti* with genotypes (groups of individuals with the same inherited map carried within the genetic code). They evaluated the *prakṛti* of 76 subjects and then determined their H.L.A. DRB1 types using low-resolution polymerase chain reaction sequence-specific primers and oligonucleotide probes. 14 alleles were examined. A reasonable correlation was observed between H.L.A. types and *prakṛti* types. H.L.A. DRB1*02 allele was completely absent in the *vāta* type and H.L.A. DRB1*13 was absent in the *kapha* type. Moreover, H.L.A. DRB1*10 allele had higher allele frequency in the *kapha* type than in the *pitta* and *vāta* types. This finding indicates that there is immunological difference between the various *prakṛti* types. Though limited by a small sample size and lack of diversity in the sample set, this study is the first attempt to understand the concept of *prakṛti* using tools of molecular biology.

11.3.4.2 Gene Polymorphism of Drug Metabolizing Enzyme

A few reports have appeared in recent years suggesting that differences in physiological responses are shown by individuals belonging to the major types of *prakṛti*. The concept of *prakṛti* revolves around discrete phenotypic groupings based on the principles of motion (*vāta*), metabolism (*pitta*) and structure (*kapha*) (Hankey 2005; Rhoda 2014). As inter-individual variability in drug response can be attributed to polymorphism in genes encoding different drug metabolizing enzymes, drug transporters and enzymes involved in D.N.A. biosynthesis (Evans 2003; Evans and McLeod 2003), distribution of these genotypes can be a good tool for studying the biochemical basis of *prakṛti*.

Ghodke et al. (2011) studied the gene polymorphism of the CYP2C19 gene involved in metabolism of many drugs (Goldstein and de Morais 1994: Daly 1995) in 132 unrelated healthy subjects. Significant association was observed between CYP2C19 genotype and major classes of *prakṛti* types. The extensive metabolizer (EM) genotype (*1/*1, *1/*2, *1/*3) was found to be predominant in *pitta prakṛti* (91%). Genotype (*1/*3) specific for EM group was present only in *pitta prakṛti*. The poor metabolizer (P.M.) genotype (*2/*2, * 2/*3, *3/*3) was highest (31%) in *kapha prakṛti* when compared with *vāta* (12%) and *pitta prakṛti* (9%). Genotype (*2/*3) which is typical of the P.M. group was significant in *kapha prakṛti*. The authors observed interesting correlations between CYP2C19 genotypes and *prakṛti*, with fast and slow metabolism being one of the major distinguishing and differentiating characteristics.

11.3.4.3 Whole Genome Expression and Biochemical Correlates

Prasher et al. (2008) carried out an interesting study to explore whether the *prakṛti* types have any molecular basis. 96 unrelated healthy individuals with predominance of *vāta* (n = 39), *pitta* (n = 29) or *kapha* (n = 28) were included in the study. Peripheral blood samples of the subjects were collected and 33 biochemical tests used in routine diagnostics were performed on the blood samples. The samples were also analyzed for genome wide expression levels. Gene ontology and pathway-based analysis were carried out on differentially expressed genes to find out if there were significant enrichments of functional categories among *prakṛti* types.

Individuals of different *prakṛti* exhibited differences in biochemical components. From the 33 biochemical parameters, 15 parameters in males and 4 in females showed significant differences with respect to *prakṛti*. Interestingly, components of lipid profile like triglycerides, total cholesterol, V.L.D.L., L.D.L., L.D.L./H.D.L. ratio, were higher in *kapha* when compared to *pitta* and *vāta* males. *Kapha prakṛti* also had lower levels of H.D.L. when compared to the *vāta* counterpart. The levels of serum uric acid, considered to be an independent predictor of cardiovascular mortality, were also found to be elevated in *kapha prakṛti*. G.G.P.T., S.G.P.T. and serum zinc were also high in the *kapha* group. Serum prolactin and prothrombin time were high in the *vāta* group when compared to the *kapha* or *pitta* groups. On the other hand, *pitta* type males showed high values of hematological parameters like hemoglobin, P.C.V. and R.B.C. count in comparison to *vāta* and/or *kapha* groups (Prasher et al. 2008).

Analysis of genome-wide expression through c.D.N.A. microarrays, using independently pooled samples of *vāta*, *pitta* and *kapha* males and females in a set of loop design experiments, unraveled a number of differentially expressed genes in each category. Among the 8416 annotated genes in the 19K array (CA Ontario), 159 in males and 92 in females were found to be differentially expressed. Amongst the differentially expressed genes there was a significant over-representation of hub genes associated with complex diseases and housekeeping genes related to basic cellular function (Prasher et al. 2008).

EGLN1 was one of the 251 differentially expressed genes among the *prakṛti* types (Prasher et al. 2008). In a follow-up study Aggarwal et al. (2010) reported a link between high-altitude adaptation and common variations rs479200 (C/T) and rs480902 (T/C) in the *EGLN1* gene. Additionally, the TT genotype of rs479200, which was more frequent in *kapha* types and correlated with higher expression of *EGLN1*, was associated with patients suffering from high-altitude pulmonary edema. However, it was present at a significantly lower frequency in *pitta* and almost absent in dwellers of high altitude. Analysis of the Human Genome Diversity Panel, Centre d'Étude du Polymorphisme Humain and Indian Genome Variation Consortium panels showed that dissimilar genetic lineages at high altitudes share the same ancestral allele (T) of rs480902 that is over-represented in *pitta* and positively correlated with altitude globally. Therefore, it can be inferred that *EGLN1* polymorphisms are associated with high-altitude adaptation (also see Yi et al. 2010; Simonson et al. 2010; Storz 2010). A genotype rare in highlanders but over-represented in the *kapha prakrti* subgroup of normal lowlanders may render them more prone to high-altitude pulmonary edema. The authors concluded that genetic analysis of healthy individuals phenotyped using the principles of Ayurveda could uncover genetic variations associated with adaptation to external environment and susceptibility to diseases.

11.3.4.4 Genetic Basis of Prakṛti

A single-nucleotide polymorphism (S.N.P.) is a variation in a single nucleotide that occurs at a specific position in the genome. Though a particular S.N.P. may not cause a disorder, some S.N.P.s are associated with certain diseases. S.N.P.s can be useful tools to evaluate an individual's genetic predisposition to develop a disease (Syvänen 2001). Govindaraj et al. (2015) reported correlation of genomic variations with the classification of *prakṛti*. They performed genome-wide S.N.P. analysis of 262 well-classified male subjects, belonging to the three *prakṛti*. These subjects were recruited from screening a population of 3416 individuals. It was found that 52 S.N.P.s were significantly different between the three *prakṛti*. On the basis of principal component analysis, these S.N.P.s could be classified into their respective *vāta*, *pitta* and *kapha prakṛti* groups. 52 clusters of differentiation (C.D.) markers were used to find the genotype-phenotype correlations. Marker rs10518915 was found to be associated with *vāta* and *pitta prakṛti*, whereas rs986846 was found to be associated with *kapha* and *vāta prakṛti*. 4 markers (rs2269241, rs2269240, rs2269239, rs2269238) of the PGM1gene were associated with *pitta prakṛti*. This study suggests that the phenotypic classification (*prakṛti*) in Ayurveda has a genetic basis.

11.3.4.5 Prakṛti and Variations in Platelet Aggregation

As platelet reactivity, which markedly influences the pathological outcome of thrombosis and sensitivity to anti-platelet drugs, has been reported to exhibit inter-individual variations (Bray 2006), Bhalerao et al. (2012) attempted to study whether platelet aggregation in response to adenosine diphosphate (A.D.P.) and its inhibition with aspirin varies in various *prakṛti* subtypes. The *prakṛti* of 137 subjects was assessed by two ayurvedic physicians. The authors observed that platelet aggregation induced by A.D.P. and the inhibitory effect of aspirin differed according to *prakṛti*. Thus, *vāta-pitta prakṛti* individuals had the highest maximal platelet aggregation compared to other *prakṛti* types. These individuals also showed highest inhibition of aggregation after exposure to 2.5 μM aspirin. A limitation of the study is the unequal number of subjects in different *prakṛti* groups and the small sample size (n = 4) in *vāta-kapha* and *kapha-vāta* types. As it is known that the inter-individual differences in platelet aggregation response to A.D.P. are more pronounced at lower concentrations of A.D.P. (1, 2 or 5 μM) (Fontana et al. 2003), the study of platelet aggregation using 1 and 2 μM A.D.P. may prove more fruitful.

11.3.4.6 Cardiovascular Responses and Prakṛti

With the aim of identifying physiological parameters that can serve as good indicators of *prakṛti*, Tripathi et al. (2011) conducted a study in 90 randomly selected clinically healthy volunteers belonging to dual constitutional types like *vāta-pitta*, *pitta-kapha* or *vāta-kapha*. They evaluated the variability of heart rate and arterial blood pressure in response to specific postural changes, exercise and the cold pressor test. The results suggest that these basic cardiovascular responses do not vary significantly according to the dual constitutional types. However, the authors noted a significant fall in diastolic blood pressure immediately after performing the isotonic exercise for five minutes in *vāta-kapha* individuals in comparison to *pitta-kapha* and *vāta-pitta prakṛti*.

11.3.4.7 Autonomic Responses and Prakrti

In a subsequent study Rapolu et al. (2015) investigated whether autonomic function tests vary according to *prakṛti* of individuals. They conducted this study in 106 healthy volunteers of both genders belonging to the age group of 17 to 35 years. The *prakṛti* of these volunteers was assessed using a validated questionnaire and also by the traditional method of interviewing. After confirmation of *prakṛti*, volunteers were grouped into three on the basis of dominant *dōṣa*. They were then subjected to various autonomic function tests like the cold pressor test, standing-to-lying ratio, Valsalva ratio and pupillary responses such as pupil cycle time and pupil size measurement in light and dark (Low et al. 2013).

The results of the study suggest that the autonomic function tests in healthy individuals may correlate linearly with the dominant *dōṣa* expressed in the *prakṛti* of the individual. Notably, individuals belonging to *kapha*-dominant *prakṛti* showed a tendency to have either a higher parasympathetic activity or a lower sympathetic activity related to their cardiovascular reactivity in comparison to individuals with *pitta*-dominant or *vāta* -dominant *prakṛti* (Rapolu et al. 2015). This study also suggests that various autonomic function tests, which are already reported to show variability in responses (Manor et al. 1981; Moodithaya and Avadhany 2009; 2012) can be used as indicators of *prakṛti*. However, their clinical suitability needs to be validated.

11.3.4.8 Isotonic Aerobic Exercise and Prakrti

Tiwari et al. (2012) studied the effect of isotonic aerobic exercise (walking) on some physiological parameters in individuals belonging to three dual-*dōṣa prakṛti* (*vāta-pitta*, *vāta-kapha* and *pitta-kapha*) groups. A total of 83 diabetics (age ranging from 35 to 65 years) were included in this study. 51 diabetics formed the experimental group and 32 formed the diabetic control group. Pulse rate, blood pressure and respiratory rate were measured before exercise, after 1 month and after 3 months. Hematological parameters like blood sugar and cholesterol were estimated before exercise, and after 3 months. Diabetic patients were encouraged to walk for 30 minutes every day for 3 months on empty stomach.

Highly significant decrease in systolic blood pressure was observed in *vāta-pitta*, *pitta-kapha* and *vāta-kapha prakṛti* individuals of the experimental group after the isotonic exercise. All the three groups showed significant decrease in mean pulse rate and respiratory rate after walking. Significant lowering of fasting blood sugar levels was observed in *vāta-pitta* and *pitta-kapha* individuals of the experimental group. However, all the three groups showed significant lowering of cholesterol and post-prandial blood sugar levels after exercise (Tiwari et al. 2012).

11.3.4.9 DNA Methylation Analysis of Prakrti Groups

Genetic factors alone cannot explain the phenotypic variations observed in the human race. It may involve many non-genetic (epigenetic) influences. Heyn et al. (2013) reported the identification of DNA methylation markers in three human populations viz., Caucasian-Americans, African-Americans and Han-Chinese Americans. Therefore, DNA methylation may contribute to human developmental plasticity, adaptation and manifestation of distinct phenotypes. Rotti et al. (2015) studied the association of natural DNA methylation variations in *prakṛti*-based grouping of populations. Whole blood genomic D.N.A. collected from 147 randomly selected male volunteers of dominant *prakṛti* (*Vāta* = 47, *Pitta* = 48, *Kapha* = 52) was used for data analysis. Their study showed that large proportions of D.N.A. methylation patterns are common between various *prakṛti* groups. The observed differences in methylation signatures between *prakṛti* suggests that different mechanisms may influence and sustain the expression of key regulator genes involved in the manifestation of distinct phenotypes. *Pitta prakṛti* showed a higher number of methylated CpG islands in the gene body region, whereas promoters' regions were more methylated in *kapha prakṛti* (Rotti et al. 2015).

Several methylated sequences of *vāta prakṛti* were found to be represented in biological processes like cell communication, transcription, signal transduction pathways and embryo morphogenesis. The associated genes are involved especially in neuronal development. Additionally, the NFIX gene was found to be associated with low body mass index (B.M.I.) which is one of the characteristic signs of *vāta prakṛti* (Rotti et al. 2015).

Pitta prakṛti showed enrichment of m.P.S.R.s (methylated *prakṛti*-specific regions) for metabolism-related pathways like regulation of hormone secretion, regulation of nucleic acid metabolism and several others. *Kapha* m.P.S.R.-associated genes were significantly enriched in cell growth and maintenance, cytoskeleton anchoring activity and cellular adhesion (Rotti et al. 2015). The 501 different m.P.S.R.s identified in the present study were sufficient to classify the study subjects on the basis of *prakṛti*, thus providing molecular evidence for one of the core concepts of Ayurveda (Rotti et al. 2015).

11.3.4.10 Immunophenotyping of Prakrti Groups

Variability in immune response is often attributed to and measured from expression of C.D. markers in lymphocytes. At present, no reports are available on the expression of C.D. markers related to *prakṛti*. Therefore, Rotti et al. (2014) evaluated a panel of lymphocyte subset C.D. markers in individuals belonging to a dominant *prakṛti*. Immunophenotyping was carried out using whole blood samples from 222 healthy subjects, grouped into *kapha* (n = 95), *pitta* (n = 57) and *vāta* (n = 70) *prakṛti*. C.D. markers such as C.D.3, C.D.4, C.D.8, C.D.14, C.D.25, C.D.56, C.D.69, C.D.71 and H.L.A.-D.R. were analyzed using the flow cytometry method. A significant difference in the expression of C.D. markers such as C.D.14 (monocytes), C.D.25 (activated B cells) and C.D.56 (natural killer cells) was observed between different *prakṛti* groups. C.D.25 and C.D.56 expression was significantly higher in *kapha prakṛti* samples. Similarly, slightly higher levels of C.D.14 were observed in *pitta prakṛti* samples. Significant difference in the expression of C.D.14, C.D.25 and C.D.56 markers between three different *prakṛti* was observed. The increased level of C.D.25 and CD.56 in *kapha prakṛti* may indicate better immune response in these individuals.

11.3.4.11 Association of Prakṛti with Diseases

Individuals belonging to various *prakṛti* are prone to diseases related to the respective element (s) of *tridōṣa* (Upadhyaya 1975c). The earliest report on the impact of *prakṛti* on the appearance of diseases is that of Venkatraghavan et al. (1987). They assessed the *prakrti* of 28 cancer patients attending the out-patient department of Government General Hospital, Madras. 57 normal, healthy individuals were considered as controls. Among the 28 patients, 1 belonged to the *vāta*-dominant *prakṛti*, 17 to the *pitta*-dominant *prakṛti* and 10 to the *kapha*-dominant *prakṛti*. The percentage of *pitta*-dominant *prakṛti* in the control group was 17.54%, whereas 60.7 % patients in the cancer group belonged to *pitta*-dominant *prakṛti*. Though limited by small sample size, this study shows that *pitta*- and *kapha*-dominant individuals are more prone to cancer.

Coronary artery disease (C.A.D.) has elements of cellular proliferation and metabolic abnormalities (Ballantyne et al. 2008). Therefore, an anomalous expression of certain elements related to the *prakṛti* of the individual will be more prevalent in C.A.D. So far, no study has demonstrated the association of risk factors (diabetes, hypertension), inflammatory markers and insulin resistance with *prakṛti* in patients suffering from cardiovascular disease. Mahalle et al. (2012) attempted to correlate constitutional types with cardiovascular risk factors, inflammatory markers and insulin resistance among subjects with established C.A.D.

The *prakṛti* of 300 patients with C.A.D (>25 years) was assessed. Biochemical parameters, inflammation markers (h.s.C.R.P., T.N.F.-alpha and I.L.-6) and insulin resistance (H.O.M.A.-I.R.) of the subjects were measured. The mean age of patients was 60.97 ± 12.5 years and 62.3% of subjects belonged to the *vata-kapha prakṛti*. Triglyceride, V.L.D.L. and L.D.L. were significantly higher and H.D.L. cholesterol was significantly lower in *vata-kapha prakṛti* in comparison with other *prakṛti* groups. *Vata-kapha prakṛti* was correlated with diabetes mellitus, hypertension and dyslipidemia. Inflammatory markers and H.O.M.A.-I.R. were also high in *vata-kapha prakṛti*. Inflammation markers were correlated positively with both *vata-kapha* and *kapha prakṛti*. There is strong relation of risk factors (diabetes, hypertension, dyslipidemia), insulin resistance, and inflammatory markers with *vata-kapha* and *kapha prakṛti*. The major limitation of the study was male predominance, and lower numbers of subjects in some groups of *prakṛti* (Mahalle et al. 2012).

A similar study was reported by Manyam and Kumar (2013). They assessed the *prakṛti* of 75 patients with established Parkinson's disease and 73 normal subjects with no known neurological disease. The results showed that the incidence of Parkinson's disease was highest in individuals belonging to *vāta prakṛti*. The incidence of the disease was higher in men than in women.

Irritable bowel syndrome (I.B.S.) is a functional gastrointestinal disorder with no known organic cause. It is characterized by chronic abdominal pain, discomfort, bloating and alteration of bowel habits. Depending on the predominant symptom, I.B.S. can be classified into subtypes such as I.B.S. with diarrhea (I.B.S.-D.), I.B.S. with constipation (I.B.S.-C.) and I.B.S. with mixed symptoms (I.B.S.-M.). Complete cure of I.B.S. cannot be achieved, as there is no pathology in the gut that can be targeted by therapeutic agents (Ford 2013). Therefore, Shirolkar et al. (2015) tried to evaluate the *prakṛti* of I.B.S. patients and their correlation with I.B.S. subtypes. Fifty I.B.S. patients were grouped into *vata*-dominant, *pitta*-dominant and *kapha*-dominant *prakrti* using a 24-item questionnaire. Of the 50 I.B.S. patients enrolled, 22 patients each were of *vāta*- and *pitta*-dominant *prakṛti*, while 6 patients belonged to the *kapha*-dominant *prakṛti*.

In the *vāta*-dominant group, I.B.S.-C. was found in 13 patients, I.B.S.-D. in 8 and I.B.S.-M. in 1. In *pitta*-dominant group I.B.S.-D. was found in 13, I.B.S.-C. in 6 and I.B.S.-M. in 3. In the *kapha*-dominant group, I.B.S.-C. was found in 5 patients and I.B.S.-M. in 1. The authors concluded that *prakṛti* examination in I.B.S. may help in detecting proneness to developing a given I.B.S. subtype (Shirolkar et al. 2015).

11.3.4.12 Prakrti in Therapeutics

Identifying phenotypic and genetic heterogeneity, gene–gene interactions and allelic spectrum is a major challenge in understanding complex diseases. Of the several factors which contribute to this, phenotypic heterogeneity is a serious limitation encountered in modern medicine (Manchia et al. 2013). Juyal et al. (2012) opine that conditioning association studies on prior risk, predictable in Ayurveda, will uncover much more variation and advance diagnostics and therapeutics. They attempted identification of genetic susceptibility markers in a rheumatoid arthritis (R.A.) cohort by combining the *prakṛti*-based grouping of individuals with genetic analysis tools. Association of 21 markers from commonly implicated inflammatory and oxidative stress pathways was tested using a case-control approach in a total cohort comprising 325 cases, 356 controls and the three subgroups separately. A few postulates of Ayurveda on the disease characteristics were also tested in the various *prakṛti* groups using clinico-genetic data (Juyal et al. 2012).

Inflammatory genes like *I.L.1β* and C.D.40 seem to be the determinants in the *vāta* subgroup, while oxidative stress pathway genes are observed in the *pitta* and *kapha* subgroups. Fixed effect analysis of the associated markers from C.D.40, S.O.D.3 and *T.N.F.-α* with genotype versus *prakṛti* interaction terms suggests heterogeneity of effects within the subgroups, suggesting disease-specific pathways. In addition to these, disease characteristics such as severity were most pronounced in the *vāta* group. The findings of this study suggest the existence of discrete causal pathways for rheumatoid arthritis etiology in *prakriti*-based subgroups, thereby validating concepts of *prakṛti* and personalized medicine in Ayurveda. This exploratory study supports the contention that subgrouping of patients based on *prakṛti* may help overcome the limitation of phenotypic heterogeneity which hampers progress in complex trait genetics research (Juyal et al. 2012).

As the molecular basis of the different *prakṛti* types has become evident, a new term, "ayurgenomics", has come into vogue. Ayurgenomics aims at integration of Ayurveda and genomics for furthering personalized nutrition, diagnosis and treatment of diseases. Validation of the fundamental concepts of Ayurveda and subsequent application of *prakṛti*-based phenotyping in complex diseases can offer the least invasive and most affordable way to assess susceptibility to diseases and prognosis. Ayurgenomics can also guide therapeutic and dietary recommendations. Research in Ayurveda had so far been confined to exploration of active principles from Ayurveda herbs indicated in the treatment of various diseases. With the burgeoning of ayurgenomics, efforts are now being made to establish the molecular basis of the principles of Ayurveda (Mukerji and Prasher 2011; Banerjee et al. 2015).

11.3.4.13 Uniform Method for Assessment of Prakṛti

Aṣṭāṅgahṛdaya describes the signs of *vata-*, *pitta-* and *kapha prakṛti* (Upadhyaya 1975c). However, various authors have employed questionnaires that do not reflect the teachings of *Aṣṭāṅgahṛdaya* (Venkatraghavan et al. 1987; Joshi 2004; Patwardhan et al. 2005; Prasher et al. 2008; Aggarwal et al. 2010; Ghodke et al. 2011; Shilpa and Murthy 2011; Tripathi et al. 2011; Bhalerao et al. 2012; Juyal et al. 2012; Mahalle et al. 2012; Suchitra and Nagendra 2012; Tiwary et al. 2012: Suchitra and Nagendra 2013; Manyam and Kumar. 2013; Rotti et al. 2014a, 2014b, 2015; Suchitra et al. 2014; Govindaraj et al. 2015; Rapolu et al. 2015; Shirolkar et al. 2015). Kurande et al. (2013a) conducted the first study to investigate comprehensively the inter-rater reliability of *prakṛti* assessment. 15 registered ayurvedic physicians having 3–15 years of experience assessed the *prakṛti* of 300 healthy subjects. Poor to substantial levels of reliability were obtained for the assessment of *prakṛti*. The authors therefore concluded that an objectively defined questionnaire should be used in ayurvedic clinical studies. Such objectivity in diagnosis will improve the confidence of physicians and these methods can be incorporated into ayurvedic clinical trials (Kurande et al. 2013b).

A major problem encountered in ayurvedic diagnosis is the determination of *prakṛti* of individuals. Based on the striking similarities between psychological somatotypes advocated by William H. Sheldon (Vertinsky 2007) and *prakṛti* types of Ayurveda, Rizzo-Sierra (2011) proposed a finite genopsycho-somatotyping of humans. This concept is based on a set of common physiological, morphological and psychological attributes related to a common basic birth constitution which remains somewhat permanent during human lifetime, as it is assumed that this primordial constitution is programmed in the D.N.A. of the individual. This approach provides a tool for classifying human population based on broad and finite phenotype clusters, cutting across barriers of ethnicity, language, geographical location or self-reported ancestry. Rizzo-Sierra (2011) proposes for males that every basic constitution has an associated identification organ, a measured property or marker, a soma and some general psychic tendencies suggesting specific behavior pattern. He proposes three basic extreme genopsycho-somatotypes or birth constitutions. They are mesomorphic or andrus (*Pitta*), endomorphic or thymus (*Kapha*) and ectomorphic or thyrus (*Vāta*). This method of genopsycho-somatotyping further postulates that male andrus constitution across races shares similarities in androgen (An) nuclear receptor behavior, while thymus constitution is mainly regulated by T-cells (Tc) nuclear receptor behavior. The thyrus constitution shares similarities in thyroxine (Th) nuclear receptor behavior. According to this hypothesis, these proposed nuclear receptors regulate the expression of specific genes, thereby controlling the embryonic development and metabolism of the human organism in very profound ways. The method predicts small differences in measured properties (An, Tc and Th nuclear receptor behavior) within a birth constitution by modulation effects in melanocyte-stimulating hormone receptor behavior (Rizzo-Sierra 2011). The endocrinological perspective proposed by Rizzo-Sierra (2011) encompassing the ayurvedic concept of *prakṛti* may provide insights for further studies in human classification. Exploratory studies combining genopsycho-somatotyping and extensive experimentation can provide objective and speedy method of assessing the *prakṛti* of individuals.

11.3.5 Taste and Drug Action

According to Ayurveda all forms of matter are derived from the combination of five primordial principles *pṛdhvi*, *ap*, *tējas*, *vāyu* and *ākāśa*, collectively called *pañcabhūta*. The *pañcabhūta* do not exist in individual forms. They always stay in combinations, the ratios being dependent on the matter in question. All substances in this universe are medicinal and they can be put to good use if one knows their medicinal value (Upadhyaya 1975d). Ayurveda states that the medicinal qualities of matter are judged on the basis of five factors viz., taste (*rasa*), qualities (*guṇa*), potency (*vīrya*), post-digestive taste (*vipāka*) and specific action (*prabhāva*) (Murthy 2017). Unlike Western science, Ayurveda considers six taste modalities sweet (*madhura*), sour (*amḷa*), salty (*lavaṇa*), pungent (*kaṭu*), bitter (*tikta*) and astringent (*kaṣāya*). These six taste modalities either increase or decrease *tridōṣa* (Upadhyaya 1975e) (Table 11.2). *Guṇa* represents qualities of matter. *Guṇa* can be innumerable. However, Ayurveda considers only 20 *guṇa* arranged in ten pairs of opposing qualities like cold (*śīta*) versus hot (*uṣṇa*), gross (*sthūla*) versus subtle (*sūkṣma*), dry (*rūkṣa*) versus moist (*snigdha*) and so on. Potency can be either hot (*uṣṇa*) or cold (*śīta*). *Vipāka* represents the post-digestive taste of each of the six taste modalities. *Prabhāva* connotes the specific action of a drug. For example, the cardiotonic property of *Arjuna* (*Terminalia arjuna*) is its *prabhāva*. Depending upon the deranged *tridōṣa*, medicinal formulations having appropriate taste (*rasa*), qualities (*guṇa*), potency (*vīrya*), post-digestive taste (*vipāka*) and specific action (*prabhāva*) are administered in various diseases (Murthy 2017).

11.3.5.1 Taste Response as Sensory Expression of Pharmacological Activity

The concept of taste influencing pharmacological actions was deeply embedded in traditional medicine of China, Greece and India. However, Ayurveda made full use of this knowledge (Beauchamp 2019). Some modern research lends support to it. Fifty years ago Roland Fischer and coworkers

TABLE 11.2
Effects of Taste Modalities on *Tridōṣa*

Taste modality	Effects on *tridōṣa*		
	Vāta	*Pitta*	*Kapha*
Madhura (Sweet)	Pacifies	Pacifies	Aggravates
Amḷa (Sour)	Pacifies	Aggravates	Aggravates
Lavaṇa (Salt)	Pacifies	Aggravates	Aggravates
Kaṭu (Pungent)	Aggravates	Aggravates	Pacifies
Tikta (Bitter)	Aggravates	Pacifies	Pacifies
Kaṣāya (Astringent)	Aggravates	Pacifies	Pacifies

had demonstrated that the bitterness of certain compounds correlated with their pharmacological activity. They called attention to a general relationship prevailing between the taste threshold of stereospecific drugs and their biological activity (Fischer and Griffin 1963, 1964). To cite an example, with l-quinine and d-quinine, as well as with d-amphetamine and l-amphetamine, the l- form of each drug pair is the biologically more potent compound. Correspondingly, humans can taste the more active compound in a lower concentration (lower taste threshold). Based on these findings they came to regard a subject's oral cavity as a pharmacological test preparation *in situ* and the taste response as a sensory expression of pharmacological activity (Fischer et al. 1965a; Fischer and Kaelbling 1967).

In a subsequent publication they reported a statistically significant positive correlation between quinine taste thresholds and the cumulative trifluoperazine dose sufficient to induce extrapyramidal side-effects like slowness of movement, tremor and jerky movements, for 48 acutely ill mental patients. Patients with high quinine taste thresholds, for instance, needed higher trifluoperazine dosage for the induction of extrapyramidal side-effects (Fischer et al. 1965b).

Ibuprofen is a non-steroidal potent modulator of inflammation and analgesia. It is a non-selective inhibitor of the cyclooxygenase enzymes COX-1 and COX-2, but not of lipoxygenase, which catalyzes steps in the biochemical inflammation pathways derived from arachidonic acid (Vane and Botting 1995). Newly pressed extra virgin olive oil contains the aldehyde oleocanthal, whose pungency induces a strong stinging sensation in the throat, very similar to that caused by solutions of ibuprofen (Breslin et al. 2001) (Figure 11.3). Like ibuprofen, oleocanthal caused dose-dependent inhibition of COX-1 and COX-2 activities. But it had no effect on lipoxygenase *in vitro* (Beauchamp et al. 2005). This finding has striking similarity with the teachings of Ayurveda and is one of the rare scientific reports noting common pharmacological activity for compounds with similar taste. As such, it is consistent with the tenet of Ayurveda that the similarity of taste of substances indicates similar pharmacological activity.

It is often asked whether it is possible to identify the *rasa*, *guṇa*, *vīrya*, *vipāka* and *prabhāva* of herbal drugs outside the formulary of Ayurveda. In addition to vegetables like cabbage (*Brassica oleracea*), carrot (*Daucus carota*), potato (*Solanum tuberosum*), lady's finger (*Abelmoschus esculentus*) and tomato (*Lycopersicon lycopersicum*), many fruits like pineapple (*Ananas sativa*), mangosteen (*Garcinia mangostana*), papaya (*Carica papaya*), custard apple (*Annona squamosa*), guava

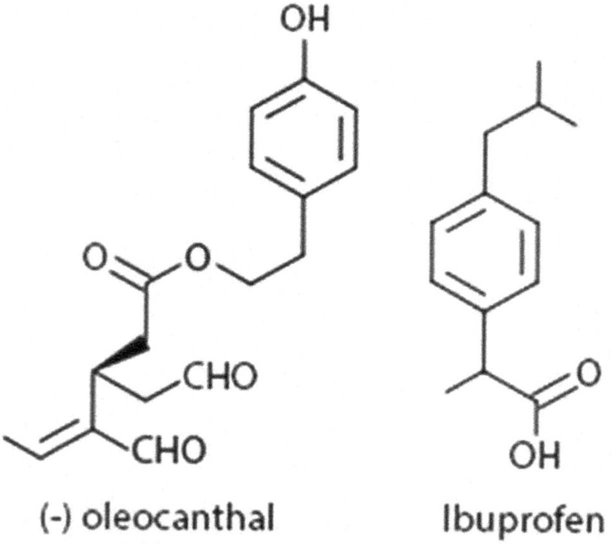

FIGURE 11.3 Though dissimilar in chemical structure, both (-) oleocanthal (left) and ibuprofen (right) are pungent and have anti-inflammatory activity.

(*Psidium guayava*), sapodilla (*Manikara achras*) and litchi (*Niphelium litchi*) were introduced into India by the Portuguese and English (Sen 1997). Though they have medicinal value, all of them are alien to Ayurveda. These herbs can be successfully incorporated into Ayurveda once the five ayurvedic features are known. Taste sensitivity studies using juices of these herbs are essential in this regard (Harris and Kalmus 1949; Fischer and Griffin 1964; Fischer et al. 1965a; Kaplan and Fischer 1965; Fischer and Kaelbling 1967; Breslin et al. 1993).

11.3.5.2 Role of Taste Preferences in Diagnosis

The concept of taste is important in diagnosis of diseases as well. As diseases spring from the derangement of *tridosa*, the patient experiences cravings for food and beverages that have tastes with the ability to restore the steady state of *tridōṣa*. The taste preferences of the patient also aid the physician in diagnosis. Table 11.3 lists the taste-related signs from the Ayurveda text *Aṣṭāṅgahṛdaya*. In addition to these, many forms of pseudogeusia are also mentioned in this ayurvedic tome. These are instances of taste sensations felt by the patient in the absence of any external stimuli. These signs are available in Table 11.4.

Most of the conditions of pseudogeusia mentioned in *Aṣṭāṅgahṛdaya* are related to diseases like *grahaṇi* (diarrhea-like conditions), *aruci* (anorexia), *chardi* (emesis), *tṛṣṇa* (excessive thirst), *kāsa* (cough), *pāṇḍu* (morbid pallor), *pratiśyāya* (rhinitis) and *rājayakṣma* (tuberculosis). It will be worthwhile to investigate whether these pseudogeusic conditions disappear following ayurvedic treatment of these diseases. Such studies are essential, as appropriate studies on the treatment of gustatory dysfunction are notably lacking (Heckmann et al. 2003).

11.4 MOLECULAR BASIS OF AYURVEDIC MEDICINES

Bioactive compounds isolated from herbs are favorable as lead structures for drug discovery. Though it is generally accepted that they show great structural diversity and could play a protagonist role for discovering new drugs, evaluating this diverse chemical space efficiently has remained a challenge for medicinal chemists and pharmacologists. Isolation of all the available bioactive compounds, followed by random screening is next to impossible (Polur et al. 2011). Therefore, experts in chemoinformatics have adopted the application of computational approaches for the identification of bioactive molecules (Harvey 2008). Systems biology approaches have been used to explore the molecular basis of Chinese medicine and to relate its terminology in the context of Western medicine (Yi et al. 2010).

TABLE 11.3
Signs Related to Taste Preferences or Aversion towards Tastes (Upadhyaya 1975f)

Sl. No.	Sign	Clinical Significance	Reference
1	Interest in sour taste	Signs of diminution of blood	*Sū.*, Ch. 11, Verse 17
2	Interest in sweet, sour and salty tastes	Sign of *vāta* constitution	*Śā.*, Ch. 3, Verses 85–89
3	Interest in sweet, astringent and bitter tastes	Sign of *pitta* constitution	*Śā.*, Ch. 3, Verses 90–95
4	Interest in bitter, astringent and pungent tastes	Sign of *kapha* constitution	*Śā.*, Ch. 3, Verses 96–102
5	Interest in sour, salty and pungent tastes	Prodrome of *jvara*	*Ni.*, Ch. 2, Verses 6–9
6	Aversion towards sweet taste	Prodrome of *jvara*	*Ni.*, Ch. 2, Verses 6–9
7	Eager longing for all tastes	Sign of *vāta grahaṇi*	*Ni.*, Ch. 8, Verses 22–25
8	Interest in jaggery	Sign of spiritual affliction	*US.*, Ch. 4, Verses 41–42

Sū = *Sūtrasthāna*, Ch. = Chapter, *Śā* = *Śārīirasthāna*, *Ni* = *Nidānasthāna*, *US* = *Uttarasthāna*, *jvara* = fever, *grahaṇi* = digestive disorder

TABLE 11.4
Taste Sensations Perceived in the Absence of Any External Stimuli (Upadhyaya 1975g)

Sl. No.	Sign	Clinical Significance	Reference
1	Sensation of sweet taste	Prodrome of *mēha*	*Ni.*, Ch. 10, Verse 38
		Sign of *kapha jvara*	*Ni.*, Ch. 2, Verses 21–22
		Prodrome of *rājayakṣma*	*Ni.*, Ch. 5, Verses 7–12
		Sign of *kapha aruci*	*Ni.*, Ch. 5, Verse 29
		Sign of *kapha chardi*	*Ni.*, Ch. 5, Verses 34–35
		Sign of *kaphaja tṛṣṇa*	Ni., Ch. 5, Verses 52–53
2	Sensation of pungent taste	Sign of *pitta pāṇḍu*	*Ni.*, Ch. 13, Verses 10–11
		Sign of *pitta jvara*	*Ni.*, Ch. 2, Verses 18–20
3	Sensation of astringent taste	Sign of *vāta jvara*	*Ni.*, Ch. 2, Verses 10–17
		Sign of *vāta aruci*	*Ni.*, Ch. 5, Verse 29
4	Sensation of bitter taste	Sign of *kapha-pitta jvara*	*Ni.*, Ch. 2, Verse 26
		Sign of *pitta aruci*	*Ni.*, Ch. 5, Verse 29
		Sign of *pitta kāsa*	*Ni.*, Ch. 3, Verses 24–25
		Sign of *pittaja tṛṣṇa*	Ni., Ch. 5, Verse 51
5	Sensation of salty taste	Prodrome of *chardi*	Ni., Ch. 5, Verse 30

Ni. = *Nidānasthāna*, Ch. = Chapter, *mēha* = polyuric disease, *jvara* = fever, *rājayakṣma* = kingly consumption or tuberculosis, *aruci* = anorexia, *chardi* = emesis, *tṛṣṇa* = excessive thirst, *pāṇḍu* = morbid pallor, *kāsa* = cough

Considering the paucity of information vis-à-vis Ayurveda, Polur et al. (2011) developed an *in silico* library of bioactive molecules from Ayurveda, coupled with structural information, source of the compounds and traditional therapeutic use. Subsequently, they utilized the ayurvedic indication of the plants, to predict biological activities of their constituent molecules, on the basis of structural similarity pairings with drug molecules. Interesting findings emerged from the study.

For example, the leaves of the neem tree (*Azadirachta indica*) are traditionally used as an anti-inflammatory drug. The leaves contain isoazadirolide, a bioactive molecule with very little information available in literature (Siddiqui et al. 1986). Isoazadirolide was linked in the network to two synthetic compounds, namely desonide and fluocinonide. Desonide is a topical agent for dermatoses, while fluocinonide is a topical glucocortisoid used in the treatment of eczema (Brogden et al. 1974; Bhankharia and Sanjana 2004).

Betulin is a triterpene present in six different plants used in Ayurveda (Dinnimath et al. 2017). It was identified to be structurally similar to calcidiol, a major circulating metabolite of vitamin D3 used in the treatment of rickets in children (Lee et al. 2013) and osteoporosis in adults (Brandi and Minisola 2013). All six plants containing betulin are traditionally used against chronic rheumatism, a disease closely related to rickets and osteoporosis. This study demonstrates that this integrated *in silico* ethnopharmacological approach would help to unravel the molecular basis of the therapeutic actions of ayurvedic medicinal plants, repurposing of drugs, and the identification of novel chemical entities with attractive scaffolds for drug discovery (Polur et al. 2011).

11.5 PROBLEMS THAT WARRANT URGENT ATTENTION

In spite of the growing interest in Ayurveda, a closer look at the system prevailing in India reveals that it is suffering from many shortcomings. Unlike Western medical texts, Ayurveda tomes are generally written in a poetic style in the Sanskrit language, using numerous synonyms and antonyms. Sound knowledge of the nuances of the language is essential for better understanding of the subject. Additionally, Ayurveda theory is based on unique concepts like *pañcabhūta* (the five primordial elements), *tridōṣa* (the three "vitiators"), *dhātu* (tissue elements) and *mala* (waste products), which are very different from those of Western medicine. Similarly, the quality of drug substances is judged on the basis of properties like *rasa* (taste), *guṇa* (quality), *vīrya* (potency), *vipāka* (post-digestive taste) and *prabhāva* (specific actions). It is nearly impossible to find equivalents for these concepts in Western science. This fundamental difference between the two systems is ignored in contemporary Ayurveda to such an extent that Ayurveda is taught and practiced along the lines of, and with the help of, Western medicine. Many stumbling blocks have to be removed if Ayurveda is to cater to the needs of a wider audience.

11.5.1 Botanical Identity of Herbs

Pharmacognosy of many ayurvedic herbs is in a chaotic state. Most of the published pharmacognostical studies are on formulations, forgetting the fact that standards for formulations can be set only when standards for the constituent herbs are fixed. Various ayurvedic lexicons such as *Abhidānamañjari*, *Bhāvaprakāśa Nighaṇṭu*, *Dhanvantari Nighaṇṭu* and *Rāja Nighaṇṭu* provide information on the botanical characteristics and medicinal properties of herbs. This information is usually given in a string of Sanskrit synonyms, composed in a metrical style, much similar to that of a religious hymn. Because of their linguistic and metrical characteristics, it is easy to memorize these quatrains. For example, *Tinospora cordifolia* (*Guḍūci* or *Amṛta*) and fruits of *Phyllanthus emblica* (*Āmalakī*) are described in the following manner:

Guḍūci kuṇḍalī sōmā chinnā chinnōtbhavāmrtā
Madhuparṇī chinnarūhā vayasthā cakṛalakṣaṇā
Āmalakī pañcarasā śṛīphalī dhātrikā śivā
Ākārāmṛtā vayasthā ca vṛṣyātiṣyāphalā tathā

Some names are common for many plants (homonyms), an example being *Vayasthā* for *Tinospora cordifolia* and *Phyllanthus emblica* (Sivarajan and Balachandran 1994). As a result, several plants are used for the same Sanskrit entity, leading to confusion. Table 11.5 lists four examples.

To clear the existing confusion in nomenclature, Krishnamurthy (1971) proposed a new method of pharmacolinguistics, which is the study of linguistic aspects of names of medicinal plants, applied in its rudimentary form by the medieval Armenian physician Amirdovlat' Amasiats' I (1420–1496) (Gueriguian 1987). Based on this method he ascertained the identities of *Haimavati* and *Kuliñjana*. From eight different plants *Haimavati* was identified as *Acorus gramineus*. *Alpinia galanga*, *Kaempferia galanga*, *Alpinia chinensis* and *Alpinia officinarum* are generally equated with the Sanskrit entity *Kuliñjana*. However, based on pharmacolinguistic study it was ascertained that *Alpinia galanga* is the correct choice (Krishnamurthy 1971).

TABLE 11.5
Some Sanskrit Entities and Their Botanical Equivalents

Sanskrit Name	Botanical Equivalents	Reference
Parpaṭaka	*Fumaria officinalis*, *Fumaria indica*, *Oldenlandia corymbosa*, *Polycarpea corymbosa*, *Justicia procumbens*, *Rungia repens*, *Rungia parviflora*, *Peristrophe bicalyculata*, *Glossocardia linaerifolia*, *Mollugo stricta*	Meulenbeld (2009)
Pāṣāṇabhēda	*Saxifraga ligulata*, *Aerva lanata*, *Kalanchoe pinnata*, *Coleus aromaticus*, *Homonoia riparia*, *Rotula aquatica*, *Ocimum basilicum*	Meulenbeld (2009)
Śaṅkhapuṣpi	*Clitoria ternatea*, *Convolvulus pluricaulis*, *Evolvulus alsinoides*, *Canscora decussata*	Sivarajan and Balachandran (1994)
Vidāri	*Pueraria tuberosa*, *Ipomoea mauritiana*, *Adenia hondala*	Sivarajan and Balachandran (1994)

In spite of being a very popular drug, there has been no authentic information available as to which plant is the true source of *Śaṅkhapuṣpi*, in place of which four plants are used in different parts of India, physicians of Kerala preferring *Clitoria ternatea* (Table 11.5). A survey of crude drug samples of *Śaṅkhapuṣpi* from various pharmaceutical companies showed that nine samples were of *Convolvulus pluricaulis*, one was of *Evolvulus alsinoides*, one was a mixture of these two, and others were comprised of some different plants (Singh and Vishwanathan 2001).

Using the pharmacolinguistic approach, Pillai (1976) ascertained the correct identity of the plant *Śaṅkhapuṣpi*. By scoring the Sanskrit synonyms of *Śaṅkhapuṣpi* against the exomorphy of these contestants, it was discovered that the name *Śaṅkhapuṣpi* is most appropriate for *Clitoria ternatea*. Out of 30 synonyms which obviously denote the morphological and pharmacological characteristics of *Śaṅkhapuṣpi*, *Clitoria ternatea* scored 30, *Canscora decussata* one, *Convolvulus pluricaulis* two and *Evolvulus alsinoides* one. Synonyms like *Vanamālini* (garland in a forest), *malavināśini* (laxative), *gōkarṇi* (shaped like the ear of a cow) and *aparājita* (sword-like), alluding to the similarity of the pod to the legendary sword, were strikingly related to *Clitoria ternatea* (Pillai 1976).

The success of the pharmacolinguistic approach is confirmed by study of chemical markers. Kaempferol, a flavonoid is reported to be present in *Clitoria ternatea*, *Convolvulus pluricaulis* and *Evolvulus alsinoides* (Kazuma et al. 2003; Gupta et al. 2007; Kumar et al. 2010; Ganie et al. 2015a). H.P.L.C. analysis showed that *Clitoria ternatea* has the highest content of kaempferol (Ali et al. 2013; Ganie and Sharma 2014; Ganie et al. 2015b). Interestingly, kaempferol, some glycosides of kaempferol and plants containing this flavonoid may have neuroprotective activity and play a protective role in Alzheimer's disease, Parkinson's disease and Huntington's disease (Calderón-Montaño et al. 2011). The pharmacolinguistic approach seems to be quite useful for ascertaining the identity of controversial medicinal plants.

11.5.2 Identification of Formulae with Lesser Ingredients

In contemporary Ayurveda, formulations with numerous ingredients are generally manufactured. At least in the province of Kerala, this trend was started in 1902, when Vaidyaratnam P.S. Variar founded *Ārya Vaidya Śāla*. All other entrepreneurs who came into this field followed suit and as a result almost all of them manufacture and sell the same set of products. However, formulae of medicines having equal or more efficacy are described in Ayurveda texts like *Cakradattam*, *Gadanigraham*, *Vaidyamanōrama* and *Cikitsāmañjari*. An example is *Guḍūci Ghṛtam* described in *Cakradattam*. It is a simple remedy prepared using juice of *Guḍūci* (*Tinospora cordifolia*), cow milk and clarified butter. This medicated clarified butter has exceptional effect on skin diseases (Sharma 1994). Equally interesting is the formulation *Piṇḍāsava* described in *Caraka Samhita* to cure of an array of digestive diseases (Sharma 1998). *Piper longum* fruits, jaggery and pericarp of *Terminalia belerica* are to be pounded together with water and kept in a heap of barley grains for a fortnight. This medicine is administered suspended in water. Healthy individuals can also consume *Piṇḍāsava* for a month in order to prevent the appearance of those disorders. Such formulations need to be popularized, so that the impact of shortages of crude drugs on medicine manufacture can be lessened to some extent.

11.5.3 Discovering New Dosage Forms

Nowadays ayurvedic companies offer tablets of *kvātha* instead of the *kvātha* itself. The justification offered is that tablets are easier to consume and carry than *kvātha*, which are invariably bitter or astringent. It is also widely known that many such *kvātha* tablets fail to bring about the desired effect of the corresponding *kvātha*. The fundamental cause of this anomaly is the disparity in dosage. The standard dose of a *kvātha*, as instructed in classical Sanskrit texts is the decoction obtained from one *pala* of herb mixture. The metric equivalent of one *pala* is 48 g (Anonymous 1978b). One *pala* of crude herb mixture will yield not less than 10% of water extract, which amounts to 4.8 g.

5 g of granulated extract, when punched, will yield ten tablets. Consuming five tablets twice a day will not be liked by patients.

To overcome the problem of dosage it is better to convert ayurvedic *kvātha* unto granules of extracts. Granules equivalent to one *pala* of crude herb powder can be easily packed into two tea-bags, which can be dipped into hot water and the reconstituted decoction consumed. This will overcome the problem of sub-therapeutic dosage of *kvātha* tablets.

11.5.4 Substitution of Plant Parts

Ayurvedic pharmacy makes use of a wide variety of plant parts or products like whole plants, flowers, fruits, seeds, leaves, galls, stems, bark, wood, roots, oil, gums and exudates (see Table 11.6). These are generally collected from the wild, though recently, attempts are being made at cultivation and conservation of some herbs. However, population growth and increasing urbanization have had their deleterious effects on the ecosystem, leading to depletion of plant wealth. A glaring example that highlights this problem is the *daśamūla* group of crude herbs, which consists of five larger roots and five smaller roots. Production of *Daśamūla kvatha*, *Daśamūlāriṣṭa* and several other formulations is adversely affected by shortage of these crude herbs. One way to solve this problem is by substitution of the roots with a renewable plant part like the leaf.

Jena et al. (2017) conducted a study to explore the substitution of root of a member of the *daśamūla* group, *Premna latifolia*, with its leaf and young stem. T.L.C. fingerprinting and monitoring of analgesic and anti-inflammatory activities of the methanol extracts of the whole root, root bark, root pith, young stem and leaves were compared. The chromatogram of the leaf extract was similar in profile to that of the whole root both under $U.V._{254}$ and after derivatization, with six major bands appearing at matching R_f values. All plant parts showed statistically significant and dose-dependent analgesic and anti-inflammatory activities. The whole root and leaf showed same level of inhibition of edema (48%) at 6 h at a dose of 400 mg/kg when compared to 59% inhibition shown by ibuprofen.

Similar results were reported earlier by Bidhan et al. (2015) who investigated the similarity of high-performance thin layer chromatographic (H.P.T.L.C.) patterns of leaf and heartwood of *Pterocarpus marsupium*, a medicinal tree exploited mainly for the heartwood, which has significant anti-diabetic activity (Ivorra et al. 1989; Kameswara et al. 2001). The study revealed that the leaf extract produced an equal number of spots in the H.P.T.L.C. system, suggesting similarity in biological activities. Similar studies need to be conducted to explore the possibility of substituting root, bark and heartwood with renewable parts like leaves and stems of many important ayurvedic herbs.

TABLE 11.6
Plant Parts Consumed in Ayurvedic Pharmacy

Plant Part	Percentage
Whole plants	13
Flowers	6.50
Fruits and seeds	25.80
Leaves	10.90
Galls	0.30
Stems and bark	12.60
Wood	1.60
Roots	26.20
Oils	0.20
Gums and exudates	2.90

Adapted from: Ved and Goraya (2007)

11.5.5 Pharmaceutics of Ayurvedic Medicines

Expressed juices and powders are rather simpler preparations, requiring only extraction of the juice in the case of *svarasa* and pulverizing and blending of the herbs, in the case of *cūrṇa*. However, chemical conversions can occur among the bioactive compounds during the process of preparation of decoctions, which are used in Ayurveda as a dosage form as well as as ingredients of medicated oils and clarified butter. The dynamics of transfer of compounds has been studied well in the case of Chinese and Kampo medicine (Takaishi and Torii 1969; Takaishi and Watanabe 1971; Noguchi et al. 1978a, 1978b; Arichi et al. 1979; Tomimori and Yoshimoto 1980; Yamaji et al. 1984). However, such information is totally lacking in the case of ayurvedic *kvātha*.

Taila and *ghṛta* are prepared by boiling juices or water extracts of herbs with vegetable oils or clarified butter, in the presence of a small quantity of paste of herbs (*kalka*). The process is carried out with frequent stirring of the boiling mixture. The function of the *kalka* is to introduce small quantities of some herbs into the preparation as well as to indicate the stage at which the *taila* or *ghṛta* is to be filtered (Lahorkar et al. 2009). The general belief is that during this process the lipid-soluble compounds in the decoction dissolve in the oil or clarified butter. However, it is possible that during the continued heating and stirring over an extended period, glycosides in the boiling mass may turn into aglycones, rendering them soluble in the lipid medium. In addition to this, all kinds of chemical transformations may occur, like isomerizations (cis/trans cinnamic acid analogues and derivatives), inter- and intra-molecular trans-esterifications, hydrolysis of esters and amides, and so on. High temperatures can also cause oxidation and decomposition. The only way to understand the chemical changes is to analyze the herbal fluids and lipid before, during and after the treatments (Dr Robert Verpoorte, Leiden University, personal communication).

Virgin coconut oil (V.C.O.) is coconut oil directly extracted from fresh coconut milk (Villariano et al. 2007). Phenolic acid fraction (caffeic acid, *p*-coumaric acid, ferulic acid and (+/–) catechin) of the coconut oil prepared by boiling coconut milk (the traditional method) is known to be more complex compared with that of coconut oil prepared by pressing copra (commercial coconut oil). The total phenol content of coconut oil produced by the traditional method is nearly seven times higher than that of commercial coconut oil (Seneviratne and Dissanayake 2008). On account of the high content of biologically active components like vitamins, phytosterols and polyphenols, V.C.O. exerts anticancer, antimicrobial, analgesic, antipyretic, and anti-inflammatory properties (Abbas et al. 2017; Intahphuak et al. 2010; Nevin and Rajamohan 2010). Considering the physico-chemical properties, biological activities and beneficial effects on health (DebMandal and Mandal 2011; Chinwong et al. 2017), V.C.O. can be a cost-effective substitute for clarified butter. Ayurvedic *ghṛta* can be prepared with V.C.O.

The traditional method of preparing ayurvedic pills is another area worthy of study. Solid-state fermentation (S.S.F.) is the growth of microorganisms on moist solid material in the absence or near-absence of free water. The microorganisms grow on a natural substrate or an inert support used as a solid base (Bhargav et al. 2008). It seems that S.S.F. is involved in the preparation of traditional ayurvedic pills. An example is *Bilvādi guḷika*, used in the treatment of bites from scorpions, rodents, insects and spiders, gastroenteritis, dyspepsia, fever, poisoning and psychological conditions (Vaidyan and Pillai 1985). It is prepared by grinding the mixture of herb powders for 1 or 2 hours daily for about 8–9 months continuously, by wetting with goat urine. On completion of the grinding process, the mass is rolled into small pills. When processed in the traditional way, the pills never dry up. The daily grinding helps the mass to acquire a particle size adequate for recharging of the microbial inocula, which may include yeasts and lactic acid bacteria. This process may lead to S.S.F. It may be possible that the biotransformation of berberine, observed elsewhere (Chandra et al. 2012) may be taking place in *Bilvādi guḷika* also under an extended period of preparation, as most of the constituent herbs are rich in berberine (Sabu and Haridas 2015). The biotransformation of herbal bioactives during the preparation of ayurvedic pills, in the presence of microflora needs to be investigated.

In Ayurveda the five-fold characteristics (*rasa, guṇa, vīrya, vipāka, prabhāva*) are indicated for whole herbs. Nevertheless, only aqueous decoctions of herbs (*kvātha*) are administered internally. All the water-insoluble compounds are left out in the spent herb as waste. Crude drugs like black pepper, long pepper, dry ginger, turmeric and others are boiled and the spent herb is discarded, along with their water-insoluble compounds. As these compounds are also biologically active, it will be interesting to judge the therapeutic efficiency of hydroalcoholic extracts as liquid preparations, instead of *kvātha*. As the hydroalcoholic extracts contain all the phytoconstituents, the medicinal preparation will have enhanced therapeutic activity. Moreover, the dose can also be reduced. The same innovation can be applied to the preparation of *taila* and *ghrta* as well. At present only water extracts are used in the preparation of *taila* and *ghrta*.

11.5.6 Research on Posology

The *rasa, guṇa, vīrya, vipāka* and *prabhāva* (factors that decide drug action) are known for all ayurvedic herbs. However, the *rasa, guṇa, vīrya, vipāka* and *prabhāva* of combinations of herbs remain unknown. These five-fold characteristics of major combinations of herbs can be determined by studying the contributions of the Arab physician Al-Kindi. Yakub ibn Ishaq al-Sabah al-Kindi (A.D. 800–866) was the first Arab philosopher. He is considered to be a "Renaissance man" of the Arab world, centuries before the Renaissance happened in Europe. Al-Kindi's most important work on medicine was *De Medicinarum Compositarum Gradibus Investigandis Libellus* (The investigation of the strength of composite medicines). The treatise is all about posology, a branch of medicine, which al-Kindi practically invented for dealing with dosages of drugs. Before al-Kindi there was very little scientific study on this subject. In this book he describes all kinds of medicines that physicians used to cure various ailments at the time. Finding out the dosages of these medicines was a guessing game in those days (Abboud 2006).

Using brilliant deduction, al-Kindi applied mathematical calculations to the earlier work on degrees of warmth and coldness of drugs done by Galen (A.D. 131–201). He formulated an easy-to-use table that pharmacists could refer to when writing prescriptions. If a drug was to be neither warm nor cold, then he would add one-part warm ingredient and one-part cold ingredient, finally cancelling out any warmth. If the formulation was to have first degree of warmth, the drug mixture should contain two-parts warm to one-part cold. For the second degree of warmth, three-parts warm were added to one-part cold, and so on. By documenting amounts of drugs with a formula which anyone could follow, al-Kindi revolutionized medicine. Drugs could now be formulated according to fixed amounts, with the result that all patients would receive standardized dosages of medicines (Abboud 2006). Compound ayurvedic formulations can be administered with precision once their *rasa, guṇa, vīrya, vipāka* and *prabhāva* are known.

11.5.7 Research on Rasaśāstra

While mainstream Ayurveda uses only herbs for the preparation of medicines, there is a distinct stream of Ayurveda which employs metals, minerals and animal products along with herbs. *Rasouṣadhi* (minerallo-metallic preparations) have the three characteristics of instant effectiveness, requiring only very small doses and extensive therapeutic utility, irrespective of constitutional variation in patients. The 8th-century Buddhist sage Nāgārjuna is said to have perfected this system (Savrikar and Ravishankar 2011). It is significant that classical Ayurveda texts like *Caraka Samhita, Suśruta Samhita, Aṣṭāṅgasamgraha* and *Aṣṭāṅgahṛdaya* do not discuss *rasaśāstra*.

Western medicine considers mercury inherently toxic, its toxicity not being due to the presence of impurities. Compounds of mercury are known to cause permanent damage to brain and kidneys (Anonymous 1999). However, exponents of *rasaśāstra* consider mercury, arsenic, zinc, tin, lead, antimony, sulfur and so on as valuable medicinal substances, if treated in the proper way. *Rasaśāstra* texts describe various elaborate procedures of *śōdhana* (detoxification-cum-potentiation) (Jagtap et al. 2013a).

Some studies have been carried out to determine the toxicity of calcined powders of metals and minerals. Jagtap et al. (2013b) studied the toxicity of *tāmra bhasma* (calcined powder of copper) using non-detoxified copper and detoxified copper. *Tāmra bhasma* prepared from non-detoxified copper caused pathological effects on several hematological parameters and cytoarchitecture of different organs, even at therapeutic dose level of 5.5 mg/kg. However, *Tāmra bhasma* prepared from detoxified copper was found to be safe even at doses five times higher than the therapeutic dose (27.5 mg/kg). These observations lend scientific evidence for effectiveness of the process of detoxification recommended in *rasaśāstra* texts.

Sidh Makardhwaj is a *rasaśāstra* formulation prepared from gold, mercury and sulfur. The safety of this preparation was reported by Kumar et al. (2014). They reported that *Sidh Makardhwaj* administered for 28 days, at doses five times higher than the human dose, did not show any toxicological effects on brain cerebrum, liver and kidney of rats. *Svarṇa bhasma*, another calcined powder of gold was also found to be safe in a 90-day toxicity study carried out in Wistar rats (Jamadagni et al. 2015)

Nevertheless, not all *rasaśāstra* medicines are free from side-effects. *Nāradīya Lakṣmīvilāsa Rasa* is an ayurvedic medicine used in tablet or powder form, prepared with mica, mercury, sulfur and 13 herbs. It is used in the treatment of sinusitis, chronic skin diseases, diabetes and urinary tract disorders (Anonymous 1978c). Hasan et al. (2016) studied the toxicological effects of this formulation following administration for 32 days. It was observed that this medicine increased the weight of heart, lungs, liver, kidneys, spleen and testes. The authors inferred that *Nāradīya Lakṣmīvilāsa Rasa* should not be administered chronically in higher doses.

It is believed that the aims of *śōdhana* procedures are removal of physical and chemical impurities, minimization of toxicity of the substance, transformation of the hard and non-homogeneous material to a soft, brittle and homogeneous state, conversion of the material to a suitable form for further processing and enhancement of therapeutic efficacy of the drug (Jagtap et al. 2013a). However, further investigations are required to confirm this. As minerallo-metallic preparations have the inherent danger of toxicity, they need to be used with extreme caution.

11.5.8 Objective Way of Teaching Ayurveda

Ayurveda is taught at present solely with the help of the Sanskrit texts and without making use of figures, tables, flowcharts or diagrams which make learning easier. Instead of encouraging them just to memorize Sanskrit quatrains, students should be taught Ayurveda objectively. Concepts like *prakṛti* (constitution) and *dōṣakōpam* (enragement of *dōṣas*) should be taught with the aid of practical classes. At present, in examinations, the knowledge of a student is judged on his ability to recite Sanskrit quatrains correctly. Instead of that, their skill in identifying *prakṛti* of individuals and assessing the degree of *dōṣakōpam* should be evaluated. Greater emphasis should be given to identification of symptoms and signs. When students are taught Ayurveda in a progressive way, they will be able to diagnose diseases on the basis of ayurvedic principles.

11.5.9 Diagnosis of Diseases

Considering the fact that the human organism is a part of the cosmos, Ayurveda insists upon the need to examine minutely ten factors for effective diagnosis and treatment of diseases (Kumar 1992). They are briefly described below.

I. Physiological constitution (*prakṛti*)

   The concept of *prakṛti* has some similarities with the Greco-Arabic concept of *mizāj* (Ahmer et al. 2015). Ayurveda considers mainly seven types of *prakṛti* (vide supra). As individuals belonging to each *prakṛti* are susceptible to diseases arising out of the destabilization of the corresponding component(s) of *tridōṣa*, knowledge of *prakṛti* helps in the selection of food, measures and medicinal substances that are to be adopted to promote the steady state of *tridōṣa* (Upadhyaya 1975c).

II. Tissue elements (*dhātu*)

An intelligent practitioner can accurately gauge the state of *dhātu* cycle by correlating symptoms and signs with the *dhātu*, *mala* and *tridōṣa*.

III. Digestive efficiency (*agni*)

Metabolic activities of the human body are regulated by the omnipresent fire (*agni*) which, for the sake of convenience, is classified into 13 varieties. The ingested food is first digested by the abdominal fire (*jaḍharāgni*), followed by five *bhūtāgni* (fires of the *pañcabhūta*), one *agni* assigned to each *bhūta* present in matter. The essential products of digestion enter the *dhātu* cycle, each step of which is catalyzed by a *dhātvagni* corresponding to each *dhātu* (Upadyaya 1975c).

Ancient authorities of Ayurveda considered *jaḍharāgni* as the prime regulator of metabolism. It is classified into four types depending on the intensity. *Viṣamāgni* (irregular fire) is observed in conditions related to increased *vāta*. Predominance of *pitta* gives rise to *tīkṣṇāgni* (sharp fire). *Mandāgni* (dull fire) results from the influence of *kapha* and *samāgni* (regular fire) is experienced by individuals in whom there is steady state of *tridōṣa* (Upadhyaya 1975b).

IV. Temporal aspects of *tridōṣa* (*kālam*)

The *tridōṣa* exhibit circadian and circannual rhythms. Day and night start with *kapha* and end with *vāta*, the intermediate period being characterized by *pitta* (Upadhyaya 1975b). Similarly, the *tridōṣa* fluctuate during the seasons of the year (Upadhyaya 1975h). The temporal aspects of *tridōṣa* are to be considered in diagnosis, selection of medicines and time of their administration.

V. Age (*vayaḥ*)

If the life span of an individual is divided into three equal parts (childhood, adulthood and old age), the first part will be dominated by *kapha*, followed by *pitta* and *vāta* respectively (Upadhyaya 1975b). The age of the patient helps the physician to decide the dosage of medicines and to make a prognosis.

VI. Physical strength (*balam*)

The season of the year and the age of the individual, by virtue of their *tridōṣa*-modulating qualities, alter the physical strength and so does the adoption of food and measures (Upadhyaya 1975i). The constitutional and temporal aspects of *balam* are to be considered in the selection of such food and measures which help the body to regain health.

VII. Place of residence (*dēśam*)

Depending on climatic and geographical features a land (*dēśam*) is divisible into *jāṅgala* (arid), *ānūpa* (wet, marshy, sylvan) and *sādhāraṇa* (mixed) varieties. *Vāta* is predominant in the first one and *kapha* in the second. The third type is favorable for the maintenance of a steady state of *tridōṣa* (Upadhyaya 1975b). The type of place of residence has its own clinical importance. For example, if a *vāta* disease manifests in one individual each of the *jāṅgala* and *ānūpa* areas, the drug should be administered to the former in the medium of clarified butter and the latter should receive only a plain decoction.

VIII. Homologation (*sātmya*)

Getting used to food, drinks and measures is called *sātmya* (homologation). Unwholesome victuals and measures will be injurious in the long run, even though they do not evoke any immediate adverse effects. The habits and addictions of a patient need to be understood for achieving homologation. The patient should achieve homologation through food, measures and medicines that bring the *tridōṣa* to steady state (Upadhyaya 1975j).

IX. Food (*āhāra*)

Ayurveda states that the physical and mental characteristics of a person are influenced by the type of food consumed. Consequentially, proper dietetics is essential to cure diseases and to maintain health (Upadhyaya 1975k).

X. Emotional status (*satvam*)

Fear, anxiety and other psychological stresses cause disease and aggravate them. The patient is therefore, expected to be optimistic and fearless to undergo treatment (Upadhyaya 1975b). Adequate counseling should be given to the patient for adopting a positive outlook.

However, in contemporary ayurvedic practice, diseases are diagnosed exclusively on the basis of Western medical jargon and technology. As ayurvedic theory correlates qualities of matter, seasons, symptoms of diseases and several other factors with *vāta*, *pitta* and *kapha*, introduction of any new parameter into ayurvedic practice calls for establishing its relationship with the *tridōṣa*. For example, clinical data obtained through instrumental techniques like spectrophotometry, electrocardiography, electromyography, computed tomography scan and the like are to be rationally correlated with *tridōṣa* before they are integrated into Ayurveda. However, as such an exercise seems to be a difficult task in the light of the present state of affairs, it will be more appropriate to use only the parameters of Ayurveda in the diagnosis of diseases. There are nearly 2500 distinct symptoms described in *Aṣṭāṅgahṛdaya*. A wise physician should collect as many symptoms as possible through *darśana* (observation), *sparśana* (palpation) and *praśna* (interrogation). Thereafter, he should classify these symptoms at the levels of *tridōṣa*, *dhātu*, *mala*, *ṣaḍkriyākāla* (the six stages of development of a disease), *pūrvarūpa* (premonitory symptoms and signs), *āvaraṇā* (envelopment) and the various diseases described in Ayurveda. Such a diagnostic approach will help the medical world to look at disease entities described in Western medicine through the Ayurveda perspective, offering the possibility of identifying lines of ayurvedic treatment for many refractory diseases. Western medical knowledge and investigation technology can be employed for evaluating the success of ayurvedic diagnosis and treatment.

11.5.10 Clinical Research on Ayurvedic Medicines

Ayurveda advocates specific protocols in the treatment of diseases. Medicinal preparations in several dosage forms, medicinal food and therapeutic measures are used in the process. However, almost all clinical studies on Ayurveda are carried out without considering the individualization of therapy, which is a salient feature of Ayurveda. For example, seven studies have been reported on the ayurvedic treatment of rheumatoid arthritis (Park and Ernst 2005). Nevertheless, none of them dealt with classical ayurvedic treatment of rheumatoid arthritis or allowed individualization of therapy. They used fixed combinations of the same formulations throughout and did not individualize the therapies. As a result, these studies show that ayurvedic treatment is ineffective (Kulkarni et al. 1992; Sander et al. 1998; Chopra et al. 2000).

The treatment of respiratory diseases (*Śvāsahidhmā cikitsa*) described in *Aṣṭāṅgahṛdaya* would serve as an example of individualized therapy. As the first step in the treatment, the patient is to be smeared with an appropriate oil mixed with powdered rock salt. Thereafter, he is to be given a bath in warm water. The *kapha* lodged inside the channels of the body get loosened and reach the alimentary canal. After sudation, the patient should consume unctuous food mixed with soup prepared with fish and meat of animals from water-logged terrains. After that, mild emetic medicine is to be administered to facilitate vomiting. The emetic should contain fruits of *Piper longum*, rock salt and honey. *Kapha* from the body will come out along with the vomitus and the patient will feel great relief. This is to be followed by inhalation through nose of the smoke emanating from powdered leaf of turmeric (*Curcuma longa*), root of *Ricinus communis*, dried grapes and heartwood of *Cedrus deodara*, all rolled into wicks, dipped in clarified butter and burnt. Fomentation of thorax and neck with sweetened warm milk, warm water, various medicinal oils and crude herbs will bestow further relief. As the final step in therapy, the patient is to be administered with appropriate decoctions (e.g., *Daśamūlakaṭutrayam*), electuaries (e.g., *Agastya rasāyanam*, *Kamsaharītaki*) and medicated clarified butter (e.g., *Amṛtaprāśam ghṛtam*, *Rāsnādaśamūla ghṛtam*) (Upadhyaya 1975l). This shows that therapy needs to be individualized and that a one disease, one medicine approach is not favored in Ayurveda.

Controlled double-blind studies of classical Ayurveda have not been conducted so far because of the lack of placebos for the traditional, individually varied, dosage forms and therapeutic measures of Ayurveda. A novel study using placebos was reported for the first time by Furst et al. (2011a). Six placebos, appearing identical to the traditional dosage forms for classical ayurvedic treatment of rheumatoid arthritis, were employed. They included decoction, powder, pills, electuaries, *āsava-ariṣṭa* and medicinal oil. The placebos were shown to effectively double-blind the trial (Furst et al. 2011b).

Forty-three seropositive rheumatoid arthritis patients with disease duration of less than 7 years were assigned to groups like methotrexate plus ayurvedic placebo, Ayurveda plus methotrexate placebo, or Ayurveda plus methotrexate. The ayurvedic physician had full freedom to prescribe any combination of medicines and therapies for all patients, based on his clinical judgment. Therefore, the patients received individualized therapy according to true ayurvedic precepts. All patients were not expected to receive all the dosage forms or corresponding placebos. But they were dispensed verum or placebo depending on their particular treatment group assignment. Forty ayurvedic formulations were used. After 36 weeks of therapy the success of treatment was evaluated using the Disease Activity Score (DAS28-CRP) (van Gestel et al. 1999), American College of Rheumatology 20/50/70 criteria for clinical improvement (Felson et al. 1995) and Health Assessment Questionnaire-Disability Index (Lubeck 2002).

All groups were comparable at baseline in disease characteristics. All three treatments were approximately equivalent in efficacy. There were no significant differences among the three groups on the efficacy measures. The methotrexate groups, however, experienced more adverse events than the Ayurveda group. This study shows that double-blind, placebo-controlled, randomized studies are possible when testing classical ayurvedic medicines against medications of the Western medical system, in a way acceptable to both Western medicine and Ayurveda (Furst et al. 2011a).

11.6 CONCLUSION

In contemporary India three viewpoints are raised when it comes to discussions on strengthening Ayurveda. The first group, composed of traditional and orthodox practitioners, vociferously argues that what Caraka, Suśruta, and Vāgbhaṭa have written is sufficient for the progress of Ayurveda. There is no need for any research on any aspect of Ayurveda. The second group, representing college-trained Ayurveda practitioners possessing B.A.M.S. and M.D. (Ayu.) degrees, favor integration of Ayurveda with Western medicine. They opine that Western medical knowledge and technology should be used for diagnosis of diseases and that ayurvedic medicines are to be used in treatment.

The third group, consisting of progressive Ayurveda physicians and researchers having firm faith in the theoretical foundation of Ayurveda, maintains a centrist attitude. They say that Ayurveda has evolved over a long span of time, accumulating centuries of knowledge. For example, therapeutic measures like *dhāra*, *pizhiccil* and all the so-called treatments of Kerala are not described in *Caraka Samhita* or *Aṣṭāṅgahṛdaya*. They were developed by the Ayurveda physicians of medieval Kerala. Similarly, medicines like *Koṭṭamcukkādi tailam*, *Veṭṭumāran guḷika*, *Kompañcādi guḷika* and *Nālpāmarādi taila* originated in Kerala. Therefore, innovation is essential in Ayurveda.

This debate brings to mind the derisive attitude of Edward Jenner's contemporaries in Gloucestershire who despised the claim of milkmaids that cowpox gave them protection from smallpox. When informed by Jenner of the curious observation and the arguments against it, John Hunter, his mentor gave his famous reply "Why think? Why not experiment?" (Jenner 1801; Valiathan and Thatte 2010). Application of modern technology, without sacrificing the principles of Ayurveda is the best approach to advancement of Ayurveda.

In spite of the advances made in Western medicine, diseases like A.I.D.S., cancer, leukemia and Alzheimer's disease continue to baffle the medical world, which is compelled to look for solutions in various alternative medical systems. Adoption of Ayurveda seems to be a good solution to such problems because of its cost-effective and safe medicines. According to Ayurveda, all diseases spring from imbalance in *tridōṣa*. *Aṣṭāṅgahṛdaya* states that a physician need not feel inferior if he cannot assign a name to a disease. Sound knowledge of the states of *tridōṣa*, *dhātu* and *mala*, gauged through careful observation and grouping of symptoms will reveal all aspects of a disease (Upadhyaya 1975g). Simple and safe ayurvedic health care can be made affordable to the deprived sections of the developing and underdeveloped countries.

REFERENCES

Abbas, A. A., B. A. Ernest, M. Akeh, P. Upla, and T. K. Tuluma. 2017. Antimicrobial activity of coconut oil and its derivative (lauric acid) on some selected clinical isolates. *International Journal of Medical Science and Clinical Inventions* 4: 3173–77.

Abboud, T. 2006. The scientist. In *Al Kindi: The father of Arab philosophy*, 68–80. New York: The Rosen Publishing Group, Inc.

Aggarwal, S., S. Negi, P. Jha et al. 2010. *EGLN1* involvement in high-altitude adaptation revealed through genetic analysis of extreme constitution types defined in Ayurveda involvement in high-altitude adaptation revealed through genetic analysis of extreme constitution types defined in Ayurveda. *Proceedings of the National Academy of Science (U.S.A.)* 107, no. 44: 18961–6.

Ahmer, S. M., F. Ali, A. W. Jamil, H. I. Ahmad, and S. J. Ali. 2015. *Mizaj*: Theory of Greco-Arabic medicine for health and disease. *Asian Journal of Complementary and Alternative Medicine* 3: 1–9.

Ali, Z., S. H. Ganie, A. Narula, M. P. Sharma, and P. S. Srivastava. 2013. Intra-specific genetic diversity and chemical profiling of different accessions of *Clitoria ternatea* L. *Industrial Crops and Products* 43: 768–73.

Anonymous. 1978a. Declaration of Alma-Ata international Conference on Primary Health Care, Alma-Ata. USSR, September, 6–12, 1978. http://www.who.int/hpr/NPH/docs/declaration_almaata.pdf (accessed June 27, 2017).

Anonymous. 1978b. Appendix IV. In *The ayurvedic formulary of India, Part I, First Edition*, 317–9. New Delhi: Ministry of Health and Family Planning.

Anonymous. 1978c. Rasayoga. In *The ayurvedic formulary of India, Part I, First Edition*, 199–219. New Delhi: Ministry of Health and Family Planning.

Anonymous. 1999. Toxicological profile for mercury. U.S. Department of Health and Human Services, Public Health Service Agency for Toxic Substances and Disease Registry, 1–676.

Anonymous. 2017. Ministry of AYUSH. http://www. http://ayush.gov.in/ (accessed June 27, 2017).

Arichi, S., T. Tani, and M. Kubo. 1979. Studies on Bupleuri radix and Saikosaponin. (3) quantitative analysis of Saikosaponin of commercial Bupleuri radix. *Medical Journal of Kinki University* 4: 235–41.

Arnold, D. 1996. The rise of western medicine in India. *Lancet* 348, no. 9034: 1075–8.

Ballantyne, C. M., R. C. Hoogeveen, A. M. McNeill et al. 2008. Metabolic syndrome risk for cardiovascular disease and diabetes in the ARIC study. *International Journal of Obesity* 32, no. Suppl 2: S21–4.

Banerjee, S., P. Debnath, and P. K. Debnath. 2015. Ayurnutrigenomics: Ayurveda-inspired personalized nutrition from inception to evidence. *Journal of Traditional and Complementary Medicine* 5, no. 4: 228–33.

Beauchamp, G. K. 2019. Basic taste: A perceptual concept. *Journal of Agricultural and Food Chemistry* 67, no. 50: 13860–9.

Beauchamp, G. K., R. S. J. Keast, D. Morel et al. 2005. Phytochemistry: Ibuprofen-like activity in extra-virgin olive oil. *Nature* 437, no. 7055: 45–6.

Bellomo, N., and L. Preziosi. 1995. *Modelling mathematical methods and scientific computation*, 1–499. Boca Raton: CRC Press.

Bhalerao, S., T. Deshpande, and U. Thatte. 2012. *Prakriti* (Ayurvedic concept of constitution) and variations in platelet aggregation. *BMC Complementary and Alternative Medicine* 12: 248. http://www.biomedcentral.com/1472-6882/12/248.

Bhankharia, D. A., and P. H. Sanjana. 2004. Efficacy of desonide 0.05% cream and lotion in steroid-responsive dermatoses in Indian patients: A post-marketing surveillance study. *Indian Journal of Dermatology, Venereology and Leprology* 70, no. 5: 288–91.

Bhargav, S., B. P. Panda, M. Ali, and S. Javed. 2008. Solid-state fermentation: An overview. *Chemical and Biochemical Engineering Quarterly* 22: 49–70.

Bidhan, M., A. B. Rema Shree, and R. Remadevi. 2015. HPTLC comparison of leaf and heartwood of *Pterocarpus marsupium* Roxb. – An endangered medicinal plant. *Journal of Scientific and Innovative Research* 4: 27–32.

Bodmer, W. F. 1997. HLA: What's in a name? A commentary on HLA nomenclature development over the years. *Tissue Antigens* 46: 293–6.

Brandi, M. L., and S. Minisola. 2013. Calcidiol [25 (OH) D3]: From diagnostic marker to therapeutical agent. *Current Medical Research and Opinion* 29, no. 11: 1565–72.

Bray, P. F. 2006. Platelet reactivity and genetics down on the Pharm. *Transactions of the American Clinical and Climatological Association* 117: 103–11.

Breslin, P. A. S., M. M. Gilmore, G. K. Beauchamp, and B. G. Green. 1993. Psychophysical evidence that oral astringency is a tactile sensation. *Chemical Senses* 18, no. 4: 405–17.

Breslin, P. A. S., T. N. Gingerich, and B. G. Green. 2001. Ibuprofen as a chemesthetic stimulus: Evidence of a novel mechanism of throat irritation. *Chemical Senses* 26, no. 1: 55–65.

Brogden, R. N., T. M. Speight, and G. S. Avery. 1974. Fluocinonide in FAPG base: A review of its therapeutic efficacy in inflammatory dermatoses. *Drugs* 7, no. 5: 337–43.

Calderón-Montaño, J. M., E. Burgos-Morón, C. Pérez-Guerrero, and M. López-Lázaro. 2011. A review on the dietary flavonoid kaempferol. *Mini-Reviews in Medicinal Chemistry* 11, no. 4: 298–44.

Chakravorty, R. C. 2008. Colonial medicine in India. In *Encyclopaedia of the history of science, technology and medicine in non-western cultures*, Volume I (A-K), 157–62. ed. H. Selin. Berlin: Springer-Verlag.

Chambers, R. B. 2000. The role of mathematical modeling in medical research: "Research without patients?" *The Ochsner Journal* 2, no. 4: 218–23.

Chandra, D. N., A. Joseph, G. K. Prasanth, A. Sabu, C. Sadasivan, and M. Haridas. 2012. Inverted binding due to a minor structural change in berberine enhances its phospholipase A2 inhibitory effect. *International Journal of Biological Macromolecules* 50, no. 3: 578–85.

Chinwong, S., D. Chinwong, and A. Mangklabruks. 2017. Daily consumption of virgin coconut oil increases high-density lipoprotein cholesterol levels in healthy volunteers: A randomized crossover trial. *Evidence-Based Complementary and Alternative Medicine: ECAM*. doi:10.1155/2017/7251562.

Chopra, A., P. Lavin, B. Patwardhan, and D. Chitre. 2000. Randomized double blind trial of an Ayurvedic plant derived formulation for treatment of rheumatoid arthritis. *Journal of Rheumatology* 27, no. 6: 1362–5.

Daly, A. K. 1995. Molecular basis of polymorphic drug metabolism. *Journal of Molecular Medicine* 73, no. 11: 539–53.

Dasgupta, S. 1997. General observations on the systems of Indian philosophy. In *A history of Indian philosophy*, Vol. 1, 62–77. Delhi: Motilal Banarsidass.

DebMandal, M., and S. Mandal. 2011. Coconut (*Cocos nucifera* L.: Arecaceae): In health promotion and disease prevention. *Asian Pacific Journal of Tropical Medicine* 4, no. 3: 241–7.

Ding, G. H., and G. T. Wan. 2008. Mathematical analysis of Yin-Yang Wu-Xing model in TCM. *Journal of Acupuncture and Tuina Science* 6, no. 5: 269–70.

Dinnimath, B. M., S. S. Jalalpure, and U. K. Patil. 2017. Antiurolithiatic activity of natural constituents isolated from *Aerva lanata*. *Journal of Ayurveda and Integrative Medicine* 8, no. 4: 226–32.

Evans, W. E. 2003. Pharmacogenomics: Marshalling the human genome to individualise drug therapy. *Gut* 52, no. Suppl 2: ii10–8.

Evans, W. E., and H. L. McLeod. 2003. Pharmacogenomics-drug disposition, drug targets, and side effects. *The New England Journal of Medicine* 348, no. 6: 538–49.

Felson, D. T., J. J. Anderson, M. Boers et al. 1995. American College of Rheumatology preliminary definition of improvement in rheumatoid arthritis. *Arthritis and Rheumatism* 38, no. 6: 727–35.

Fischer, R., and F. Griffin. 1963. Quinine dimorphism: A cardinal determinant of taste sensitivity. *Nature* 200: 343–7.

Fischer, R., and F. Griffin. 1964. Pharmacogenetic aspects of gustation. *Drug Research* (*Arzneimittel-Forschung*) 14: 673–86.

Fischer, R., F. Griffin, R. C. Archer, S. C. Zinsmeister, and P. S. Jastram. 1965a. The weber ratio in gustatory chemoreception: An indicator of systemic (drug) reactivity. *Nature* 207, no. 5001: 1049–53.

Fischer, R., and R. Kaelbling. 1967. Increase in taste acuity with sympathetic stimulation the relation of a just-noticeable taste difference to systemic psychotropic drug dose. In *Recent advances in biological psychiatry*, ed. J. Wortis, 183–95. New York: Plenum Press.

Fischer, R., W. Knopp, and F. Griffin. 1965b. Taste sensitivity and the appearance of trifluoperazine-tranquilizer induced extrapyramidal symptoms. *Drug Research* (*Arzneimittel-Forschung*) 15: 1379–82.

Fontana, P., A. Dupont, S. Gandrille et al. 2003. Adenosine diphosphate-induced platelet aggregation is associated with P2Y12 gene sequence variations in healthy subjects. *Circulation* 108, no. 8: 989–95.

Ford, A. C. 2013. Definitions and classifications of irritable bowel syndrome. In *Irritable bowel syndrome-diagnosis and clinical management*, ed. A. Emmanuel, and E. M. M. Quigley, 3–21. West Sussex: John Wiley & Sons Ltd.

Furst, D. E., M. M. Venkatraman, M. McGann et al. 2011. Double-blind, randomized, controlled, pilot study comparing classic ayurvedic medicine, methotrexate, and their combination in rheumatoid arthritis. *Journal of Clinical Rheumatology* 17, no. 4: 185–92.

Furst, D. E., M. M. Venkatraman, B. G. K. Swamy et al. 2011. Well controlled, double-blind, placebo-controlled trials of classical Ayurvedic treatment are possible in rheumatoid arthritis. *Annals of Rheumatic Diseases* 70: 392–3.

Ganie, S. H., Z. Ali, S. Das, P. S. Srivastava, and M. P. Sharma. 2015a. Identification of Shankhpushpi by morphological, chemical and molecular markers. *European Journal of Biotechnology and Bioscience* 3: 1–9.

Ganie, S. H., Z. Ali, S. Das, P. S. Srivastava, and M. P. Sharma. 2015b. Molecular characterization and chemical profiling of different populations of *Convolvulus pluricaulis* (Convolvulaceae); an important herb of ayurvedic medicine. *3 Biotech* 5, no. 3: 295–302.

Ganie, S. H., and M. P. Sharma. 2014. Molecular and chemical profiling of different populations of *Evolvulus alsinoides* (L.) L. *International Journal of Agriculture and Crop Science* 7: 1322–31.

Ghodke, Y., K. Joshi, and B. Patwardhan. 2011. Traditional medicine to modern pharmacogenomics: Ayurveda *prakriti* type and CYP2C19 gene polymorphism associated with the metabolic variability. *Evidence-Based Complementary and Alternative Medicine: ECAM* 2011. doi:10.1093/ecam/nep206.

Goldstein, J. A., and S. M. F. de Morais. 1994. Biochemistry and molecular biology of the human CYP2C subfamily. *Pharmacogenetics* 4, no. 6: 285–99.

Govindaraj, P., S. Nizamuddin, A. Sharath et al. 2015. Genome-wide analysis correlates Ayurveda *Prakriti*. *Scientific Reports* 5: 15786. doi:10.1038/srep15786.

Gueriguian, J. L. 1987. Amirdovlat' Amasiats' I : His life and contributions. *Journal of the Society for Armenian Studies* 3: 63–91.

Gupta, P., Siripurapu K. B. Akanksha, K. B. Siripurapu et al. 2007. Anti-stress constituents of *Evolvulus alsinoides*: An ayurvedic crude drug. *Chemical and Pharmaceutical Bulletin* 55, no. 5: 771–5.

Hall, J. J., and R. Taylor. 2003. Health for all beyond 2000: The demise of the Alma-Ata Declaration and primary health care in developing countries. *Global Health* 178, no. 1: 17–9.

Han, J., and J. Huang. 2012. Mathematical model of biological order state or syndrome in traditional Chinese medicine: Based on electromagnetic radiation within the human body. *Cell Biochemistry and Biophysics* 62, no. 2: 377–81.

Hankey, A. 2005. A test of the systems analysis underlying the scientific theory of Ayurveda's tridosha. *Journal of Alternative and Complementary Medicine* 11, no. 3: 385–90.

Harris, H., and H. Kalmus. 1949. The measurement of taste sensitivity to phenylthiourea (PTC). *Annals of Eugenics* 15, no. 1: 24–31.

Harvey, A. L. 2008. Natural products in drug discovery. *Drug Discovery Today* 13, no. 19–20: 894–901.

Hasan, S., M. M. Sikder, M. Ali et al. 2016. Toxicological studies of the ayurvedic medicine *naradiya laksmi-vilasa rasa* used in sinusitis. *Biology and Medicine* 8: 7. doi:10.4172/0974-8369.1000359.

Heckmann, J. G., S. M. Heckmann, C. J. G. Lang, and T. Hummel. 2003. Neurological aspects of taste disorders. *Archives of Neurology* 60, no. 5: 667–71.

Heyn, H., S. Moran, I. Hernando-Herraez et al. 2013. DNA methylation contributes to natural human variation. *Genome Research* 23, no. 9: 1363–72.

Intahphuak, S., P. Khonsung, and A. Panthong. 2010. Anti-inflammatory, analgesic, and antipyretic activities of virgin coconut oil. *Pharmaceutical Biology* 48, no. 2: 151–7.

Ivorra, M. D., M. Paya, and A. Villar. 1989. A review of natural products and plants as potential antidiabetic drugs. *Journal of Ethnopharmacology* 27, no. 3: 243–75.

Jagtap, C. Y., B. K. Ashok, B. J. Patgiri, P. K. Prajapati, and B. Ravishankar. 2013b. Acute and sub-chronic toxicity study of *Tamra bhasma* (incinerated copper) prepared from *ashodhita* (unpurified) and *shodhita* (purified) *Tamra* in rats. *Indian Journal of Pharmaceutical Sciences* 75, no. 3: 346–52.

Jagtap, C. Y., B. J. Patgiri, and P. K. Prajapati. 2013a. Purification and detoxification procedures for metal *Tamra* in medieval Indian medicine. *Indian Journal of History of Science* 48: 383–404.

Jamadagni, P. S., S. B. Jamadagni, A. Singh et al. 2015. Toxicity study of *Swarna Bhasma*, an Ayurvedic medicine containing gold, in Wistar rats. *Toxicology International* 22: 11–7.

Jena, A. K., M. Karan, and K. Vasisht. 2017. Plant parts substitution- based approach as a viable conservation strategy for medicinal plants: A case study of *Premna latifolia* Roxb. *Journal of Ayurveda and Integrative Medicine* 8, no. 2: 68–72.

Jenner, E. 1801. *On the origin of the vaccine inoculation*, 1–8. Soho (London): D.N. Shury.

Jin, P., and E. Wang. 2003. Polymorphism in clinical immunology – From HLA typing to immunogenetic profiling. *Journal of Translational Medicine* 1, no. 1: 8. doi:10.1186/1479-5876-1-8.

Joshi, R. R. 2004. A biostatistical approach to Ayurveda: Quantifying the *tridosha*. *Journal of Alternative and Complementary Medicine* 10, no. 5: 879–89.

Juyal, R. C., S. Negi, P. Wakhode, S. Bhat, B. Bhat, and B. K. Thelma. 2012. Potential of ayurgenomics approach in complex trait research: Leads from a pilot study on rheumatoid arthritis. *PLoS One* 7, no. 9: e45752. doi:10.1371/journal.pone.0045752.

Kameswara, B., R. Giri, M. M. Kesavulu, and C. H. Apparao. 2001. Effect of oral administration of bark extracts of *Pterocarpus marsupium* L. on blood glucose level in experimental animals. *Journal of Ethnopharmacology* 74: 69–74.

Kaplan, A. R., and R. Fischer. 1965. Taste sensitivity for bitterness: Some biological and clinical implications. In *Recent advances in biological psychiatry*, ed. J. Wortis, 183–96. New York: Plenum Press.

Kazuma,, K., N. Noda, and M. Suzuki. 2003. Malonylated flavonol glycosides from the petals of *Clitoria ternatea*. *Phytochemistry* 62, no. 2: 229–37.

Kessler, C., M. Wischnewsky, A. Michalsen, C. Eisenmann, and J. Melzer. 2013. Ayurveda: Between religion, spirituality, and medicine. *Evidence-Based Complementary and Alternative Medicine: ECAM* 2013. doi:10.1155/2013/952432.

Keswani, N. H. 1974. Medical heritage of India. In *The science of medicine and physiological concepts in ancient and medieval India*, ed. N. H. Keswani, 3–49. New Delhi: All India Institute of Medical Sciences.

Kim, J., M. Song, J. Kang et al. 2014. A mathematical model for the deficiency-excess mechanism of yin-yang in five viscera. *Chinese Journal of Integrative Medicine* 20, no. 2: 155–60.

Krishnamurthy, K. H. 1971. Botanical identification of ayurvedic medicinal plants: A new method of pharmacolinguistics- A preliminary report. *Indian Journal of Medical Research* 59, no. 1: 90–103.

Kulkarni, R. R., P. S. Patki, V. P. Jog, Patwardhan Gadage, and B. 1992. Efficacy of an ayurvedic formulation in rheumatoid arthritis: A double blind, placebo controlled, cross-over study. *Indian Journal of Pharmacology* 24: 98–101.

Kumar, D. S. 1992. Elucidation of the factors to be considered in ayurvedic clinical practice. *Ancient Science of Life* 12, no. 1–2: 292–98.

Kumar, D. S. 2016. Medical herbalism through the ages. In *Herbal bioactives and food fortification: Extraction and formulation*. Boca Raton: CRC Press: 1–28.

Kumar, G., A. Srivastava, S. K. Sharma, and Y. K. Gupta. 2014. Safety evaluation of mercury - Based Ayurvedic formulation (*Sidh Makardhwaj*) on brain cerebrum, liver and kidney in rats. *Indian Journal of Medical Research* 139, no. 4: 610–18.

Kumar, M., A. Ahmad, P. Rawat et al. 2010. Antioxidant flavonoid glycosides from *Evolvulus alsinoides*. *Fitoterapia* 81, no. 4: 234–42.

Kurande, V., A. E. Bilgrau, R. Waagepetersen, E. Toft, and R. Prasad. 2013a. Interrater reliability of diagnostic methods in traditional Indian ayurvedic medicine. *Evidence-Based Complementary and Alternative Medicine: ECAM* 2013. doi:10.1155/2013/658275.

Kurande, V. H., R. Waagepetersen, E. Toft, and R. Prasad. 2013b. Reliability studies of diagnostic methods in Indian traditional Ayurveda medicine: An overview. *Journal of Ayurveda and Integrative Medicine* 4, no. 2: 67–76.

Lahorkar, P., K. Ramitha, V. Bansal, and D. B. A. Narayana. 2009. A comparative evaluation of medicated oils prepared using ayurvedic and modified processes. *Indian Journal of Pharmaceutical Sciences* 71, no. 6: 656–62.

Lee, J. Y., T. Y. So, and J. Thackray. 2013. A review on vitamin D deficiency treatment in pediatric patients. *The Journal of Pediatric Pharmacology and Therapeutics: JPPT: The Official Journal of PPAG* 18, no. 4: 277–91.

Low, P. A., V. A. Tomalia, and K. L. Park. 2013. Autonomic function tests: Some clinical applications. *Journal of Clinical Neurology* 9, no. 1: 1–8.

Lubeck, D. 2002. Health-related quality of life measurements and studies in rheumatoid arthritis. *The American Journal of Managed Care* 8, no. 9: 811–20.

Mahalle, N. P., M. V. Kulkarni, N. M. Pendse, and S. S. Naik. 2012. Association of constitutional type of Ayurveda with cardiovascular risk factors, inflammatory markers and insulin resistance. *Journal of Ayurveda and Integrative Medicine* 3, no. 3: 150–7.

Manchia, M., J. Cullis, G. Turecki, G. A. Rouleau, R. Uher, and M. Alda. 2013. The impact of phenotypic and genetic heterogeneity on results of genome wide association studies of complex diseases. *PLoS One* 8, no. 10: e76295. doi:10.1371/journal.pone.0076295.

Manor, R. S., Y. Yassur, R. Seigal, and I. Ben. 1981. The pupil cycle time test: Age variations in normal subjects. *British Journal of Ophthalmology* 65, no. 11: 750–3.

Manyam, B. V., and A. Kumar. 2013. Ayurvedic constitution (*prakruti*) identifies risk factor of developing Parkinson's disease. *Journal of Alternative and Complementary Medicine* 19, no. 7: 644–9.

Ming, L., M. Zengchun, L. Qiande et al. 2010. Mathematical analysis of Si-Wu-Tang model in TCM. *World Science and Technology* 12, no. 5: 684–90.

Moodithaya, S., and S. T. Avadhany. 2009. Pupillary autonomic activity by assessment of pupil cycle time; Reference value for healthy men and women. *Science and Medicine* 1: 1–6.

Moodithaya, S., and S. T. Avadhany. 2012. Gender differences in age - Related changes in cardiac autonomic nervous function. *Journal of Aging Research* 2012. doi:10.1155/2012/679345.

Mukerji, M., and B. Prasher. 2011. Ayurgenomics: A new approach in personalized and preventive medicine. *Science and Culture* 77: 10–7.

Murthy, K. R. S. 2017. *Prathama Khaṇḍa (First section).* In *Śārṅgadhara Samhita*, 10–4. Varanasi: Chaukhamba Orientalia.

Nevin, K. G., and T. Rajamohan. 2010. Effect of topical application of virgin coconut oil on skin components and antioxidant status during dermal wound healing in young rats. *Skin Pharmacology and Physiology* 23, no. 6: 290–7.

Noguchi, M., M. Kubo, T. Hayashi, and M. Ono. 1978a. Studies on the pharmaceutical quality evaluation of the crude drug preparations used in oriental medicine kampo. Part 1, precipitation reaction of the components of coptis rhizome and these of Glycyrrhiza root or rheum rhizome in decoction solution. *Shoyakugaku Zaashi* 32: 104.

Noguchi, M., M. Kubo, and Y. Naka. 1978b. Studies on the pharmaceutical quality evaluation of crude drug preparations used in Orient medicine "Kampo" IV. Behavior of alkaloids in ephedra herb mixed with other crude drugs under decoction processes. *Yakugaku Zaashi* 98, no. 7: 923–8.

Pandey, V. N., and A. Pandey. 1988. A study of the *Nāvanītaka*: The Bower manuscript. *Bulletin of the Indian Institute of History of Medicine* 18: 1–46.

Park, J., and E. Ernst. 2005. Ayurvedic medicine for rheumatoid arthritis: A systematic review. *Seminars in Arthritis and Rheumatism* 34, no. 5: 705–13.

Patwardhan, B., K. Joshi, and A. Chopra. 2005. Classification of human population based on HLA gene polymorphism and the concept of *prakriti* in Ayurveda. *Journal of Alternative and Complementary Medicine* 11, no. 2: 349–53.

Pillai, N. G. 1976. On the botanical identity of *Śaṅkhapuṣpi*. *Journal of Research in Indian Medicine, Yoga & Homoeopathy* 11: 67–76.

Polur, H., T. Joshi, C. T. Workman, G. Lavekar, and I. Kouskoumvekaki. 2011. Back to the roots: Prediction of biologically active natural products from Ayurveda traditional medicine. *Molecular Informatics* 30, no. 2–3: 181–7.

Prabhakar, Y. S., and D. S. Kumar. 1993. A model to quantify disease state based on the ayurvedic concept of tridōṣa. *Bulletin of the Indian Institute of History of Medicine* 23: 1–19.

Prasher, B., S. Negi, S. Aggarwal et al. 2008. Whole genome expression and biochemical correlates of extreme constitutional types defined in Ayurveda. *Journal of Translational Medicine* 6: 48. doi:10.1186/1479-5876-6-48.

Rapolu, S. B., M. Kumar, G. Singh, and K. Patwardhan. 2015. Physiological variations in the autonomic responses may be related to the constitutional types defined in Ayurveda. *TANG (Humanitas Medicine)* 5, no. 1: 7.1–7.

Rhoda, D. 2014. Ayurvedic psychology: Ancient wisdom meets modern science. *International Journal of Transpersonal Studies* 33, no. 1: 158–71.

Rizzo-Sierra, C. V. 2011. Ayurvedic genomics, constitutional psychology, and endocrinology: The missing connection. *The Journal of Alternative and Complementary Medicine* 17, no. 5. doi:10.1089/acm.2010.0412.

Rotti, H., K. P. Guruprasad, J. Nayak et al. 2014a. Immunophenotyping of normal individuals classified on the basis of human *dosha prakriti*. *Journal of Ayurveda and Integrative Medicine* 5, no. 1: 43–9.

Rotti, H., S. Mallya, S. P. Kabekkodu et al. 2015. DNA methylation analysis of phenotype specific stratified Indian population. *Journal of Translational Medicine* 13: 151. doi:10.1186/s12967-015-0506-0.

Rotti, H., R. Raval, S. Anchan et al. 2014b. Determinants of *Prakriti*, the human constitution types of Indian traditional medicine and its correlation with contemporary science. *Journal of Ayurveda and Integrative Medicine* 5, no. 3: 167–75.

Sabu, A., and M. Haridas. 2015. Fermentation in ancient Ayurveda: Its present implications. *Frontiers in Life Science* 8, no. 4: 324–31.

Sander, O., G. Herborn, and R. Rau. 1998. Is H15 (resin extract of *Boswellia serrata*, "incense") a useful supplement to established drug therapy of chronic polyarthritis? Results of a double-blind pilot study. *Zeitschrift für Rheumatologie* 57, no. 1: 11–16.

Savrikar, S. S., and B. Ravishankar. 2011. Introduction to rasashastra- the iatrochemistry of Ayurveda. *African Journal of Traditional, Complementary and Alternative Medicines* 8, no. 5: 66–82.

Sen, C. T. 1997. The Portuguese influence on Bengali cuisine. In *Food on the move*, ed. H. Walker, 288–96. Devon: Prospect Books.

Seneviratne, K. N., and D. M. S. Dissanayake. 2008. Variation of phenolic content in coconut oil extracted by two conventional methods. *International Journal of Food Science and Technology* 43, no. 4: 597–602.

Sharma, P. V. 1994. Treatment of *Vātarakta*. In *Cakradatta*, 215–23. Varanasi: Chaukhamba Orienalia.

Sharma, P. V. 1998. Treatment of *grahaṇi* disorder. In *Caraka Samhita*, 249–72. Varanasi: Chaukhamba Orientalia.

Shilpa, S., and C. G. V. Murthy. 2011. Development and standardization of Mysore *tridosha* scale. *Ayu* 32, no. 3: 308–14.

Shirolkar, S. G., R. K. Tripathi, and N. N. Rege. 2015. Evaluation of *prakṛti* and quality-of-life in patients with irritable bowel syndrome. *Ancient Science of Life* 34, no. 4: 210–15.

Siddiqui, S., B. S. Siddiqui, S. Faizi, and T. Mahmood. 1986. Isoazadirolide, a new tetranortriterpenoid from *Azadirachta indica* A. Juss. (Meliaceae). *Heterocycles* 24, no. 11: 3163–7.

Siepmann, J., and F. Siepmann. 2008. Mathematical modeling of drug delivery. *International Journal of Pharmaceutics* 364, no. 2: 328–43.

Simonson, T. S., Y. Yang, C. D. Huff et al. 2010. Genetic evidence for high-altitude adaptation in Tibet. *Science* 329, no. 5987: 72–5.

Singh, H., and M. Vishwanathan. 2001. Need for authentication of crude drug shankpushpi *Convolvulus microphyllus*. *Journal of Medicinal and Aromatic Plant Sciences* 23: 612–8.

Sivarajan, V. V., and I. Balachandran. 1994. Introduction. In *Ayurvedic drugs and their plant sources*, 1–14. New Delhi: Oxford & IBH Publishing Co. Pvt Ltd.

Storz, J. F. 2010. Evolution. Genes for high altitudes. *Science* 329, no. 5987: 40–1.

Suchitra, S. P., A. Jagan, and H. R. Nagendra. 2014. Development and initial standardization of Development and initial standardization of Ayurveda child personality inventory. *Journal of Ayurveda and Integrative Medicine* 5, no. 4: 205–8.

Suchitra, S. P., and H. R. Nagendra. 2012. Effects of yoga on *prakrti* in children – A pilot study. *SENSE* 2: 293–8.

Suchitra, S. P., and H. R. Nagendra. 2013. A self-rating scale to measure *tridōṣa* in children. *Ancient Science of Life* 33, no. 2: 85–91.

Svejgaard, A. 1978. The clinical and biological significance of the HLA system. *Proceedings of the European Dialysis and Transplant Association* 15: 273–82.

Syvänen, A. C. 2001. Accessing genetic variation: Genotyping single nucleotide polymorphisms. *Nature Reviews. Genetics* 2, no. 12: 930–42.

Takaishi, K., and Y. Torii. 1969. Studies on the decoction of Chinese medicines. I. On the interaction of some chemicals with starch in aqueous solution. *Yakugaku Zasshi: Journal of the Pharmaceutical Society of Japan* 89, no. 4: 538–43.

Takaishi, K., and Y. Watanabe. 1971. Studies on the decoction of Chinese medicines. II. The extraction of some drugs by starch aqueous solution and the property of Gegen Tang. *Yakugaku Zasshi: Journal of the Pharmaceutical Society of Japan* 91, no. 10: 1092–7.

Tiwari, S., S. Gehlot,S. K. Tiwari, and G. Singh. 2012. Effect of walking (aerobic isotonic exercise) on physiological variants with special reference to *Prameha* (diabetes mellitus) as per *Prakriti*. *Ayu* 33, no. 1: 44–9.

Tomimori, T., and M. Yoshimoto. 1980. Quantitative variation of glycyrrhizin in the decoction of Glycyrrhizae Radix mixed with other crude drugs. *Shoyakugaku Zaashi* 34: 138.

Tripathi, P. K., K. Patwardhan, and G. Singh. 2011. The basic cardiovascular responses to postural changes, exercise, and cold pressor test: Do they vary in accordance with the dual constitutional types of Ayurveda? *Evidence-Based Complementary and Alternative Medicine: ECAM* 2011. doi:10.1155/2011/251850.

Upadhyaya, Y. 1975a. *Dōṣādivijñānīyam* (Knowledge of *dōṣa*). In *Aṣṭāṅgahṛdaya*, 85–90. Varanasi: Chowkhamba Sanskrit Sansthan.

Upadhyaya, Y. 1975b. *Āyuṣkāmīyam* (Desire for long life). In *Aṣṭāṅgahṛdaya*, 1–17. Varanasi: Chowkhamba Sanskrit Sansthan.

Upadhyaya, Y. 1975c. *Aṅgavibhāgam* (Different parts of the body). In *Aṣṭāṅgahṛdaya*, 184–96. Varanasi: Chowkhamba Sanskrit Sansthan.

Upadhyaya, Y. 1975d. *Dravyādivijñānīyam* (Properties of matter). In *Aṣṭāṅgahṛdaya*, 79–82. Varanasi: Chowkhamba Sanskrit Sansthan.

Upadhyaya, Y. 1975e. *Rasabhēdīyam* (Types of taste). In *Aṣṭāṅgahṛdaya*, 82–85. Varanasi: Chowkhamba Sanskrit Sansthan.

Upadhyaya, Y. 1975f. *Aṣṭāṅgahṛdaya*, 1–614. Varanasi: Chowkhamba Sanskrit Sansthan.

Upadhyaya, Y. 1975g. *Nidānasthāna* (Classification of diseases). In *Aṣṭāṅgahṛdaya*, 216–84. Varanasi: Chowkhamba Sanskrit Sansthan.

Upadhyaya, Y. 1975h. *Dōṣabhēdīyam* (Types of *dōṣa*). In *Aṣṭāṅgahṛdaya*, 90–6. Varanasi: Chowkhamba Sanskrit Sansthan.

Upadhyaya, Y. 1975i. *Ṛtucarya* (Seasonal regimen). In *Aṣṭāṅgahṛdaya*, 26–34. Varanasi: Chowkhamba Sanskrit Sansthan.

Upadhyaya, Y. 1975j. *Annasamrakṣaṇīyam* (Protection of food). In *Aṣṭāṅgahṛdaya*, 67–73. Varanasi: Chowkhamba Sanskrit Sansthan.

Upadhyaya, Y. 1975k. *Annasvarūpavijñāṇīyam* (Knowledge of food). In *Aṣṭāṅgahṛdaya*, 50–67. Varanasi: Chowkhamba Sanskrit Sansthan.

Upadhyaya, Y. 1975l. *Śvāsahidhma cikitsa* (Treatment of respiratory diseases). In *Aṣṭāṅgahṛdaya*, 316–20. Varanasi: Chowkhamba Sanskrit Sansthan.

Vaidyan, K. V. K., and A. S. G. Pillai. 1985. *Guḷikāyōgaṅgaḷ* (Formulae of pills). In *Sahasrayōgam*, 133–66. Alleppey: Vidyarambham Publishers.

Valiathan, M. S., and U. Thatte. 2010. Ayurveda: The time to experiment. *International Journal of Ayurveda Research* 1, no. 1: 3.

Van Gestel, A. M., J. J. Anderson, P. L. van Riel et al. 1999. ACR and EULAR improvement criteria have comparable validity in rheumatoid arthritis trials. *Journal of Rheumatology* 26, no. 3: 705–11.

Vane, J. R., and R. M. Botting. 1995. New insights into the mode of action of anti-inflammatory drugs. *Inflammation Research* 44, no. 1: 1–10.

Ved, D. K., and G. S. Goraya. 2007. Profiles of botanicals in trade. In *Demand and supply of medicinal plants in India*, 15–20. New Delhi-Bangalore: N.M.P.B.- F.R.L.H.T.

Venkatraghavan, S., T. P. Sundaresan, V. Rajagopalan, and K. Srinivasan. 1987. Constitutional study of cancer patients – Its prognostic and therapeutic scope. *Ancient Science of Life* 7, no. 2: 110–5.

Vertinsky, P. 2007. Physique as destiny: William H. Sheldon, Barbara Honeyman heath and the struggle for hegemony in the science of somatotyping. *Canadian Bulletin of Medical History* 24, no. 2: 291–316.

Villariano, B. J., L. Dy, and C. C. Lizada. 2007. Descriptive sensory evaluation of virgin coconut oil and refined, bleached, and deodorized coconut oil. *LWT - Food Science and Technology* 40, no. 2: 193–9.

Wang, Y., N. Wang, J. Wang, Z. Wang, and R. Wu. 2013. Delivering systems pharmacogenomics towards precision medicine through mathematics. *Advanced Drug Delivery Reviews* 65, no. 7: 905–11.

Yamaji, A., Y. Mareda, M. Oishi, et al. 1984. Determination of saikosaponins in Chinese medicinal extracts containing Bupleuri radix. *Yakugaku Zaashi* 104, no. 7: 812–5.

Yi, X., Y. Liang, E. Huerta-Sanchez et al. 2010. Sequencing of 50 human exomes reveals adaptation to high altitude. *Science* 329, no. 5987: 75–78.

Zhao, J., P. Jiang, and W. Zhang. 2010. Molecular networks for the study of TCM pharmacology. *Briefings in Bioinformatics* 11, no. 4: 417–30.

Index

Abdominal fire (*jaḍharāgni*), 263
Abelmoschus esculentus, 212
Abhayāriṣṭa, 136–137
Abhidānamañjari, 106, 257
Abhijñāna Śākuntaḷam (Kāḷidāsa), 213
Acacia arabica, 214
Acacia concinna, 214
Aconitum heterophyllum, 181, 183, 193
Acorus calamus, 102, *104*
  anti-convulsant action, 104
  CNS depressant action, 102, 104
  sedative action of, 102
Acorus gramineus, 257
Activator protein 1 (A.P.1), 205
Adenosine diphosphate (A.D.P.), 248
Adhatoda vasica, 3, 144
Adil Shah II, Ibrahim, 10
A.D.P., *see* Adenosine diphosphate
Adulterants, fingerprinting in detection of, 84
Adulteration and substitution, 76, *77*
Agadatantra (toxicology), 15
Agastya harītaki rasāyana, 146
Agilent 7700X inductively coupled mass spectrometer (ICP-MS), 74
Aglaia roxburghiana, 211
Agni, 114, 171, 209, 263
Agnimāndya (lowering of abdominal fire), 114
Agnivēśa, 16, 165
Agnivēśa Tantra (Agnivēśa), 4, 16
Āhāra (food), 263–264
Air classifiers, 46
Aitihyamāla (Garland of legends) (Sankunny), 4
Ājasṛk Rasāyana, 203
Akapporuḷ (Tamil text), 7
Akattiyar (*Agastya*), 7
Albizia lebbeck, 211
Albuquerque, 242
Al-Kindi, Abu Yusuf Ya'qub bin Ishaq, 9, 261
Allium cepa, 210
Allium sativum, 104–105, 210, 212, 214
Alloxan, 117
Al-Ma'mun, 9
Alma Ata Declaration, 242
Aloe, 206
Aloe-gel, 206
Aloe vera, 214
Aloin, 206
Alpinia galanga, 214, 257
Al-Rashid, Harun, 9
Al-Razi, Abu Bakr Muhammed ibn Zakariyya, 9
Alstonia scholaris, *81*
Āma, 171, 209, 211
Āmalakī, see *Phyllanthus emblica*
Āmalakī rasāyana, 149
Āmla, see *Emblica officinalis*
Amṛtā, see *Tinospora cordifolia*
Amṛtā ghṛta, 56
Amṛtāriṣṭa, 137
Amṛtōttaram kvātha, 50
biological effects of, 144
  master formula to produce, *51*
  quality check of, 85
Andhra Ayurveda Pharmacy Ltd, 42
Andrographis paniculata, 78
Anilaghnaka, see *Terminalia belerica*
Animals, learnings from, 2
Anthocyanins, 206
Anti-aging properties, in Ayurveda, 213
Aṇu taila, 147–148
A.P.1, *see* Activator protein 1
Aparājita, see *Evolvulus alsinoides*
Apigenin, 206
Āpta (documented evidence), 165
Aqueous extracts, 116
Arachidonic acid, 206
Āragvadhāriṣṭa, 137
Arbuda neoplasms, 211
Ariṣṭa and *āsava* (fermented liquids), 11, 56–57, 72
  alcohol percentage in commercial samples of, *91*
  biological effects of, 136–138
  bottling of, *58*
  preparation of, 79–80
  proposed protocol to brew, *92*
  souring of, 89
    brewing area hygienic design, 90
    brewing vessel, 89–90
    disinfection, 90
    herbs quality, 89
    improvised fermentation, 90–93
    temperature, 90
    water quality, 89
    yeast use, 90
Arjuna cūrṇa, 85, *86*
Arjuna Kṣīrapāka, 203
Arjun tree, *see Terminalia arjuna*
Arka, 8–9
Arkaprakāśa (King Rāvaṇa), 8, 9
Arthritis
  Ayurveda in, 211–212
  nutraceuticals in, 206–207
Art of Living International Center, The, 242
Arya Vaidya Sala (Kottakkal), 42, 188, 189
Arzani, Muhammed Akbar, 11
Asili, Muhammed bin Ismail Asavale, 10
Aśōka, see *Saraca asoca*
Aśōka ariṣṭa, 116, 137
Asparagus adscendens, 193
Aspilia mossambicensis, 2
Aspilia pluriseta, 2
Aṣṭāṅga Ayurveda healing system, 201
Aṣṭāṅgahṛdaya (Vāgbhaṭa), 4, 5, 7, 10, 15, 17, 48, 142, 165, 261, 264, 266
  Dhānvantaram kvātha in, 144
  Māṇibhadra guḍa in, 147
  nidānasthāna of, 15, 16
  on *prakṛti*, 252

respiratory diseases treatment in, 264
single herb remedies in, 102
taste in, 255
Aṣṭāṅgasaṃgraha (Vāgbhaṭa), 4, 5, 17, 48, 261
Aṣṭavaidya tradition, of Kerala, 5
Aśvagandhādi avalēha, 59
master formula to produce, *61*
Aśvagandha ghṛta, 140
Aśvagandhāriṣṭam, 92
Atees, see *Aconitum heterophyllum*
Ātrēya, 16
Aurangazeb, 11
Austria, 242
Avalēha/lēha (confections/electuaries), 57–59, *61*, 72
biological effects of, 146–148
bottling of, *60*
production of, 80
Āvaraṇā (enveloping/encompassing), 16, 264
Avascular necrosis, *see* Osteonecrosis
Avipattikara cūrṇa, 139
Ayurgenomics, 252
Ayurveda and Panchakarma (Joshi), 233
Ayurveda industry, early days of, 22
"Ayurvedic biology", 163
Ayurvedic Formulary of India (Part I & II), The, 29, 48, 73, 142
Ayurvedic Pharmacopoeia of India (Part I & II), The, 29, 32, 33, 48, 73, 90
A.Y.U.S.H. Department, 25, 28, 33, 34, 73, 200, 228, 230–231
G.C.P. guidelines monitoring, 35, 36
ineffectiveness of, 226
Azadirachta indica (neem tree), 79, 124, 210, 214, 256
Azad Khan, Hakim, 11
Azardichta indica, 204

Babar, 11
Bacopa monnieri, 214
Balāguḷūcyādi kvātha, 144
Balam (physical strength), 263
Balanites aegyptiaca, 2
Balā taila, 148
Barleria prionitis, 146, 214
Basella rubra, 211
Batch manufacturing record, 49
Bayt-al-Hikma (House of Wisdom), 9
Beauty pills, *see* Nutricosmetics
Benchmarks for Training in Ayurveda (W.H.O.), 227, 231, 233, 234, 242
Berberine, 204, 260
Berberis aristata, 193, 211
Berberis asiatica, 193
Berberis chitria, 193
Berberis lycium, 193
Berberis tinctoria, 193
β-boswellic acid, 211
β-cell mass, reduction of, 119
Betulin, 256
Bhagavad Gīta, 163
Bhaiṣajya kalpana (ayurvedic pharmacy), 93
Bhaiṣajya Ratnāvali, 48
Bhasmas, 36–37
Bhatta, Krishnarama, 13
Bhāvamiśra, 17, 193, 211
Bhāvaprakāśa (Bhāvamiśra), 17, 193
Bhāvaprakāśa Nighaṇṭu, 257
Bhore, J. W., 242
Bhore Committee, 242
Bhṛṅgarāj (Eclipta alba), 51
Bhṛṅgarājāsavam, 57
master formula to produce, *58*
Bhūmikūśmāṇḍa, see *Ipomoea mauritiana*
Bhūtāgni (fires of the *pañcabhūta*), 171, 263
Bhūtavidya (knowledge of subtle beings or spirits), 15
Bilvādi Gutika, 61
biological effects of, 142
grinding of, *62*
master formula to produce, *63*
solid-state fermentation and, 260
Bilvādi lēhya, 147
Bioactives, content of, 76–77, 79, 119, 255
BioCAP, 190
Biodiversity hotspots, 179
Biofilms, 89
Biogeographic zones, in India, 178–179, *178*
Biological Diversity Act (2002), The, 25
Biological effects, of Ayurvedic formulations, *see individual entries*
Biologics Control Act (1902) (United States), 22
Biomarkers, 82
Black cotton soil stabilization, 65–66
Bleomycin, 106
Boswellia serrata, 183, 206, 208, 211, 214
Bradycardia, 120
Brahma Rasāyana, 203
Brahmī ghṛta, 140–141
Briquetting of solid wastes, 64–65
British Medical Journal, 110
Brix value, 91
Bronchial asthma (*tamaka śvāsa*), 145
Bupleurum root, 77
Burman, S.K., 42
Butein, 208

C.A.D., *see* Coronary artery disease
Caesalpinia sappan, 211
Cakradatta, 102
Cakradattam, 258
Cakrapāṇi Datta, 211
Calcidiol, 256
Calebin A, 208
C.A.M.P., *see* Conservation Assessment and Management Plan
Cancer
Ayurveda in, 211
nutraceuticals in, 207–208
Capsaicin, 208
Caraka, 72, 146, 168, 214
Caraka Samhita (Caraka), 4, 7, 15, 16, 48, 93, 261
āpta (documented evidence) in, 165
on arthritis, 212
on cancer, 211
French translation of, 231
on massage therapy, 227
on mental disorders, 227
on oil massage, 228

organizational order of, *162*
Pañcakarma in Europe and, 232
Piṇḍāsava in, 258
significance of, 4, 167
Vimānasthāna of, 4
Carbamazepine, 126
Cardiovascular diseases
Ayurveda in, 211
nutraceuticals in, 204–205
Carum roxburghianum, 140
Caṭṭaimuni Nikaṇṭu-1200 (*Caṭṭaimuni*), 7
Cavya, see *Piper chaba*
Caya, 15
C.C.F., *see* Congestive cardiac failure
C.D.40 gene, 252
C.D.S.C.O., *see* Central Drugs Standard Control Organization
Cedrus deodara, 144, 146, 214, 264
Celastrus paniculata, 214
Centella asiatica, 79, 214
Center of Excellence program, Ministry of Environment, Forest and Climate Change, 192
Central Council, 24
Central Drugs Standard Control Organization (C.D.S.C.O.), 34
Centre d'Étude du Polymorphisme Humain, 248
Chandraprabhā vaṭi, 143
Chemoinformatics, 255
Chemoprevention, 114
Chiang Mai Declaration, 176
Chimpanzees, 2
Chinese medicine, revalidation for, 168
Chinmaya Mission, 242
Chinnarūha, see *Tinospora cordifolia*
Chlorhexidine, 148
Chloroform, 116
Chlorophytum borivilianum, 183
Chopra, D., 223, 225
Chromatographic techniques, 82
Chronic diseases
Ayurveda in, 209
in arthritis, 211–212
in cancer, 211
in cardiovascular disease, 211
in chronic respiratory diseases, 211
in diabetes, 210
in digestive diseases, 212
nutraceuticals in, 203
Chronic obstructive pulmonary disease (C.O.P.D.), 208
Chronic respiratory diseases
Ayurveda in, 211
nutraceuticals in, 208–209
Chronic rheumatism, 256
Chronic stable angina, 120–121
Cikitsāmañjari, 35, 38, 258
Cikitsāsamgraha (Vaṅgasēna), 17
Cilappatikāram (Tamil treatise), 7
Cinnamomum tamala, 204
Cirupañcamūlam (Tamil treatise), 7
Citraka, see *Agni*
Citrus aurantium, 212
Citrus limon, 212
Clinical research, on Ayurvedic medicines, 264–265
Clitoria ternatea, 84, 258
Clonazepam, 143
Cochlospermum religiosum, 183
Code of Criminal Procedure (1973), 30
Comminution, *see* Size reduction
Commiphora berryi, 193
Commiphora wightii, 185, 193, 212
Congestive cardiac failure (C.C.F.), 119–120
Conophylline, 119
Conservation, of herbs, 176
and biodiversity measurement and monitoring
geographic information system (G.I.S.), 190
pilot-scale measurements, 189
botanical families, *180*
classical way of, 193
conservation
botanic gardens, 188
cultivation, 189
ex situ conservation, 183–184
in situ conservation, 183
medicinal plant conservation areas (M.P.C.A.s), 184–185, 187
medicinal plants development areas (M.P.D.A.s), 187–188
seed banks, 188–189
success stories, 185–186
Indian flora and, 176–177
biodiversity hotspots, 179
medicinal plants botanical profile, 177
medicinal plants distribution, 177–179
medicinal plants in trade and, 179
species traded in high volume, 181
substitutes and adulterants, 179–181, *181*
supply sources, 182
plant distribution mapping and, 191–192
sustainable harvesting and, 192
success story, 192–193
threat assessment, 182–183, *183*
Conservation Assessment and Management Plan (C.A.M.P.), 182
Conservation Breeding Specialist Group, 182
Contaminant limits, for Ayurveda, Siddha, and Unani products, *76*
Contamination, 74–77
Convolvulus pluricaulis, 84, 85, 258
C.O.P.D., *see* Chronic obstructive pulmonary disease
Coptis rhizome, 77
Coptis teeta, 176
Coronary artery disease (C.A.D.), 121, 251
Coscinium fenestratum, 76, 185, 191, 193
Cosmeceuticals
in Ayurveda, 213
history of, 212–213
important, 213–214
significance of, 212
Crateva magna (Varuna tree), 105–106
Curcuma longa, 78, 106, 139, 214, 264
Curcumin, 200, 205–208
Cūrṇa (medicinal powders), 50–51, 57, 72
biological effects of, 139–140
herbs pulverized for, 79
Cutting mills, 79
Cyavanaprāśa, 76, 80, 146–147, 203

Cyclooxygenase enzymes, 254
Cyclosporin, 37
CYP2C19 gene, 247
Cyperus rotundus, 106–107, 193, 214
CYPs, *see* Cytochrome P-450 isozymes
Cysteine, 104
Cystine, 104
Cytochrome P-450 isozymes (CYPs), 126

D.A.D., *see* Diode array detection
Dāḍimāṣṭaka cūrṇa, 139
Dahana, see *Agni*
Daljit Singh, Vaidyaraj Hakim, 13
Daphniphyllum himalayense, 193
Darśana (observation), 264
Dāruharidra, see *Berberis aristata*
Daśamūla, 212, 259
Daśamūla kaṭutraya kvātha, 144
Daśamūla kvātha, 144, 259
Daśamūlāriṣṭa, 67, 137–138, 259
Dastūr-ul-Atibbā/ Ikhtiyārāt-e-Qāsimi (Firishta), 10
Datura stramonium, herbarium sheet image of, *80*
Daucus crinitus, 76
D.A.Y., *see* Dried activated yeast
Decalepis arayalpathra, 185
Decalepis hamiltonii, 186
De Medicinarum Compositarum Gradibus Investigandis Libellus (al-Kindi), 261
Dēśam (place of residence), 263
Desmodium gangeticum, 107–108
Desonide, 256
Deva, Mahadeva, 13
Dēvadāru, see *Cedrus deodara*
Dhānvantaram guṭika, 143
Dhānvantaram kvātha, 144
Dhanvantari Nighaṇṭu, 257
Dhātakī puṣpa (*Woodfordia fruticosa* flowers), 57
Dhātu (tissue elements), 15, 243, 257
  relationship with *tridōṣa* and *mala*, *244*, 263
Dhātu kṣaya (degenerative state), 167
Dhātvagni (fire at tissue), 171
Dhātvagni māndya (hypometabolic state), 166
Diabetes, 165–166
  Ayurveda in, 210
  heterogenic pathogenesis in, *166*
  nutraceuticals in, 203–204
Diabetic ulcers, of lower extremity, 123
Digestive diseases, Ayurveda in, 212
Digital Atlas of Medicinal Plants, 192
Digoxin, 37
Diltiazem, 125, 126
Dinacarya (diet and lifestyle), 227
Dincarya, 169
Diode array detection (D.A.D.), 83
Dōṣa, 16, 166
Dosage forms
  adoption, new, 13
test conducted on traditional Ayurvedic, *82*Dosage forms, manufacture of
  āsava and *ariṣṭa* (fermented liquids), 56–57
  avalēha/lēha (confections), 57–59
cūrṇa (medicinal powders), 50–51
  ghṛta and taila (medicinal lipids), 51–52, 54–56
  gulika (pills), 59–61
  kvātha (decoctions), 50
  lēpa (paste), 61–63
Dōṣakōpam (enragement of *dōṣas*), 262
Drākṣādi kvātha, 145
Drākṣāriṣṭa, 138
Dṛḍhabala, 4, 16
Dried activated yeast (D.A.Y.), 91
Drōṇa (12.288 kg), 56
Drug Importation Act (1848) (United States), 22
Drugs (Price Control) Order (1995), The, 24
Drugs and Cosmetics Act (1940), 22–23, 25, 28, 30
  on *Bhasmas*, 36–37
  on cosmetics, 213
  effective implementation of, 37–38
  First Schedule of, 36
    revision of, 35
  need to amend, 67
  punishment for violation of, 31–32
  Rule 161 of, 32–33
  Rule 161-A of, 32
  Rule 161-B of, 33
  Schedule E1 revision of, 37
Drugs and Cosmetics Act 1940, 29
Drugs and Cosmetics Rules (1945), 23–24
  Form 17 of, 31
  Form 18-A of, 31
  rules, 23, 29
  schedules, 23
Drugs and Magic Remedies Act (1954), The (objectionable advertisements), 24
Dry cleaning, 46
Dry ginger, significance of, 102

Earthworm (red) (*Eisenia foetida*), 64
Eclipta alba, 214
Eco-distribution maps, 191–192
EGLN1 gene, 248
Elaeocarpus tuberculatus, 211
Ēlāti (Tamil treatise), 7
Elephantopus scaber, 211
Ellagic acid, 206
Embelia ribes, 136, 193, 211
Embelia tsjeriam-cottam, 193
Emblica officinalis, 146, 212, 214
Endoglin (CD105), 115
Enterotoxins, 107
Ēraṇḍa taila (castor oil), 212
Escherichia coli, 89, 107, 112, 124
Essential Commodities Act (1955), 24
Ethanol extracts, 116
Eucalyptus oil, 187
Evidence building, in Ayurveda, 161–163
  assembling, 163
    āpta (documented evidence), 165
    direct and indirect evidence, 163–164
    yukti (experiment), 164–165
  clinical medicine and research designs and, 171
  clinical research priority areas, 168–170
  generation of new, 168–169
  journey, 165
    etiopathogenesis and disease presentation, 165–167
  optimization of old, 167

re-appropriation, 167–168
research synthesis, 168
revalidation, 168
research designs required for, 170
Evolvulus alsinoides, 84, 85, 108, 258

fAbeta (fAβ), 205
F.A.O., *see* Food and Agriculture Organization
Fawāid-ul-Akhyār (Benefits of the Best) (Muhammed bin Yusuf), 11
FDA, *see* Food and Drug Administration
Fenugreek, 204
Ficus *glomerata*, 211
Ficus racemosa, 204
Ficus religiosa, 210
Fingerprinting, 82–83
application to quality check, 84–85
data, application of, 83
high-performance liquid chromatography (H.P.L.C.), 83
principal-component analysis (P.C.A.) and, 84
similarity evaluation of, 84
thin-layer chromatography (T.L.C.), 83
Firishta, 10–11
Flavonoids, 79, 118, 208
Flavopiridol, 208
Fluocinonide, 256
Folk medicine, *see* Traditional medicine
Food and Agriculture Organization (F.A.O.), 176
Food and Drug Administration (FDA) (United States), 78
Forest Development Agency, 193
Forest produce, sources of, *46*
Forest types, of India, *180*
Form 24-E, 26
Form 25-D, 26, 27
Form 25-E, 26, 27
Form 26-D, 26
Form-35, 26, 27
Foundation for Revitalization of Local Health Traditions (F.R.L.H.T.), 182, 184, 185, 188, 192
Franklin, B., 237
F.R.L.H.T., *see* Foundation for Revitalization of Local Health Traditions
Futuristic provisions, in Ayurvedic pathogenesis, 15

G.A.C.P.s, *see* Good Agricultural And Collection Practices
G.A.D., *see* Generalized anxiety disorder
Gadanigraham, 258
Galen, 261
Gallic acid, 207
Gandharvahastādi ēraṇḍa taila, 148
Gandharvahastādi kvātha, 145
Garcinia cambogia, 212
Garcinia pedunculata, 193
Garlic, *see Allium sativum*
GCP-ASU, see *Good clinical practice guidelines for clinical trials in Ayurveda, Siddha and Unani medicine*
Gender bias, 14–15
General Guidelines for Drug Development of Ayurvedic Formulations, 79
Generalized anxiety disorder (G.A.D.), 143
Genistein, 207, 208
Genopsycho-somatotyping, of humans, 252–253
Geographic information system (G.I.S.), 190
Germany, 242
Gestational diabetes, 203–204
Ghṛta and taila (medicinal lipids), 3, 15, 51–52, 54–56, 72, 260
biological effects of, 140–142, 147–149
preparation of, 79
G.I.S., *see* Geographic information system
Glycosides, 200
Glycyrrhiza glabra, see Licorice (*Glycyrrhiza glabra*)
Glycyrrhizin, 77
G.M.P, *see* Good Manufacturing Practices
Gobal and Indian distribution maps, 191
Good Agricultural And Collection Practices (G.A.C.P.s), 74, 189
Good Agricultural Practices for Medicinal Plants, 74
Good clinical practice guidelines for clinical trials in Ayurveda, Siddha and Unani medicine (GCP-ASU), 34
Good Manufacturing Practices (G.M.P.), 23
Greco-Arabic medicine
Ayurveda benefiting from, 12–14
Indian influence on, 9–10
origin of, 9
Guḍa, see *Guḷika/guṭika*
Guḍūci, see *Tinospora cordifolia*
Guḍūci Ghṛtam, 258
Guggulu, see *Commiphora wightii*
Guidelines on Good Field Collection Practices for Indian Medicinal Plants (2009), 74
Guineensine, 113
Guḷika/guṭika (pills), 59–61, 72
biological effects of, 142–144
heavy metals in, 85, 88–89
production of, 80
Guḷkhaṇḍ, 13
Guṇa (qualities), 3, 253, 257, 261
Gymnema sylvestre, 76, 183, 210

Haimavati, see *Acorus gramineus*
Hakims (Muslim physicians), 4, 12, 14, 25
Hammer mills, 47–48, 79
Haridrākhaṇḍa, 139
"Health for All by the Year 2000" initiative, 176
Heavy metals, contamination with, 74–75
Hedychium spicatum, 214
Hemidesmus indicus, 186
Hemorrhoids, 114
Herbs, fingerprinting in authentication of, 84–85
Hētu (etiological description), 165
Hibiscus rosasinensis, 214
High-performance liquid chromatography (H.P.L.C.), 83, 84, 93, 258
High-performance thin-layer liquid chromatography (H.P.T.L.C.), 83, 85, *88*, 93, 259
Hikmatpradīpa (Deva), 13
Hikmatprakāśa (Deva), 13
Himalaya biodiversity hotspot, 179
Hippocrates, 201
History of medicine and Ayurveda, 201
H.L.A., *see* Human leukocyte antigen
Holarrhena alkaloids, 200

Homeostasis, 236
Honey, as *yōgavāhi* (bioavailability enhancer), 126
Honokiol, 208
Hospital Real, 242
H.P.T.L.C., *see* High-performance thin-layer liquid chromatography
Hṛdayaviśuddhi, 211
Hṛdgraha, 211
Human Genome Diversity Panel, 248
Human leukocyte antigen (H.L.A.), 246–247
  DRB1, 247
Humboldtia vahliana, 186
Hutāśa, see *Agni*
Hydroxycitric acid, 212
Hygrophila auriculata, 110–111
Hygrophila spinosa, 110

Ibn Dhan, 9
Ibn Wadih-al-Yaqubi, 9
I.B.S., *see* Irritable bowel syndrome
Ibuprofen, 254, 259
I.C.A.R.-N.B.P.G.R. National Gene Bank, 188
ICP-MS, *see* Agilent 7700X inductively coupled mass spectrometer
Ilāj-ul-Amrāz (Muhammed bin Yusuf), 11
I.L.1β gene, 252
Impact mill, *see* Pin mill
I.M.P.L.A.D, *see* Indian Medicinal Plant Database
India's gift to world, Ayurveda as, 3–6
Indian cuisine and Ayurveda, 3
Indian Desert, 179
Indian Genome Variation Consortium, 248
Indian Medicinal Plant Database (I.M.P.L.A.D.), 177, 179
Indian Medicine Central Council Act (1970), 28
Indian System of Medicine and Homeopathy (I.S.M. & H.), 243
Indinavir, 37
Indo-Burma biodiversity hotspot, 179
Indukāntam kvātha, 36
Industrial manufacture, of traditional Ayurvedic medicines, 42
  dosage forms and
    āsava and *ariṣṭa* (fermented liquids), 56–57
    avalēha/lēha (confections), 57–59, *60*
    cūrṇa (medicinal powders), 50–51
    ghṛta and taila (medicinal lipids), 51–52, 54–56
    guṭika (pills), 59–61
    kvātha (decoctions), 50
    lēpa (paste), 61–63
  herbal raw material procurement and, 42
  leading manufacturers, *43*
  master formula for production, 49–50
  medicinal formulae sources, 48
  raw material analysis, 48
  raw material pre-processing
    herbs cleaning, 45–46
    size reduction of herbs, 47–48
  raw material storage and, 42–45
  standard operating procedures, 48–49
  waste processing, 64
    black cotton soil stabilization, 65–66
    briquetting of solid wastes, 64–65
    vermicomposting, 64
    wastewater recycling, 66
Inflammatory diseases, nutraceuticals in, 205–206
Influenza, 105
International Conference on Primary Health Care (W.H.O.), 35, 242
International Consultation on Conservation of Medicinal Plants (Thailand), 176
International Sivananda Yoga Vedanta Center, 242
International Society for Krishna Consciousness (ISKCON), 242
Ipomoea digitata, 111
Ipomoea mauritiana, 111
Irimēdādi taila, 148
Irritable bowel syndrome (I.B.S.), 251
Isha Foundation, 242
ISKCON, *see* International Society for Krishna Consciousness
Islamic medicine and Ayurveda, interaction between, 9–11; *see also* Greco-Arabic medicine
I.S.M. & H., *see* Indian System of Medicine and Homeopathy
Isoazadirolide, 256
Isosorbide mononitrate, 120–121
I.U.C.N., 182, 194

Jackknife test, 190
Jami-ul-Fawāid (Collection of Benefits) (Muhammed bin Yusuf), 11
Janakia arayalpathra, 191
Jatharāgni (gastrointestinal fire), 171
Jātyādi ghṛta, 141
Jawaharlal Nehru Tropical Botanical Garden (Thiruvananthapuram), 188
Jīrakāriṣṭa, 138
Joshi, S., 233

Kaempferol, 258
Kaiśōra guggulu, 143
Kalka (paste of herbs), 51, 52, 260
Kalyāṇaka ghṛta, 141
Kampo medicine, 77
Kaṇa, see *Piper longum*
Kanakāsava, 138
Kapha, 16, 166; see also *Prakṛti;* Renaissance; *Tridōṣa*
Kaumārabhṛtya (pediatrics, obstetrics and gynecology), 15
Kāyacikitsa (internal medicine), 14
Kerala, 5, *77*, 228, 265
Kesari Kuteeram Ayurveda Ouṣadhaśāla, 42
Khair-Andesh Khan, Nawab, 14
Khamīra (medicated spirituous liquors), 11
Khamīra-e-Abrēṣam, 11
Khaṇḍakādyō lauham, 3
Khara (very hard sediment), 52, *55*
Khāranādi (ancient text), 4
Khari Baoli (Delhi), 42
Khawas Khan, Behwa bin, 10
Kitāb Firdaus al-Hikma (Paradise of Wisdom) (Rabban-al-Tabari), 10
Kiṭibha kuṣḍha, *see* Psoriasis
Knife mill, 47
Knowledge gain, factors limiting, *164*
Kōkilākṣa, see *Hygrophila auriculata*
Koṅkaṇar Nāṭi (*Koṅkaṇar*), 7

Kṛṣṇa, see *Piper longum*
Kṣārasūtra, 163
Kṣīrabalā taila, 66, 148–149
Kṣīravidāri, see *Ipomoea mauritiana*
Kṣōda, 50
Kṣudrapatrī, see *Acorus calamus*
Kuḷantai Vākaṭam (Book of diseases of children), 7
Kuliñjana, see *Alpinia galanga*
Kumāryāsava, 138
Kuṅkumādi lēpa, 61
  packing of, *63*
Kuṅkumādi taila, 61
Kuṣṭā (calcined mineral/metal), 12
Kuṭajāriṣṭa, 93
Kvātha (decoctions), 50, 57
  biological effects of, 144–146
  contamination in, 75
  drug boiler in manufacture of, *52*
  herbs pulverized for, 79
  quality check of, 85
  tablets of, 258–259

Lawsonia alba, 214
Lēpa (paste), 61–63
Leucorrhoea, 116
L.H.T.s, *see* Local health traditions
Licensing Authority, 32, 33
Licorice (*Glycyrrhiza glabra*), 37, 75, 77, 108–109, 214
Lipoxygenase, 254
Lippia plicata, 2
Loan licensing, 26
  conditions of, 27
Local health traditions (L.H.T.s), 182
Lumbrifiltration, *see* Vermifiltration

Madan-ul-Shifā Sikander Shāhi (Khawas Khan), 10
Maddhyama (hard wax), 52, *55*, 79
Mādhavanidāna (Madhava), 9
Madhuca indica, 136
Madhudrava, see Licorice (*Glycyrrhiza glabra*)
Madhūkam, see Licorice (*Glycyrrhiza glabra*)
Madhumēha, *see* Diabetes
Madhuyaṣṭi, see Licorice (*Glycyrrhiza glabra*)
Madin-Darby Canine Kidney cells, 105
Maesa indica, 193
Māgadhi, see *Piper longum*
Mahāmañjiṣṭādi kvātha, 145
Mahāmāṣa taila, 149
Mahāmāyūra ghṛta, 15
Mahānārāyaṇa taila, 149
Mahārāsnādi kvātha, 67, 145
Maharishi Foundation, 242
"Maharishi Vedic Approach to Health" (M.V.A.H.), 225
Mahātiktaka ghṛta, 141
Mahāyōgarāja Guggul, 212
Mahesh Yogi, M., 225
Mahmud Shah of Gujarat, 10
Mahonia leschenaultia, 193
Mahonia napalensis, 193
Majmā-e-Ziāyi (Collections of Zia), 10
Major histocompatibility complex (M.H.C.), 246
Mā'jun (electuaries), 11
Mā'jun Murawwāh al-Arwāh, 11
Makoi, see *Solanum nigrum*
Mala (feces), 243, 257
  relationship with *tridōṣa* and *dhātu*, *244*, 263
Malassez cell, 92
Malham (malahara), 13
Mallotus nudiflorus, 186
Mānasamitra vaṭaka, 143–144
Mandāgni (dull fire), 263
Mandis (large yards), 42, *44*, *45*
Mangifera indica, 210
Māṇibhadra guḍa, 147
Mañjiṣṭa, see *Rubia sikkimensis*
Manka, 9
Manufacture, of Ayurvedic medicines
  applying, 26
    license conditions, 26–27
    loan license conditions, 27
  G.M.P. certification, 28
    duties of inspectors, 29–30
    guideline principles, 28
    guidelines, 28–29
    inspectors for conducting inspection, 29
    joint inspection of testing laboratories, 31
    procedure to collect drug sample by inspector, 30–31
  good clinical practices and, 34–35
  labeling of products and, 32
  licensing, 25–26
  loan license and, 26
  proof of effectiveness
    classical formulations, 27
    proprietary formulations, 27
  punishment for violation of Act and, 31–32
  raw materials record, 28
  regulatory problems requiring solution and
    A.Y.U.S.H. G.C.P. guidelines monitoring, 35
    change regulation in dosage form, 36
    complementary products regulation, 36
    first schedule revision, 35
    labeling of products, 36–37
    Schedule E1 revision, 37
  shelf life of Ayurvedic medicines and, 32–33
  suspension/cancellation of licenses and, 32
Mārutāpahā, see *Crateva magna* (Varuna tree)
Massage therapy, 227–228
Mass spectrometry (M.S.), 83
MaxEnt software, 190
Maximal electroshock (M.E.S.), 104
Medical systems, evolution of, 2–3
Medicinal plant conservation areas (M.P.C.A.s), 184–185, 187
Medicinal plants
  botanical profile, 177
  distribution of, 177–179
  in trade, 179
    species traded in high volume, 181
    substitutes and adulterants, 179–181, *181*
    supply sources, 182
Medicinal Plants Conservation Parks (M.P.C.P.s), 184
Medicinal Plants Development Agency, 187
Medicinal plants development areas (M.P.D.A.s), 187–188
Mēgha Sandēśam (Kāḷidāsa), 213
Menorrhagia, 116
Menstrual disorders, 116
M.E.S., *see* Maximal electroshock

Mesua ferrea, 193, 214
Methanol, 116, 117
Methionine, 104
Metrorrhagia, 116
M.H.C., *see* Major histocompatibility complex
Micro-organisms, contamination with, 75
Millennium Seed Bank Project (Kew, Britain), 188
Mīmāmsa school, 201, 243
Mimusops elengi, 214
Mitochondria, 114
Mōdaka, see *Guḷika/guṭika*
Mohammed-bin-Qasim, 4
Momordica charantia, 204, 210
Moringa oleifera, 111, 214
Mouktika, 13
M.P.C.A.s, *see* Medicinal plant conservation areas
M.P.C.P.s, *see* Medicinal Plants Conservation Parks
M.P.D.A.s, *see* Medicinal plants development areas
Mṛdu (soft wax-like sediment), 52, *55*, 79
M.S., *see* Mass spectrometry
Mubarak, Zia Muhammed, 10
Mucuna pruriens, 200
Muhammed bin Yusuf, Yusuf bin, 11
Muhit-i-Āzam (Azad Khan), 11
Munnātakhyārōga, 12
Musa paradisiaca, 211
M.V.A.H., *see* "Maharishi Vedic Approach to Health"
Mycotoxins, 75
Myrobalan, *see Terminalia chebula*
Myrsine africana, 193

Nadvi, Sayyid Sulayman, 10
Nāgakēsar, see *Mesua ferrea*
Nāgārjuna, 261
Najab-al-Din-Samarqandi, 11
N.A.M., *see* National A.Y.U.S.H. Mission
Nāradīya Lakṣmīvilāsa Rasa, 262
Nārasimha rasāyana, 141
Narcotic Drugs and Psychotropic Substances Rules Act (1985), The, 24
Nasya (errhine therapy), 149
Nāthasiddhas, 7
National A.Y.U.S.H. Mission (N.A.M.), 189
National Botanical Research Institute (N.B.R.I.) Botanic Garden (Lucknow), 188
National Formulary (United States), 22
National Medicinal Plant Board (N.M.P.B.), 176, 189, 200
National Mission on Medicinal Plants (N.M.M.P.), 189
Nāvanītakam, 242
Nayōpāya kvātha, 145
N.B.R.I., *see* National Botanical Research Institute Botanic Garden
N.C.M., *see* Non-Conventional Medicine
Nelumbium speciosum, 193
Nelumbo nucifera, 193
Nepal, 237
Neurodegenerative diseases, nutraceuticals in, 205
Neurological diseases (*vāta vyādhi*), 145
"New Age", significance of, 225
N.F.-κB, *see* Nuclear factor kappa-B
Nicobar Islands biodiversity hotspot, 179
Nidāna (etiology), 15
Nīlapuṣpā, see *Evolvulus alsinoides*
Nimbāriṣṭa, 90
Nimitantram (ancient text), 4
N.M.M.P., *see* National Mission on Medicinal Plants
N.M.P.B., *see* National Medicinal Plant Board
N.M.R., *see* Nuclear magnetic resonance
N.N. Sen and Company, 22, 42
Non-Conventional Medicine (N.C.M.), 234
Non-steroidal anti-inflammatory drugs (N.S.A.I.D.s), 171
Non-timber forest produce (N.T.F.P.), 192, 193
North-East India biodiversity hotspot, 179
N.S.A.I.D.s, *see* Non-steroidal anti-inflammatory drugs
N.T.F.P., *see* Non-timber forest produce
Nuclear factor kappa-B (N.F.-κB), 205, 206, 208, 211
Nuclear magnetic resonance (N.M.R.), 83
Nutraceuticals, 202
  in arthritis, 206–207
  in Ayurveda, 203
  in cancer, 207–208
  in cardiovascular diseases, 204–205
  in chronic diseases, 203
  in chronic respiratory diseases, 208–209
  in diabetes, 203–204
  in inflammatory diseases, 205–206
  in neurodegenerative diseases, 205
  in obesity, 207
Nutricosmetics, 212
Nyāya school, 201, 243

Obesity, nutraceuticals in, 207
Ochratoxin A, 75
Ocimum sanctum, 204, 210, 214
Ōjas, 243
Oleocanthal, 254
Oral cosmetics, *see* Nutricosmetics
Oregano, 77
Oroxylum indicum, 183
Osho Foundation, 242
Osteoarthritis, 206–207
Osteonecrosis, 145

Pañcabhūta (five primordial elements), 3, 227, 243, 253, 256
Pañcakarma, 163, 168–170, 202
  in Europe, 232–235
Pañca Karma Therapy (Singh), 233
Pañcamahābhūta (five basic elements), 14
Papio hamadryas, 2
Parpola, A., 7
Pascātkarma (post-treatment), 202
P.A.S.I., *see* Psoriasis Area and Severity Index
Pathyāṣaḍaṅgam kvātha, 85, *88*
Paṭōla Kaṭurōhiṇyādi kvātha, 145
Pātrapāka, 56
Payasvini, see *Ipomoea mauritiana*
P.C.A., *see* Principal-component analysis
Pesticides, contamination with, 75–76
P-glycoprotein efflux pumps, 37
Phala Ghṛta, 203
Phala sarpis, 142
Pharmaceutics of ayurvedic medicines, 260–261
Pharmacognosy, 257
Pharmacolinguistics, 257–258
Pharmacopeial Laboratory for Indian Medicine (P.L.I.M.), 33, 73
The Pharmacy Act (1948), 24
Pharmacy Council of India, 24
Phellodendrum bark, 77

Phenolics, 79
Phenols, 118
Phenotypic variability
in osteoarthritis, *166*, 167
significance of, 166
Phenytoin, 104
Phyllanthus amarus, 211
Phyllanthus emblica, 257
Phytochemicals, 208
Piceatannol, 208
Pillai, T. V. S., 7
Piḷḷaippiṇi Vākaṭam (Book of diseases of children), 7
Pills rolling machine, *62*
Pimpinella anisum, 214
Piṇḍāsava, 258
Piṇḍi, see *Guḷika/guṭika*
Pin mill, 47
Piper chaba, 193
Piperine (1-Piperoylpiperidine), 114
Piper longum, 57, 112, 258, 264
antibacterial activity of, 113
anti-cancer effect of, 113, 211
anti-inflammatory activity of, 113
effect on digestive efficiency, 112–113
Pippalī cūrṇa, see *Piper longum*
Pippalyāsava, 138
Pitta, 166; see also *Prakṛti;* Renaissance; *Tridōṣa*
Pittaja arbuda tumors, 211
Plant distribution mapping, 191–192
Plant parts, consumed in ayurvedic pharmacy, *259*
Plant sterols, 204
Plīha (splenomegaly), 114
P.L.I.M., *see* Pharmacopeial Laboratory for Indian Medicine
Plumbagin, *see Plumbago zeylanica*
Plumbago zeylanica, 93, 114–115
Poisons Act (1919), The, 24
Polyalthia longifolia, 84
Polyphenols, 205
Polyvalence, 77
Pongamia glabra, 214
Posology, research on, 261
Prabhāva (specific action), 3, 14, 253, 257, 261
Pradhānakarma (primary treatment), 202
Prakōpa, 15
Prakṛti (constitutional types), 163, 171, 209, 223, 227, 262
association, with diseases, 250–251
autonomic responses and, 249
cardiovascular responses and, 249
DNA methylation analysis of, 250
gene polymorphism of drug metabolizing enzyme and, 247
genetic basis of, 248
immunophenotyping of, 250
isotonic aerobic exercise and, 249
significance of, 246
in therapeutics, 251–252
understanding of, 246–247
uniform method for assessment of, 252–253
and variations, in platelet aggregation, 248–249
whole genome expression and biochemical correlates and, 247–248
Prakṣēpa dravya, 57
Pramēha (polyuria), 210
Praśama, 15
Praśna (interrogation), 264
Pratinidhi Dravya, 193
Premna latifolia, 259
Primary amenorrhea, 116
Principal-component analysis (P.C.A.), 84
Procyanidins, 214
Prostaglandins, 116
Protocol for Testing Ayurvedic, Siddha & Unani Medicines (P. L.I.M), 73
Pseudogeusia, 255
Pseudomonas aeruginosa, 112, 113
Psoralea corylifolia, 214
Psoriasis, 123, 137
Psoriasis Area and Severity Index (P.A.S.I.), 123
Pterocarpus marsupium, 183, 204, 210, 259
Pterocarpus santalinus, 191, 193
Pulamanthole Mooss family, of Kerala, 5
Pulse examination (*nāḍiparīkṣa*), 12
Pulverizing equipment, 47–48, 79
Punarnavādi kvātha, 146
Pure Food and Drug Act (1906) (United States), 22
Pūrvakarma (pre-treatment), 202
Pūrvarūpa (premonitory symptoms), 15, 264
Puṣyānuga cūrna, 139

Qasidā fi Hifz-ul-Sihhat, Riyāz -ul-Adwiya (Garden of Remedies) (Muhammed bin Yusuf), 11
Q.S.E., *see* Quality, safety and efficacy
Quality, safety and efficacy (Q.S.E.), 78
Quality control, of Ayurvedic medicines, 72
in ancient times, 72–73
The Ayurvedic Pharmacopoeia of India and, 73
factors influencing, 73–74
contamination, 74–77
Good Agricultural And Collection Practices, 74
processing methods, 79–80
stability, 78–79
methods
fingerprinting, 82–85
finished formulations, 85
macroscopic evaluation, 80–81
microscopic evaluation, 81
physicochemical analysis, 82
present state of, 73
problems related to
ariṣṭa and *āsava* souring, 89–93
heavy metals in pills, 85, 88–89
irrational substitution, 93
Quantum Healing (Chopra), 223, 225
Quercetin, 206

Rabban-al-Tabari, Ali ibn Sahl, 10
Rāja Nighaṇṭu, 257
Rajanyādi cūrna, 139–140
Rajas, 50
Raktacandana, see *Pterocarpus santalinus*
Raktadhātu (hemoglobin), 226, 227
Raktavāhasrōta (blood vessels), 226
Randomized controlled trials (R.C.T.s), 34–35
Rasa (taste), 3, 253, 257, 261
Rasadhātu (blood plasma), 226
Rasakriya, 58
Rasaśāstra (treatment using mercurials), 35
research on, 261–262
Rasa sindūram, 149

Rasāyana (promotive/ rejuvenation therapy), 15, 104, 121, 122, 140, 146, 169, 201, 209
classification of, *202*
nutraceuticals and, 203
significance of, 202
Rasāyana cikitsa, 202
Rāsnādi cūrna, 140
Rasōna Kṣīrapāka, 203
Rasouṣadhi (minerallo-metallic preparations), 261
Rauvolfia serpentina, 176, 183
Rauwolfia alkaloids, 200
Ray, G., 22, 42
R.C.T.s, *see* Randomized controlled trials
Reactive oxygen species (R.O.S.), 147
Recurrent aphthous ulcers, 109
Recycling, of wastewater, 66
Reductionism, 3
Regional distribution maps, 191
Regulatory activity, emergence of, 22
Related specialists, significance of, 232
Renaissance, 242
molecular basis of Ayurvedic medicines and, 255–256
problems warranting urgent attention and, 256–257
botanical identity of herbs, 257–258
clinical research on Ayurvedic medicines, 264–265
disease diagnosis, 262–264
formulae with lesser ingredients, 258
new dosage forms discovery, 258–259
objective way to teach Ayurveda, 262
pharmaceutics of ayurvedic medicines, 260–261
plant parts substitution, 259
research on posology, 261
research on *rasaśāstra*, 261–262
renewed interest in Ayurveda and, 242–243
theoretical constructs of Ayurveda and, 243
dhatu, 243
prakṛti, 246–253
taste and drug action, 253–255
tridōṣa, 243–246
Resveratrol, 204, 205, 208
Retinoids, 208
Rhododendron arboreum, 190
Ricinus communis, 264
Rōganidāna (etiopathogenesis), 15
R.O.S., *see* Reactive oxygen species
Rosa damascene, 214
Royal Botanic Gardens (Kew, Britain), 188
Rtucarya, 169
Rubia cordifolia, 76, 193, 214
Rubia sikkimensis, 193
Rumex vesicarius, 193
Rūpa (symptoms), 15

ṣaḍkriyākāla, 209, 264
Sage, 77
Sahacarādi kvātha, 146
Sahasrayōga, 48, 142
Saikosaponins, 77
Salacia reticulata, 210
Śālākya (knowledge of diseases of supra-clavicular region), 14–15
Saleh-bin-Bhela, 9
Śāliparṇ, see *Desmodium gangeticum*
Salmonella gallinarum, 112
Salvadora persica, 214
Śalya (surgery), 14
Samāgni (regular fire), 263
Śamana (palliative care), 233
Sāmkhya Darśana, 127
Sāmkhya school, 201, 227, 243
Samprāpti (pathogenesis), 15, 164
Samprāpti vighaṭana, 171
Sandhāna kalpana, 56
Sandhivāta (osteoarthritis), 166
Sanguinarine, 208
Śaṅkhapuṣpi, see *Clitoria ternatea*
Sankunny, K., 4
Sanskrit entities and botanical equivalents, *257*
Sapindus trifoliatus, 214
Sapogenins, *87*
Sapta dhātu ōja, 171
Saraca asoca, 84, 115–117, 185, 186, 191
Saraca declinata, 84
Saraca indica, 116
Sarbat, 13
Śārṅgadhara Samhita (medieval text), 8, 12, 17, 36, 48, 50, 52, 102, 136
on quality of medicinal plants, 72–73
Śataparvikā, see *Acorus calamus*
Śatāvari (*Asparagus racemosus*), 51
Śatāvarī Ghṛta, 203
ṣaṭkriyākāla (six specific stages of disease development), 15
Sātmya (homologation), 263
Satvam (emotional status), 264
Savīryatā avadhi (expiry date), 36
Self-regulating versus government-regulated bodies, 237
Semecarpus anacardium, 93, 211
Septicemia, 123
Sesamum indicum, 214
Sētuvṛkṣa, see *Crateva magna* (Varuna tree)
Shannon-Wiener index, 189, 190
Shifa-e-Mahmūdi (Cure of Mahmud), 10
Shigella flexneri, 107
Shikakai, see *Acacia concinna*
Shorea robusta, 186
Shredding, 79
Sida aordifolia, 229
Sida cordifolia, 144
Sida retusa, 144
Sida rhombifolia, 144, 148
Sida species, 76
Siddhabhēṣajamaṇimāla (Bhatta), 13
Siddha medicine, *see* Tamil medicine and Ayurveda
Siddha Vaidya Agastya Āśramam (Tanjore), 42
Siddhayōga (Vṛnda), 10, 12
Sidh Makardhwaj, 262
Sieving/screening, 46
Śīghraphala, see *Moringa oleifera*
Signal transducer and activator of transcription 3 (S.T.A.T.3), 205
Śigru, see *Moringa oleifera*
Śikhi, see *Agni*
Silymarin, 205
Simpson index, 189
Singh, R.H., 233
Single herb remedies, 102
co-administered adjuvants and, 126

evidence in favor of, 102, 104–125
recommendations of, *103*
Single-nucleotide polymorphism (S.N.P.), 248
Śirōdhāra (medicated oil dripping over forehead), 143
Śirōlēpa (application of herbal paste on the scalp), 140
Sivagnana Siddha Vaidya Salai (Koilpatti), 42
Size reduction, 47, 79
Skin photoaging, 147
S-methyl cysteine sulphoxide, 210
Snāyukarōga (dracunculiasis), 12
S.N.P., *see* Single-nucleotide polymorphism
Śōbhāñjana, see *Moringa oleifera*
Śōdhana (reducing) procedures, significance of, 37, 93, 232, 233, 261–262
Solanum nigrum, 181
Solid-state fermentation (S.S.F.), 260
Sparśana (palpation), 264
Species richness measure, 189
Species Survival Commission, 182
Sri Lanka, 237
Srōtāmsi, 226
S.S.F., *see* Solid-state fermentation
Standard and strange drugs, comparison of, 14
Staphylococcus aureus, 111, 113
S.T.A.T.3, *see* Signal transducer and activator of transcription 3
St John's Wort (*Hypericum perforatum*), 37
State Licensing Authority, 33
State Pharmacy Council, 24
Sterculia urens, 183
Stereospermum chelonoides, 183
Steroidal lactones, 200
Sthirā, see *Desmodium gangeticum*
Strengthening the Medicinal Plants Resource Base in Southern India in the Context of Primary Health Care, 187
Streptococcus sanguinis, 109
Streptozotocin (S.T.Z.), 118, 119, 210
Strychnos potatorum, 117–118
S.T.Z., *see* Streptozotocin
Śukra dhatu, 15
Sukumāra ghṛta, 142
Sundaland biodiversity hotspot, 179
Suśruta, 15, 72, 168
Suśruta Samhita (*Suśruta*), 4, 7, 48, 165, 167, 201, 261
on cancer, 211
Kalyāṇaka ghṛta in, 141
on skin layers, 227
Sustainable harvesting, 192
success story, 192–193
Svarasa (fresgh herb juice), 54, 260
Svarṇa bhasma, 262
Śvāsahidhmā cikitsa (respiratory diseases), 264
Śvētadāru, see *Moringa oleifera*
Śvēta śāriva, see *Decalepis hamiltonii*
Switzerland, 237, 238, 242
Symplocos racemosa, 211, 214
Synergy, significance of, 77
Syzygium aromaticum, 214
Syzygium cumini, 204

Tabernaemontana Divaricata, 118–119
Tailabindu parīkṣa (pouring an oil drop into urine), 168
Tamil medicine and Ayurveda, 7–8
Tāmra bhasma (calcined powder of copper), 262
Taste, 253
preferences, in diagnosis, 255
response, as sensory expression of pharmacological activity, 253–255
sensations, in absence of external stimuli, 255, *256*
Taxus wallichiana, 183
T.C.M., *see* Traditional Chinese medicine
T.D.U., *see* Trans-disciplinary University
Tectona grandis, 211
Teeth mill, *see* Pin mill
Tephrosia purpurea, 84, 85
Terminalia arjuna, 85, 119–121, 211, 253
Terminalia belerica, 121–122, 214, 258
fruits, seeds, and pericarp of, *122*
Terminalia chebula, 85, 136, 140, 144, 192, 214
Thin-layer chromatography (T.L.C.), 77, 83, 85, 93
Threatened medicinal plants of Kerala, adulterants of, *77*
Thymus, 77
Tibb-e-Firus Shāhi (Medicine of Firus Shah), 10
Tibb-e-Hindi (Medicine of the Hindus) (Arzani), 11
Tibb-e-Mahmūdi (Medicine of Mahmud), 10
Tibb-e-Yūsufi (Medicine of Yusuf) (Muhammed bin Yusuf), 11
Tīkṣṇāgni (sharp fire), 263
Tinospora cordifolia, 51, 85, 122, 144, 204
antibacterial activity, 124
anti-cancer effects of, 211
anti-inflammatory activity, 124
in clinical studies, 123
description of, 257
effects on immunity, 123–124
in ethnomedicine, 123
Guḍūci Ghṛtam and, 258
Tinospora ordifolia, 124
Tirikaṭukam (Tamil treatise), 7
Tirukkuraḷ (Tamil philosophical treatise), 7
Tirumantiram (Tamil text), 7
T.L.C., *see* Thin-layer chromatography, 77
Tocotrienol, 208
Tōla (4.800 kg), 56
Trachyspermum ammi, 140
The Trade Marks Act (1999), 24–25
Traditional Chinese medicine (T.C.M.), 225, 226, 236
Traditional medicine, 2–3
Trans-disciplinary University (T.D.U.), 182, 188, 192, 194
Tridōṣa, 3, 12, 16, 17, 102, 144, 166–167, 171, 209, 213, 225, 227, 236, 243, 253, 257, 262; *see also Prakṛti;* Renaissance
mathematical modeling of, 244–246
relationship with *mala* and *dhatu*, *244*
significance of, 263
taste and, 255
Trigonella foenum graceum, 204
Trikaṭu, 144
Trimalla, 12
Triphala, 121
Triphalā Cūrṇam, 51, 212
master formula to produce, *54*
Triterpenoids, 208
Triticum sativum, 214
Tṛparṇi, see *Desmodium gangeticum*
Tṛphalā ghṛta, 142
Tughlaq, Firus Shah, 10

Tughlaq, Muhammed, 10
Turmeric, *see Curcuma longa*

Ugragandha, see *Acorus calamus*
Ūnāni Dravyaguṇavijñān (Daljit Singh), 13
Ūnāni *Dravyaguṇādarś* (Daljit Singh), 13
Ūnāni medicine
    Ayurvedic drugs borrowed from, *13*
emergence of, 10–11
    influence of Ayurveda on, 11–12
    interest in, 13
Ūnāni Siddhayōgasamgraha (Daljit Singh), 13
Unani-Tibb, 225
Upaśaya (trial and error), 15
Ūrjaskara (promotive) therapy, 213
U.S.A., 237
Usaybia, Ibn Abi, 9
U.S. Pharmacopeia, 22
Utleria salicifolia, 185
Uttaravasti (enema), 142
Uyūn-al-anbā fi tabāqat-al-atibbā (Ibn Abi Usaybia), 9
U.V.B. irradiation, 147

Vaca, see *Acorus calamus*
Vāgbhaṭa, 5, 15–17, 142
    omission of tenets by, 15
    sepulcher (believed) of, *7*
Vahni, see *Agni*
Vaidyamanōrama, 35, 38, 102, 258
Vaidyamuktāvali (Mouktika), 13
*Vaidyā*s, 16, 22, 25, 28, 167
Vaidyavinōda Samhita (Śamkarā), 12
Vaiśēṣika school, 201, 243
Vaiśvānara cūrna, 140
Vaittiyacintāmaṇi-800 (Yūkimuni), 8
    ayurvedic origins of, *8*
Vājīkaraṇa (virilification), 15, 169
Vaṅgasēna, 17
Vara Asanādi kvātha, 91
Vardhma, 12
Variar, P.S., 22, 42, 258
Varti, see *Guḷika/guṭika*
Vāśā, see *Adhatoda vasica*
Vāśā ghṛtam, 3
Vāśā khaṇḍa kūśmāṇḍakam, 3
Vāśā khaṇḍam, 3
Vāśāriṣṭa, 138
Vāśāriṣṭam, 3
Vāstu Śāstra (Indian architecture), 229
Vāta, 16, 166, 263; *see also Prakṛti;* Renaissance; *Tridōṣa*
Vātarakta, 144, 168
Vātaśōṇita, 16, 110
Vātavyādhi, 16
Vaṭi, see *Guḷika/guṭika*
Vaṭika, see *Guḷika/guṭika*
Vayaḥ (age), 263
V.C.O., *see* Virgin coconut oil
Vēdānta school, 201, 243
Vermicomposting, 64
Vermifiltration, 66
Vermouth, 89
Vernonia amygdalina, 2

Vibhītaki, see *Terminalia belerica*
Vibrio cholerae, 113
Vicitrapratyayārabdha dravya (strange drugs), 14
Viḍaṅga, see *Embelia ribes*
Vidārigandhā, see *Desmodium gangeticum*
Vidāryādi ghṛta, 142
Vikṛti (new biological entity), 164
Vipāka (post-digestive taste), 3, 253, 257, 261
Virgin coconut oil (V.C.O.), 260
Vīrya (potency), 3, 253, 257, 261
Viṣamāgni (irregular fire), 263
Viṣṇukrānta, see *Evolvulus alsinoides*
Vitex negundo, 124–125
Vitis vinifera, 136
V:P:K code, 17, 244
Vyādhihara (curative) therapy, 213

Warfarin, 37
Waste processing, 64
    black cotton soil stabilization, 65–66
    briquetting of solid wastes, 64–65
    vermicomposting, 64
    wastewater recycling, 66
"Wellness Ayurveda", 223, 227–228
West, Ayurveda in, 223–225
    fundamental problems, 226
        B.A.M.S. syllabus and western courses, 228–229
        different rules in different western countries for
            T.M/C.A.M., 234–238
        lack of support from Indian government, 226–228
        western Ayurveda versus educational priorities,
            229–234
    schools teaching, currently, *224*
    turning point of, 225
Western Ghats–Sri Lanka biodiversity hotspot, 179
Wet cleaning, 46
WHO, *see* World Health Organization
Wild ginger, *see Zingiber zerumbet*
Withania somnifera, 212, 229
Withanolides, 200
Wooden vats, 89–90
Woodfordia fruticosa, 90
World Conservation Strategy, 194
World Health Organization (WHO), 25, 79, 170, 200, 223
Wound Severity Score, 123

Xanthium strumarium, 211

Yakṛt (hepatomegaly), 114
Yaṣṭimadhu, see Licorice (*Glycyrrhiza glabra*)
Yaṣṭimadhūkam, see Licorice (*Glycyrrhiza glabra*)
Yōgāmṛtam, 35, 38
Yōgarāja Guggul, 212
Yōgaratnākara, 13, 138
Yoga school, 201, 243
Yōgataraṅgiṇi (Trimalla), 12
Yōni picu, 142
Yukti (experiment), 164–165

Zerumbone, 208
Zingiber officinale, 85, 140, 144, 146, 211
Zingiber zerumbet, 204